RNA–Protein Interaction Protocols

METHODS IN MOLECULAR BIOLOGY™

John M. Walker, SERIES EDITOR

129. **Integrin Protocols**, edited by *Anthony Howlett, 1999.*
122. **Confocal Microscopy Methods and Protocols**, edited by *Stephen W. Paddock, 1999*
121. **Natural Killer Cell Protocols:** *Cellular and Molecular Methods*, edited by *Kerry S. Campbell and Marco Colonna, 1999*
120. **Eicosanoid Protocols**, edited by *Elias A. Lianos, 1999*
119. **Chromatin Protocols**, edited by *Peter B. Becker, 1999*
118. **RNA–Protein Interaction Protocols**, edited by *Susan R. Haynes, 1999*
117. **Electron Microscopy Methods and Protocols**, edited by *M. A. Nasser Hajibagheri, 1999*
116. **Protein Lipidation Protocols**, edited by *Michael H. Gelb, 1999*
115. **Immunocytochemical Methods and Protocols (2nd ed.)**, edited by *Lorette C. Javois, 1999*
114. **Calcium Signaling Protocols**, edited by *David Lambert, 1999*
113. **DNA Repair Protocols:** *Eukaryotic Systems*, edited by *Daryl S. Henderson, 1999*
112. **2-D Proteome Analysis Protocols**, edited by *Andrew J. Link 1999*
111. **Plant Cell Culture Protocols**, edited by *Robert Hall, 1999*
110. **Lipoprotein Protocols**, edited by *Jose M. Ordovas, 1998*
109. **Lipase and Phospholipase Protocols**, edited by *Mark H. Doolittle and Karen Reue, 1999*
108. **Free Radical and Antioxidant Protocols**, edited by *Donald Armstrong, 1998*
107. **Cytochrome P450 Protocols**, edited by *Ian R. Phillips and Elizabeth A. Shephard, 1998*
106. **Receptor Binding Techniques**, edited by *Mary Keen, 1998*
105. **Phospholipid Signaling Protocols**, edited by *Ian M. Bird, 1998*
104. **Mycoplasma Protocols**, edited by *Roger J. Miles and Robin A. J. Nicholas, 1998*
103. **Pichia Protocols**, edited by *David R. Higgins and James M. Cregg, 1998*
102. **Bioluminescence Methods and Protocols**, edited by *Robert A. LaRossa, 1998*
101. **Mycobacteria Protocols**, edited by *Tanya Parish and Neil G. Stoker, 1998*
100. **Nitric Oxide Protocols**, edited by *Michael A. Titheradge, 1998*
99. **Human Cytokines and Cytokine Receptors**, edited by *Reno Debets and Huub Savelkoul, 1999*
98. **Forensic DNA Profiling Protocols**, edited by *Patrick J. Lincoln and James M. Thomson, 1998*
97. **Molecular Embryology:** *Methods and Protocols*, edited by *Paul T. Sharpe and Ivor Mason, 1999*
96. **Adhesion Proteins Protocols**, edited by *Elisabetta Dejana, 1999*
95. **DNA Topoisomerases Protocols:** *II. Enzymology and Drugs*, edited by *Mary-Ann Bjornsti and Neil Osheroff, 1998*
94. **DNA Topoisomerases Protocols:** *I. DNA Topology and Enzymes*, edited by *Mary-Ann Bjornsti and Neil Osheroff, 1998*
93. **Protein Phosphatase Protocols**, edited by *John W. Ludlow, 1998*
92. **PCR in Bioanalysis**, edited by *Stephen J. Meltzer, 1998*
91. **Flow Cytometry Protocols**, edited by *Mark J. Jaroszeski, Richard Heller, and Richard Gilbert, 1998*
90. **Drug–DNA Interaction Protocols**, edited by *Keith R. Fox, 1998*
89. **Retinoid Protocols**, edited by *Christopher Redfern, 1998*
88. **Protein Targeting Protocols**, edited by *Roger A. Clegg, 1998*

87. **Combinatorial Peptide Library Protocols**, edited by *Shmuel Cabilly, 1998*
86. **RNA Isolation and Characterization Protocols**, edited by *Ralph Rapley and David L. Manning, 1998*
85. **Differential Display Methods and Protocols**, edited by *Peng Liang and Arthur B. Pardee, 1997*
84. **Transmembrane Signaling Protocols**, edited by *Dafna Bar-Sagi, 1998*
83. **Receptor Signal Transduction Protocols**, edited by *R. A. John Challiss, 1997*
82. **Arabidopsis Protocols**, edited by *José M Martinez-Zapater and Julio Salinas, 1998*
81. **Plant Virology Protocols:** *From Virus Isolation to Transgenic Resistance*, edited by *Gary D. Foster and Sally Taylor, 1998*
80. **Immunochemical Protocols (2nd. ed.)**, edited by *John Pound, 1998*
79. **Polyamine Protocols**, edited by *David M. L. Morgan, 1998*
78. **Antibacterial Peptide Protocols**, edited by *William M. Shafer, 1997*
77. **Protein Synthesis:** *Methods and Protocols*, edited by *Robin Martin, 1998*
76. **Glycoanalysis Protocols (2nd. ed.)**, edited by *Elizabeth F. Hounsell, 1998*
75. **Basic Cell Culture Protocols (2nd. ed.)**, edited by *Jeffrey W. Pollard and John M. Walker, 1997*
74. **Ribozyme Protocols**, edited by *Philip C. Turner, 1997*
73. **Neuropeptide Protocols**, edited by *G. Brent Irvine and Carvell H. Williams, 1997*
72. **Neurotransmitter Methods**, edited by *Richard C. Rayne, 1997*
71. **PRINS and *In Situ* PCR Protocols**, edited by *John R. Gosden, 1996*
70. **Sequence Data Analysis Guidebook**, edited by *Simon R. Swindell, 1997*
69. **cDNA Library Protocols**, edited by *Ian G. Cowell and Caroline A. Austin, 1997*
68. **Gene Isolation and Mapping Protocols**, edited by *Jacqueline Boultwood, 1997*
67. **PCR Cloning Protocols:** *From Molecular Cloning to Genetic Engineering*, edited by *Bruce A. White, 1997*
66. **Epitope Mapping Protocols**, edited by *Glenn E. Morris, 1996*
65. **PCR Sequencing Protocols**, edited by *Ralph Rapley, 1996*
64. **Protein Sequencing Protocols**, edited by *Bryan J. Smith, 1997*
63. **Recombinant Protein Protocols:** *Detection and Isolation*, edited by *Rocky S. Tuan, 1997*
62. **Recombinant Gene Expression Protocols**, edited by *Rocky S. Tuan, 1997*
61. **Protein and Peptide Analysis by Mass Spectrometry**, edited by *John R. Chapman, 1996*
60. **Protein NMR Techniques**, edited by *David G. Reid, 1997*
59. **Protein Purification Protocols**, edited by *Shawn Doonan, 1996*
58. **Basic DNA and RNA Protocols**, edited by *Adrian J. Harwood, 1996*
57. **In Vitro Mutagenesis Protocols**, edited by *Michael K. Trower, 1996*
56. **Crystallographic Methods and Protocols**, edited by *Christopher Jones, Barbara Mulloy, and Mark R. Sanderson, 1996*

METHODS IN MOLECULAR BIOLOGY™

RNA–Protein Interaction Protocols

Edited by

Susan R. Haynes

Uniformed Services University of the Health Sciences
Bethesda, Maryland

CHOW H. LEE

Humana Press ✷ Totowa, New Jersey

© 1999 Humana Press Inc.
999 Riverview Drive, Suite 208
Totowa, New Jersey 07512

All rights reserved. No part of this book may be reproduced, stored in a retrieval system, or transmitted in any form or by any means, electronic, mechanical, photocopying, microfilming, recording, or otherwise without written permission from the Publisher. Methods in Molecular Biology™ is a trademark of The Humana Press Inc.

All authored papers, comments, opinions, conclusions, or recommendations are those of the author(s), and do not necessarily reflect the views of the publisher.

This publication is printed on acid-free paper. ∞
ANSI Z39.48-1984 (American Standards Institute) Permanence of Paper for Printed Library Materials.

Cover design by Patricia F. Cleary.
Cover illustration: Fig. 15-3; *see* page 193.

For additional copies, pricing for bulk purchases, and/or information about other Humana titles, contact Humana at the above address or at any of the following numbers: Tel: 973-256-1699; Fax: 973-256-8341; E-mail: humana@humanapr.com, or visit our Website at www.humanapress.com

Photocopy Authorization Policy:
Authorization to photocopy items for internal or personal use, or the internal or personal use of specific clients, is granted by Humana Press Inc., provided that the base fee of US $10.00 per copy, plus US $00.25 per page, is paid directly to the Copyright Clearance Center at 222 Rosewood Drive, Danvers, MA 01923. For those organizations that have been granted a photocopy license from the CCC, a separate system of payment has been arranged and is acceptable to Humana Press Inc. The fee code for users of the Transactional Reporting Service is: [0-89603-568-9/99 $10.00 + $00.25].

Printed in the United States of America. 10 9 8 7 6 5 4 3 2 1

Library of Congress Cataloging-in-Publication Data

RNA-protein interaction protocols / edited by Susan R. Haynes.
      p. cm. -- (Methods in molecular biology ; vol. 118)
   Includes bibliographical references and index.
   ISBN 0-89603-568-9 (alk. paper)
 1. RNA-protein interactions--Laboratory manuals. I. Haynes, Susan R. II. Series: Methods in molecular biology (Totowa, N.J.) ; vol. 118.
QP623.8.P75R63   1999
572.8'8--dc21   98-55324
                                                         CIP

Preface

The molecular characterization of RNA and its interactions with proteins is an important and exciting area of current research. Organisms utilize a variety of RNA–protein interactions to regulate the expression of their genes. This is particularly true for eukaryotes, since newly synthesized messenger RNA must be extensively modified and transported to the cytoplasm before it can be used for protein synthesis. The realization that posttranscriptional processes are critical components of gene regulation has sparked an explosion of interest in both stable ribonucleoprotein (RNP) complexes and transient RNA–protein interactions.

RNA is conformationally flexible and can adopt complex structures that provide diverse surfaces for interactions with proteins. The fact that short RNA molecules (aptamers; *see* Chapter 16) can be selected to bind many different types of molecules is evidence of the structural variability of RNA. RNA molecules are rarely entirely single- or double-stranded, but usually contain multiple short duplexes interrupted by single-stranded loops and bulges; in some RNAs, such as tRNAs, the short duplexes stack on each other. Further variability is generated by the presence of non-Watson-Crick base pairs, modified nucleotides, and more complex structures, such as pseudoknots and triple-strand interactions.

The techniques described in *RNA–Protein Interaction Protocols* cover a wide range of approaches to studying RNA–protein interactions. They address several broad methodological questions: How do I analyze the structural details of an RNA–protein interaction—what residues contact each other and how strongly do the components interact? If I know one component of a suspected RNA–protein interaction, how do I identify the other(s)? How do I purify RNP complexes from cells? How do I assay the effects of proteins and RNP complexes on mRNA metabolism?

Many of the techniques in this book require in vitro synthesized RNA, either unlabeled, uniformly labeled, or labeled at specific sites. The first chapter describes methods for generating and purifying large quantities of labeled or unlabeled RNAs by in vitro transcription from bacteriophage promoters. Chapter 2 covers the use of ligase to introduce a labeled phosphate at a specific internal location within a long RNA for label-transfer studies.

An important aspect of understanding the structural details of RNA–protein interactions is determining what portions of the two molecules are in close contact. A variety of crosslinking techniques, described in the next four chapters, have been developed for this purpose. Each has advantages and disadvantages, and multiple approaches must often be used to analyze an RNP complex. Chapter 3 describes the use of photoactivatable nucleotide analogs, which are incorporated into RNA during transcription and form covalent crosslinks with adjacent amino acids after irradiation. This method is optimal for crosslinking proteins to single-stranded regions of RNA, but crosslinking protein to double-stranded (ds) RNA is much less efficient. Chapter 4 details a procedure using the intercalating dye methylene blue to generate efficient crosslinks to dsRNA. Although the dye facilitates the reaction, it does not itself form part of the crosslink. Crosslinking methods rely on the close apposition of reactive species, but not all residues form adducts efficiently. The use of photoactivatable nucleotide analogs addresses this problem for the RNA. A similar approach for protein, described in Chapter 5, uses site-directed mutagenesis to introduce cysteine at a specific location in the protein, which is subsequently derivatized with psoralen. Upon irradiation, the psoralen will crosslink with adjacent pyrimidines. The procedures in these chapters have generally been applied to the interaction of a single protein with a single RNA. Obviously, the analysis is considerably more complicated for multi-component RNP complexes. However, mass spectrometry techniques offer new tools for analyzing crosslinking in such RNP complexes. Chapter 6 describes the use of N-terminal sequencing and matrix-assisted laser desorption/ionization mass spectrometry (MALDI-MS) as applied to analyzing RNA–protein crosslinks in the native 30 S ribosomal subunit.

The absence of a crosslink does not imply the lack of close contact between RNA and protein in a particular region, and it is necessary to use additional methods for a complete understanding of the structure of an RNP complex. RNA footprinting and modification interference techniques (Chapter 7) provide a way of assessing all residues in an RNA for their association with protein. RNA footprinting yields information on how the binding of a protein affects the accessibility of an RNA to enzymatic and chemical probes. Modification interference studies allow one to identify bases that are essential for protein binding. In an approach that is conceptually similar to footprinting, short DNA oligonucleotides can be used to compare the regions of an RNA that are available for hybridization and cleavage by RNAse H in the presence and absence of protein binding (Chapter 8).

Several techniques have been developed to provide quantitative measures of the affinity or kinetics of RNA–protein interactions, as well as allowing one to assay for the presence of an RNA-binding protein in a complex mixture. The most commonly used are nitrocellulose filter binding (Chapter 9) and gel retardation assays (Chapter 10). These assays are most effective for relatively strong, stable interactions. The polyacrylamide coelectrophoresis (PACE; Chapter 11) procedure has been developed to study weaker interactions that may not be kinetically stable under the conditions of gel retardation assays. More sophisticated technology is also available: Biosensors employing surface plasmon resonance measurements can provide data on real-time interactions, their stoichiometry, and the effects of other molecules on those interactions. The applications of these instruments to RNA–protein interactions are discussed in Chapter 12.

The analysis of variant protein or RNA sequences, either naturally occurring or experimentally-induced, provides important information on the contributions of different residues to an interaction. Methods that allow screening of populations of variant molecules are described in the next few chapters. Two of the methods utilize bacterial genetic assays, in which the binding of a protein to its RNA target results in either repression of translation (Chapter 13) or transcription antitermination (Chapter 14). In both systems, the level of β-galactosidase expression provides a readout of the extent of interaction. Other procedures utilize in vitro selection of populations of variant molecules. Chapter 15 describes the use of phage display technology to generate a pool of mutated proteins that can be assayed for binding to a specific RNA target. The opposite approach, involving preparation of combinatorial RNA libraries and selection of molecules that bind a particular protein, is discussed in Chapter 16.

Identifying an unknown, biologically relevant partner in an RNA–protein interaction can be a major technical challenge. For example, although selection of a target RNA sequence from a combinatorial library of short RNAs may provide a consensus binding sequence, that information alone may not be sufficient to identify an in vivo target mRNA. Chapter 17 describes the use of libraries of natural RNA sequences to identify such RNAs. Two approaches for identifying unknown proteins are described: screening of expression libraries for binding to an RNA target by a Northwestern assay (Chapter 18), and by a solution-based assay (Chapter 19). The methods in these three chapters assume that the interaction is bimolecular, but that is not always true. Purification of RNP complexes by immunoprecipitation and the identification of component RNAs by PCR-based approaches (Chapter 20) should be particularly useful in situations in which the RNA binding specificity requires multiple proteins, only one of which is known.

Studies of RNP complex composition and function depend critically on the ability to purify the complex from contaminating cellular components.

The remainder of the book provides methods for purification and analysis of RNP complexes, and assays for some of the major RNA–protein interactions that occur in the cell. Chapter 21 describes the use of biotinylated antisense oligoribonucleotides to either purify or deplete specific RNP complexes from cell extracts by affinity chromatography. Immunoaffinity chromatography, either alone or in combination with other separation methods, has also proved to be a useful technique for purification of RNP complexes. A protocol describing its use in purification of the small nuclear RNP (snRNP) complexes required for splicing is given in Chapter 22; immunoaffinity purification of heterogeneous nuclear RNP complexes, which are composed of nuclear proteins and newly synthesized mRNA, is described in Chapter 23. Analyses of the mechanism and regulation of constitutive and alternative splicing are major topics of current research in the RNA field. The next eight chapters provide protocols for such studies in mammalian tissue culture cells and yeast. Chapters 24 and 25 cover the preparation and use of HeLa cell extracts for in vitro splicing assays, and Chapter 26 provides similar protocols for studying splicing in yeast. In vivo, splicing occurs in large multicomponent complexes termed spliceosomes, and Chapters 27 and 28 describe methods for their purification and analysis from mammalian cells and yeast, respectively. The regulation of splicing of mRNAs transcribed from complex transcription units depends on a combination of *cis* elements and *trans* factors. Chapter 29, while not a protocol chapter in the strict sense, discusses strategies and pitfalls in defining the *cis* elements that regulate alternative mRNA splicing. A protocol for selecting functional *cis* elements in vivo from a randomized pool is given in Chapter 30. The SR protein family constitutes a major class of non-snRNP splicing factors that regulate splicing in *trans*, and Chapter 31 describes their purification from tissue culture cells or organs.

The final chapters cover protocols for analyzing mRNA polyadenylation, translation, and turnover. These processes can be crucial control points for regulating the expression of specific mRNAs during cellular metabolism and development. The preparation and use of extracts for studying mRNA 3' end cleavage and polyadenylation reactions are described in Chapter 32. The length of the poly(A) tail on a particular mRNA plays an important role in its stability and translatability; Chapter 33 describes a PCR-based assay for measuring poly(A) tail lengths. The ability to prepare translation extracts (Chapter 34) from different cell types has been important for studying the regulation of translation of specific mRNAs, particularly the cap-independent translation of certain viral and cellular mRNAs. Finally, regulation of mRNA stability can be a crucial aspect of gene regulation, and Chapter 35 provides assays for studying mRNA turnover in cell-free extracts.

The protocols in *RNA–Protein Interaction Protocols* were written with the novice RNA researcher in mind, although a knowledge of basic molecular biology, biochemistry, and cell culture is assumed. They should be particularly useful for the scientist who discovers that the expression of his or her favorite gene is posttranscriptionally regulated, and who wants to know how to proceed in analyzing that regulation. Although every effort has been made to provide a comprehensive protocol book, it is impossible to include all the variations and applications that have been developed for studying RNA-protein interactions. However, a broad range of methods is covered in this book, and many of the protocols can be readily adapted to other systems. These protocols distill a great deal of wisdom and experience, and provide an excellent starting point for investigating many types of RNA–protein interactions.

Susan R. Haynes

Contents

Preface v
Contributors xv

1 Labeling and Purification of RNA Synthesized
   by In Vitro Transcription 1
   Paul A. Clarke

2 Joining RNA Molecules with T4 DNA Ligase 11
   Melissa J. Moore

3 RNA–Protein Crosslinking with Photoreactive Nucleotide Analogs ... 21
   **Michelle M. Hanna, Lori Bentsen, Michael Lucido,
   and Archana Sapre**

4 The Methylene Blue Mediated Photocrosslinking Method for
   Detection of Proteins that Interact with Double-Stranded RNA 35
   Zhi-Ren Liu and Christopher W. J. Smith

5 Probing RNA–Protein Interactions by Psoralen Photocrosslinking 49
   Zhuying Wang and Tariq M. Rana

6 Identification and Sequence Analysis of RNA–Protein Contact Sites
   by N-Terminal Sequencing and MALDI-MS 63
   **Bernd Thiede, Henning Urlaub, Helga Neubauer,
   and Brigitte Wittmann-Liebold**

7 RNA Footprinting and Modification Interference Analysis 73
   Paul A. Clarke

8 Oligonucleotide-Targeted RNase H Protection Analysis
   of RNA–Protein Complexes 93
   Arthur Günzl and Albrecht Bindereif

9 Nitrocellulose Filter Binding for Determination
   of Dissociation Constants 105
   Kathleen B. Hall and James K. Kranz

10 Measuring Equilibrium and Kinetic Constants
   Using Gel Retardation Assays 115
   David R. Setzer

11 PACE Analysis of RNA–Peptide Interactions 129
   Christopher D. Cilley and James R. Williamson

12	Detection of Nucleic Acid Interactions Using Surface Plasmon Resonance 143
	Robert J. Crouch, Makoto Wakasa, and Mitsuru Haruki
13	An *Escherichia coli*-Based Genetic Strategy for Characterizing RNA Binding Proteins 161
	Chaitanya Jain
14	Screening RNA-Binding Libraries Using a Bacterial Transcription Antitermination Assay 177
	Kazuo Harada and Alan D. Frankel
15	In Vitro Genetic Analysis of RNA-Binding Proteins Using Phage Display 189
	Ite A. Laird-Offringa
16	In Vitro Selection of Aptamers from RNA Libraries 217
	Daniel J. Kenan and Jack D. Keene
17	Identification of Specific Protein–RNA Target Sites Using Libraries of Natural Sequences 233
	Lucy G. Andrews and Jack D. Keene
18	Northwestern Screening of Expression Libraries 245
	Paramjeet S. Bagga and Jeffrey Wilusz
19	Screening Expression Libraries with Solution-Based Assay 257
	Philippa J. Webster and Paul M. Macdonald
20	An Immunoprecipitation-RNA:rPCR Method for the In Vivo Isolation of Ribonucleoprotein Complexes 265
	Edward Chu, John C. Schmitz, Jingfang Ju, and Sitki M. Copur
21	Purification and Depletion of RNP Particles by Antisense Affinity Chromatography 275
	Benjamin J. Blencowe and Angus I. Lamond
22	Purification of U Small Nuclear Ribonucleoprotein Particles 289
	Berthold Kastner and Reinhard Lührmann
23	Preparation of Heterogeneous Nuclear Ribonucleoprotein Complexes 299
	Maurice S. Swanson and Gideon Dreyfuss
24	Preparation of Hela Cell Nuclear and Cytosolic S100 Extracts for In Vitro Splicing 309
	Akila Mayeda and Adrian R. Krainer
25	Mammalian In Vitro Splicing Assays 315
	Akila Mayeda and Adrian R. Krainer

Contents

26 Yeast Pre-mRNA Splicing Extracts *323*
    Stephanie W. Ruby
27 Prespliceosome and Spliceosome Isolation and Analysis *351*
    Laura A. Lindsey and Mariano A. Garcia-Blanco
28 A Yeast Spliceosome Assay *365*
    Stephanie W. Ruby
29 Defining Pre-mRNA *cis* Elements that Regulate
    Cell-Specific Splicing *391*
    Thomas A. Cooper
30 In Vivo SELEX in Vertebrate Cells *405*
    Thomas A. Cooper
31 Purification of SR Protein Splicing Factors *419*
    Alan M. Zahler
32 Processing mRNA 3' Ends In Vitro *433*
    Michael J. Imperiale
33 Analysis of Poly(A) Tail Lengths by PCR: *The PAT Assay* *441*
    Fernando J. Sallés and Sidney Strickland
34 In Vitro Translation Extracts from Tissue Culture Cells *449*
    Kazuko Shiroki and Akio Nomoto
35 Messenger RNA Turnover in Cell-Free Extracts
    from Higher Eukaryotes *459*
    Jeff Ross
    Index *477*

Contributors

LUCY G. ANDREWS • *Department of Biology, The University of Alabama at Birmingham, Birmingham, Alabama, USA*
PARAMJEET S. BAGGA • *Department of Biology, Montclair State University, Upper Montclair, New Jersey, USA*
LORI BENTSEN • *Department of Microbiology and Immunology, University of Oklahoma Health Sciences Center, Oklahoma City, Oklahoma, USA*
ALBRECHT BINDEREIF • *Institut für Biochemie, Humboldt-Universität/Charité, Berlin, Germany*
BENJAMIN J. BLENCOWE • *C.H. Best Institute, University of Toronto, Toronto, Ontario, Canada*
EDWARD CHU • *Department of Medicine and Pharmacology, Yale Cancer Center, Yale University School of Medicine and VA Connecticut Healthcare System, New Haven, Connecticut, USA*
CHRISTOPHER D. CILLEY • *Department of Molecular Biology and Skaggs Institute of Chemical Biology, The Scripps Research Institute, La Jolla, California, USA*
PAUL A. CLARKE • *Cancer Research Campaign (CRC) Centre for Cancer Therapeutics, Institute of Cancer Research, Royal Marsden Hospital, Surrey, United Kingdom*
THOMAS A. COOPER • *Departments of Pathology and Cell Biology, Baylor College of Medicine, Houston, Texas, USA*
SITKI M. COPUR • *St. Francis Cancer Center, St. Francis Memorial Health Center, Grand Island, Nebraska, USA*
ROBERT J. CROUCH • *Laboratory of Molecular Genetics, National Institute of Child Health and Human Development, National Institutes of Health, Bethesda, Maryland, USA*
GIDEON DREYFUSS • *Howard Hughes Medical Institute, Department of Biochemistry and Biophysics, University of Pennsylvania School of Medicine, Philadelphia, Pennsylvania, USA*

ALAN D. FRANKEL • *Department of Biochemistry and Biophysics, University of California San Francisco, San Francisco, California, USA*
MARIANO A. GARCIA-BLANCO • *Departments of Pharmacology and Cancer Biology, Microbiology, and Medicine, Levine Science Research Center, Duke University Medical Center, Durham, North Carolina, USA*
ARTHUR GÜNZL • *Abteilung Zellbiologie, Zoologisches Institut der Universität Tübingen, Tübingen, Germany*
KATHLEEN B. HALL • *Department of Biochemistry and Molecular Biophysics, Washington University School of Medicine, St. Louis, Missouri, USA*
MICHELLE M. HANNA • *Department of Chemistry and Biochemistry, University of Oklahoma, Norman, Oklahoma, USA*
KAZUO HARADA • *Department of Life Sciences, Tokyo Gakugei University, Tokyo, Japan*
MITSURU HARUKI • *Department of Material and Life Sciences, Graduate School of Engineering, Osaka University, Osaka, Japan*
MICHAEL J. IMPERIALE • *Department of Microbiology and Immunology and Comprehensive Cancer Center, University of Michigan Medical School, Ann Arbor, Michigan, USA*
CHAITANYA JAIN • *Skirball Institute of Biomolecular Medicine, New York University Medical Center, New York, New York, USA*
JINGFANG JU • *Department of Medicine and Pharmacology, Yale Cancer Center, Yale University School of Medicine and VA Connecticut Healthcare System, New Haven, Connecticut, USA*
BERTHOLD KASTNER • *Institut für Molekularbiologie und Tumorforschung, Philipps Universität Marburg, Marburg, Germany*
JACK D. KEENE • *Department of Microbiology, Combinatorial Sciences Center, Duke University Medical Center, Durham, North Carolina, USA*
DANIEL J. KENAN • *Departments of Pathology and Microbiology, Combinatorial Sciences Center, Duke University Medical Center, Durham, North Carolina, USA*
ADRIAN R. KRAINER • *Cold Spring Harbor Laboratory, Cold Spring Harbor, New York, USA*

Contributors

JAMES K. KRANZ • *Department of Biochemistry and Molecular Biophysics, Washington University School of Medicine, St. Louis, Missouri, USA*

ITE A. LAIRD-OFFRINGA • *Departments of Surgery and Biochemistry and Molecular Biology, Norris Cancer Center, University of Southern California, Los Angeles, California, USA*

ANGUS I. LAMOND • *Department of Biochemistry, University of Dundee, Dundee, Scotland*

LAURA A. LINDSEY • *Department of Pharmacology and Cancer Biology, Levine Science Research Center, Duke University Medical Center, Durham, North Carolina, USA*

ZHI-REN LIU • *Department of Animal and Dairy Science, Auburn University, Auburn, Alabama, USA*

MICHAEL LUCIDO • *Department of Chemistry and Biochemistry, University of Oklahoma, Norman, Oklahoma, USA*

REINHARD LÜHRMANN • *Institut für Molekularbiologie und Tumorforschung, Philipps Universität Marburg, Marburg, Germany*

PAUL M. MACDONALD • *Department of Biological Sciences, Stanford University, Stanford, California, USA*

AKILA MAYEDA • *Department of Biochemistry and Molecular Biology, University of Miami School of Medicine, Miami, Florida, USA*

MELISSA J. MOORE • *Howard Hughes Medical Institute, Department of Biochemistry, Brandeis University, Waltham, Massachussetts, USA*

HELGA NEUBAUER • *Max-Delbrück-Center for Molecular Medicine, Berlin-Buch, Germany*

AKIO NOMOTO • *Department of Microbiology, Institute of Medical Science, The University of Tokyo, Tokyo, Japan*

TARIQ M. RANA • *Department of Pharmacology, Robert Wood Johnson (Rutgers) Medical School, University of Medicine and Dentistry of New Jersey, Piscataway, New Jersey, USA*

JEFF ROSS • *McArdle Laboratory for Cancer Research, Madison, Wisconsin, USA*

STEPHANIE W. RUBY • *Department of Molecular Genetics and Microbiology, University of New Mexico Health Sciences Center, Albuquerque, New Mexico, USA*

FERNANDO J. SALLÉS • *Department of Cellular and Molecular Pharmacology, University Medical Center at Stony Brook, Stony Brook, New York, USA*
ARCHANA SAPRE • *Department of Chemistry and Biochemistry, University of Oklahoma, Norman, Oklahoma, USA*
JOHN C. SCHMITZ • *Department of Medicine and Pharmacology, Yale Cancer Center, Yale University School of Medicine and VA Connecticut Healthcare System, New Haven, Connecticut, USA*
DAVID R. SETZER • *Department of Molecular Biology and Microbiology, School of Medicine, Case Western Reserve University, Cleveland Ohio, USA*
KAZUKO SHIROKI • *Department of Microbiology, Institute of Medical Science, The University of Tokyo, Tokyo, Japan*
CHRISTOPHER W. J. SMITH • *Department of Biochemistry, University of Cambridge, Cambridge, United Kingdom*
SIDNEY STRICKLAND • *Department of Cellular and Molecular Pharmacology, University Medical Center at Stony Brook, Stony Brook, New York, USA*
MAURICE S. SWANSON • *Department of Molecular Genetics and Microbiology and Centers for Gene Therapy and Mammalian Genetics, University of Florida College of Medicine, Gainesville, Florida, USA*
BERND THIEDE • *Wittman Institute of Technology and Analysis of Biomolecules, WITA GmBH, Teltow, Germany*
HENNING URLAUB • *Institut für Molekularbiologie und Tumorforschung, Philipps Universität Marburg, Marburg, Germany*
MAKOTO WAKASA • *Department of Research and Development, Kishimoto Clinical Laboratory Group 2-3-9, Hiyoshi-cho, Tomakomai-shi, Hokkaido, Japan*
ZHUYING WANG • *Department of Pharmacology, Robert Wood Johnson (Rutgers) Medical School, University of Medicine and Dentistry of New Jersey, Piscataway, New Jersey, USA*
PHILIPPA J. WEBSTER • *ZymoGenetics, Inc., Seattle, Washington, USA*
JAMES R. WILLIAMSON • *Department of Molecular Biology and Skaggs Institute of Chemical Biology, The Scripps Research Institute, La Jolla, California, USA*

Contributors

JEFFREY WILUSZ • *Department of Microbiology and Molecular Genetics, University of Medicine and Dentistry of New Jersey—New Jersey Medical School, Newark, New Jersey, USA*

BRIGITTE WITTMANN-LIEBOLD • *Max-Delbrück-Center for Molecular Medicine, Berlin-Buch, Germany*

ALAN M. ZAHLER • *Department of Biology and Center for Molecular Biology of RNA, Sinsheimer Laboratories, University of California Santa Cruz, Santa Cruz, California, USA*

1

Labeling and Purification of RNA Synthesized by In Vitro Transcription

Paul A. Clarke

1. Introduction

The problems of isolating sufficient quantities of rare RNAs for detailed biochemical analysis can be circumvented by synthesis of the desired RNA in vitro *(1–3)*. Early methods of in vitro transcription included the use of eukaryotic cell extracts or *Escherichia coli* RNA polymerase to transcribe DNA templates containing the appropriate promoter. Ideally, however, the optimal in vitro transcription system should require simple buffer components without need for preparation of extracts and should precisely initiate and terminate transcription at definable sites. In vitro bacteriophage transcription systems fulfill these criteria. Single-stranded RNA of the desired sequence can now be synthesized using commercially available SP6, T3, and T7 bacteriophage RNA polymerases that have a very high specificity for their respective promoters. Large quantities of RNA can be synthesized and used as substrates in assays involving translation, RNA processing, microinjection, or transfection. The RNA can be end-labeled for structural analysis or for examining RNA–protein interactions. Alternatively, the RNA can be internally labeled during transcription and used as a riboprobe for Southern/Northern blots, for RNase protection assays, or in the assays described previously.

1.1. Principle of the Procedure

SP6, T3, and T7 bacteriophage RNA polymerases and plasmid vectors containing multiple cloning sites flanked by bacteriophage promoters are now available from a large number of commercial sources. The DNA to be transcribed is inserted into the polylinker site of the plasmid vector using a restriction endonuclease site downstream of the promoter element. The cloned

From: *Methods in Molecular Biology, Vol. 118: RNA-Protein Interaction Protocols*
Edited by: S. Haynes © Humana Press Inc., Totowa, NJ

plasmid is then used as the DNA template for transcription. In preparation for transcription, the template is linearized by restriction endonuclease digestion (*see* **Note 1**). The template is then purified and transcribed. Uniform RNA of a single defined length is produced because the RNA polymerase initiates at the defined promoter site and "runs off" the end of the linear template. The 3' end of the RNA is defined by the choice of restriction endonuclease used to linearize the template. There are, however, several potential drawbacks to the plasmid-based approach. In most instances it is difficult to insert the DNA fragment exactly at the site of transcription initiation, while at the 3' end of the template, convenient restriction endonuclease sites located exactly at the 3' end of the sequence to be transcribed are often lacking. This results in RNAs containing additional vector sequences at both the 5' and 3' ends of the RNA (**Fig. 1**). In some circumstances these extra sequences will not influence the outcome of the experiment. However, if the RNA is used for structural analysis, protein binding, or functional studies the extra sequences may influence both structure and function. This is especially critical in the case of small RNAs in which the extra sequence can exert a large influence. In these circumstances it is crucial that the transcribed sequence is free of vector sequences. This can be achieved easily by using the polymerase chain reaction (PCR) to place the template exactly at the transcription initiation site by using a PCR primer containing the core bacteriophage promoter sequence (**Fig. 1**). The 3' end of the RNA is defined by the position of the downstream PCR primer. The PCR fragment can be used directly in the transcription reaction or if desired can be cloned into a plasmid vector by the addition of restriction endonuclease recognition sequences to the PCR primers (*see* **Note 2**). It is also possible to produce small RNAs using a single-stranded oligonucleotide as the template, provided the core promoter sequence is double-stranded *(4)*.

The protocols outlined in this chapter describe the in vitro synthesis and gel purification of unlabeled, internally labeled, or end-labeled RNA.

2. Materials

1. RNase inhibitor (RNasin or equivalent).
2. T3, T7, or SP6 bacteriophage RNA polymerases.
3. T4 polynucleotide kinase.
4. T4 RNA ligase.
5. Calf intestinal phosphatase.
6. RNase-free DNase I.
7. RNase-free bovine serum albumin.
8. Distilled phenol, phenol/chloroform (1:1 v/v), and chloroform.
9. Acrylamide, *bis*-acrylamide, 10% ammonium persulfate (w/v) and TEMED (*N,N,N',N'*-tetramethylethylenediamine).

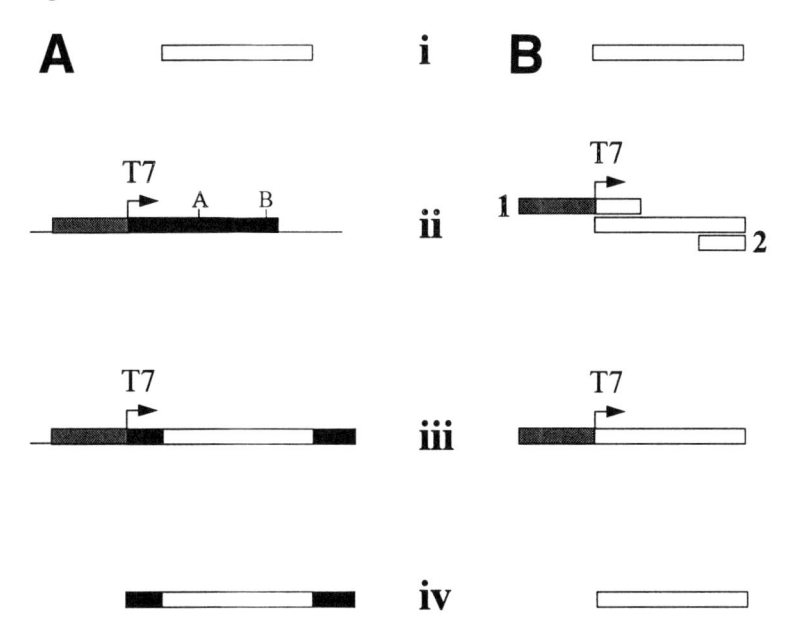

Fig. 1. The DNA template (**i**) to be transcribed is either (**A; ii**) cloned into site A of the multiple cloning site (black box) of a vector containing a bacteriophage RNA polymerase promoter (gray box; transcription initiates at the arrow) or (**B; ii**) is used as a PCR template with primers 1 and 2. Primer 1 also contains a T7 RNA polymerase promoter sequence (gray box). The templates for transcription (**iii**) are prepared either by (**A**) linearizing the vector by restriction endonuclease digestion at site B or (**B**) by PCR. The templates (**iii**) are transcribed to give RNA products (**iv**). The RNA product from (**A**) will contain additional sequences from the multiple cloning site sequences (black boxes), whereas the RNA product from the PCR-generated template (**B**) will contain no extra sequences.

10. Nucleotide triphosphates (ATP, CTP, GTP, and UTP). Prepare a stock solution containing each NTP at 25 mM.
11. Dichlorodimethylsilane.
12. Dimethyl sulfoxide (DMSO).
13. 10 mM CaCl$_2$.
14. 5 M NaCl.
15. Rad tape (Sigma-Aldrich Ltd, Poole, Dorset, BH12 4QH, UK).
16. Siliconized 0.5- and 1.5-mL tubes.
17. Scotch 3M electrical tape (Life Technologies, Paisley, PA4 9RF, UK).
18. [α-^{32}P]NTP (800 Ci/mmol; 10 mCi/mL).
19. [γ-^{32}P]ATP (3000 Ci/mmol; 10 mCi/mL).
20. [^{32}P]pCp (3000 Ci/mmol; 10 mCi/mL).
21. DEPC-H$_2$O: add 1 mL of diethyl pyrocarbonate (DEPC) to 1 L of double-distilled H$_2$O, mix vigorously, and autoclave (*see* **Note 3**).

22. 0.5 *M* EDTA, pH 8.0: dissolve 186 g of ethylenediaminetetraacetic acid·2H$_2$O in 800 mL of DEPC-H$_2$O and adjust to pH 8.0 with NaOH (approx 20 g). Adjust to 1 L with DEPC-H$_2$O, autoclave, aliquot, and store at room temperature.
23. 10× TBE: dissolve 54 g of Tris base and 27.5 g of boric acid in 20 mL of EDTA, pH 8.0, and add DEPC-H$_2$O to 1 L. Autoclave and store at room temperature.
24. 30% Acrylamide/*bis*-acrylamide mix: dissolve 145 g of acrylamide and 5 g of *bis*-acryla-mide in DEPC-H$_2$O to 500 mL, filter sterilize, and store at 4°C in a light-tight bottle.
25. Gel mix: for 500 mL of a denaturing 10% gel mix dissolve 240 g of urea in 25 mL of 10× TBE and 167 mL of 30% acrylamide/*bis*-acrylamide mix and add DEPC-H$_2$O to 500 mL. Some gentle heating may be required to dissolve the urea. Filter sterilize and store at 4°C in a light-tight bottle. Occasionally the urea will crystallize out of solution during storage, but can be redissolved by gentle warming to room temperature. Care must be taken to pour the gel quickly, as warming the gel mix will significantly increase the rate of polymer-ization. This is more of a problem with high percentage gels. The percentage of poly-acrylamide in the gel is adjusted according to the size of product being purified (*see* **Subheading 3.5.**).
26. 5× Transcription reaction buffer: 200 m*M* Tris-HCl, pH 7.8, 20 m*M* dithiothreitol (DTT), 100 m*M* spermidine, 7 m*M* MgCl$_2$; filter sterilize and store at –20°C.
27. TE pH 7.4: 10 m*M* Tris-HCl, pH 7.4, 1 m*M* EDTA, pH 8.0.
28. 10× Phosphatase reaction buffer: 0.5 *M* Tris-HCl, pH 8.0, 1 m*M* EDTA, pH 8.0; autoclave and store at –20°C.
29. 10× T4 polynucleotide kinase buffer: 0.5 *M* Tris-HCl, pH 7.6, 100 m*M* MgCl$_2$, 50 m*M* DTT, 1 m*M* spermidine, 1 m*M* EDTA, pH 8.0; filter sterilize and store at –20°C.
30. 1.5× gel loading buffer: 10 *M* urea, 1.5× TBE, 0.015% (w/v) bromophenol blue, 0.015% (w/v) xylene cyanol; filter sterilize and store at –20°C.
31. 10× T4 RNA ligase reaction buffer: 0.5 *M* HEPES-KOH, pH 7.5, 33 m*M* DTT, 150 m*M* MgCl$_2$; filter sterilize and store at –20°C.
32. TE/sodium dodecyl sulfate (SDS): add 2.5 mL of 20% (w/v) SDS to 97.5 mL of TE, pH 7.4.

3. Methods
3.1. In Vitro Transcription

The following protocol has been optimized for a 160-nucleotide RNA using T7 bacteriophage RNA polymerase *(5,6)*. If large amounts of RNA are required the reaction can be scaled up accordingly.

1. Either digest 10 µg of the transcription vector containing the sequence to be transcribed with the appropriate restriction endonuclease or amplify the appropriate sequence by PCR as outlined in **Subheading 1.1.** (*see* **Notes 1** and **2**). Purify the DNA template on an aga-rose gel following digestion or PCR (*see* **Note 4**).
2. At *room temperature* (*see* **Note 5**) combine 4 µL of 5× transcription reaction buffer, 3.2 µL of 25 m*M* NTPs, 0.8 µL (25 U) of RNasin, 1 µL of 1 mg/mL gel-

purified DNA, 2 µL (20 U) of T7 bacteriophage RNA polymerase, and 9 µL of DEPC-treated H_2O. Incubate at 37°C for 2 h (*see* **Note 6**). To avoid problems with condensation use a cabinet incubator or overlay with RNase-free mineral oil and incubate in a water bath or heating block.

3. The DNA template is removed by the addition of 2 µL of 10 mM $CaCl_2$ and RNase-free DNase I to 20 µg/mL followed by incubation at 37°C for 15 min. Add 1 µL of 0.5 M EDTA, pH 8.0, 3 µL of 5 M NaCl, and TE, pH 7.4, to a final volume of 100 µL (*see* **Note 7**).
4. Add an equal volume of phenol/chloroform. Vortex the mixture vigorously for 1–2 min and microcentrifuge at room temperature for 10 min at full speed (approx 16,000g). Recover the upper aqueous phase and extract with an equal volume of chloroform as before.
5. Collect the upper aqueous phase and recover the RNA by the addition of 2.5 vol of ethanol. Incubate in a dry ice/ethanol bath for 10–20 min. Precipitate the RNA by microcentrifugation at 4°C for 15 min at 16,000g. Wash the RNA pellet once with ice-cold 70% ethanol and pellet as before.
6. Resuspend the pellet in DEPC-treated H_2O and store at –80°C. A number of factors will influence the yield; 25–50 µg/reaction is usual, but can be as high as 100 µg/reaction (*see* **Notes 6** and **8**).
7. The quality of the RNA should be checked by electrophoresis on a 2% agarose minigel (*see* **Note 9**). If required, the RNA can be purified on a polyacrylamide gel (*see* **Notes 4** and **10**) (**Subheading 3.5.**).

3.2. Internal Labeling of Substrate RNA

The RNA can be internally labeled during the transcription process by incorporation of a radiolabeled precursor (*see* **Note 11**).

1. Perform the in vitro transcription exactly as described under **Subheading 3.1.** in the presence of 5 µL of [α–^{32}P]UTP or CTP (800 Ci/mmol; 10 mCi/mL). Omit the unlabeled UTP or CTP or keep at a concentration of 10–50 µM (*see* **Note 10**). Incubate for a maximum of 1 h.
2. Following DNase digestion load the RNA on a denaturing polyacrylamide gel and purify the full-length labeled RNA as described in **Subheading 3.5.** (*see* **Note 7**).

3.3. 5' End-Labelling of Substrate RNA

RNA is labeled at its 5' end by T4 polynucleotide kinase. This enzyme transfers ^{32}P from [γ-^{32}P]ATP onto the 5' end of the RNA molecule. However, to 5' end-label one has to first remove the 5' phosphate using calf intestinal phosphatase.

1. Dilute 1–2 µg of RNA to a final volume of 16 µL in DEPC-treated H_2O. Denature the RNA by incubation at 95°C for 2 min followed by quenching on ice for 5 min.
2. Add the remaining components: 2 µL of 10× phosphatase reaction buffer and 2 µL of calf intestinal phosphatase (1 U/µL); incubate at 37°C for 30 min.

3. Dilute the RNA to 200 µL with TE, pH 7.4, add 6 µL of 5 M NaCl, and extract/precipitate the RNA as described in **Subheading 3.1**.
4. Resuspend the RNA to 0.5 µg/µL in TE, pH 7.6. Denature 2 µL of the RNA at 95°C for 2 min and quench on ice for 5 min (*see* **Note 12**).
5. Add 1 µL of 10× T4 polynucleotide kinase buffer, 4 µL of [γ-^{32}P]ATP (3000 Ci/mmol; 10 mCi/mL), and 1 µL of T4 polynucleotide kinase (3–6 U/µL). Incubate for 30 min at 37°C.
6. Add 20 µL of 1.5× gel loading buffer and purify the labeled RNA by denaturing polyacrylamide gel electrophoresis (**Subheading 3.5.**).

3.4. 3' End-Labeling of Substrate RNA

The addition of label to the 3' ends of RNA uses T4 RNA ligase and an excess of [^{32}P]pCp that is ligated to the 3' end of the RNA.

1. Denature 1–2 µg of RNA by heating to 68°C for 2 min and then quenching on ice for 5 min (*see* **Note 13**).
2. Add 4 µL of 10× T4 RNA ligase reaction buffer, 4 µL of DMSO, 1 µL of 20 µM ATP, 1 µL of 10 µg/mL RNase-free bovine serum albumin, 1–2 µg of RNA, 5–10 µL of [^{32}P]pCp (3000 Ci/mmol; 10 mCi/mL), and DEPC-treated H_2O to a final volume of 40 µL. Incubate overnight at 4°C (*see* **Note 13**).
3. Add 80 µL of 1.5× gel loading buffer and purify the labeled RNA by denaturing polyacrylamide gel electrophoresis (**Subheading 3.5.**).

3.5. Polyacrylamide Gel Purification of Labeled RNA

For almost all purposes gel purification of the full-length RNA is essential. The majority of commercially available or homemade sequencing systems can be used. The protocol described here is for the Gibco-BRL S2 system using 31 cm × 38.5 cm × 0.4 mm gels.

1. Clean the gel plates scrupulously with soap and water and rinse in deionized water. Treat the gel plates by wiping with 5% dichlorodimethylsilane (in chloroform), allow to dry in a fume hood, and finally wipe with ethanol and acetone. Assemble the plates with two spacers and secure using bulldog clips or Scotch 3M electrical tape.
2. Mix 75 mL of gel mix (**Subheading 2.5.**) with 75 µL of TEMED and 450 µL of fresh 10% ammonium persulfate. Immediately pour the mix between the plates. A number of alternate techniques for pouring gels can be employed according to personal preference. Insert the gel comb and allow 15–45 min for the gel to polymerize. To accommodate the larger volumes of the labeling reactions use a comb that forms a 2–4 cm well (usually cut from an old sharkstooth comb or spacer).
3. Remove the comb and immediately remove unpolymerized acrylamide by flushing the wells with deionized water or running buffer (0.5× TBE). Assemble the apparatus using 0.5× TBE (**Subheading 2.5.**) as the running buffer and prerun the gel at a constant 55 W power setting until the external gel plate is hot to touch (30–60 min). Incubate the sample for 10 min at 68°C and load immediately. Elec-

trophorese the sample at 55 W constant power until the full-size product migrates to approximately a third to a halfway down the gel. The time of running depends on the size of the product and percentage of gel used. On 5%, 10%, and 20% gels RNAs of 135, 60, and 25 nucleotides, respectively, migrate with the xylene cyanol dye front, while RNAs 35, 10, and 5 nucleotides in length migrate with the bromophenol blue dye front.

4. Disassemble the apparatus, taking care with the bottom buffer tank (containing unincorporated radionucleotides), and separate the plates using a thin spatula or razor blade. Cover the gel/plate with plastic wrap. ^{32}P labeled RNA can be detected by direct autoradiography as detailed in **step 5**, whereas unlabeled RNA is detected by UV shadowing as described in **step 6**.

5. Stick several fluorescent markers (*see* **Note 14**) around the edge of the gel and autoradiograph briefly (generally 1–5 min) to locate the intact RNA species. Mark the band corresponding to the full-length RNA on the autoradiograph, align the film with the gel using the image of the fluorescent markers, and secure firmly with tape. Cut out the band using a sterile scalpel, remove the plastic wrap from the gel slice, and transfer the gel slice to a 1.5-mL sterile screw-capped Eppendorf tube. The gel can be reautoradiographed to confirm that the correct band has been cut.

6. To UV shadow, turn the plastic wrap covered plate over. Carefully remove the other gel plate and cover the gel with plastic wrap. Place the plastic wrapped gel on a fluorescent thin-layer chromatography plate or an intensifying screen. The RNA can be visualized using a hand-held UV light set at a short wavelength (254 nm). The RNA band will be seen as a dark purplish band and is cut using a sterile scalpel. Remove the plastic wrap from the gel slice and place the gel slice in a 1.5-mL sterile screw-capped Eppen-dorf tube.

7. Elute the RNA from the gel slice by the addition of 400 µL of TE/SDS and incubate for 2–4 h at room temperature on a rotator. Remove the supernatant to a fresh screw-capped tube and store. Add an additional 400 µL of TE/SDS elution buffer to the gel slice and elute for an additional 2–4 h.

8. Extract the supernatants with equal volumes of phenol/chloroform and chloroform as described in **Subheading 3.1**.

9. Precipitate the RNA by the addition of 12.5 µL of 5 *M* NaCl and 2.5 vol of ethanol as described in **Subheading 3.1**. Resuspend the RNA in a small volume of DEPC-treated H_2O. The cpm of labeled RNA can be quickly estimated by Cerenkov counting the aqueous solution in a scintillation counter.

4. Notes

1. When linearizing the transcription vector it is often preferable to use a double digest; this minimizes copurifying undigested plasmid that can give rise to high molecular weight RNAs. In addition, Schenborn and Mierindorf *(7)* have reported the production of complementary RNA from templates that have 3' overhanging ends (e.g., those produced by *Kpn*I, *Pst*I, and *Sac*I). To avoid this problem it is best to choose restriction endonucleases that give a blunt

end or a 5' overhang. Mellits et al. *(6)* have also reported the presence of low levels of double-stranded RNAs (dsRNAs) that are produced during in vitro transcription with bacteriophage RNA polymerases. Their presence may complicate analysis of proteins with dsRNA binding motifs (dsRBMs) *(5)*. In this case sequential purification through denaturing and native polyacrylamide gels is required for their complete removal.

2. The following promoter sequences should be added to the PCR primer that is located upstream of the sequence to be transcribed. The core promoter sequence is in the bold type-face.

    T3 — 5'-GCATGC**AATTAACCCTCACTAAA**GGG-3'
    T7 — 5'-GCATGC**TAATACGACTCACTATA**GGG-3'
    SP6 — 5'-GCATGC**ATTTAGGTGACACTATA**GAA-3'

    The first six nucleotides downstream of the promoter are important for transcription efficiency *(4)*. This is especially true for the first three nucleotides, GGG for T3 and T7 and GAA for SP6, that should not be changed.

3. All stock solutions or reaction buffers should be made with DEPC-treated double-distilled H_2O and sterilized by autoclaving or filtration.

4. It is not always necessary to gel purify templates generated by PCR, provided the PCR reaction gives a single product. However, before transcription we generally remove the PCR primers using one of a number of commercially available PCR "clean-up" kits.

5. The transcription reaction should be set up at room temperature, as the reaction buffer contains spermidine which will precipitate the template DNA at low temperatures.

6. The yield of RNA is somewhat template dependent and is also affected by the choice of polymerase, the incubation temperature, and incubation time. Two hours at 37°C is suggested as a starting point and will usually give a yield sufficient for most purposes. The yield of transcription can be improved by increasing either the incubation time or temperature (up to 40°C) or both (if a large yield is crucial; *see* **ref. 3** for further details).

7. During polyacrylamide gel purification some template DNA may copurify with the RNA; therefore DNase digestion is absolutely necessary prior to gel purification.

8. The yields of RNA can be estimated by measuring absorbance at A_{260} ($1OD_{260}$ = 40 µg/mL). Unincorporated nucleotides may coprecipitate with the full-length RNA and will influence the A_{260}. If accurate quantitation is required it is best to either gel purify the full-length RNA or remove unincorporated nucleotides by gel filtration through Sephadex G-25 or G-50. Alternatively, two sequential precipitations with 1/10 vol of 5 M NH_4OAc and 3 vol of ethanol will eliminate most of the unincorporated nucleotides. If radioactive nucleotides are included, incorporation can be estimated by trichloroacetic acid precipitation and scintillation counting.

9. Agarose minigels are usually run under nondenaturing conditions. RNAs analyzed in this way will occasionally give multiple bands, probably due to the single RNA species adopting a number of different structural conformations. These can usually be resolved to a single band by heating the RNA sample to 68°C and quenching on ice prior to analysis. If there is still concern, the RNA should be

analyzed on denaturing urea polyacrylamide gels or denaturing formaldehyde agarose gels. If prematurely terminated RNA products are detected, the most likely cause is the presence of sequences resembling factor-independent transcription termination sequences. These consist of a GC-rich hairpin loop followed by a poly(U) tract. In most cases sufficient full-length RNA can be still be recovered by gel purification. If larger yields are required, premature termination can sometimes be overcome by increasing nucleotide concentrations or lowering the incubation temperature while increasing the reaction time *(8)*.
10. RNAs can either be trace labeled for gel purification purposes or labeled at high specific activity for other uses. The concentration of nonradioactive limiting nucleotide will depend on the required specific activity. The greater the concentration of unlabeled limiting nucleotide the lower the specific activity of the RNA. For trace labeling full concentrations of all nucleotides and 1–10 µCi of an [α-^{32}P]NTP are used. For very high specific activities the nonradioactive nucleotide should be omitted. At nucleotide concentrations <5 µ*M* the yield of full-length RNA will decline dramatically (especially true for longer RNAs). Generally the final concentration of limiting nucleotide (nonradioactive plus radioactive) should be between 10 and 50 µ*M*. The specific activity of the labeled nucleotide should be around 800 Ci/mmol; higher specific activity label is generally less concentrated, requiring the addition of unlabeled nucleotide to get the concentration above 10 µ*M*. If the limiting nucleotide is found within approximately the first 10 nucleotides of the substrate there may be problems with premature termination. This can be avoided by increasing the limiting nucleotide concentration or by using a nucleotide not found in the first 10 nucleotides. We have found this is occasionally a problem with [α-^{32}P]CTP and T3 bacteriophage RNA polymerase and can also be a problem with [α-^{32}P]UTP *(9)*. Another problem with limiting nucleotide concentrations is premature termination of transcription that is template dependent. This can be overcome by increasing nucleotide concentrations or by lowering the incubation temperature, but increasing the reaction time *(8)*.
11. This method can also be used for nonisotopic labeling of RNA. Biotin-14-CTP, biotin-16-UTP, digoxigenin-11-CTP, and fluorescein-12-UTP can all be incorporated into RNA by in vitro transcription with bacteriophage RNA polymerases.
12. T4 polynucleotide kinase requires single-stranded 5' ends for efficient labeling and does not work efficiently on recessed 5' ends. To avoid this problem the RNA should be denatured prior to labeling. If 5' labeling is still a problem it is possible to "cap" in vitro transcripts using [α-^{32}P]GTP and guanylyltransferase (GTP transferase) *(10)*.
13. T4 RNA ligase prefers single-stranded 3' termini, and prior denaturation of the RNA may improve labeling. If necessary the efficiency of 3' labeling can also be improved by increasing the time of incubation to a maximum of 96 h.
14. After being marked with an ordinary pen and briefly exposed to light, the fluorescent labels will give a unique image on the autoradiograph (within 1–2 min). An alternative is to cut the fluorescent label into unique shapes that can be easily identified.

Acknowledgments

The author is supported by the Cancer Research Campaign.

References

1. Melton, D. A., Krieg, P. A., Rebagliati, M. R., Maniatis, T., Zinn, K., and Green, M. R. (1984) Efficient *in vitro* synthesis of biologically active RNA and RNA hybridization probes from plasmids containing a bacteriophage SP6 promoter. *Nucleic Acids Res.* **12,** 7035–7056.
2. Krieg, P. A. and Melton, D. A. (1987) In vitro synthesis with SP6 RNA polymerase. *Methods Enzymol.* **155,** 397–415.
3. Gurevich, V. V. (1996) Use of bacteriophage RNA polymerase in RNA synthesis. *Methods Enzymol.* **275,** 382–397.
4. Milligan, J. F., Groebe, D. R., Witherell, G. W., and Uhlenbeck, O. C. (1987) Oligoribo-nucleotide synthesis using T7 RNA polymerase and synthetic DNA template. *Nucleic Acids Res.* **15,** 8783–8798.
5. Clarke, P. A. and Mathews, M. B. (1995) Interactions between the double-stranded RNA binding motif and RNA: definition of the binding site for the interferon-induced protein kinase (PKR) on adenovirus VA RNA. *RNA* **1,** 7–20.
6. Mellits, K. H., Pe'ery, T., Manche, L., Robertson, H. D., and Mathews, M. B. (1990) Removal of double-stranded contaminants from RNA transcripts: synthesis of adenovirus VA RNA_I from a T7 vector. *Nucleic Acids Res.* **18,** 5401–5406.
7. Schenborn, E. T. and Mierindorf, R. C. (1985) A novel transcription property of SP6 and T7 RNA polymerases: dependence on template structure. *Nucleic Acids Res.* **13,** 6223–6236.
8. Krieg, P. A. (1991) Improved synthesis of full length RNA probe at reduced incubation temperatures. *Nucleic Acids Res.* **18,** 6463
9. Ling, M. L., Risman, S. S., Klement, J. P., McGraw, N., and McAllister, W. T. (1989) Abortive initiation by bacteriophage T3 and T7 polymerase under conditions of limiting substrate *Nucleic Acids Res.* **17,** 1605–1618.
10. Knapp, G. (1989) Enzymatic approaches to probing of RNA secondary and tertiary structure. *Methods Enzymol.* **180,** 192–212.

2

Joining RNA Molecules with T4 DNA Ligase

Melissa J. Moore

1. Introduction

Researchers interested in studying RNA structure and function or RNA–protein interactions are increasingly using site-specifically modified RNAs to probe sites of interest *(1)*. In addition to expanding the repertoire of functional groups beyond the very limited set available through standard mutagenesis, current synthesis techniques allow for inclusion of radioactive labels and/or various crosslinking reagents at single internal sites in RNA. The latter probes are particularly useful for determining what proteins or other RNAs interact with a site of interest on the test RNA. This chapter deals mainly with recently developed methodology for introducing such modifications into long RNAs.

Although it is relatively straightforward to generate, either by complete chemical synthesis *(2)* or transcription *(3,4)*, short RNA oligomers containing internal site-specific modifications, incorporating such modifications into longer RNAs requires additional steps. The general strategy is first to make the RNA in segments, one of which contains the desired modification, and then join these fragments together via some sort of ligation reaction. RNA ligations can be performed either enzymatically or chemically. During the ligation reaction, a single radioactive label can be readily introduced at the ligation junction. The following section presents the advantages and disadvantages of each ligation option. The remainder of the chapter focuses on the use of T4 DNA ligase to incorporate a single site-specific radioactive phosphate into long RNAs for label transfer experiments.

1.1. Techniques for Ligation of RNA

Prior to 1992 *(5)*, RNA ligations were performed either chemically or with T4 RNA ligase. Although both of these approaches have advantages in particular situations, both also have important limitations. Chemical coupling with cyanogen bromide or water-soluble carbodiimides *(6,7)* allows for synthesis of both natural and unnatural internucleotide linkages (such as 2'-5' phosphodiesters, pyrophosphates, and 5'-N-P or 3'-N-P phosphoramides). However, these reactions can be quite slow (0.2–6 d) and are prone to side reaction of the coupling reagents with base functional groups in single-stranded regions. T4 RNA ligase *(8,9)* has been used to prepare numerous site-specifically modified RNAs, particularly modified oligonucleotides and tRNAs with anti-codon loop modifications. This enzyme is also commonly used for 3' end-labeling of RNA with [5'-^{32}P]pCp (*see*, e.g., **ref. *10***). But its very high K_m (>1 m*M*) for polynucleotides and its propensity for production of side products (*see* **ref. *11***) limit its general usefulness for ligating multiple long RNA fragments.

The third option is T4 DNA ligase. This enzyme (once known as polynucleotide ligase *[12,13]*) joins nicks in double-stranded duplexes, including (RNA:RNA)/DNA hybrids *[14,15]*. T4 DNA ligase presents several advantages over the other methods for ligating RNA *[11]*. First, its low K_m for polynucleotide duplexes (from 10^{-8} to 10^{-7} *M*) means that the ligation reactions can proceed in reasonable times even when substrate concentrations are in the submicromolar to micromolar range. Second, because this enzyme is specific for nicks in double-stranded duplexes, production of undesired side products is kept to a minimum. An added advantage of this feature is that three- and even four-way ligations can be performed simultaneously in a single reaction, as long as the RNAs do not crosshybridize with each other's DNA templates. For these reasons, T4 DNA ligase is often the best starting place for attempting to join long RNA molecules, where in many cases only picomole amounts of starting material can be obtained and production of side products must be kept to a bare minimum.

1.2. Choice and Preparation of RNA Ligation Substrates

The RNA segments to be joined in a ligation reaction can be prepared in any number of ways. Short segments (generally fewer than 25 nts) can be either synthesized chemically *(2)* or transcribed from partially double-stranded deoxyoligonucleotide templates with T7 RNA polymerase *(3)*. Longer fragments can be transcribed from PCR products or an appropriate linearized plasmid template (*4; see* **Notes 1** and **3**). Alternatively, if a natural RNA is required (i.e., one that contains any post-transcriptional modifications normally added in vivo), the RNA of interest can be purified from a cellular extract and, if necessary, cleaved site-specifically with RNase H and an appropriate complementary oligonucleotide *(16,17)*.

No matter how the RNAs to be ligated are initially prepared, they must have the appropriate 5' and 3' ends. T4 DNA ligase absolutely requires a 3'-OH on the acceptor substrate and a 5' monophosphate on the donor substrate — a 5' triphosphate or 5'-OH will not work. If the donor RNA is synthesized chemically, a 5' monophosphate can either be incorporated during the synthesis or added later using labeled or unlabeled ATP and T4 polynucleotide kinase. If the donor RNA is prepared by transcription with T7 polymerase, it would normally start with 5'-pppG; in this case the RNA must be completely dephosphorylated with a phosphatase and then rephosphorylated with ATP and T4 polynucleotide kinase. Alternatively, if a labeled phosphate at the ligation junction is not required, then the RNA can be transcribed with a ready-made 5'-monophosphate by simply including GMP at a four- to five-fold excess over GTP in the transcription reaction. Another approach is to initiate the RNA with a 5'-NpG-3' dinucleotide primer (again at four- to five-fold excess over GTP in the transcription reaction) and then kinase the product. This latter approach is often the easiest for incorporating a labeled phosphate at the ligation junction *(11)*.

1.3. Choice and Preparation of cDNA Template

Generally, a cDNA oligonucleotide template that extends at least 10 nucleotides on either side of the ligation junction is sufficient for efficient ligations, provided there is neither significant RNA secondary structure in the area nor a high percentage of Us in the duplex region. In these cases, lengthening the template can often dramatically increase the ligation efficiency. There seems to be no upper limit to the length of the cDNA template (*see* **Notes 3** and **4**).

2. Materials
2.1. T4 DNA Ligase

T4 DNA ligase is available from any number of commercial sources, including (but not limited to) Amersham/USB (Arlington Heights, IL, USA), New England Biolabs (Beverly, MA, USA) and Promega (Madison, WI, USA). It is important to note that because T4 DNA ligase is relatively inefficient at joining RNA molecules, it is required at near stoichiometric concentrations in the ligation reactions. This means that if you want to ligate 10 pmol of RNA, then you need to include 10 pmol of enzyme in the reaction. A good rule of thumb for calculating how much enzyme you'll need is that 1 Weiss Unit of T4 DNA ligase corresponds to approx 1 pmol of enzyme. Note, however, that some companies (e.g., New England Biolabs) use different activity units.

Because so much enzyme is required, the cost of doing large-scale ligations (or even multiple small-scale reactions) using commercially obtained enzyme can become prohibitive. In this case, a His-tagged version of the protein that you can prepare yourself is also available *(18)*.

T4 DNA ligase is quite stable for at least a year at $-20°C$, but should not be stored at any lower temperature.

2.2. Other Reagents for Phosphorylation and Ligation Reactions

1. RNAs to be joined (in water; see **Note 1**).
2. T4 polynucleotide kinase (T4 PNK) (usually 10 U/µL).
3. 10× Ligase buffer: 500 mM Tris-HCl, pH 7.5, 100 mM MgCl$_2$, 200 mM dithiothreitol (DTT), 0.5 mg/mL bovine serum albumin (BSA) (for ligase from New England Biolabs, Beverly MA, USA), or 660 mM Tris-HCl, pH 7.6, 66 mM MgCl$_2$, 100 mM DTT (for ligase from Amersham/USB, Arlington Heights, IL, USA).
4. [γ-^{32}P]ATP (10 mCi/mL, 3000 Ci/mmol = 3.33 µM)
5. Oligonucleotide cDNA template (in water or TE).
6. RNase-free water (either DEPC-treated or from a MilliQ [Millipore, Bedford, MA, USA] or similar deionization system).
5. 10 mM ATP.
6. RNasin (Promega, Madison, WI, USA) — optional.
7. RQ1 RNase-free DNase (Promega, Madison, WI, USA).

2.3. Other Items

1. SpeedVac™ concentrator.
2. 5× TBE: 54 g of Tris base, 27.5 g of boric acid, and 20 mL of 0.5 M EDTA, pH 8.0, per liter.
3. 2× Denaturing gel loading buffer: 10 M urea, 1× TBE; or 80% formamide, 1× TBE.
4. Denaturing polyacrylamide gel (*see* Chapter 1).
5. TE: 10 mM Tris-HCl, pH 8.0, 1 mM EDTA.

3. Methods

One of the most common uses for joining of RNAs with T4 DNA ligase is to incorporate a single radiolabeled phosphate having extremely high specific activity at a single internal site. One use of such singly labeled RNAs is for short-wave ultraviolet (UV) crosslinking experiments to identify proteins interacting with the site of interest (*see*, e.g., Chapters 3 and 6, and **ref.** *19* and references therein). To increase crosslinking efficiencies, the label can be juxtaposed with a modified base superactivated for photocrosslinking. Selected examples of such bases include 4-thiouracil *(20)*, 6-thio-guanosine *(21)*, 2,6-diaminopurine *(22)*, and thiol-containing nucleotides derivatized with benzophenone *(23,24)* or azidophenacyl bromide *(25,26)*. Proteins or other RNA molecules reacting with the photocrosslinking group thus become "tagged" with the adjacent labeled phosphate. This section describes a general protocol for incorporating a single labeled phosphate of high specific activity at the ligation junction between two RNA segments. It assumes that the 3' ligation piece initially has a free 5'-hydroxyl. The individual steps of the protocol are schematized in **Fig. 1**.

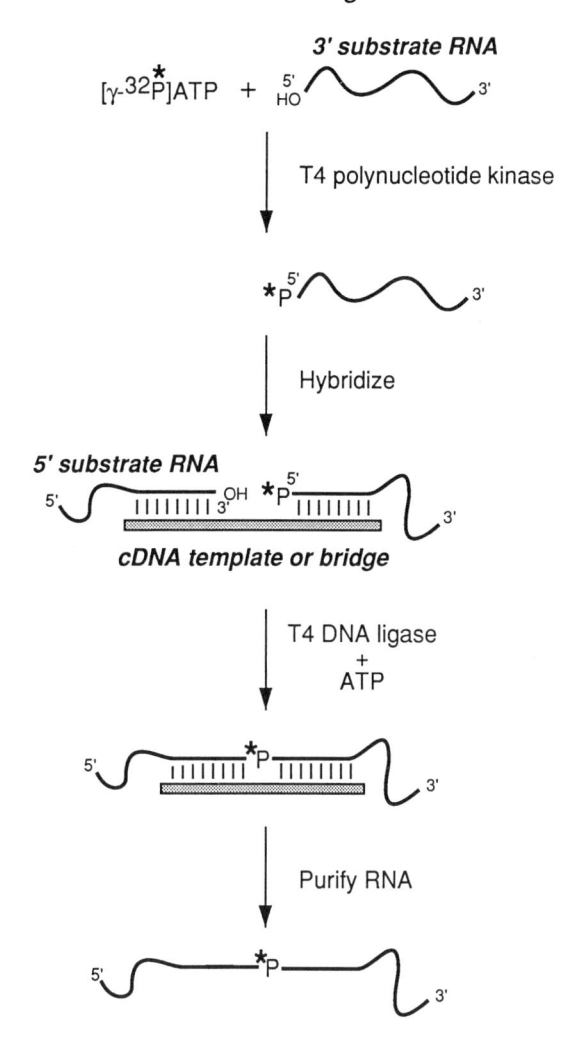

Fig. 1. Overview of strategy for incorporating a single radiolabeled phosphate at an internal site in an RNA molecule using T4 polynucleotide kinase and T4 DNA ligase.

3.1. Phosphorylation of the 3' RNA

1. In the Speed-Vac™, dry down 5 µL of [γ-^{32}P]ATP (50 µCi, 16 pmol) in a 0.5- or 1.5-mL tube. This is easily done in a closed microcentrifuge tube with a hole made in the top using a 25-gauge needle to minimize chances of accidental radioactive contamination.
2. To the tube, add in the following order: 0.5 µL of 10× ligation buffer, 3.0 µL of 20 µM 3' RNA, 1.0 µL of water, and 0.5 µL of T4 PNK. The final volume is 5.0 µL.

3. Incubate at 37°C for 30–60 min.
4. Inactivate the kinase by incubating at 75°C for 10 min or 92–95°C for 2 min. Store on ice.

3.2. Ligation Reaction

1. To the inactivated kinase reaction, add 1.0 µL of 10× ligation buffer, 3.0 µL of 20 µ*M* 5' RNA, and 3.0 µL of 20 µ*M* cDNA bridge (*see* **Note 2**).
2. Hybridize the RNA and DNA pieces by heating at 75°C for 2 min, followed by 5 min at room temperature.
3. To the hybridized reaction, add 1.0 µL of 10 m*M* cold ATP and 2.0 µL of T4 DNA ligase. The total volume should be 15 µL.
4. Incubate at 30°C for 2–4 h.
5. Add 1 µL of RQ1 RNase-free DNase. Incubate at 37°C for 15–30 min.
6. Add 16 µL of 2× gel loading buffer and load into a large well on a denaturing polyacrylamide gel. Labeled 5' and 3' RNAs, as well as full-length RNA that corresponds to the ligated product, should be run as markers.

4. Notes

1. Runoff transcription reactions often generate substantial amounts of both short aborted transcripts and full-length RNAs with one or more extra nucleotides at their 3' ends (so-called "N + 1" products). Because the short transcripts that aborted soon after initiation have the same 5' end sequence as the full-length transcript, these oligonucleotides can significantly interfere with hybridization of a 3' substrate to the cDNA template during the ligation reaction. For this reason it is advisable to gel-purify transcribed 3' substrates prior to ligation reactions; desalting columns (e.g., Sephadex G-50) are not particularly effective at eliminating abortive transcripts. In contrast, the 5' ligation substrate need only be gel filtered prior to ligation, as its abortive initiation products should be too short to interfere (unless of course the 5' end of the 5' substrate is significantly similar in sequence to the 5' end of the 3' substrate). At the opposite end of the RNA, little can often be done about the "N + 1" transcripts since they are usually almost identical in size to the desired transcript. Luckily, T4 DNA ligase discriminates against such longer 5' substrates and does not incorporate them into ligated products *(5)*.
2. For efficient ligation to occur, it is crucial that the concentration of the bridging cDNA template be equal to or intermediate between those of the RNA ligation substrates. If excess template is present, then individual substrate molecules can hybridize to different template molecules and not have a ligation partner. For this reason, any attempt to "drive" the hybridization by adding excess template is futile. However, if one RNA substrate is much more "precious" than the other, then increasing the concentration of the less precious RNA over that of the more precious RNA and the cDNA can lead to higher ligation efficiencies. If this is contemplated, it is advisable to titrate the concentration of the more abundant RNA to determine what conditions work best. We generally perform ligation reactions with RNA substrate concentrations between 0.1 and 10 µ*M*.

3. There seems to be no upper limit to the length of the RNA substrates joinable by T4 DNA ligase, as long as there is no strong propensity for either RNA substrate to form secondary or tertiary structure in the region that could interfere with the ability of the cDNA template to hybridize. If such problems are encountered, however, remedies include lengthening the cDNA template to improve its hybridization or the inclusion of a second "disrupter" oligonucleotide to break up tertiary interactions (27). At the opposite extreme, RNA substrates as short as six nucleotides can be ligated quantitatively in the standard incubation time of 2 h. Ligation products can also be obtained with four- and five-nucleotide substrates, but only with longer incubation times. We have not been able to detect ligation products for substrates of 3 nts or fewer (Suntharalingam, M., Dulude, E., and Moore, M. J., unpublished data).
4. Often experimental design necessitates three-way (and sometimes even four-way [28]) ligations. If the middle piece is short, then both ligation junctions can be bridged by a single long cDNA oligonucleotide. Alternatively individual ligation sites can be bridged with different cDNAs, as long as their basepairing regions do not overlap. Since the overall yield of any multipart ligation is the product of the yields of the individual ligations, it is advisable to ensure first that each individual ligation reaction is proceeding as efficiently as possible before attempting the multipart ligation.

Further Reading

1. Moore, M. J. and Query, C. C. (1998) Uses of site-specifically modified RNAs constructed by RNA ligation, in *RNA–Protein Interactions: A Practical Approach* (Smith, C., ed.), Academic, New York, pp. 75–108.
2. Zimmermann, R. A., Gait, M. J. and Moore, M. J. (1998) Incorporation of modified nucleotides into RNA for studies on RNA structure, function and intermolecular interactions, in *Modification and Editing of RNA: The Alteration of RNA Structure and Function* (Grosjean, H. and Benne, R., eds.), ASM Press, Washington, D.C., pp. 59–84.

References

1. Zimmermann, R. A., Gait, M. J. and Moore, M. J. (1998) Incorporation of modified nucleotides into RNA for studies on RNA structure, function and intermolecular interactions, in *Modification and Editing of RNA: The Alteration of RNA Structure and Function* (Grosjean, H. and Benne, R., eds.), ASM Press, Washington, D.C., pp. 59–84.
2. Eckstein, F., ed. (1991) *Oligonucleotides and Analogues: A Practical Approach.* IRL Press, Oxford.
3. Milligan, J. F. and Uhlenbeck, O. C. (1989) Synthesis of small RNAs using T7 RNA polymerase. *Methods Enzymol.* **180**, 51–62.
4. Chabot, B. (1994) Synthesis and purification of RNA substrates, in *RNA Processing, Vol. I: A Practical Approach,* (Higgins, S. J. and Hames, B. D.), IRL Press, Oxford, pp. 1–30.

5. Moore, M. J. and Sharp, P. A. (1992) Site-specific modification of pre-mRNA: the 2' hydroxyl groups at the splice sites. *Science* **256**, 992–997.
6. Dolinnaya, N. G., Sokolova, N. I., Ashirbekova, D. T., and Shabarova, Z. A. (1991) The use of BrCN for assembling modified DNA duplexes and DNA-RNA hybrids; comparison with water-soluble carbodiimide. *Nucleic Acids Res.* **19**, 3067–3072.
7. Shabarova, Z. A. (1988) Chemical development in the design of oligonucleotide probes for binding to DNA and RNA. *Biochimie* **70**, 1323–1334.
8. Uhlenbeck, O. C. and Gumport, R. I. (1982) T4 RNA Ligase, in *The Enzymes* (Boyer, P. D., ed.), Academic, New York, pp. 31–58.
9. Romaniuk, P. J. and Uhlenbeck, O. C. (1983) Joining of RNA molecules with RNA ligase. *Methods Enzymol.* **100**, 52–59.
10. Enright, C. and Sollner-Webb, B. (1994) Ribosomal RNA processing in vertebrates, in *RNA Processing, Vol. II: A Practical Approach* (Higgins, S. J. and Hames, B. D., eds.), IRL Press, Oxford, pp. 135–172.
11. Moore, M. J. and Query, C. C. (1998) Uses of site-specifically modified RNAs constructed by RNA ligation, in *RNA-Protein Interactions: A Practical Approach* (Smith, C., ed.), IRL Press, Oxford, pp. 75–108.
12. Fareed, G. C., Wilt, E. M., and Richardson, C. C. (1971) Enzymatic breakage and joining of deoxyribonucleic acid. VIII. Hybrids of ribo- and deoxyribonucleotide homopolymers as substrates for polynucleotide ligase of bacteriophage T4. *J. Biol. Chem.* **246**, 925–932.
13. Kleppe, K., Van de Sande, J. H., and Khorana, H. G. (1970) Polynucleotide ligase-catalyzed joining of deoxyribo-oligonucleotides on ribopolynucleotide templates and of ribo-oligonucleotides on deoxyribopolynucleotide templates. *Proc. Natl. Acad. Sci. USA* **67**, 68–73.
14. Higgins, N. P. and Cozzarelli, N. C. (1979) DNA-joining enzymes: a review. *Methods Enzymol.* **68**, 50–71.
15. Engler, M. J. and, C. C. Richardson (1982) DNA Ligases, in *The Enzymes* (Boyer, P. D., ed.) Academic, New York, pp. 3–29.
16. Inoue, H., Hayase, Y., Iwai, S., and Ohtsuka, E. (1987) Sequence-dependent hydrolysis of RNA using modified oligonucleotide splints and RNase H. *FEBS Lett.* **215**, 327–330.
17. Lapham, J. and Crothers, D. M. (1996) RNase H cleavage for processing of in vitro transcribed RNA for NMR studies and RNA ligation. *RNA* **2**, 289–296.
18. Strobel, S. A. and Cech, T. R. (1995) Minor groove recognition of the conserved G. U pair at the Tetrahymena ribozyme reaction site. *Science* **267**, 675–679.
19. Chiara, M. D., Gozani, O., Bennett, M., Champion-Arnaud, P., Palandjian, L., and Reed, R. (1996) Identification of proteins that interact with exon sequences, splice sites, and the branchpoint sequence during each stage of spliceosome assembly. *Mol. Cell. Biol.* **16**, 3317–3326.
20. Sontheimer, E. J. (1994) Site-specific RNA crosslinking with 4-thiouridine. *Mol. Biol. Rep.* **20**, 35–44.

21. Ping, Y. H., Liu, Y., Wang, X., Neenhold, H. R., and Rana, T. M. (1997) Dynamics of RNA-protein interactions in the HIV-1 Rev-RRE complex visualized by 6-thioguanosine-mediated photocrosslinking. *RNA* **3**, 850–860.
22. Query, C. C., Strobel, S. A., and Sharp, P. A. (1996) Three recognition events at the branch site adenine. *EMBO J.* **15**, 1392–1402.
23. MacMillan, A. M., Query, C. C., Allerson, C. R., Chen, S., Verdine, G. L., and Sharp, P. A. (1994) Dynamic association of proteins with the pre-mRNA branch region. *Genes Dev.* **8**, 3008–3020.
24. Musier-Forsyth, K. and Schimmel, P. (1994) Acceptor helix interactions in a class II tRNA synthetase: photoaffinity cross-linking of an RNA miniduplex substrate. *Biochemistry* **33**, 773–779.
25. He, B., Riggs, D. L., and Hanna, M. M. (1995) Preparation of probe-modified RNA with 5- mercapto-UTP for analysis of protein-RNA interactions. *Nucleic Acids Res* **23**, 1231–1238.
26. Burgin, A. B. and Pace, N. R. (1990) Mapping the active site of ribonuclease P RNA using a substrate containing a photoaffinity agent. *EMBO J.* **9,** 4111–4118.
27. Strobel, S. A. and Cech, T. R. (1993) Tertiary interactions with the internal guide sequence mediate docking of the P1 helix into the catalytic core of the Tetrahymena ribozyme. *Biochemistry* **32**, 13,593–13,604.
28. Moore, M. J. and Sharp, P. A. (1993) The stereochemistry of pre-mRNA splicing: evidence for two active sites in the spliceosome. *Nature* **365**, 364–368.

3

RNA–Protein Crosslinking with Photoreactive Nucleotide Analogs

Michelle M. Hanna, Lori Bentsen, Michael Lucido, and Archana Sapre

1. Introduction

Photochemical crosslinking is a powerful technique for characterization of RNA–protein interactions in ribonucleoprotein complexes. Intermolecular crosslinks can be generated without chemical modification of either the RNA or the protein by irradiation of native complexes with short wavelength ultraviolet (UV) light (254 nm). Covalent crosslinks form when the nucleotides or amino acids in the RNA and protein are photochemically converted to reactive species. This method has the advantage of allowing the study of interactions in which neither macromolecule has been modified, which can, in some cases, alter the way that the protein and RNA make contact. However, this approach produces very low crosslinking yields and often leads to nicking of the RNA and degradation of some proteins. This photodamage to the RNA and protein makes subsequent mapping of the position(s) of the crosslink(s) at the nucleotide and amino acid level difficult. Crosslinking by this method is also relatively nonspecific, making localization of the protein contacts to a single nucleotide or region in the RNA fairly difficult.

Here we describe an alternative approach to RNA–protein crosslinking that utilizes nucleotide analogs that can be incorporated into RNA during transcription with bacteriophage T7 or *E. coli* RNA polymerases. The RNA produced after incorporation of these analogs can be photoactivated by irradiation with ultraviolet light having a wavelength greater than 300 nm, where both the RNA and protein sustain minimal photodamage. The nucleotide analogs contain functional groups that are chemically inert in the absence of light but can be converted to chemically reactive species upon irradiation. The yield of

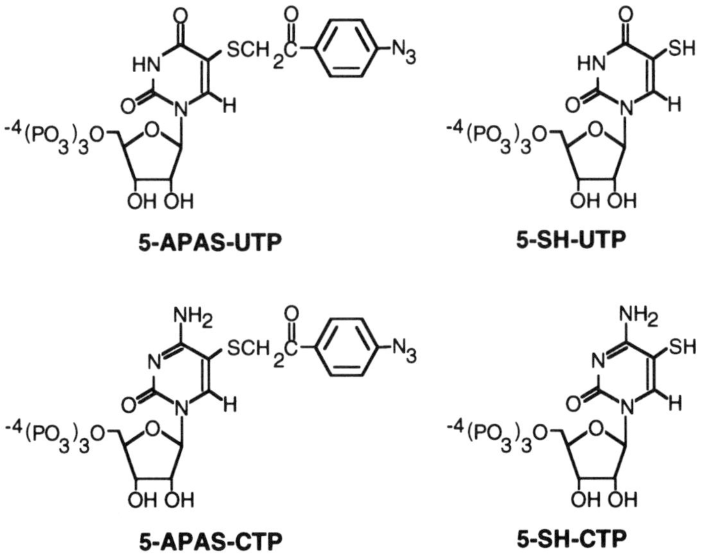

Fig. 1. Nucleotide analogs for analysis of protein–nucleic acid interactions. Four ribonucleoside triphosphate analogs are described. 5-azido(phenacylthio)-uridine-5'-triphosphate [5-APAS-UTP] and 5-azido(phenacylthio)-cytidine-5'-triphosphate [5-APAS-CTP] are photocrosslinking ribonucleotide analogs that are substrates for *E. coli* and T7 RNA polymerases. These analogs can be incorporated directly at internal or 3' terminal positions in RNA during transcription. The precursors to these analogs, 5-mercapto-UTP [5-SH-UTP] and 5-mercapto-CTP [5-SH-CTP], are not themselves photoreactive. After incorporation of these precursors into RNA during transcription, the RNA can be posttranscriptionally tagged with a photoreactive crosslinking group by alkylation with *p*-azidophenacylbromide (APB) or any of several other commercially available photocrosslinking reagents. Some of these other reagents have longer linker arms between the base and the crosslinker and/or have cleavable groups within the linker arm.

RNA–protein crosslinks is higher than that obtained by direct irradiation with short wavelength ultraviolet light. In addition, neither the RNA nor the protein is significantly nicked by the irradiation process, thereby producing covalently linked ribonucleoprotein complexes that are suitable for precise mapping of the crosslink at the nucleotide or peptide/amino acid level.

Although a variety of photocrosslinking nucleotide analogs have been developed, methods for incorporation and use of only two are described here. However, these methods are generally applicable for use with several other photocrosslinking ribonucleoside-5'-triphosphates. The two described here, 5-azido(phenacylthio)-UTP (5-APAS-UTP, **refs.** *1–3*) and 5-azido(phenacylthio)-CTP (5-APAS-CTP, **refs.** *4* and *5*), are modified on the C-5 position of a pyrimidine base (**Fig. 1**), which is not involved in normal Watson–Crick base pairing. They are therefore capable of specific hydrogen bonding in both RNA–DNA and RNA–RNA hybrids, and both function well as substrates for *E. coli* and T7 RNA polymerases in transcription elongation. They both contain a reactive aryl azide group approx 13 Å from the base that can be photoactivated for crosslinking upon irradiation with long wavelength UV light. Upon photoactivation of analog-tagged RNA, it can be covalently attached to proteins, or other macromolecules, with which it may have direct interactions. Therefore, nucleic acid binding domains in complexes can be characterized, nucleic acid binding proteins in extracts can be identified, and determinants involved in these specific interactions can be characterized at the level of individual nucleotides and amino acids.

The precursors to these analogs, 5-mercapto-UTP (5-SH-UTP, **ref. *4***) and 5-mercapto-CTP (5-SH-CTP, **ref. *5***) are also substrates for *E. coli* and T7 RNA polymerases. They are not themselves photoreactive. However, if they are incorporated into RNA, the reactive thiol groups can be posttranscriptionally alkylated to attach a photoreactive crosslinker. By attaching the crosslinking groups after the RNA has been synthesized and has adopted its folded tertiary structure, only nucleotides that are exposed to the surface of the RNA molecule will be modified.

The methods are divided into two parts (**Fig. 2**). The conditions for preparation and isolation of the analog-tagged RNA are described first (**Fig. 2A**), followed by the methods for photocrosslinking and detection of crosslinked proteins (**Fig. 2B**). The appropriate ratio of analog to the normal nucleotide substrate must be determined empirically for each RNA. This ratio depends upon both the sequence and length of the RNA being made. RNA synthesized with 5-APAS-UTP or 5-APAS-CTP can be used directly for protein–RNA crosslinking. RNA synthesized with 5-SH-UTP or 5-SH-CTP must be posttranscriptionally alkylated for attachment of the photocrosslinking group.

2. Materials

2.1. Synthesis of Full-Length RNA Containing Photocrosslinking Nucleotide Analogs Throughout

1. Distilled, deionized, autoclaved water is to be used in the preparation of all buffers and solutions (*see* **Note 1**).
2. Buffer A: 20 mM Tris-OAc, pH 8.0, 10 mM Mg(OAc)$_2$, 50 mM potassium glutamate, 5% (v/v) glycerol, 40 mM Na$_2$EDTA, pH 8.0, 40 µg/mL of acetylated bovine serum albumin (AcBSA); store at –20°C (*see* **Note 2**).
3. Buffer B: 1.5 M NH$_4$OAc, 37.5 mM Na$_2$EDTA, pH 8.0, 50 µg/mL RNase-free tRNA. Store at –20°C. To prepare RNase-free tRNA, extract tRNA 5× as described in **Subheading 3.1.**, **steps 6–9**, and then dialyze into distilled, deionized, autoclaved water.

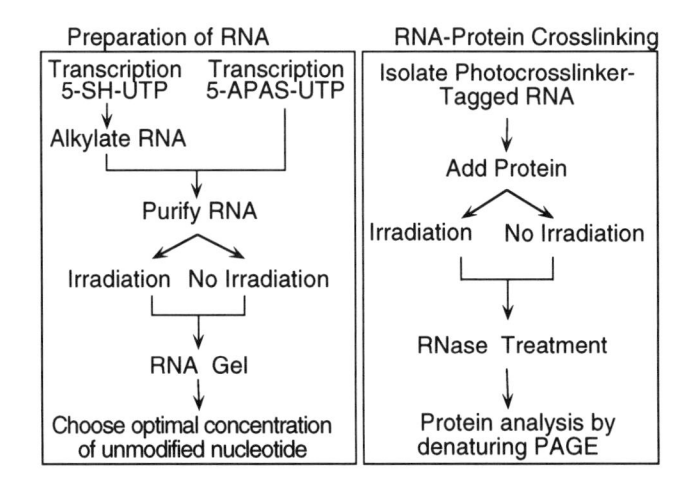

```
Preparation of RNA                          RNA-Protein Crosslinking
┌─────────────────────────────────┐  ┌──────────────────────────────┐
│ Transcription    Transcription  │  │ Isolate Photocrosslinker-    │
│  5-SH-UTP         5-APAS-UTP    │  │      Tagged RNA              │
│      ↓                          │  │           ↓                  │
│  Alkylate RNA                   │  │       Add Protein            │
│      └──────────┬─────────┘     │  │         ↙     ↘              │
│                 ↓               │  │  Irradiation  No Irradiation │
│             Purify RNA          │  │         └────┬───┘           │
│              ↙     ↘            │  │              ↓               │
│      Irradiation  No Irradiation│  │      RNase Treatment         │
│         └────┬────┘             │  │              ↓               │
│              ↓                  │  │      Protein analysis by     │
│           RNA Gel               │  │      denaturing PAGE         │
│              ↓                  │  └──────────────────────────────┘
│  Choose optimal concentration   │
│     of unmodified nucleotide    │
└─────────────────────────────────┘
```

Fig. 2. Preparation of analog-tagged RNA. (**A**) RNA that is labeled throughout with photoreactive crosslinking groups can be prepared two different ways. When it is desirable to look at crosslinking between proteins and only those residues on the surface of the RNA after it has adopted its tertiary structure, RNA is synthesized with a precursor nucleotide, 5-SH-UTP or 5-SH-CTP, and the crosslinker is added to the RNA after synthesis. Alternatively, photocrosslinker-tagged analogs can be incorporated throughout the RNA before secondary and tertiary structures are adopted by using 5-APAS-UTP or 5-APAS-CTP during transcription. In either case, the amount of UTP or CTP that must also be present to produce full-length, photoreactive RNA must be determined empirically for each RNA by analyzing its mobility on a gel before and after irradiation. The band corresponding to RNA that contains photocrosslinkers will smear or disappear completely from the gel after irradiation. RNA that contains no crosslinker will run as a sharp band in both the irradiated and unirradiated lanes. (**B**) Once photocrosslinker-tagged RNA has been made, by either method, the RNA is isolated and incubated with the protein(s). An aliquot is removed and kept in the dark, and the remainder of the sample is irradiated to crosslink the protein to the RNA. The RNA is digested with RNase T1 prior to analysis of the proteins by denaturing polyacrylamide gel electrophoresis (*see* **Fig. 3**).

4. Buffer C (2×): 8 M urea, 0.1% (w/v) xylene cyanol, 0.1% (w/v) bromophenol blue, in autoclaved water.
5. FPLC-purified ribonucleotides, ATP, CTP, GTP, and UTP (Pharmacia LKB Biotechnologies, Inc., Newark, NJ, USA); store at –80°C.
6. 5-SH-UTP, 5-SH-CTP, 5-APAS-UTP, and 5-APAS-CTP are synthesized from UTP or CTP (*1,2,4,5*) or are commercially available from Oklahoma Biolabs, Inc., Norman, OK, USA. All analogs should be stored in water at concentrations of 1–3 M and kept in the dark at –80°C (*see* **Notes 3** and **4**).
7. T7 bacteriophage RNA polymerase or *E. coli* RNA polymerase (Epicenter Technologies, Madison, WI, USA).
8. [α-^{32}P] GTP, used at between 10^6 and 10^7 cpm/pmol in the reactions (*see* **Note 5**).

9. 1 M NH$_4$OAc dissolved in 100% ethanol, store at room temperature.
10. Heparin: 0.5 mg/mL in water; store at –20°C.
11. Buffer F: 40 mM Tris-HCl, pH 7.6, 15 mM MgCl$_2$, 10 mM β-mercaptoethanol, 50 µg/mL of AcBSA. Use autoclaved stocks of 1 M Tris-HCl, pH 7.6, and 1 M MgCl$_2$; store in 1-mL aliquots at –20°C.
12. Phenol:chloroform:isoamyl alcohol (25:24:1).
13. 80% (v/v) Ethanol.
14. RQ1 RNase-free DNase (Promega, Madison, WI, USA).
15. Spectroline model XX-15B UV light source (Spectroline, Westbury, NY, USA), 1800 µW/cm^2 at 15 cm (or a comparable UV light source, λ_{max} = 302 nm).
16. 1-mL Polystyrene colorless microcentrifuge tubes (Robbins Scientific, Sunnyvale, CA, USA; cat. no. 10100-00-0).

2.2. Synthesis of Full-Length RNA Containing Photocrosslinking Nucleotide Analogs Only on the Surface of the RNA

1. 100 mM Azidophenacyl bromide (APB, Sigma) dissolved in dimethyl sulfate (*see* **Note 6**).
2. Sephadex G-50 DNA grade resin (Pharmacia).
3. Buffer D: 30 mM Tris-HCl, pH 7.0; 10 mM KCl; 0.5 mM Na$_2$EDTA, pH 8.0.

2.3. Crosslinking of RNA to Protein

1. Buffer C: Refer to **Subheading 2.1.4**.
2. 10× TBE buffer: 1 M Tris base, 0.97 M boric acid, 0.63 M EDTA in autoclaved water; store at room temperature.
3. For a 40 cm × 0.75 cm gel, prepare 75 mL of a 10% (w/v) polyacrylamide/7 M urea (acrylamide:methylene *bis*-acrylamide, 19:1) solution: Dissolve 7.1 g of acrylamide, 0.4 g of *bis*-acrylamide, and 30.5 g of urea in 10–20 mL of distilled, deionized, autoclaved water and 7.5 mL of 10× TBE. After the urea dissolves, bring the volume to 75 mL with distilled water (*see* **Note 7**). Degas the solution under vacuum for 10 min, and then add 375 µL of 10% (w/v) ammonium persulfate (make fresh daily) and 37.5 µL *of N,N,N',N'*-tetramethylethylenediamine (TEMED). Pour the gel immediately.
4. X-ray film (Hyperfilm-MP, Cronex), Intensifying screen, Autoradiography Cassette.
5. RNase T1 digestion buffer: 25 mM sodium citrate, 1 mM Na$_2$EDTA, pH 8.0, stored at –20°C.
6. RNase T1.
7. 2× Protein loading buffer: 60 mM Tris-HCl, pH 8.0, 60 mM dithiothreitol (DTT), 3.4% (w/v) sodium dodecyl sulfate (SDS), 17% (v/v) glycerol, 0.02% (w/v) bromophenol blue. Add DTT just before use. Use an autoclaved stock solution of 1 M Tris-HCl, pH 8.0, and store the buffer in 1-mL aliquots at –20°C.
8. Denaturing Tris-glycine SDS polyacrylamide gel (Novex; San Diego, CA, USA). The appropriate percentage of acrylamide to use depends on the protein size, as follows: proteins over 100 kDa, 4% or 6%; 50–100 kDa, 8%; 25–50 kDa, 10% or 12%; below 25 kDa, 16%.

9. Transfer buffer: 25 mM Tris, 192 mM glycine, 20% (v/v) methanol in a final volume of 1 L, in water; store at 4°C.
10. Nitrocellulose membrane.
11. Electroblotting apparatus (such as the Fisher semidry blotting apparatus).
12. Silver staining solution: 2% (w/v) sodium citrate, 0.8% (w/v) ferrous sulfate, 0.1% (w/v) silver nitrate (compound is light sensitive and should be stored in foil); store at 4°C.

3. Methods

3.1. Synthesis of Full-Length RNA Containing Photocrosslinking Nucleotide Analogs Throughout (Fig. 2A)

1. In a final reaction volume of 100 µL, mix 100 nM template DNA (calculated in terms of promoter concentration, with one promoter per template) and 100 nM *E. coli* RNA polymerase or T7 RNA polymerase (in buffer A for *E. coli* RNA polymerase or buffer F for T7 RNA polymerase), and incubate for 5 min at 37°C (*5*).
2. In the dark, add reagents to a final concentration of: 100 µM ATP, 100 µM CTP or UTP (depending on the analog used), 20 µM [α-^{32}P] GTP, 150 µM nucleotide analog (5-APAS-UTP or 5-APAS-CTP) and in separate reactions include increasing concentrations of UTP or CTP (0 µM, 1 µM, 5 µM, 10 µM, 20 µM) (*see* **Note 8**). Incubate at 37°C for 15 min.
3. Add 1.0 µL of heparin and incubate for 10 min at 37°C (*see* **Note 9**).
4. Add RQ1 RNase-free DNase I (5 U) and incubate at 37°C for 10 min.
5. Add MgCl$_2$ to 100 mM (final) and add 20 µg of RNase-free tRNA.
6. Add an equal volume (100 µL) of phenol:chloroform:isoamyl alcohol (25:24:1), vortex, and centrifuge at 14,000g for 5 min in a microcentrifuge.
7. Remove the top, aqueous layer and place in a clean microcentrifuge tube. Add 3 vol (300 µL) of ethanol/1 M NH$_4$OAc, mix, and place in a dry ice/ethanol bath for 10 min. Centrifuge at 14,000g for 5 min.
8. Remove the supernatant fraction, without disturbing the RNA pellet. Rinse the pellet with 100 µL of 80% ethanol. Centrifuge at 14,000g for 5 min.
9. Remove the supernatant fraction, without disturbing the RNA pellet. Dry the pellet in a Speed-Vac with the heat on until no liquid remains on the pellet (approx 10 min).
10. Resuspend the RNA pellet in 100 µL of distilled, deionized, autoclaved water or buffer.
11. Split the reaction in half and place one half in a 1-mL colorless, polystyrene microcentri-fuge tube. Place the polystyrene tube 1.5 cm above the UV light source (302 nm max for these analogs) and irradiate the sample for 2 min at room temperature (*see* **Subheading 2.1.**, **step 15**). Keep the other half of the reaction in the dark at room temperature during the irradiation process.
12. Add DTT to a final concentration of 100 mM to all reactions to reduce any remaining azide (irradiated and nonirradiated), and leave them in the dark for at least 5 min. Manipulations after this step can be carried out with the lights on.

13. Add buffer C to a final concentration of 1× and heat the samples for 2 min at 95°C. Quick-chill the samples on ice and then load them on a 7 M urea/polyacrylamide gel (loading 5 µL of sample is sufficient). The percentage of polyacrylamide, as well as the length of the gel, depends upon the size of RNA being analyzed (e.g., 6% for more than 100 nt, 15% for 30–100 nt, 25% for fewer than 30 nt). Gels should be run so that they are slightly warm to the touch, and electrophoresis wattage and duration depends upon the gel length and RNA size.
14. Gels that are 4–15% polyacrylamide should be dried before autoradiography. Higher percentage gels often crack upon drying and can be exposed without drying. Place the gel in a film cassette on a piece of filter paper. Place a piece of plastic wrap on the gel, a piece of X-ray film on the plastic wrap, and an intensifying screen on the film. Expose the film overnight at –80°C.
15. Determine from the autoradiogram which UTP (or CTP) concentration used in the titration is optimal for production of full-length RNA that is still photoreactive (*see* **Note 10**).
16. Proceed to **Subheading 3.3.** for protein–RNA crosslinking.

3.2. Synthesis of Full-Length RNA Containing Photocrosslinking Nucleotide Analogs Only on the Surface of the RNA

1. Prepare and purify RNA containing 5-SH-UMP or 5-SH-CTP (*4*) as described in **Subheading 3.1., steps 1–10**, using 5-SH-UTP or 5-SH-CTP as the nucleotide analog. Adjust the pH to between 6.0 and 7.0.
2. In the dark, add APB (dissolved in dimethyl sulfate) to a final concentration of 5 m*M* (*see* **Note 11**).
3. Incubate at room temperature for 2 h in the dark.
4. Prepare a Sephadex G-50 column in the following manner: Weigh out 7 g of Sephadex G-50 (DNA grade) and add 100 mL of distilled, deionized water. Autoclave the slurry formed. Add 1 mL of the G-50 slurry to a 1-mL minicolumn with a filter at the bottom and centrifuge in an IEC model clinical centrifuge at approx 3500*g* (setting 5) for 3 min. Repeat as necessary until the column contains approx 750 µL of packed resin after centrifugation (*see* **Note 12**).
5. Add 200 µL of 1× buffer D to equilibrate the column and centrifuge at 3500*g* for 4 min. Repeat this step three times. Discard the buffer washes.
6. In the dark, carefully load 200 µL of the reaction directly onto the resin without touching the walls of the column. It is important that the ratio of resin volume to sample volume is 750:200.
7. Centrifuge the column in a clinical centrifuge at approx 3000*g* (setting 4) for 3 min.
8. Collect the flow through and proceed to **Subheading 3.3.** for RNA–protein crosslinking.

3.3. Crosslinking of RNA to Protein (Fig. 2B)

1. Using the optimal UTP or CTP concentration determined in the previous section for protein–RNA crosslinking, repeat **steps 1–9** of **Subheading 3.1.** Resuspend the RNA in distilled, deionized, autoclaved water or the buffer appropriate for

the specific RNA–protein interaction being characterized. Add the putative RNA-binding protein (at a concentration appropriate for that interaction) to the RNA (*see* **Note 13**) *(6,7)*.
2. Split the reaction in half and place one half in a 1-mL colorless polystyrene microcentri-fuge tube. Place the polystyrene tube 1.5 cm above the UV light source and irradiate for 2 min at room temperature, unless the temperature must be controlled for the interaction being examined. Keep the other half of the reaction in the dark at room temperature as the minus irradiation control.
3. Add DTT to all reactions to a final concentration of 100 mM and leave the samples in the dark for at least 5 min (see **Note 4**). After this step, manipulations can be carried out with the lights turned on.
4. Withdraw a small aliquot from the nonirradiated sample to verify that the correct RNA length has been made (as in **Subheading 3.1., step 13**).
5. To the remaining samples (both the irradiated and nonirradiated), add an equal volume of T1 digestion buffer and add 10 U/µL of RNase T1 (*see* **Note 14**). Incubate at 37°C for 10 min.
6. Add an equal volume of 2× protein loading buffer to the remaining irradiated samples, heat at 94°C for 2–3 min, and load the samples onto a denaturing SDS-polyacrylamide gel. Run the gel until the bromophenol blue dye reaches the bottom.
7. Cut a piece of the nitrocellulose membrane and six pieces of filter paper exactly the same size as the gel (*see* **Note 15**).
8. Soak the membrane and the filter paper in the transfer buffer for 3 min. Make a "sandwich" by stacking three pieces of filter paper, the gel, the membrane, and then three more pieces of filter paper.
9. Put this in the blotting apparatus, apply a current of 500 mA for 2 h, (*see* **Note 16**), and then remove the membrane from the gel and filter paper.
10. Put the membrane in silver staining solution for 5 min or until bands are visible. Air dry the membrane.
11. Expose the membrane to X-ray film with an intensifying screen at –80°C. Crosslinked proteins can be identified as described in **Fig. 3**. Visualization of crosslinked proteins can take several days if the crosslinking yield is low.

4. Notes

4.1. Synthesis of Full-Length RNA Containing Photocrosslinking Nucleotide Analogs Throughout

1. If a problem with RNase contamination develops, make all buffers and solutions with water that has been treated with diethylpyrocarbonate (DEPC). Prepare DEPC-treated water in a fume hood by adding 1 mL of DEPC/L of distilled, deionized water. Cap the bottle loosely and let it stand at room temperature overnight in the hood. Autoclave the water to remove unreacted DEPC. If a fruity smell is detectable after autoclaving, the water contains residual DEPC and should be autoclaved again. DEPC inactivates RNase by modifying amine groups, and any residual DEPC in your buffers may also cause inhibition of transcription.

2. Commercially available bovine serum albumin (BSA) can be contaminated with RNase. Acetylation of the BSA also inactivates any contaminating RNase. Gloves should be worn when preparing any solutions and handling any materials that will be used during transcription. RNase that is on skin can be transferred to materials, such as pipet tips, that can then contaminate the reaction.
3. Store nucleotide analogs at –80°C in brown microcentrifuge tubes, wrapped in aluminum foil to minimize exposure to light. All manipulations with analog should be done in reduced light (indirect illumination from another lamp or light from an adjacent room).
4. Azides are reduced rapidly by DTT. If thiol must be present in buffers, substitute 10 mM β-mercaptoethanol for the DTT.
5. Care must be taken when using [α-^{32}P] GTP. It is a strong β emitter and standard safety precautions must be taken. A dosimeter and safety glasses must be worn at all times. Once the use of isotope has begun, all work must be done behind a Plexiglas shield.
6. Photocrosslinker is light sensitive and must be handled in reduced light.
7. Unpolymerized acrylamide is a known neurotoxin that can be absorbed through unbroken skin. Gloves, eye protection, and a dust mask must be worn when working with dry acrylamide, and only gloves and eye protection must be worn when working with acrylamide in solution. The addition of urea will increase the volume of the acrylamide solution greatly, and care must be taken with the initial volume when the urea is added. Keep this volume at a minimum until all of the urea has dissolved and then bring the solution to its final volume.
8. Although the analogs described here are substrates for both *E. coli* and T7 RNA polymerases, only the unmodified mercapto analogs can be incorporated at two adjacent positions in the RNA by the polymerases. When using 5-APAS-UTP, a small amount of UTP must be present for the polymerase to transcribe through regions of sequential "Us" in the transcript. The amount of UTP that must be added depends on the RNA sequence and length, and the optimal ratio of UTP to APAS-UTP required for production of full-length RNA must be determined empirically for each RNA. The same is true for 5-APAS-CTP, except that CTP must be included in the reaction. In the absence of UTP or CTP, the polymerases will terminate transcription after incorporation of one analog if the next encoded nucleotide is the same. To determine the concentration of the unmodified nucleotide that must be added, RNA is synthesized in the presence of increasing amounts of the unmodified nucleotide and then analyzed by gel electrophoresis. By analyzing the RNA made with or without irradiation, the concentration necessary for production of full-length RNA that is still photoreactive can be determined. The concentration required depends upon both the RNA sequence and the RNA length.
9. Heparin is a polyanion that binds to free RNA polymerase. RNA polymerase binds to heparin after it reaches a transcription termination signal and is released from the template. Addition of heparin prevents reinitiation and ensures that all RNA transcripts are full-length.

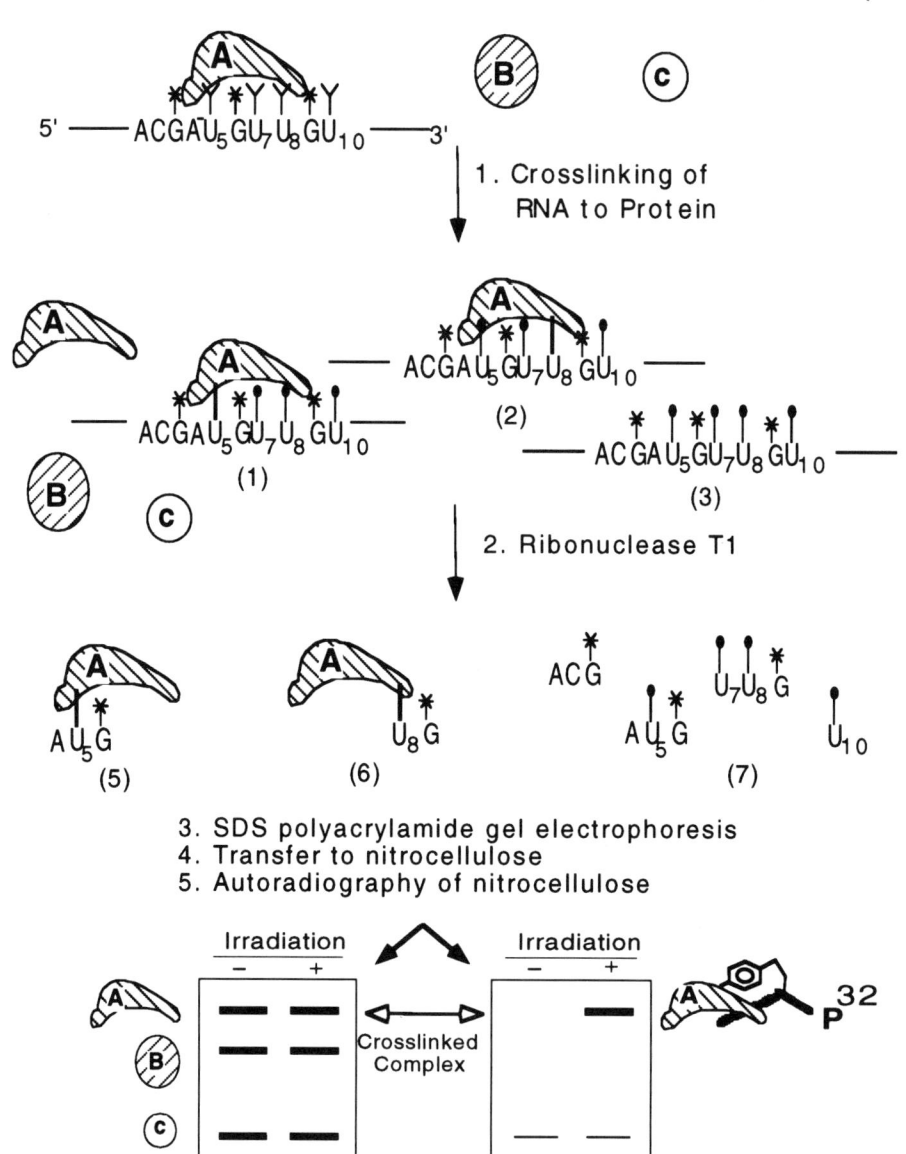

Fig. 3. Strategy for identification of proteins crosslinked to long RNA. When the RNA is modified with more than one crosslinker, it is possible for bound protein to be crosslinked to the RNA through different nucleotides. In the example shown here, a large piece of RNA contains a region with four crosslinkers. If protein binding occurred as shown in this example, upon irradiation, crosslinking could occur, e.g., between the protein and nucleotides at positions 5 and/or 8 (step 1). Analogs at positions 7 and 10 do

10. This will be the reaction in which full-length RNA is present in the minus irradiation lane, but in which irradiation results either in the disappearance (*see* **ref. 3**) or smearing of the bands (*see* **refs. 5** and **8**). The sample that is not irradiated should give a sharp, full-length band. In general, upon irradiation of larger RNA (more than approx 20 nt), the RNA band disappears because the RNA forms intramolecular crosslinks or reacts with solvent to produce a heterogeneous population of RNAs, all having different mobilities. A smear, rather than complete loss of bands, is observed with short RNAs that may not form intramolecular crosslinks, but only crosslink to solvent molecules, producing RNA with slightly different molecular weights and mobilities.

4.2. Synthesis of Full-Length RNA Containing Photocrosslinking Nucleotide Analogs Only on the Surface of the RNA

11. APB alkylates the 5-SH-UMP or 5-SH-CMP in the RNA posttranscriptionally to form 5-APAS-UMP and 5-APAS-CMP, respectively. The APB reacts with the mercapto group and attaches a photoreactive azide group for crosslinking. Because the 5-SH-UMP is incorporated during transcription and then the APB is

---

not contact the protein and therefore react with solvent. Two different ribonucleoprotein complexes would be generated for single crosslinking events [labeled (1) and (2) in the figure]. When RNA becomes covalently crosslinked to a protein during irradiation, the resulting ribonucleoprotein complex has a larger molecular weight than the protein itself does. This will cause a retardation of the mobility of the protein in SDS-polyacrylamide gels. To leave only a small radioactively tagged RNA oligonucleotide on the protein, the ribonucleoprotein complex is treated with the ribonuclease T1. T1 cleaves 3' to the GMP residues, so that one ^{32}P group will remain in the oligonucleotide between the position of the U-analog crosslink and the T1 cleavage site (step 2). To identify the proteins that are crosslinked to the RNA, the ribonuclease-treated complex is fractionated by denaturing polyacrylamide gel electrophoresis (step 3). The material from the gel is then transferred to nitrocellulose electrophoretically (step 4). Under the conditions used, both proteins and protein–nucleic acid complexes are retained by the nitrocellulose, but free RNA and DNA are not. This step removes the uncrosslinked RNA from the reaction so that any remaining radioactive components on the nitrocellulose are RNA–protein complexes. Because the RNA has been treated with ribonuclease, it retards the mobility of the protein to which it is crosslinked only negligibly. Except in the case of relatively small proteins (under 10 kDa) the ribonucleoprotein complex will comigrate in SDS-polyacrylamide gels with the uncrosslinked protein. The positions of proteins are determined by silver-staining the nitrocellulose. Proteins will be visible on both the irradiated and nonirradiated control reactions. After staining, the nitrocellulose is subjected to autoradiography (step 5). Only proteins that produce a band in the irradiated reactions are crosslinked to RNA (Protein A). Bands that occur in both lanes are not due to photocrosslinking (Protein C), but to some other form of radioactive labeling of the protein *(9)*. The asterisk (*) indicates the position of [α-^{32}P] GMP and the (V) indicates the position of 5-APAS-UMP crosslinking analog. Azides that react with solvent rather than producing covalent crosslinks to protein are shown with a closed circle (●).

added, only those mercapto groups that are accessible (analogs on surface) will be modified and have the ability to crosslink. This approach allows one to distinguish between interactions with the surface binding sites and interactions with proteins that occur when the RNA is buried within a complex.
12. The Sephadex G-50 column is a size-exclusion column. The large RNA will flow through the column, while the smaller APB will be retained on the column.

4.3. Crosslinking of RNA to Protein

13. Pure protein, cell extracts, or a mixture of several purified proteins may be used in this step. A positive control reaction using a known RNA binding protein and a negative control with a protein that does not bind to RNA may be useful to facilitate interpretation of the results.
14. T1 RNase digestion is done when the RNA is long (> 20 nucleotides). The mobility of a small protein is affected by attachment to long RNA *(6)*. T1 RNase digestion is not needed when the RNA synthesized is smaller than 20 nucleotides.
15. The proteins are transferred from the SDS-polyacrylamide gel to the nitrocellulose membrane because the nitrocellulose binds proteins tightly, thereby capturing the protein–RNA complexes, while allowing free RNA to pass. This can be important when the protein being examined is very small and the RNA has considerable secondary structure which makes it somewhat resistant to the RNase treatment. The longer pieces of RNA will comigrate at the bottom of the gels with the small proteins, making identification of the crosslinked protein difficult or impossible.
16. The conditions for electroblotting vary according to the apparatus, the protein size, etc., and have to be determined empirically.

Acknowledgments

This work was support by NIH Grant RO1 GM47493, NSF Grant MCB-9509132, and an OCAST award for project no. HR4-072, from the Oklahoma Center for the Advancement of Science and Technology.

References

1. Hanna, M. M. (1989) Photoaffinity crosslinking methods for studying RNA – protein interactions. *Methods Enzymol. RNA Proc.* **180**, 383–409.
2. Hanna, M. M., Dissinger, S., Williams, B. D., and Colston, J. E. (1989) Synthesis and characterization of 5-((4-azidophenacyl)thio)-uridine-5'-triphosphate: a cleavable photocrosslinking nucleotide analog. *Biochemistry* **28**, 5814–5820.
3. Dissinger, S. and Hanna, M. M. (1990) Active site labeling of *E. coli* transcription complexes with 5-((4-azidophenacyl)thio)-uridine-5'-triphosphate. *J. Biol. Chem.* **265**, 7662–7668.
4. He, B. and Hanna, M. M. (1995) Preparation of probe-modified RNA with 5-mercapto-UTP for analysis of protein-RNA Interactions. *Nucleic Acids Res.* **73**, 1231–1238.

5. Hanna, M. M., Zhang, Y., Reidling, J. C., Thomas, M. J., and Jou, J. (1993) Synthesis and characterization of a new photocrosslinking CTP analog and its use in photoaffinity labeling *E. coli* and T7 RNA polymerases. *Nucleic Acids Res.* **21**, 2073–2079.
6. Liu, K. and Hanna, M. M. (1995) NusA interferes with interactions between the nascent RNA and the C-terminal domain of the alpha subunit in *E. coli* transcription complexes. *Proc. Natl. Acad. Sci. USA* **92**, 5012–5016.
7. Liu, K. and Hanna, M. M. (1995) NusA contacts nascent RNA in *E. coli* transcription complexes. *J. Mol. Biol.* **247**, 547–558.
8. Zhang, Y. and Hanna, M. (1994) NusA changes the conformation of *E. coli* RNA polymerase at the binding site for the 3' end of the nascent RNA. *J. Bacteriol.* **176**, 1787–1789.
9. Schmidt, M. and Hanna, M. M. (1986) Nonenzymatic radiolabeling of protein by ^{32}P-containing nucleotides. *FEBS Lett.* **194**, 305–308.

4

The Methylene Blue Mediated Photocrosslinking Method for Detection of Proteins that Interact with Double-Stranded RNA

Zhi-Ren Liu and Christopher W. J. Smith

1. Introduction

Photochemical crosslinking is an approach that is widely used to detect and analyze interactions between proteins and nucleic acids. The approach involves illuminating a mixture of protein and nucleic acid with light of a suitable wavelength to induce photochemical reactions that result in covalent linkages between the nucleic acid and the protein. The protein can be a purified or a recombinant RNA binding protein, but perhaps a more common application is the identification of proteins within a complex cell extract that bind to a specific RNA. Radiolabeled RNA is incubated with the extract, illuminated with suitable wavelength light, the RNA digested and proteins analyzed by sodium dodecyl sulfate-polyacrylamide gel electrophoresis (SDS-PAGE) and autoradiography. Only proteins that were initially bound to the RNA at the time of illumination should have acquired radiolabel. The simplest method is to use short-wave ultraviolet (UV) light (254 nm), which causes direct crosslinking between the bases and adjacent proteins. However, this can result in large amounts of nonspecific damage to bases not actually involved in crosslinking; this limits, e.g., the ability to subsequently map the sites involved in crosslinking. More refined approaches involve the substitution of modified bases that can be activated at longer wavelengths at which the usual bases will remain intact. 4-thiouridine (activated at 330–350 nm) is one of the more common modified bases. Such bases can be incorporated randomly by RNA polymerases (*see* Chapters 1 and 3) *(1)*, site specifically within chemically synthesized RNA oligonucleotides *(2,3)*, or site specifically into longer RNAs

From: *Methods in Molecular Biology, Vol. 118: RNA-Protein Interaction Protocols*
Edited by: S. Haynes © Humana Press Inc., Totowa, NJ

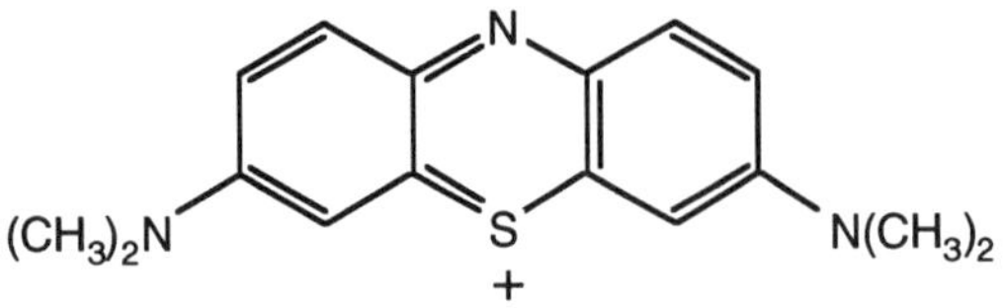

Fig. 1. Structure of methylene blue. Note that the planar three-ring structure facilitates intercalation between nucleic acid basepairs, while the cationic properties are consistent with electrostatic binding to nucleic acid phosphodiester backbones.

using the oligonucleotide-mediated ligation technique pioneered by Moore and Sharp (*see* Chapter 2) (*4,5*).

Detection of RNA binding proteins by this general approach relies entirely on a fortuitous apposition of the reacting bases to the protein, and upon the presence of a labeled residue within the fragment of digested RNA that is covalently bound to the protein. Lack of a detectable crosslink does not allow one to conclude that there is no RNA–protein interaction. This caveat becomes more pronounced when UV crosslinking is used with highly structured dsRNA. Here there are two factors that may operate to decrease the likelihood of successful crosslinking:

1. The close association of the hydrogen-bonded bases means that efficient crosslinking between the bases themselves may compete with RNA-protein crosslinking. Indeed UV crosslinking using site-specific 4-thiouridine incorporation has been used extensively to physically map sites of RNA–RNA interaction (*6,7*).
2. In continuous A-form dsRNA, the major groove is too deep and narrow to allow access of amino-acid side chains to the bases (*8*). Proteins that bind to purely dsRNA are therefore likely to contact the bases through the more accessible minor groove side. If a photochemical reaction generates reactive species only on the major groove side it would be unlikely to detect such interactions.

Methylene blue (MB)-mediated photocrosslinking provides a useful addition to the repertoire of RNA–protein crosslinking methods, having a specificity that is complementary to that of traditional UV crosslinking methods. MB is a member of the phenothiazinium group of dyes and has for a long time been used as a dye in microscopy (for review *see* [*9*]). Owing to its planar ring structure (*see* **Fig. 1**) MB is able to bind to nucleic acids at relatively low stoichiometry and high affinity by intercalation between the bases. At much higher concentrations its cationic nature also allows binding to the surface and phosphodiester backbone by electrostatic interactions. MB has an absorbance maximum in the visible spectrum at 665 nm. When illuminated with visible light it becomes photosensitized and can cause damage to a wide array of biological macromolecules including proteins, lipids, nucleic acids, and nucleotides *(9)*. These reactions can be direct or they can be mediated by the production of reactive species such as singlet oxygen, hydrogen peroxide, superoxide anion, and hydroxyl radicals *(9)*. Despite this tendency to behave as a "molecular hooligan," causing wanton and widespread damage, under suitable conditions MB can be used to induce efficient and specific crosslinks between dsRNA or DNA and bound proteins in response to illumination with visible light *(10–12)*. Experiments with dsDNA indicate that the MB mediates crosslinking by intercalating between the DNA basepairs but does not actually form part of the crosslinked structure *(11)*. MB crosslinking therefore resembles the simple forms of UV crosslinking with the important distinction that MB crosslinking shows marked preference for inducing crosslinks between dsRNA and bound proteins rather than between ssRNA and proteins. This specificity is complementary to that of UV crosslinking, which is more efficient for single-stranded RNA. In our experience, MB and UV crosslinking are excellent complementary approaches that detect distinct arrays of RNA–protein interactions (*see*, e.g., **Fig. 2**).

We originally demonstrated that MB crosslinking could be used to detect interactions of dsRNA with either purified proteins or with dsRNA binding proteins in crude cell extracts *(12)*. However, various other applications are possible. We have already used MB crosslinking in two further kinds of experiments. In the first we have been investigating a proposed scanning process during step 2 of splicing *(13,14)*. It was already known that stable hairpin structures placed between the branchpoint and 3' splice site could specifically block step 2 of splicing. It was proposed that the hairpins interfered with the scanning process. Consistent with this, we have detected a specific spliceosome component that MB crosslinks to these structures after step 1 of splicing and identified it as the 116-kDa subunit of U5 snRNP *(15,16)*. In this case MB crosslinking was used as a crude form of site-specific crosslinking to the only stable structured region of a pre-mRNA. In the second application we have used MB crosslinking to look for proteins that interact with the pre-mRNA–snRNA duplexes that form transiently during pre-mRNA splicing. We have detected a ~65-kDa protein (**Fig. 2**) that interacts with the duplex formed between U1 snRNA and the 5' splice site *(17)*. Again, in this case the entire pre-mRNA was labeled, but we were able to focus upon proteins that interacted with a short region of dsRNA, in this case an intermolecular duplex. In both of these examples various control experiments had to be carried out to verify the functional specificity of the interaction. Both the proteins detected, U5 116-kDa and the unidentified 65-kDa protein, were undetectable by UV crosslinking. Thus these examples show that MB crosslinking can be harnessed to extend the range of RNA–protein interactions that are accessible to analysis by crosslinking. Finally, it is worth noting that although this method makes use

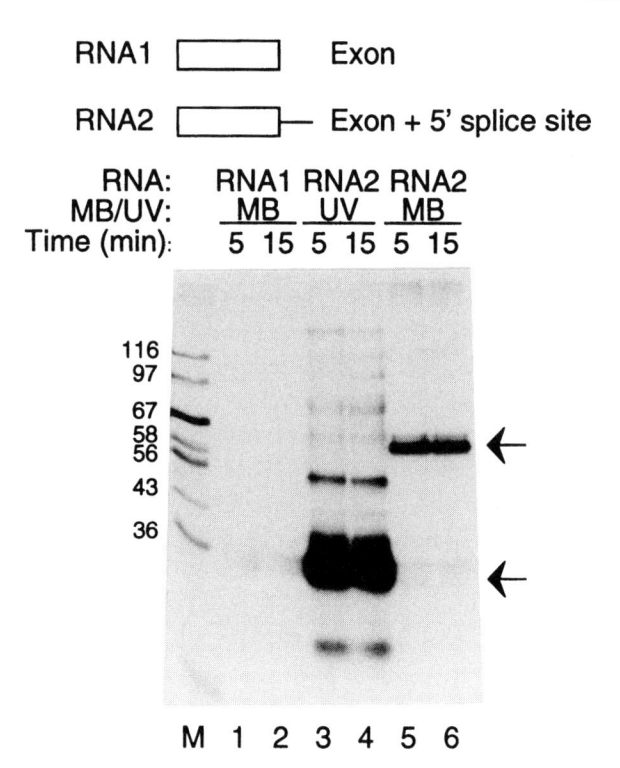

Fig. 2. Comparison of MB and UV crosslinking. The two RNAs shown schematically were incubated in HeLa cell nuclear extract under conditions for pre-mRNA splicing. RNA 1 contains rat α-tropomyosin exon 2. RNA 2 also contains the 5' splice site and adjacent intron sequence. After the times indicated the samples were MB or UV crosslinked, digested with RNases, and the proteins analyzed by SDS-PAGE and autoradiography. MB crosslinking detects mainly a band of ~65 kDa. This band requires the presence of the 5' splice site (compare *lanes 1* and *2* with *5* and *6*) and also the 5' end of U1 snRNA. Various data indicate that it crosslinks to the duplex formed between the 5' splice site and U1 snRNA *(17)*. Note that the UV crosslinking pattern of RNA 2 (*lanes 3* and *4*) is complementary to the MB crosslinking. The major band is a doublet that is much fainter in the MB lanes. This doublet probably corresponds to SR proteins which bind to a single-stranded splicing enhancer region in the exon (they are detected faintly by MB crosslinking with RNA 1). In contrast, the 65-kDa band is not detected by UV.

of the binding of MB to conventional Watson–Crick basepaired RNA, given its mode of binding by intercalation and general reactivity, it is likely that MB crosslinking could be used to probe the interactions of proteins with other RNA structures. These could include RNA duplexes containing noncanonical basepairs or other irregularities, or even structures such as four-stranded G quartets *(18)*.

We detail in the following sections the straightforward procedure used in our laboratory for MB crosslinking including measures that can be used to curtail undesirable side reactions.

2. Materials

2.1. Equipment

1. Light. We use a 60-W fluorescent tube light, 1.2 m long (but *see* **Note 2**).
2. Standard apparatus for SDS-PAGE, autoradiography, and/or phosphorimaging.
3. Microtiter plates.

2.2. Reagents

1. Methylene blue. Dissolved in water at a concentration of approx 2.5 mg/mL. The concentration is determined accurately by the absorbance at 665 nm; ε_{665nm} = 81,600 $M^{-1}cm^{-1}$. The stock solution is stored in aliquots (~ 1 mL) in light-tight tubes at −20°C.
2. RNA labeled to high specific activity by transcription (*see* Chapter1) using one or more [α-^{32}P] NTPs (*see* **Notes 7** and **8**). Note that if the transcription template is not removed by DNase digestion, then the dsDNA may bind a significant amount of MB by intercalation. This should not usually present a problem, although it may increase the MB concentration for optimal crosslinking (*see* **Note 4**). If the studies are carried out with two complementary ssRNAs the two RNAs should be first be annealed. In these circumstances, use one strand radiolabeled and mix with two to five times molar excess of unlabeled complementary RNA. The two RNAs are heated to 65°C for a short time (~5 min) and allowed to cool slowly to room temperature. We usually carry out the annealing immediately before each set of experiments.
3. 10× Binding buffer (*see* **Note 1**). Composition will vary according to the interaction being investigated. Typically a 1× binding buffer might contain 10 mM Tris-HCl, pH 7.5, 20–100 mM KCl, 2 mM MgCl$_2$, and 0.5 mM dithiothreitol (DTT).
4. Ribonuclease stocks (*see* **Note 8**).
   a. RNase A: Solid dissolved at 10 mg/mL in 10 mM Tris-HCl, pH 7.5, 15 mM NaCl, boiled for 15 min, slow cooled and stored in aliquots at −20°C.
   b. RNase T1: (100,000 U/mL).
   c. RNase V1: (700 U/mL).
5. 5× SDS sample buffer: 10% SDS, 1 M Tris-HCl, pH 6.8, 50% glycerol, 0.03% bromophenol blue. Store at room temperature. Add 2-mercaptoethanol to 5% just before use.
6. Ascorbic acid (*see* **Note 5**). Prepared as small volume of stock 1 M aqueous solution by dissolving directly in RNase-free water (double-distilled, optionally diethylpyrocarbonate [DEPC]-treated, autoclaved). The pH of the solution is not adjusted because aqueous ascorbate is rapidly oxidized and this process is accelerated at higher pH. The solution is either made fresh each time or it can be stored at −70°C for short periods (less than a month).

3. Method

1. Assemble the components of the binding reaction in a volume of ~10 µL in the wells of a microtiter plate. Typical quantities are: 1 µL of 10× binding buffer, ~5–10 ng [^{32}P]-labeled RNA, ~100 ng of pure protein or ~10 µg of protein in a crude extract, and H$_2$O to a final volume of 10 µL. Incubate for a sufficient time to allow equilibration of the binding reaction. Conditions may vary considerably according to the system under investigation. For most simple binding reactions 5–10 min at room temperature is sufficient. For more complex reactions, e.g., detection of dsRNA–protein interactions at specific stages of the spliceosome cycle, longer incubations may be necessary.
2. Place the microtiter plate on a bed of ice. To maximize subsequent illumination, we usually first cover the bed of ice with a sheet of aluminum foil. Alternatively, the microtiter plate can be placed on an aluminum foil covered adjustable platform in a cold room. Add MB to a final concentration of 0.1–2 ng/µL from a 20× stock solution (*see* **Note 4**). Mix the MB into the binding reaction by pipetting up and down. Then add ascorbic acid or other quenching reagent to an appropriate concentration, if desired (*see* **Note 5**). Add ascorbic acid along with a twofold molar excess of Tris-HCl, pH 7.8, to maintain the pH of the sample. Mix the quenching reagent into the binding reaction by pipetting up and down.
3. Place the microtiter plate 1–5 cm below the light source (*see* **Note 2**). Illuminate for 5–20 min (*see* **Note 3**).
4. Prepare a 10× RNase mix containing 10 µg/µL RNase A, 3 U/µL RNase T1, and 0.35 U/µL RNase V1 (*see* **Note 8**). Transfer each sample to a 0.5-mL microcentrifuge tube, add 1 µL of the RNase mix and incubate at 37°C for 30 min. This step can be omitted to visualize the intact RNA–protein complex on SDS-PAGE if a pure protein is being used (*see* **Note 8**).
5. Add 4 µL of 5× SDS sample buffer to each tube. Samples can either be applied immediately to an SDS polyacrylamide gel or stored temporarily at –20°C.
6. Before loading on an SDS gel, denature the samples by incubating in a boiling water bath for 3 min or a heat block set to 85°C for 5 min. Load the samples onto an SDS gel with a suitable percentage polyacrylamide. After electrophoresis it is often useful to stain the gel with Coomassie blue and then partially destain before drying. This allows a convenient visual inspection for equal loading of samples and also for any gross protein–protein crosslinking that may occur at higher MB concentrations (*see* **Notes 4** and **5**). Expose the dried gel to X-ray film or to a phosphorimager screen if quantitation of data is required.

4. Notes

As should be apparent, the basic procedure for MB crosslinking is very straightforward. However, a number of variables may need to be optimized before embarking on a series of experiments.

1. Binding buffer. MB crosslinking appears to be compatible with the usual conditions of pH and ionic strength (up to 10 mM Mg^{2+}, and 150 mM K$^+$) used in most

experiments on RNA–protein interactions *(12)*. Note, however, that EDTA can reduce the excited triplet state of MB *(9)*. We have not studied the effects of EDTA on crosslinking experiments, but it may decrease the efficiency of crosslinking as well as potentially causing unwanted side reactions due to the production of free radicals (*see* **Note 5** on the use of radical scavengers).

2. Light source. The light source that we have routinely used is a simple domestic fluorescent tube light with an output of 40–60 W. Because the tube is itself about 1.2 m in length, the power delivered to the samples is considerably less. We have not systematically investigated the efficacy of different light sources. The tube lights used were those already mounted at the back of our laboratory benches. We suspect that one major advantage of fluorescent lights is that they have a low heat output and so are less likely to cause overheating of the samples with the attendant possibility of protein and/or RNA denaturation before crosslinking has occurred.

3. Time. In our experience the intensity of crosslinking usually reaches a maximum after approx 5–6 min. However, this may vary according to other parameters (such as the light source, distance of sample from light) so it is worth checking that maximal crosslinking has been achieved.

4. MB concentration. The optimal MB concentration should be determined. This will depend to a large extent upon the amount and degree of double-strandedness of the RNA being used. In addition, if crosslinking experiments are being carried out with crude cell extracts multiple MB binding species may be present such as rRNA, tRNAs, or snRNAs. Optimal crosslinking efficiency presumably depends upon partial occupancy of MB intercalation binding sites. However, binding of intercalating agents distorts the structure of double-stranded nucleic acids, causing partial unwinding. At high concentrations this eventually disrupts RNA–protein interactions as shown by gel shift assays. Thus, high levels of MB eventually lead to decreased crosslinking *(12)*. In general, the optimal concentration of MB, expressed in mass per unit volume, is of the same order of magnitude as that of the RNA. Because the concentrations of both MB and RNA are probably well below the K_d for binding by intercalation *(19)*, this probably represents relatively low occupancy of potential binding sites. In addition to optimizing the efficiency of the specific crosslink, attention should also be paid to minimizing other unwanted side reactions (*see* **Note 5**).

5. Quenching competing side reactions and enhancing specific crosslinking. As mentioned in the Introduction, MB can mediate a wide range of reactions in response to illumination with visible light (reviewed in *[9]*). It is likely that only a subset of these reactions result in specific crosslinking of proteins to dsRNA. Many other reactions may be positively harmful. For example, protein–protein crosslinking can occur as a result of photooxidation reactions *(20,21)*. We have noticed some MB-induced protein–protein crosslinking in our experiments. For instance, in the case of the *staufen* dsRBD binding to the adenovirus VAI RNA, we noted small amounts of apparent protein dimers and trimers *(12)*. In this case, the small amounts of protein–protein crosslinking were not a major

problem. Indeed, such observations may give useful information on the stoichiometry of binding if, as appears, the protein–protein crosslinking is between protein monomers bound to the same RNA molecule.

However, the more common application of photocrosslinking is to identify proteins that interact with a specific RNA within a complex cell extract. In these cases MB-induced protein–protein crosslinking causes large aggregates of protein to become stuck in the wells and at the bottom of the stacking gel upon SDS-PAGE (**Fig. 3**). These are detectable both by autoradiography and by Coomassie blue staining, which also reveals the complete disappearance of some high molecular weight proteins, probably due to protein–protein crosslinking. In addition, the fuzzy appearance of some bands may be due to variable amounts of amino acid modification without crosslinking (*see* also **Note 8** on RNases).

Reasoning that the chemistry of specific dsRNA-protein crosslinking may be quite distinct from that of protein–protein crosslinking, we have tested a number of compounds that are expected to act as radical scavengers and/or reducing reagents (e.g., ascorbic acid), singlet oxygen quenchers (e.g., sodium azide, semicarbazide *[21]*), or as competitors for amino acid photooxidation (histidine) *(21)*. We found that ascorbic acid was particularly potent at quenching the protein–protein crosslinking *(22)*, as indicated by the reduction in aggregates stuck in the wells, and enhanced dsRNA–protein crosslinking, probably by release of material previously stuck in the wells (*see* **Fig. 3**). Coomassie blue staining also revealed release of material from the wells and reappearance of high molecular weight bands. Histidine was less effective, while semicarbazide reduced both the protein aggregates and the specific crosslinking in parallel. Another point about the specificity of ascorbate quenching is evident from **Fig. 3**. The doublet of bands at ~30 kDa that crosslink more efficiently with UV than with MB are probably the splicing factors ASF/SF2 and SC-35 *(23)*, most probably binding to a single-stranded exon splicing enhancer sequence within the RNA probe (*see also* **Fig. 2**). Note that MB crosslinking of these bands, which is already inefficient compared with UV crosslinking, is further decreased by ascor-bic acid. Thus ascorbic acid apparently reduces both protein–protein and protein–ssRNA crosslinking. It therefore appears that ascorbic acid should be considered as a routine component of MB crosslinking reactions unless dictated by some other overriding consideration.

The optimal ascorbic acid concentration probably depends upon various factors including the amounts of MB used and the type and concentration of extracts. We have found the optimal concentration of ascorbic acid in our HeLa extract experiments to be around 2 mM. At higher concentrations (10 mM) it leads to abolition of all crosslinking so it is advisable to titrate to find the optimal concentration for a given experiment. However, 0.5–2 mM is probably a good starting range. In our experiments we have added ascorbic acid along with twice the molar concentration of Tris-HCl, pH 7.8, to maintain the pH of the sample. Ascorbate is very unstable in neutral aqueous solution, so it is added immediately before the illumination.

Ascorbic acid (mM)

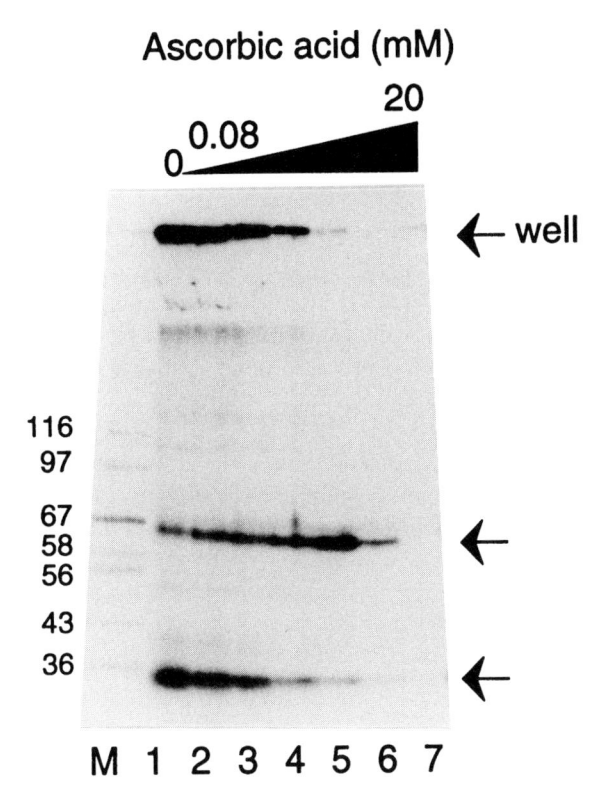

Fig. 3. Quenching of side reactions by ascorbic acid. MB crosslinking was carried out with RNA 2 from **Fig. 2**. Reactions contained 40% HeLa nuclear extract and 5 ng of [α-^{32}P]GTP labeled RNA 2 and were incubated for 15 min, followed by addition of 1 ng of methylene blue and the following final concentrations of ascorbic acid: 0, 0.08, 0.25, 0.75, 2.2, 6.7, and 20 mM, in lanes 1–7 respectively. Each ascorbic acid addition was accompanied by addition of a twofold molar excess of Tris-HCl, pH 7.8. Note that the large amount of material stuck in the wells and at the bottom of the stacking gel in the absence of ascorbic acid. This material, which probably results from protein–protein crosslinking, decreases in intensity with increasing ascorbic acid concentrations. However, the 5' splice site–U1 snRNA duplex binding p65 increases to a maximum at 2.2 mM. Finally, note the different response of the ~30 kDa bands which are thought to bind to single-stranded exon sequences (*see legend to* **Fig. 2**). The intensity of these bands decreases over the ascorbic acid concentration range in which p65 is increasing, suggesting that the p30 crosslinks occur via a distinct photochemical mechanism from the dsRNA–p65 crosslinking.

6. Testing specificity of crosslinking. In many experiments on RNA–protein interactions it is routine to test the sequence specificity of an interaction by testing the relative effects of adding increasing molar excesses of unlabeled specific self-

competitor and of various other unlabeled nonspecific competitors. This is routine in testing the specificity of gel-shift complexes (Chapter 10) and of UV-crosslinked proteins (Chapters 3 and 6). In theory this can be done for MB crosslinking reactions too, although we have not carried out this procedure. In many cases, such competition experiments may not be particularly informative because many proteins that interact with dsRNA are expected to do so without any particular sequence specificity *(24)*. One potential problem is that competitor nucleic acids, especially dsRNA or dsDNA, would be expected to bind MB. In an extreme case this could lead to a reduction in the crosslink signal, not because the unlabeled nucleic acid competitor binds the protein but because it significantly reduces the free concentration of MB (although, this may not be a major effect as most experiments are carried out well below the K_d for MB binding; *see* **Note 4** *[9]*). In any case, one would expect that a specific competitor would still reduce the crosslinking signal more rapidly than a nonspecific competitor. However, one potential way to circumvent this problem would be to carry out MB titrations at each concentration of specific and nonspecific competitor. The effectiveness of the competitors could then be ascertained by comparing the maximum signals achieved at the optimal MB concentrations for each concentration of each competitor.

7. Base preference for MB crosslinking. For many dsRNA-protein interactions, binding is relatively nonspecific with respect to the RNA sequence so any dsRNA duplex is likely to be able to crosslink successfully even if there is some differential reactivity between the different bases. However, in some instances where short duplexes are involved the particular configuration of bases in juxtaposition to an interacting protein potentially could affect the efficiency of crosslinking. Previous reports indicate that guanine is the most common target for damage by MB *(9)*. We have recently tested the base preference of MB mediated crosslinking using both recombinant RED-1 editing enzyme, which is a double-stranded RNA specific adenine deaminase, and a GST fusion protein with a dsRNA binding domain from the *Drosophila staufen* protein *(22,25)*. Both proteins show sequence nonspecific binding to dsRNA. The proteins were MB crosslinked to hairpins consisting solely of A–U or G–C base pairs and to a model hairpin in which one arm of the stem contained only purines and the other only pyrimidines (Pu:Py hairpin). In general we saw only slight variations (< twofold) in crosslinking efficiency between the different hairpins. In particular with the Pu:Py hairpin, crosslinking efficiency was not markedly different between RNAs containing radiolabel only in the purine or pyrimidine arm of the stem. These data suggest that MB crosslinking shows little base specificity and can be mediated by any of the four common bases *(16)*. The technique should therefore be quite versatile and widely applicable even with specific short duplexes. Nevertheless, in the latter case attention should be paid to the appropriate combinations of ribonucleases and radiolabeled nucleotides (*see* **Note 8**).

8. Ribonucleases and [α-^{32}P]NTPs. In most of our experiments we have used a cocktail of single- and double-stranded specific RNases. In parallel UV and MB crosslinking experiments we have noticed that RNA becomes slightly more

resistant to RNase in the presence of MB, leading to slightly fuzzy bands if digestion is incomplete. The digestion for 30 min is twice the duration routinely used in UV crosslinking. If sharp bands are not obtained, it may be worthwhile extending the RNase digestion.

In some circumstances it may be appropriate to vary the choice of RNase. First, it is possible in some circumstances to omit the ribonuclease treatment entirely. This would usually only be the case when a pure protein preparation is being used. In this case free and crosslinked RNA could be visualized directly after SDS-PAGE. This type of experiment can be used to measure the efficiency of a crosslinking reaction and is preferable to one in which the sample has been treated by RNases, because the exact number of radiolabeled residues in the crosslinked complex is known. After RNase digestion, assumptions have to be made about the degree of digestion and hence the number of radiolabeled nucleotide residues attached to the crosslinked protein.

In most cases the experiment is being carried out with a complex cellular extract in an attempt to identify particular proteins. In these situations RNases must be used after the crosslinking step. Because the RNA is double stranded it is appropriate to use cobra venom nuclease V1 which digests dsRNA with no base specificity, leaving 5' phosphates. However, this nuclease is expensive and for complete digestion we usually also incorporate RNases A and T1. These are both specific for single stranded RNA and leave 3' phosphates. RNase T1 cleaves adjacent to G, while RNase A cleaves 3' to C and U. Inclusion of these enzymes usually gives more complete digestion with sharper bands on autoradiographs. Possibly they complete digestion of RNA fragments that have first been cleaved by nuclease V1. For most experiments this gives suitable results. Note, however, that the RNase digestion results in an ill-defined fragment of RNA crosslinked to the protein. First, as with UV crosslinking, it is likely that digestion immediately adjacent to crosslinked bases may not occur. Second, depending upon whether it was generated by the RNase V1 or one of the other RNases, the crosslinked fragment may carry either P or OH groups at both the 5' and 3' ends. With small duplexes where there may only be one or two base positions that are able to form a crosslink, the precise combination of nucleases and choice of [α-^{32}P]NTP may have a major influence on the ability to detect a crosslink. Note that cleavage by RNases A and T1 results in transfer of label to the adjacent base on the 5' side. In contrast, nuclease V1 cleavage results in retention of label by the originally labeled nucleotiode. These considerations may not be applicable in many instances with larger RNA duplexes. However, for rather specific interactions it may be worth considering whether the combination of RNases and labeling [α-^{32}P]NTP is optimal.

Acknowledgments

The work on developing this method has been supported by a grant from the Wellcome Trust (040375/Z/94/Z/PMG/RB). We thank Justine Southby for critically reading the manuscript.

References

1. Milligan, J. F. and Uhlenbeck, O. C. (1989) Synthesis of small RNAs using T7 RNA polymerase. *Methods Enzymol.* **180**, 51–62.
2. Grasby, M. J. and Gait, M. J. (1994) Synthetic oligoribonucleotides carrying site-specific modifications for RNA structure–function analysis. *Biochimie* **76**, 1223–1234.
3. Gait, M. J., Earnshaw, D. J., Farrow, M. A., Fogg, J. H., Grenfell, R. L., Naryshkin, N. A., and Smith, T. V. (1998). Applications of chemically synthesized RNA, in *RNA:Protein Interactions. A Practical Approach* (Smith, C. W. J., ed.), Oxford University Press, Oxford and New York, pp. 1–36.
4. Moore, M. J. and Sharp, P. A. (1992) Site-specific modification of pre-mRNA: the 2'-hydroxyl groups at the splice site. *Science* **256**, 992–997.
5. Moore, M. J. and Query, C. C. (1998) Uses of site specifically modified RNAs constructed by RNA ligation, in *RNA:Protein Interactions. A Practical Approach* (Smith, C. W. J., ed.), Oxford University Press, Oxford and New York, pp. 75–108.
6. Sontheimer, E. J. and Steitz, J. A. (1993) The U5 and U6 small nuclear RNAs as active site components of the spliceosome. *Science* **262**, 1989–1995.
7. Newman, A. J., Teigelkamp, S., and Beggs, J. D. (1995) snRNA interactions at 5' and 3' splice sites monitored by photoactivated crosslinking in yeast spliceosomes. *RNA* **1**, 968–980.
8. Weeks, K. M. and Crothers, D. M. (1993) Major groove accessibility of RNA. *Science* **261**, 1574–1577.
9. Tuite, E. M. and Kelly, J. M. (1993) Photochemical interactions of methylene blue and analogues with DNA and other biological substrates. *J. Photochem. Photobiol. B: Biol.* **21**, 103–124.
10. Lalwani, R., Maiti, S., and Mukherji, S. (1990) Visible light induced DNA-protein crosslinking in DNA-histone complex and sarcoma-180 chromatin in the presence of methylene blue. *J. Photochem. Photobiol. B: Biol.* **7**, 57–73.
11. Lalwani, R., Maiti, S., and Mukherji, S. (1995) Involvement of H1 and other chromatin proteins in the formation of DNA-protein crosslinks induced by visible light in the presence of methylene blue. *J. Photochem. Photobiol. B: Biol.* **27**, 117–122.
12. Liu, Z.-R., Wilkie, A. M., Clemens, M. J., and Smith, C. W. J. (1996) Detection of double-stranded RNA-protein interactions by methylene blue-mediated photocrosslinking. *RNA* **2**, 611–621.
13. Smith, C. W. J., Porro, E. B., Patton, J. G., and Nadal-Ginard, B. (1989) Scanning from an independently specified branch point defines the 3' splice site of mammalian introns. *Nature* **342**, 242–247.
14. Smith, C. W. J., Chu, T. T., and Nadal-Ginard, B. (1993) Scanning and competition between AGs are involved in 3' splice site selection in mammalian introns. *Mol. Cell. Biol.* **13**, 4939–4952.
15. Fabrizio, P., Laggerbauer, B., Lauber, J., Lane, W. S., and Lührmann, R. (1997) An evolutionarily conserved U5 snRNP-specific protein is a GTP binding factor closely related to the ribosomal translocase EF-2. *EMBO J.* **16**, 4092–4106.

16. Liu, Z.-R., Laggerbauer, B., Lührmann, R., and Smith, C. W. J. (1997) Cross-linking of the U5 snRNP specific 116 kDa protein to RNA hairpins that block step 2 of pre-mRNA splicing. *RNA* **3,** 1207–1219.
17. Liu, Z.-R., Sargueil, B., and Smith, C. W. J. (1998) Detection of a novel ATP-dependent cross-linked protein at the 5' splice site:U1 snRNA duplex by methylene blue mediated photo-crosslinking. *Mol. Cell. Biol.* **18,** 6910–6920.
18. Saenger, W. (1984) *Springer Advanced Texts in Chemistry: Principles of Nucleic Acid Structure* (Cantor, C. R., ed.), Springer-Verlag, New York, p. 12.
19. Atherton, S. J. and Harriman, A. (1993) Photochemistry of intercalated methylene blue. Photoinduced hydrogen-atom abstraction from guanine and adenine. *J. Am. Chem. Soc.* **115,** 1816–1822.
20. Girotti, A. W., Lyman, S., and Deziel, M. R. (1979) Methylene blue-sensitized photo-oxidation of hemoglobin: evidence for cross-link formation. *Photochem. Photobiol.* **29,** 1119–1125.
21. Van Steveninck, J. and Dubbelman, T. M. A. R. (1984) Photodynamic intramolecular crosslinking of myoglobin. *Biochim. Biophys. Acta* **791,** 98–101.
22. Liu, Z.-R., Sargueil, B., and Smith, C. W. J. (1997) Methylene blue mediated cross-linking of proteins to double-stranded RNA. *Methods Enzymol.*, in press.
23. Fu, X.-D. (1995) The superfamily of arginine/serine-rich splicing factors. *RNA* **1,** 663–680.
24. Bass, B. L., Hurst, S. R., and Singer, J. D. (1994) Binding properties of newly identified Xenopus proteins containing dsRNA-binding motifs. *Curr. Biol.* **4,** 301–314.
25. Bycroft, M., Grunert, S., Murzin, A. G., Proctor, M., and Johnston, D. S. (1995) NMR solution structure of a dsRNA binding domain from *Drosophila* staufen protein reveals homology to the N-terminal domain of ribosomal protein S5. *EMBO J.* **14,** 3563–3571.

5

Probing RNA–Protein Interactions by Psoralen Photocrosslinking

Zhuying Wang and Tariq M. Rana

1. Introduction

RNA molecules can fold into extensive structures containing regions of double-stranded duplex, hairpins, internal loops, bulged bases, and pseudoknotted structures *(1)*. Owing to the complexity of RNA structure, the rules governing sequence-specific RNA–protein recognition are not well understood. RNA–protein interactions are vital for many regulatory processes, especially in gene regulation where proteins specifically interact with binding sites found within RNA transcripts. In the absence of high-resolution crystallographic and nuclear magnetic resonance data, new methods are needed to determine the topology of RNA–protein complexes under physiological conditions. We have devised a new method based on psoralen photochemistry to identify specific contacts in RNA–protein complexes *(2,3)*.

1.1. Description of Psoralen Photochemistry

Psoralens are bifunctional photoreagents that form covalent bonds with the pyrimidine bases of nucleic acids *(4)*. The primary reaction is cyclobutane ring formation between the 5,6 double bond of uridine in RNA (or thymidine in DNA) and either the 4',5' or 3,4 double bond of the psoralen (**Fig. 1**). Reactions at the 4',5' double bond create a furan-side monoadduct (MAf), which can further react at a site with a flanking pyrimidine on the opposite strand to create an interstrand crosslink (XL). There are four steps involved in the mechanism of crosslink formation *(5)*: (a) intercalation of the psoralen into the nucleic acid, (b) absorption of a long wavelength ultraviolet photon (320–410 nm; UVA) and bond formation to create the MAf, (c) conformational changes induced by the adduct formation during a 1-µs period in which the absorption of an addi-

From: *Methods in Molecular Biology, Vol. 118: RNA-Protein Interaction Protocols*
Edited by: S. Haynes © Humana Press Inc., Totowa, NJ

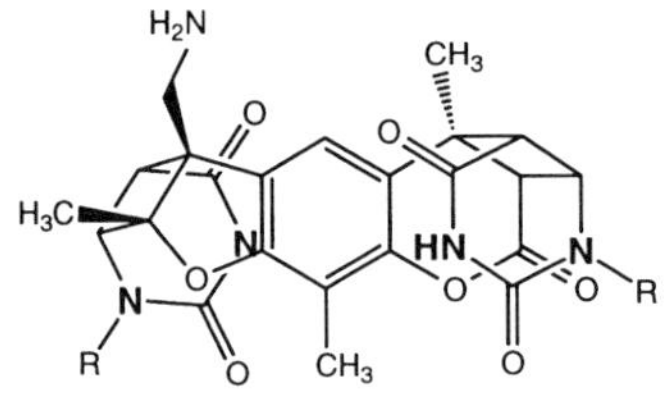

Fig. 1. Structures of the two types of 4'-aminomethyl-4,5',8-trimethylpsoralen (AMT)–uridine adducts formed by the photoreaction of AMT with RNA: *(1)* AMT; *(2)* furan-side monoadduct; *(3)* furan-side and pyrone-side diadduct.

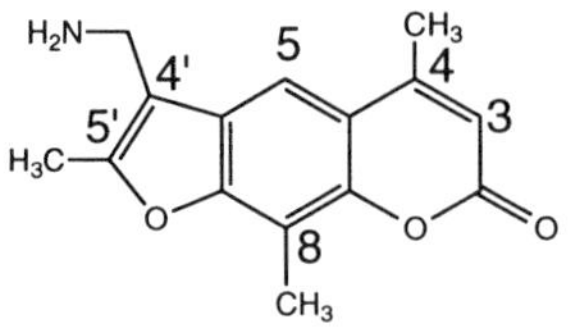

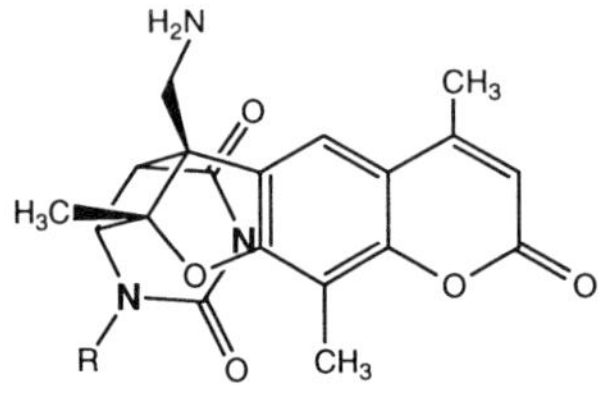

tional UVA photon does not form a XL, and (d) absorption of a second UVA photon by the conformationally relaxed MAf and formation of the XL.

1.2. Principle of the Procedure

Attachment of psoralen to a nucleic acid binding molecule creates an efficient nucleic acid crosslinking molecule. Oligonucleotides carrying psoralen at a specific site in the sequence are powerful tools for molecular biology and could provide a new class of therapeutic agents *(6)*. Photochemical properties of psoralens can be exploited to generate DNA probes containing psoralen monoadducts at specific sites, and these probes can form site-specific crosslinks to the complementary target sequences *(7,8)*. Psoralen-derivatized oligodeoxyribonucleotides and oligoribonucleoside methylphosphonates can specifically crosslink to double-helical DNA and viral mRNA targets, respectively *(9–11)*.

We reasoned that, as with nucleic acids, sequence-specific nucleic acid binding peptides can be converted into sequence-specific crosslinking peptides by introducing a psoralen at a specific site within the peptide sequence. In a protein–nucleic acid complex, site-specifically placed psoralen would photoreact with adjacent pyrimidines. The protein–nucleic acid crosslinked products can be purified and characterized. Isolation and identification of the crosslinked products could provide important information about the three-dimensional structure of the protein–nucleic acid complex, revealing, e.g., which base(s) of the nucleic acid are close to the psoralen site.

2. Materials

1. Standard apparatus for polyacrylamide gel electrophoresis (PAGE): plates, combs, spacers, and a power supply.
2. A photochemical reactor with long-wavelength (360 nm) tubes. We use Rayonet RPR-100 photochemical reactor (The Southern New England Ultraviolet Company, Branford, CT, USA). Other commercially available ultraviolet sources (e.g., Stratalinker) can also be used. One can also use a hand-held UV-lamp (Model UVGL-25, San Gabriel, CA, USA) with long-wavelength to perform the crosslinking, but longer irradiation time are required.
3. Sep-pak C_{18} cartridges from Waters (Milford, MA, USA).
4. Speed-Vac.
5. Sigmacote (Sigma, St. Louis, MO, USA).
6. Denaturing gel electrophoresis running buffer (TBE): 0.09 M Tris-borate, 0.002 M EDTA. Make a 5× TBE stock solution by dissolving 54 g of Tris base and 27.5 g of boric acid in 900 mL of double-distilled water (double-distilled water is used in all procedures), add 20 mL of 0.5 M EDTA, pH 8.0, and water to make a final volume of 1 L. Sterilize by autoclaving for 20 min at 15 lb/in.^{2} on liquid cycle.
7. Native gel electrophoresis running buffer: TBE buffer containing 0.1% Triton X-100.
8. 1× Binding buffer: 50 mM Tris-HCl, pH 7.5, 20 mM KCl, 0.1% Triton X-100.
9. 10× Hydrolysis buffer: This buffer is prepared by mixing equal volumes of 0.5 M Na_2CO_3 and 0.5 M $NaHCO_3$. Final pH of this buffer is pH 9.2.
10. Ribonuclease T1, and ribonuclease T1 buffer: 16 mM sodium citrate, pH 5.0, 0.8 mM EDTA, 0.5 mg/mL carrier tRNA and 3.5 M urea.
11. Ribonuclease *B. cereus* and ribonuclease *B. cereus* buffer: 16 mM sodium citrate, pH 5.0, 0.8 mM EDTA, 0.5 mg/mL carrier tRNA.
12. Loading buffer: For denaturing gels, either a solution of 9 M urea and 1 mM EDTA supplemented with 0.1% bromophenol blue or a solution of 98% formamide, 1 mM EDTA, and 0.1% bromophenol blue can be used. For native gels, the loading buffer is the same as the binding buffer but supplemented with 40% (w/v) glycerol and 0.1% bromophenol blue.
13. Gel stock solution: Make up a 40% (w/v) acrylamide (19:1 acrylamide: N',N'-methylene *bis*-acrylamide) stock solution by dissolving 380 g of acrylamide and 20 g of *bis*-acrylamide in 900 mL of water, and adding water to make a final volume of 1 L. Filter and store at 4°C. The denaturing gel working solution contains 7 M urea and 1× TBE. The native gel working solution contains 1× TBE and 0.1% Triton X-100.

14. RNA elution buffer: TBE buffer, pH 7.2. Adjust the pH of TBE buffer with 0.1 *M* HCl.
15. *N,N,N',N'*-tetramethylethylenediamine (TEMED); Ammonium persulfate.
16. Psoralen-modified peptide (or protein) and RNA labeled with ^{32}P at its 5'- or 3'-end (*see* Chapter 1).
17. AMV reverse transcriptase and buffer from Promega (Madison, WI, USA). Deoxynucleotides and dideoxynucleotides from Sigma (St. Louis, MO, USA).
18. For autoradiography, cassettes, X-ray films, and developer are required. A phosophorImager or a scanning densitometer is required for quantification.
19. Proteinase K, and proteinase K digestion buffer: 10 m*M* Tris-HCl, pH 7.8, 5 m*M* EDTA, and 0.5% SDS.
20. 8-Hydroxypsoralen, 1,3-diiodopropane, potassium carbonate, acetone, petroleum ether, ethyl acetate, sodium sulfate, and silica gel were obtained from Aldrich (Milwaukee, WI, USA).
21. 0.1 *M* Sodium phosphate, pH 7.4.
22. 10% Trifluoracetic acid in H$_2$O.
23. Dimethylformamide.

3. Methods

3.1. Synthesis of Psoralens Suitable for Cysteine Modification

3.1.1. Synthesis of 8-([3-Iodopropyl-1]oxy)psoralen, Compound 1 (Fig. 2).

1. Dissolve a mixture of 8-hydroxypsoralen (0.176 g, 1 mmol), 1,3-diiodopropane (1.15 mL, 2.96 g, 10 mmol), and potassium carbonate (1.38 g, 10 mmol) in acetone (15 mL). This reaction is carried out in a 100-mL round-bottom glass flask.
2. Stir the reaction mixture for 10 h at room temperature. Monitor the reaction by thin-layer chromatography (petroleum ether:ethyl acetate, 4:1). A complete reaction shows the absence of starting material.
3. Concentrate the reaction mixture to dryness under reduced pressure.
4. Dissolve the residue in 30 mL of water and extract 3× with 20 mL of ethyl acetate.
5. Wash the extract with 10 mL of water and then with 10 mL of saturated sodium chloride solution.
6. Add approx 2 g of sodium sulfate to the extract and let it stand at room temperature for 4–6 h.
7. Filter and evaporate ethyl acetate under reduced pressure.
8. Chromatograph the residue on silica gel and elute with petroleum ether:ethyl acetate, 4:1. Pure compound 1 (0.31 g, 85%) is obtained as a pale yellow solid. ^{1}H NMR (CDCl$_3$): d 7.78 (1H, d, *J* = 9.8), 7.71 (1H, d, *J* = 2.2), 7.39 (1H, s), 6.84 (1H, d, *J* = 2.2), 6.39 (1H, d, *J* = 9.8), 4.57 (2H, t, *J* = 5.74), 3.54 (2H, t, *J* = 6.8), 2.36 (2H, m). Mass spectrum 371 (M + H).

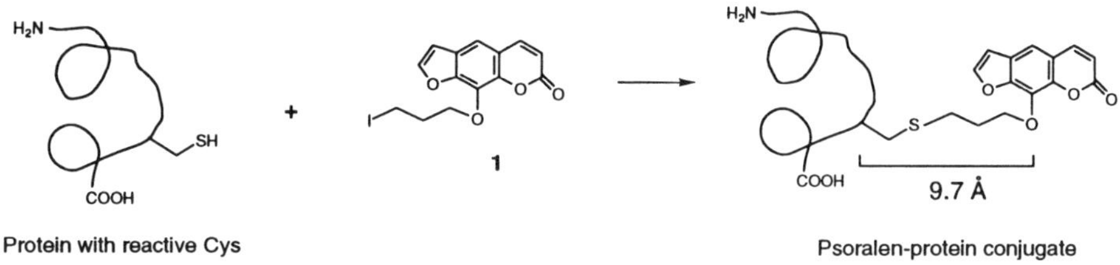

Fig. 2. Schematic representation for site-specific chemical modification of a protein with psoralen. The structure of 8-([3-Iodopropyl-1]oxy)psoralen, compound **1** is shown. The distance between the psoralen molecule and the α-carbon of a cysteine amino acid in the protein sequence is 9.7 Å.

3.2. Conjugation of Psoralen to Proteins

The experimental strategy for site-specific psoralen conjugation of a protein is outlined in **Fig. 2**.

1. Introduce a unique reactive (solvent-accessible) cysteine at a predetermined position into the protein sequence by site-directed mutagenesis or during solid-phase peptide synthesis according to established procedures.
2. After purification and characterization of the protein, dissolve the protein (e.g., 1 nmol) and 8-([3-Iodopropyl-1]oxy)psoralen **1** (10 nmol) in 100 µL of dimethylformamide.
3. Adjust the final pH to 7.0 by adding 50 µL of 0.1 M sodium phosphate buffer, pH 7.4.
4. Incubate for 16 h at room temperature.
5. Prepare 10% trifluoroacetic acid solution in water. Add approx 10 µL of this solution to the reaction mixture to bring the pH between 4.0 and 5.0.
6. Extract unreacted psoralen three times with 0.5 mL of chloroform.
7. Wash the organic phase three times with 0.2 mL of 1% trifluoroacetic acid in water to recover any dissolved peptide.
8. Combine the aqueous layers and concentrate to ≈200 µL in a Speed-Vac.
9. Purify the psoralen–protein conjugate by HPLC on a Zorbax 300 SB-C$_8$ column or other columns appropriate for the protein under investigation. The final yield of psoralen–peptide conjugates is usually ≈80%.

3.3. Photocrosslinking

3.3.1. Analytical Scale Crosslinking Reaction

1. Prepare a denaturing polyacrylamide gel in advance of performing the crosslinking assay. To prepare a 20% gel, add 1 mL of a fresh 10% (w/v) ammonium persulfate solution and 30 µL of TEMED to 100 mL of 20% (w/v) acrylamide and 7 M urea in 1× TBE buffer while stirring. Pour the gel (*see* **Notes 1** and **2**), set the comb, and let the gel polymerize.

2. Remove the comb as soon as the gel has set, and flush the wells with TBE buffer from a washing bottle.
3. Mount the gel in the gel tank and fill the two chambers with TBE buffer. Remove the air bubbles from the bottom of the gel, and prerun at 30 W for 3–4 h.
4. Dilute the ^{32}P-end-labeled RNA to approx 2.5 μM in 20 μL of binding buffer, heat the solution at 85°C for 3 min to denature RNA, and then let it cool down slowly to room temperature to refold the RNA (*see* **Note 3**).
5. Use sterile Eppendorf tubes for the crosslinking assay. In each tube, place 2 μL of refolded RNA, 2 μL of 5× binding buffer, 1 μL of 0.1 μg/μL yeast tRNA as competitor, 1–5 μL of ddH$_2$O, and finally protein or psoralen-modified protein (final volume of 10 μL). The amount of protein to be added can be determined by gel retardation analysis (*see* **Note 4** and Chapter 10).
6. Pick the RNA/protein ratio where only specific RNA–protein complexes are formed. Thoroughly mix the reaction mixtures and incubate on ice for 20 min.
7. Meanwhile, clean a small plate (glass or polystyrene) with 95% ethanol and coat the plate with Sigmacote for better recovery of samples after crosslinking. Place the plate on ice.
8. Transfer the samples to be irradiated to the plate and UV (360 nm) irradiate in a photoreaction chamber or by hand-held UV lamp for 10–20 min. The optimal time can be found by a time-course experiment.
9. After irradiation, transfer the samples to new tubes, rinse the sample spots on the plate with 10 μL of loading buffer, and combine with the original samples.
10. Add 10 μL of loading buffer to nonirradiated samples. Add 1 μL of 20 μg/μL yeast tRNA to all the samples to compete with RNA–protein binding (*see* **Note 5**).
11. Heat all the samples at 90°C for 3 min and then load onto the gel that has been prerun for 3–4 h. Run the gel at 30 W for 2–4 h until the bromophenol blue band is 5 cm from the bottom of the gel (*see* **Note 6**).
12. Turn off the power supply, remove the gel from the plates, and wrap in a piece of plastic wrap. Expose it to X-ray film at –80°C.

3.3.2. Large-Scale Crosslink Preparation

1. To map the crosslink site on an RNA sequence, a large-scale crosslink reaction is required. The crosslink reaction as described above can be scaled up to 100× or more, and the final reaction volume can be reduced to half without affecting the crosslink reaction significantly. When transferring the sample to a coated plate for UV irradiation, the sample is placed on the plate in a small volume (< 20 μL) for efficient UV irradiation and easy handling. If the drop is more than 20 μL, it will move on the plate and may mingle with other drops when the plate is moved around.
2. Prepare a gel as described in **Subheading 3.3.1**. Pour a two-well gel with a narrow well for ^{32}P-labeled RNA as a control, and a wide well for large-scale crosslink separation. Prerun the gel as described.
3. After UV irradiation, transfer the sample to a new tube. Add 1/10 vol of 3 M NaOAc, pH 5.2, then add 2.5 vol of absolute ethanol. Let it stand at –80°C for 30 min to precipitate RNA and RNA crosslinks.

4. Spin it down at 16,000g for 20 min, aspirate the supernatant, rinse the pellet with a small volume of ice-cold 75% ethanol, centrifuge again, and take off the supernatant. Dry the pellet briefly in a Speed-Vac.
5. Redissolve the pellet in loading buffer, and add yeast tRNA. Heat the solution at 90°C for 3 min and load onto a denaturing gel.
6. Run the gel for the same time as in the crosslink assay.
7. Take off the gel, wrap it in plastic wrap, and expose to X-ray film (label both the gel and film with a marker for later match) at room temperature for a short time.
8. After developing the film, place the gel on the film according to the mark, cut the RNA and the RNA–protein crosslink bands, crush the gel and elute in the elution buffer, recover RNA with a cartridge or by ethanol precipitation (*see* **Note 7**).

3.4. Mapping the Crosslink Sites

Mapping of the crosslink site on the RNA to single nucleotide resolution can be carried out by partial RNase digestion and alkaline hydrolysis of the gel-purified RNA–protein crosslink. Fragment sizes are determined by comparison with RNA oligonucleotides of defined sequence and length generated by digesting RNA with RNases T1 and *B. cereus* (*see* **Note 8**).

In the case of large or highly unstable RNA structures that cannot be sequenced by alkaline hydrolysis, the exact site of the crosslink on the RNA is determined by primer extension analysis of the gel-purified RNA–protein crosslink *(12,13)*. Reverse transcriptase synthesizes cDNA copies from an RNA template and this enzyme stops at a point where RNA has been modified or crosslinked *(12–17)*. The stops are mapped by comparing their cDNA length with standard chain termination sequencing.

3.4.1. Partial Alkaline Hydrolysis Sequencing of RNA

1. Add ^{32}P-labeled RNA to two sterile Eppendorf tubes (no. 1 and no. 2) and the RNA–protein crosslink to a third tube (no. 3). Use labeled RNAs of approximately equal specific activity.
2. To the first tube (no. 1), add ddH$_2$O and 1 µL of 10× ribonuclease buffer to a final volume of 9 µL, then add 1 µL of 0.1 U/µL of ribonuclease T1 or ribonuclease *B. cereus* (*see* **Note 8**). Incubate at 55°C for 6–12 min. The reaction products will provide size markers for the hydrolysis reactions *(3)*.
3. To the other two tubes, add ddH$_2$O to a final volume of 9 µL, then add 1 µL of 10× hydrolysis buffer. Incubate at 85°C for 8 min to obtain alkaline hydrolysis ladders of the RNA.
4. Add 10 µL of loading buffer (*see* **Note 9**) to each tube, and place it on ice. Heat at 95°C for 2 min and then load onto a denaturing acrylamide gel that has been prerun for 4 h (*see* **Note 10**). Run the gel at 30 W until the bromophenol blue band is 12 cm from the gel bottom to resolve all the bands. Bromophenol blue can be loaded in a blank lane.
5. Expose to film.

3.4.2. Reverse Transcriptase (or Primer Extension) Sequencing of RNA

Usually 0.1–1 pmol of RNA or RNA–protein crosslink is used for this method, depending upon the specific activity of the ^{32}P-labeled DNA primer.

1. Treat RNA–protein crosslink with proteinase K to digest the protein. This will produce an RNA crosslinked to a small peptide containing a few amino acids. Resuspend the gel-purified RNA–protein crosslink product in 50 µL of proteinase K digestion buffer and add 1 µL of proteinase K stock solution (2.5 mg/mL in ddH$_2$O). Incubate the reaction mixture at 37°C for 30 min.
2. Precipitate the RNA–peptide crosslink with ethanol as described in **Subheading 3.3.2.** and purify the RNA products by denaturing gel electrophoresis (*see* **Subheading 3.3.1.**).
3. Place 4 pmol of uncrosslinked RNA, 4 pmol of ^{32}P-labeled DNA primer, 2 µL of 5× RT buffer in a microcentrifuge tube, and add ddH$_2$O to a final volume of 10 µL. In a second tube, place 1 pmol of digested RNA–protein crosslink, 1 pmol of ^{32}P-labeled primer, 0.5 µL of 5× RT buffer, and add ddH$_2$O to a final volume of 2.5 µL. Heat the two annealing reactions at 75°C for 3 min and allow to cool down slowly to room temperature on the benchtop. Then spin down briefly to recover the contents.
4. Label five microcentrifuge tubes G, A, C, U, and X. Add 3 µL of the appropriate RT nucleotide mix to each tube (*see* **Note 11**). Add 1 µL of 5× RT buffer and 2.5 µL of the annealing mix (from above) to each tube. Add the annealing mix for the RNA–protein crosslink in tube X.
5. Add 1 µL of RT enzyme (5 U) to each tube to initiate the extension.
6. Incubate for 10 min at room temperature, then 50 min at 42°C.
7. Stop the reactions by adding 12 µL of loading buffer. Heat the reactions for 5 min at 95°C before loading onto a denaturing gel.
8. Expose to film.

3.5. Analysis of Results

Alkaline hydrolysis of RNA and crosslinked RNA–protein complex generates a ladder of RNA degradation products. Bands of crosslinked RNA–peptide complexes migrate much slower than corresponding free RNA (**Fig. 3**). For example, a Tat protein fragment containing a cysteine modified with psoralen at position 57 forms a crosslink with TAR RNA at nucleotide U31. Base hydrolysis of the 5'-end-labeled crosslinked complex results in an RNA ladder in which all fragments up to C30 are resolved. There is an obvious gap in the hydrolysis ladder after C30 that is not seen with the uncrosslinked RNA. This indicates that the fragments above C30 from the 5'-end are linked to the psoralen–protein conjugate. Thus, U31 is the 5'-end crosslink site. To define the 3'-end boundary of the crosslink site, 3'-end-labeled RNA–protein crosslink is purified and subjected to partial alkaline hydolysis. Base hydrolysis of the 3'-end-labeled crosslink produces a ladder in which the fragments from the

Psoralen RNA–Protein Crosslinking

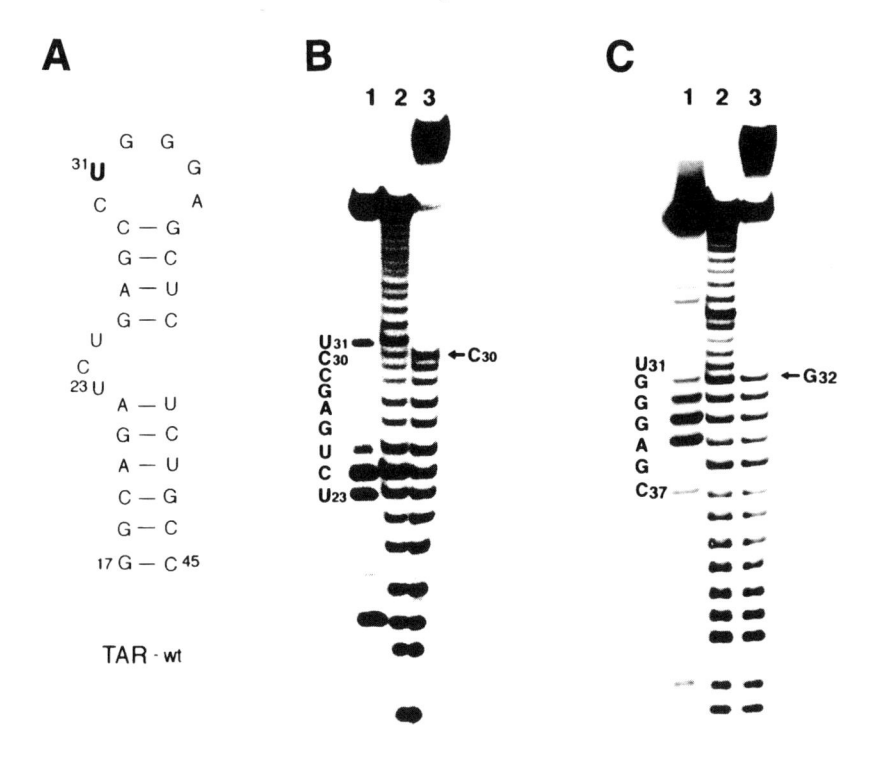

A
```
        G  G
    ³¹U    G
     C     A
      C — G
      G — C
      A — U
      G — C
     U
     C
   ²³U
      A — U
      G — C
      A — U
      C — G
      G — C
  ₁₇G — C⁴⁵

    TAR · wt
```

B 1 2 3
U₃₁
C₃₀ ←C₃₀
C
G
A
G
U
C
U₂₃

C 1 2 3
U₃₁
G ←G₃₂
G
G
A
G
C₃₇

Fig. 3. Mapping of the crosslinked base in the crosslinked RNA–protein complexes by alkaline hydrolysis. A specific cysteine is introduced into the Tat protein and subsequently modified with psoralen. This psoralen–Tat conjugate photocrosslinks to TAR RNA at a single site. **(A)** Secondary structure of wild-type TAR RNA. Uridine 31, a psoralen addition site, is highlighted in the TAR RNA. The numbering of nucleotides in the RNA corresponds to their positions in wild-type TAR RNA. **(B)** Analysis of the RNA–protein crosslink containing 5'-^{32}P end-labeled TAR RNA: *B. cereus* ladder of TAR RNA (lane 1); hydrolysis ladder of TAR RNA (lane 2); hydrolysis ladder of RNA–protein crosslinked complex (lane 3). The sequence of TAR RNA from U23 to U31 is labeled, and a gap in the sequence is obvious after the C30 residue, indicating that U31 is the crosslinked base. **(C)** Analysis of RNA–protein crosslink containing 3'-^{32}P end-labeled TAR RNA: RNase T1 ladder of TAR RNA (lane 1); hydrolysis ladder of TAR RNA (lane 2); hydrolysis ladder of crosslinked RNA–protein complex (lane 3). The sequence of TAR RNA from C37 to U31 is labeled, and a gap in the sequence is obvious after residue G32 indicating that U31 is the only crosslinked base. Reprinted in part with permission from **ref. 3**. Copyright [1996] American Society for Biochemistry and Molecular Biology.

crosslinked RNA–peptide complex match those from free RNA until G32 from 3'-end. After G32, a clear gap is observed in the hydrolysis ladder of the 3'-end-labeled crosslink, whereas alkaline digestion of 3'-end-labeled RNA re-

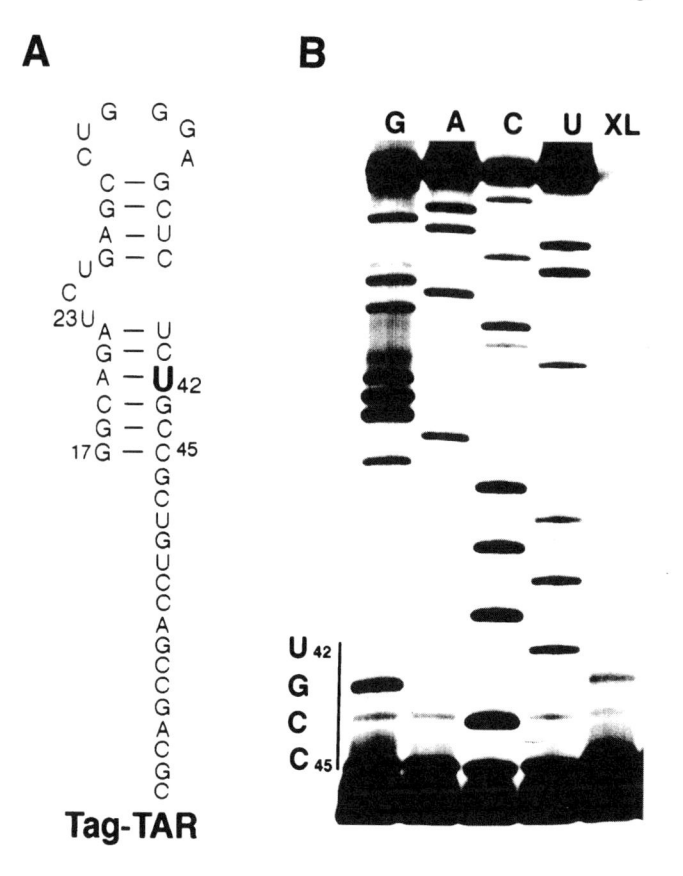

Fig. 4. Primer extension analysis of an RNA–protein crosslink. Psoralen is attached to the N-terminus of a Tat peptide and this peptide is crosslinked to TAR RNA. (**A**) Secondary structure of Tag TAR RNA. Wild-type TAR RNA is shown in **Fig. 3**. Tag TAR RNA contains an extra 15 nucleotides at the 3'-end of TAR RNA to allow the hybridization of a DNA primer for reverse transcriptase analysis of the crosslink. Uridine 42, a psoralen addition site, is highlighted in Tag TAR RNA. (**B**) Mapping the exact position of the psoralen crosslink to TAR RNA by primer extension analysis. Gel-purified crosslinked Tag TAR RNA was primer extended by hybridizing an oligonucleotide complementary to the 15 nucleotides at the 3'-end (lane XL). Lanes of sequencing reactions are labeled as G, A, C, and U. Sequencing lanes are presented as their representative complementary RNA position rather than the cDNA itself. Sequence of TAR RNA in the crosslinked region is shown on left. Reprinted in part with permission from **ref. 2**. Copyright [1995] American Chemical Society.

sults in a standard ladder. This result indicates that the fragments above G32 from the 3'-end contain Tat peptide. Based on these results, one can conclude that U31 of TAR RNA is the only site at which crosslinking occurs.

During primer extension of psoralen-crosslinked RNA, AMV reverse transcriptase displays a characteristic pattern by pausing at 1 nucleotide from the 3'-end of the crosslink site. However, there are some instances where the transcriptase will advance to the adducted site *(12–14,17–19)*. In this case, a characteristic doublet is produced, corresponding to stops at 1 nucleotide from the 3'-end of the crosslink and up to the crosslink site. **Figure 4** outlines a typical primer extension mapping of a crosslink.

3.6. Limitations and Modification of the Method

Although the psoralen–peptide conjugates offer a new class of probes for sequence-specific protein–nucleic acid interactions, they also have their limitations. Specificity of the psoralen photoreactions limits the use of this approach. Psoralen primarily reacts with thymidine in DNA and uridine in RNA, although a minor reaction with cytosine also occurs *(4)*. Therefore, this limitation of psoralen photoreactivity provides a cautionary note for the interpretation of crosslinking results: the presence of a crosslink indicates proximity, but absence of a crosslink could be due to unfavorable photochemistry rather than lack of proximity.

The psoralen analogs shown in **Fig. 2** possess a chemically reactive functional group that modifies the cysteine side chains of a protein. Unfortunately, these reagents cannot distinguish a specific amino acid when it is present more than once in the sequence of the protein. When only one reactive cysteine is present in the protein sequence, the protein–psoralen conjugate can be used for crosslinking reactions and the results provide structural information about RNA–protein complexes. However, when more than one cysteine is present, the crosslinking of RNA gives complicated results. This problem can be overcome by using a protected amino acid analog of psoralen that can be incorporated at any internal position in a peptide sequence during solid-phase peptide synthesis. A psoralen amino acid analog, N-α-Fmoc-L-aspartic acid-β-(4'-[aminomethyl]-4,5',8-trimethylpsoralen), has recently been synthesized and used for site-specific labeling of a peptide by standard solid phase peptide synthesis methods *(2)*. For details of the synthesis of psoralen amino acid and application of this psoralen–peptide conjugate, *see* **ref. 2**.

4. Notes

1. Electrophoresis plates should be cleaned thoroughly with water, and then washed with 95% ethanol. Because 0.8 mm acrylamide gels are used, silanization is good but not necessary. Also, gels do not need to be dried; they are wrapped in plastic wrap and autoradiographed with X-ray film at –80°C, or placed on a phosphor image screen for phosphor image analysis.
2. Acrylamide is a potent neurotoxin. Always wear gloves when working with psoralen and acrylamide.

3. In most cases, RNA refolding is required because gel conditions and cartridge purification may denature RNA structure. Denatured RNA will not bind or will nonspecifically bind to the protein under investigation. This can be easily tested on a native gel: one will see more than one band in the RNA lane and many bands in the RNA plus protein lane. One can expect to see mainly one band in the RNA lane if the RNA is correctly folded. Different RNAs may need different refolding conditions and the proper conditions can be found by mutiple trials.
4. Protein modification with psoralen may alter the protein structure or RNA-binding capabilities of the protein. To characterize and evaluate the binding capabilities of a psoralen–protein conjugate, dissociation constants for psoralen–protein:RNA complexes should be determined and compared with the wild-type protein:RNA complexes. Equilibrium dissociation constants of the protein:RNA complexes can be measured using direct and competition electrophoretic mobility assays *(20)*. For direct mobility shift assays, the fractional saturation of 5'-^{32}P-end-labeled RNA is measured as a function of wild-type and psoralen modified protein *(2)*. Binding constants are calculated from multiple sets of experiments. In the case of a Tat:TAR complex, results of gel shift experiments show that psoralen attachment to a Tat fragment does not significantly alter the structure of the Tat and preserves the TAR-binding affinities of the peptide *(2,3)*.
5. Yeast tRNA is added to the crosslinking reaction before samples are loaded on a denaturing gel to compete out binding. Without the addition of yeast tRNA, one can see protein–RNA binding bands even in a denaturing gel. Usually 100–200 times (w/w) yeast tRNA or other unlabeled competitor RNA of radiolabeled RNA is added to a crosslink reaction.
6. Make sure to prerun the gels for the proper time. Usually a vertical 20 cm × 45 cm gel is prerun at 30 W for about 4 h before samples are loaded. If the gel is prerun for only a short time, short RNA fragment bands (one base up to several bases long) will not be separated cleanly. Optimum prerunning time can be determined by a few trials.
7. RNA is very sensitive to hydrolysis by acids and bases, especially bases. To protect the RNA from hydrolysis, the RNA is eluted out of gels with TBE buffer whose pH is adjusted to 7.0–7.2. The RNA is further purified with a C_{18} cartridge to get rid of any salts and redissolved in ddH_2O. Up to 1 µmol of RNA in ≈1 mL of buffer can be loaded onto a 1-mL cartridge. Depending on the properties of protein–RNA crosslink, sometimes it is not suitable to purify the crosslink product on a cartridge, because the hydrophilicity of protein decreases the binding affinity of the crosslink product to the cartridge matrix. In this case, the crosslink product can be recovered by ethanol precipitation. Always wear gloves and use double distilled autoclaved water when handling RNA.
8. Ribonuclease *B. cereus* cleaves RNA predominantly at UpN and CpN bonds. Ribonuclease T1 cleaves RNA only at GpN bonds.
9. The loading buffer for RNA sequencing gels should contain very little (<0.05%) or no bromophenol blue or other dyes. Too much dye will affect the band separation around the dye front. If a dye band is needed for monitoring the gel running, dye can be loaded in a blank lane.

10. The acrylamide percentage of an RNA sequencing gel is determined by the size of RNA. For a small RNA fragment of < 30 bases, 20% gels are used for optimal resolution.
11. For primer extension sequencing lanes, the nucleotide mix (RT mix) solution contains 500 µM of each deoxynucleotide (dGTP, dATP, dCTP, and dTTP) and one dideoxynucleotide. U, G, C, and A nucleotide mix solutions contain ddATP, ddCTP, ddGTP, and ddTTP, respectively. The concentration of dideoxynucleotide is from one tenth of the deoxynucleotide concentration (for longer RNA sequencing) to one third of the deoxynucleotide concentration (15–30 nucleotides long RNA). The nucleotide mix for RNA–protein crosslink sequencing contains only the four deoxynucleotides.

References

1. Tinoco, I., Jr., Puglisi, J. D., and Wyatt, J. R. (1990) RNA folding. *Nucleic Acids Mol. Biol.* **4,** 205–226.
2. Wang, Z. and Rana, T. M. (1995) Chemical conversion of a TAR RNA-binding fragment of HIV-1 Tat protein into a site-specific crosslinking agent. *J. Am. Chem. Soc.* **117,** 5438–5444.
3. Wang, Z., Wang, X., and Rana, T. M. (1996) Protein orientation in the Tat-TAR complex determined by psoralen photocross-linking. *J. Biol. Chem.* **271,** 16,995–16,998.
4. Cimino, G. D., Gamper, H. B., Isaacs, S. T., and Hearst, J. E. (1985) Psoralens as photoactive probes of nucleic acid structure and function: organic chemistry, photochemistry, and biochemistry. *Ann. Rev. Biochem.* **54,** 1151–1193.
5. Spielmann, H. P., Dwyer, T. J., Sastry, S. S., Hearst, J. E., and Wemmer, D. E. (1995) DNA structural reorganization upon conversion of a psoralen furan-side monoadduct to an interstrand cross-link: implications for DNA repair. *Proc. Natl. Acad. Sci. USA* **92,** 2345–2349.
6. Hearst, J. E. (1988) A photochemical investigation of the dynamics of oligonucleotide hybridization. *Annu. Rev. Phys. Chem.* **39,** 291–315.
7. Gamper, H. B., Cimino, G. D., Isaacs, S. T., Ferguson, M., and Hearst, J. E. (1986) Reverse southern hybridization. *Nucleic Acids Res.* **14,** 9943–9954.
8. Van Houten, B., Gamper, H. B., Hearst, J. E., and Sancar, A. (1986) Construction of DNA substrates modified with psoralen at a unique site and study of the action of ABC excinuclease on these uniformly modified substrates. *J. Biol. Chem.* **261,** 14,135–14,141.
9. Takasugi, M., Guendouz, A., Chassingnol, M., Decout, J. L., Lhomme, J., Thuong, N. T., and Helene, C. (1991) Sequence-specific photo-induced cross-linking of the two strands of double-helical DNA by a psoralen covalently linked to a triple helix-forming oligonucleotide. *Proc. Natl. Acad. Sci. USA* **88,** 5602–5606.
10. Lee, B. L., Murakami, A., Blake, K. R., Lin, S.-B., and Miller, P. S. (1988) Interaction of psoralen-derived oligodeoxyribonucleoside methylphosphonates with single-stranded DNA. *Biochemistry* **27,** 3197–3203.

11. Kean, J. M. and Miller, P. S. (1994) Effect of target structure on cross-linking by psoralen-derivatized oligonucleoside methylphosphonates. *Biochemistry* **33**, 9178–9186.
12. Youvan, D. C., and Hearst, J. E. (1981) A sequence from *Drosophila melanogaster* 18S rRNA bearing the conserved hypermodified nucleoside amψ: analysis by reverse transcription and high-performance liquid chromatography. *Nucleic Acids Res.* **9**, 1723–1741.
13. Youvan, D. C. and Hearst, J. E. (1982) Sequencing psoralen photochemically reactive sites in *Escherichia coli* 16 S rRNA. *Anal. Biochem.* **119**, 86–89.
14. Ericson, G. and Wollenzien, P. (1988) Use of reverse transcription to determine the exact locations of psoralen crosslinks in RNA. *Anal. Biochem.* **174**, 215–223.
15. Burgin, A. B. and Pace, N. R. (1990) Mapping the active site of ribonuclease P RNA using a substrate containing a photoaffinity agent. *EMBO J.* **9**, 4111–4118.
16. Harris, M. E., Nolan, J. M., Malhotra, A., Brown, J. W., Harvey, S. C., and Pace, N. R. (1994) Use of photoaffinity crosslinking and molecular modeling to analyze the global architecture of ribonuclease P RNA. *EMBO J.* **13**, 3953–3963.
17. Nolan, J. M., Burke, D. H., and Pace, N. R. (1993) Circularly permuted tRNAs as specific photoaffinity probes of ribonuclease P RNA structure. *Science* **261**, 762–765.
18. Ericson, G. and Wollenzien, P. (1989) An RNA secondary structure switch between the active and inactive conformations of the *Escherichia coli* 30S ribosomal unit. *J. Biol. Chem.* **264**, 540–545.
19. Watkins, K. P., Dungan, J. M., and Agabian, N. (1994) Identification of a small RNA that interacts with the 5' splice site of the trypanasoma brucei spliced leader RNA in vivo. *Cell* **76**, 171–182.
20. Fried, M. and Crothers, D. M. (1981) Equilibria and kinetics of lac repressor-operator interactions by polyacrylamide gel electrophoresis. *Nucleic Acids Res.* **9**, 6505–6525.

6

Identification and Sequence Analysis of RNA–Protein Contact Sites by N-Terminal Sequencing and MALDI-MS

Bernd Thiede, Henning Urlaub, Helga Neubauer, and Brigitte Wittmann-Liebold

1. Introduction

Crosslinking techniques have been used widely to obtain meaningful structural information on RNA–protein interactions. Exact data on the contact sites of RNA–protein complexes at the molecular level are required for a detailed modeling of three-dimensional structures within the complexes and for the identification of recognition motifs. For this purpose an approach was developed to determine the contact sites of RNA–protein crosslinks by combining N-terminal sequencing and matrix-assisted laser desorption/ionization mass spectrometry (MALDI-MS). This approach allowed the precise localization of the contact sites within the complex ribonucleoprotein particle of the native 30S ribosomal subunit *(1–4)*. Crosslinking experiments are performed either by ultraviolet (UV)-irradiation or by combining 2-iminothiolane and UV treatment. The generated RNA–protein crosslinks are digested with ribonuclease T_1 and endoprotease (Lys-C or Glu-C) and purified by size-exclusion chromatography and reverse phase–high-performance liquid chromatography (RP-HPLC). The purified oligoribonucleotide-peptide complexes are then subjected to N-terminal sequencing and MALDI-MS to identify the position of the crosslink in the protein and RNA (**Fig. 1**). The amino acid sequence of the peptide enabled the identification of the corresponding ribosomal protein and a gap in the sequence defined the crosslinking position. The mass spectrometrical analysis of the complexes prior to and after partial alkaline hydrolysis and treatment with 5'→3' phosphodiesterase led to the identification of the composition, the sequence, and crosslinking position of the oligoribonucleotide moiety.

From: *Methods in Molecular Biology, Vol. 118: RNA-Protein Interaction Protocols*
Edited by: S. Haynes © Humana Press Inc., Totowa, NJ

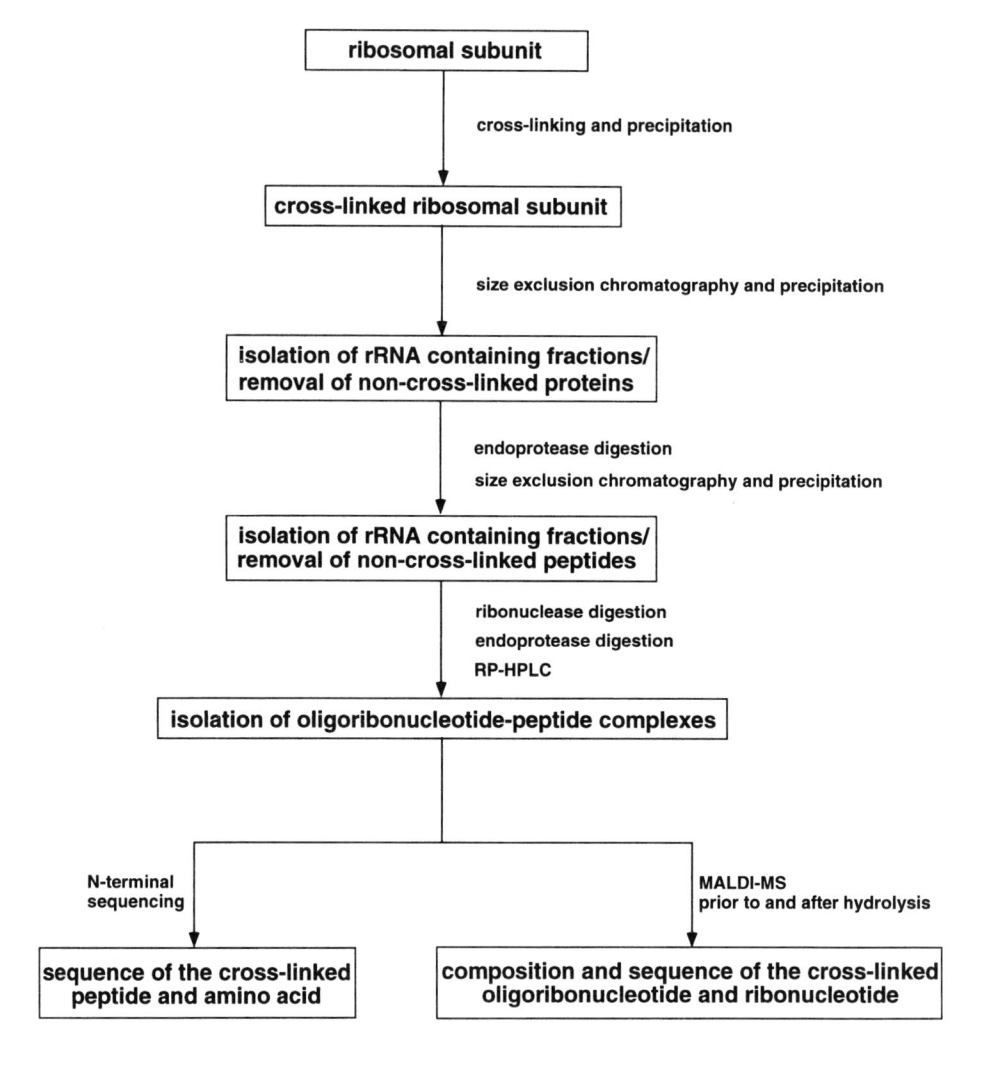

Fig. 1. Strategy for the analysis of crosslinked RNA–protein complexes.

The sequence analysis is achieved by ladder sequencing: the nucleotides from the 5' and 3' end are partially hydrolyzed from the oligonucleotide–peptide complex after ammonium hydroxide treatment. The masses of the resulting fragments in the mixture are determined by mass analysis. Starting from the entire nucleotide complex, the mass differences are calculated and yield the oligoribonucleotide sequence information, as shown in **Fig. 2**. In addition, the 5'→3' phosphodiesterase digest with subsequent mass analysis defines the 5' end.

N-Terminal Sequencing and MALDI-MS

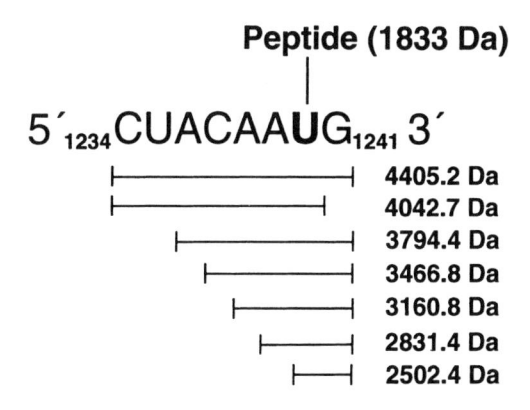

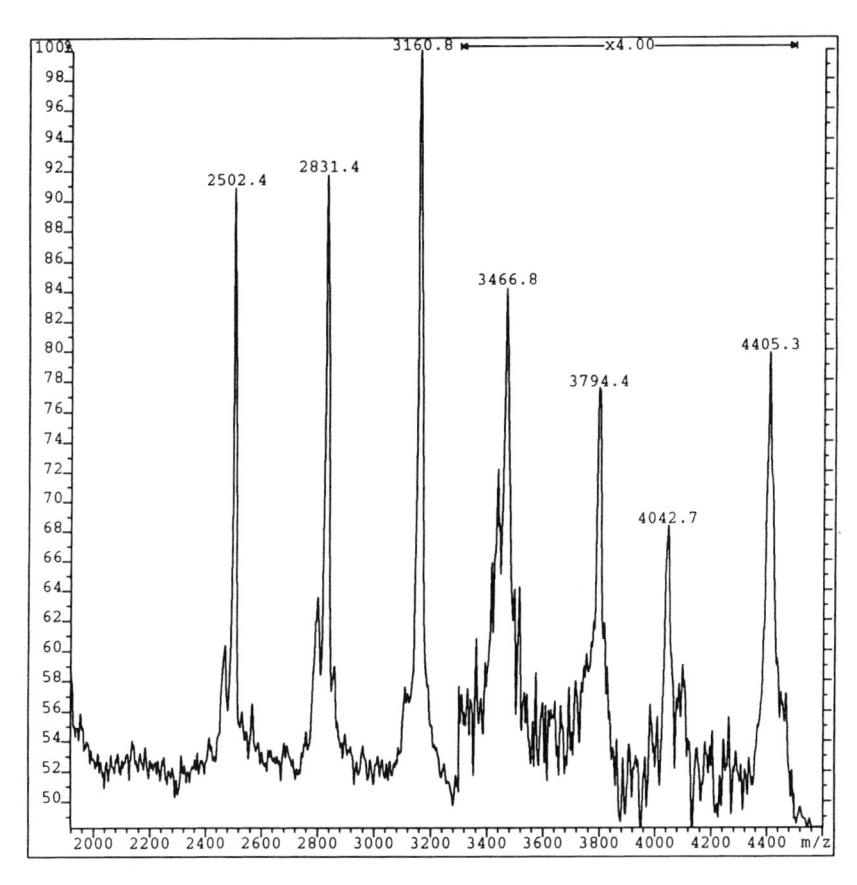

Peptide (1833 Da)
|
5´ ₁₂₃₄CUACAAUG₁₂₄₁ 3´

|————————————| 4405.2 Da
|————————————| 4042.7 Da
  |——————————| 3794.4 Da
   |—————————| 3466.8 Da
    |————————| 3160.8 Da
     |———————| 2831.4 Da
      |——————| 2502.4 Da

Fig. 2. MALDI mass spectrum of an oligoribonucleotide–peptide complex (Met-114 in ribosomal protein S7 to U-1240 of the 16S rRNA) after partial hydrolysis with aqueous ammonium hydroxide.

The approach described here to determine contact sites between RNA and proteins within a native cell complex promises to be applicable for other known RNA–protein complexes as well. For instance, the method may be applied for topographical studies of small nuclear RNPs. However, considerable amounts of isolated ribosomes were used for one crosslinking experiment. Three thousand A_{260} units were used as starting material for the ribosomal subunits to perform the various experiments and to reproduce the results. It is obvious that for other RNP complexes these amounts are difficult to prepare. However, these amounts may be reducible by using new mass spectrometers (delayed extraction MALDI-MS; nanoelectrospray-MS). Simultaneously with the mass analysis of intact molecules, fragment ions can be generated by these methods (postsource decay; tandem-MS). These techniques have not yet been applied to the entire sequence analysis of short oligoribonucleotide–peptide complexes. Although these complexes are not as well-studied as peptides by mass analysis, this approach will likely be possible in the future.

2. Materials

2.1. Crosslinking

1. UV-lamp G8T58W GERMICIDAL at 254 nm (Herolab, Wiesloch, Germany).
2. 20 mM 2-Iminothiolane solution (Pierce, Rockford, IL, USA): 500 mg of 2-iminothiolane dissolved in 10 mL of 500 mM triethanolamine.
3. Crosslinking buffer: 25 mM triethanolamine-HCl, pH 7.8, 5 mM magnesium acetate, 50 mM KCl.
4. Glass dishes with an inner diameter of 12.5 cm.
5. 98% Ethanol (store at –20°C).
6. 1 M Sodium acetate, pH 5.5.
7. Sorvall® RC-5B centrifuge with HB 6 rotor (DuPont, Wilmington, DE, USA).
8. 80% Ethanol (store at –20°C).
9. 2-Mercaptoethanol.

2.2. Enzymatic Cleavages

1. Endoproteases (Lys-C, Glu-C) and 5'→3' phosphodiesterase from calf spleen (Boehringer Mannheim, Mannheim, Germany).
2. Ribonuclease T_1 (Calbiochem, San Diego, CA, USA).
3. 1 M Tris-HCl, pH 7.8.
4. 2 mM Tris-HCl, pH 7.8.
5. 0.5 M EDTA.
6. RNasin (Promega, Madison, WI, USA).

2.3. Size-Exclusion Chromatography

1. Sephacryl S-300 column 100 × 2.6 cm (Pharmacia, Uppsala, Sweden).
2. Running buffer: 25 mM Tris-HCl, pH 7.8, 2 mM EDTA, 0.1% SDS, 6 mM 2-mercapto-ethanol.

3. RNase A (Sigma, Deisenhofen, Germany). Stock solution: 1 mg RNase A in 1 mL water.
4. SDS-polyacrylamide gel electrophoresis (SDS-PAGE) apparatus.
5. Sample buffer: 125 mM Tris-HCl, pH 6.8, 0.1% SDS, 2% dithioerythritol, 20% glycerol, 0.02% bromophenol blue.
6. Water, pH 6.5.

2.4. RP-HPLC

1. HPLC system (Shimadzu Corporation, Kyoto, Japan): SCL-6a system controller, LC-6a liquid chromatograph, SPD-7AV UV-VIS spectrometric detector, and SPD-M6A photodiode array UV-VIS detector.
2. Vydac C-18 column (250 × 4 mm, 300 Å, The Separation Group, Hesperia, CA, USA).
3. Buffer A: 0.1% trifluoroacetic acid in water. Buffer B: 0.085% trifluoroacetic acid in acetonitrile.
4. Microcentrifuge.

2.5. N-Terminal Sequencing

1. Procise™ microsequencer (Applied Biosystems, Foster City, CA, USA).
2. BioBrene (Applied Biosystems, Foster City, CA, USA).
3. Micro TFA filter (Applied Biosystems, Foster City, CA, USA).

2.6. MALDI Mass Spectrometry

1. MALDI-TOF mass spectrometer.
2. Matrix: saturated solution of α-cyano-4-hydroxy-cinnamic acid (Sigma, Deisenhofen, Germany) in 60% water, 0.1% trifluoroacetic acid, and 40% acetonitrile (stable at 4°C for 1 wk).
3. Aqueous 0.1% trifluoroacetic acid/acetonitrile (1:1).

2.7. Alkaline Hydrolysis

1. Aqueous ammonium hydroxide, pH 10.0; made with 1 mL of water and 10 µL of ammonium hydroxide (30%).

3. Methods
3.1. Crosslinking with 2-Iminothiolane

1. Dialyze 3000 A_{260} U of ribosomal subunits (108 nmol 50S and 216 nmol 30S, respectively) against crosslinking buffer plus 6 mM 2-mercaptoethanol for 16 h (*see* **Notes 1–5**).
2. Dilute with the same buffer plus 6 mM 2-mercaptoethanol to a concentration of 20 A_{260} U/mL (*see* **Note 6**).
3. Incubate for 20 min at room temperature under gentle stirring with 7.3 mL of 20 mM 2-iminothiolane solution (*see* **Note 6**).
4. Precipitate at –20°C for at least 2 h by adding 2 vol of chilled 98% ethanol and 0.1 vol of 1 M sodium acetate, pH 5.5.

5. Centrifuge for 30 min at 16,500g.
6. Wash the pellet with 2 mL of chilled 80% ethanol.
7. Centrifuge for 15 min at 16,500g.
8. Dissolve the pellet in 30 mL of crosslinking buffer under gentle stirring.
9. Adjust the solution to a concentration of 5 A_{260} U/mL with crosslinking buffer and perform UV irradiation (*see* **Subheading 3.2.3.**).

3.2. Crosslinking by UV Irradiation

1. Dialyze 3000 A_{260} U of ribosomal subunits against the crosslinking buffer.
2. Dilute with the same buffer to a concentration of 5 A_{260} U/mL.
3. Pipet the ribosomal solution to 1 mm depth in glass dishes (volume = $\pi \cdot r^2 \cdot h = \pi \cdot (6.25 \text{ cm})^2 \cdot 0.1 \text{ cm} = 12.3 \text{ mL}$) and perform UV crosslinking for 5 min at 254 nm at a distance of 3–5 cm to the surface of the solution.
4. Adjust the solution to a final concentration of 3% 2-mercaptoethanol.
5. Incubate for 30 min at 37°C.
6. Precipitate at –20°C for at least 2 h by adding 2 vol of 98% chilled ethanol and 0.1 vol of 1 *M* sodium acetate, pH 5.5.
7. Centrifuge for 30 min at 16,500g.
8. Wash the pellet with 2 mL of chilled 80% ethanol.
9. Centrifuge for 15 min at 16,500g.
10. Dissolve the pellet in running buffer (*see* **Subheading 2.3.2.**) to a final concentration of 300 A_{260} U/mL ribosomal subunit.
11. Store in 1-mL aliquots at –80°C.

3.3. Size-Exclusion Chromatography

1. Centrifuge 300 A_{260} U of crosslinked ribosomal subunit for 15 min at 17,000g and load the supernatant onto the size-exclusion column.
2. Wash with running buffer at a flow rate of 0.8 mL/min. Collect fractions of 1.5 mL and monitor the absorbance at 254 nm.
3. Check the crosslinking reaction by SDS-PAGE (*1*; *see* Chapters 23 and 31). Take 100-µL aliquots of every fifth eluted fraction and add 1 µg of RNase A to each aliquot. Incubate for 2 h at 50°C. Evaporate the solution in the vacuum. Dissolve the dried material in 20 µL of sample buffer. Incubate for 5 min at 95°C and perform SDS-PAGE (*5*), with 4% stacking gel and 15% separation gel. Stain with Coomassie blue. The comparison between a control (noncrosslinked) and the crosslinked sample indicates the success of the crosslinking reaction.
4. Pool the RNA-containing fractions.
5. Precipitate at –20°C for at least 2 h by adding 2 vol of 98% chilled ethanol and 0.1 vol of 1 *M* sodium acetate, pH 5.5.
6. Centrifuge for 30 min at 16,500g.
7. Wash the pellet with 2 mL of chilled 80% ethanol.
8. Centrifuge for 15 min at 16,500g.
9. Dissolve the pellet in 2 mL of water.

N-Terminal Sequencing and MALDI-MS

3.4. Digestion with Endoprotease

1. Adjust the solution to 25 mM Tris-HCl, pH 7.8, and 2 mM EDTA.
2. Add 40 U of RNasin/mL.
3. Add 3 µg/300 A_{260} U of endoprotease (Lys-C or Glu-C, dissolved in water) (*see* **Note 7**).
4. Incubate for 16 h at 37°C.
5. Evaporate the solution in the vacuum.
6. Dissolve the dried material in 2 mL of running buffer to perform a second size-exclusion chromatography as described under **Subheading 3.3.** (without step 3).

3.5. Digestion with Ribonuclease T_1 and Endoprotease

1. Adjust the solution to 1 mM EDTA.
2. Add 10 µg/300 A_{260} U of ribonuclease T_1 (dissolved in water).
3. Incubate for 2 h at 50°C.
4. Adjust the solution to 25 mM Tris-HCl, pH 7.8, and 2 mM EDTA.
5. Add 3 µg/300 A_{260} U of the same endoprotease as in **Subheading 3.4.3.** (Lys-C or Glu-C, dissolved in water) (*see* **Note 8**).
6. Incubate for 16 h at 37°C.
7. Divide into two aliquots of 150 A_{260} U each.
8. Store at –80°C until performing RP-HPLC.

3.6. RP-HPLC

1. Centrifuge aliquots from **Subheading 3.5., step 8** (150 A_{260} U) for 10 min at 12,500g and load onto the RP-HPLC column.
2. Perform an isocratic elution with 10% buffer B at a flow rate of 0.5 mL/min until the absorbance at 220 nm returns to baseline (about 45 min).
3. Perform a linear gradient elution within 240 min from 10% buffer B to 45% buffer B, at a flow rate of 0.5 mL/min; monitor the absorbance at 220 and 254 nm.
4. Collect fractions of 0.3 mL; avoid pooling (*see* **Note 9**).
5. Evaporate the fractions containing material that absorbs at 220 and 254 nm in the vacuum.
6. Store the fractions at –80°C until they are used for N-terminal sequencing and MALDI-MS. At that time, dissolve the sample in 5 µL of aqueous 0.1% trifluoroacetic acid (buffer A)/acetonitrile (1:1).

3.7. N-terminal Sequencing

1. Apply 2.5 µL of the sample from **Subheading 3.6., step 6** onto a BioBrene precoated filter.
2. Perform N-terminal sequencing as described by the manufacturer (*see* **Note 10**).

3.8. MALDI-MS

1. Add 0.7 µL of the matrix onto the sample holder.
2. Mix immediately with 0.5 µL of the sample from **Subheading 3.6., step 6** and air dry.
3. Record a spectrum for the mass analysis of the total complex (*see* **Notes 11–13**).

3.9. Alkaline Hydrolysis

1. Evaporate 1.0 µL of the sample from **Subheading 3.6.**, **step 6** in the vacuum.
2. Add 20 µL of aqueous ammonium hydroxide, pH 10.0.
3. Incubate for 15 min at 95°C.
4. Evaporate in the vacuum.
5. Dissolve the dried sample in 0.5 µL of aqueous 0.1% trifluoroacetic acid/acetonitrile (1:1).
6. Use 0.5 µL for the mass analysis of the hydrolyzed complex as described in **Subheading 3.8.** (*see* **Notes 14** and **15**).

3.10. Digestion with 5'→3' phosphodiesterase

1. Evaporate 1.0 µL of the sample from **Subheading 3.6.**, **step 6** in the vacuum.
2. Add 20 µL of 2 m*M* Tris-HCl, pH 7.8.
3. Add 0.002 U of phosphodiesterase.
4. Incubate for 1 h at 37°C.
5. Evaporate in the vacuum.
6. Dissolve the dried sample in 0.5 µL of aqueous 0.1% trifluoroacetic acid/50% acetonitrile (1:1).
7. Use 0.5 µL for the mass analysis of the hydrolyzed complex as described in **Subheading 3.8.** (*see* **Note 14**).

4. Notes

1. If not indicated otherwise all procedures have to be performed at 4°C.
2. Use sterile glass equipment.
3. All buffers have to be sterile.
4. Use gloves in all experiments.
5. The starting 3000 A_{260} are measured exactly; other A_{260} values assume full recovery.
6. RNA–protein crosslinking with the heterobifunctional reagent 2-iminothiolane is a two-step reaction. The first step involves the reaction of lysine residues on the proteins with the imidate function of the reagent and formation of free sulfhydyl groups. Therefore, a reducing agent such as 2-mercaptoethanol must be present to avoid S–S bridge formation. In the second step, the crosslinking to RNA is achieved by UV irradiation.
7. The use of a second endoprotease in a second crosslinking experiment is highly recommended. Thereby, many of the results of the first crosslinking experiments can be confirmed and other RNA–protein contact sites may be found.
8. Two endoprotease digests are performed to obtain higher yields of crosslinked products. The second endoprotease digest allows complete cleavage of the proteins due to the fact that the large RNA is cleaved by ribonuclease T_1.
9. For further purification of RP-HPLC fractions, precipitation at –20°C for at least 2 h with 2 vol of chilled 98% ethanol, 0.1 vol of 1 *M* sodium acetate, pH 5.5, and 40 µg of glyco-gen (Boehringer Mannheim, Mannheim, Germany) and an additional RP-HPLC run may be performed.

10. Usually, lysines are found crosslinked by 2-iminothiolane, whereas mainly tyrosines and methionines are crosslinked by UV treatment.
11. Adduct ions from sodium (+22 Daltons) and magnesium (+26 Daltons) are visible within the mass spectra. The sodium ions can be reduced by dissolving the sample in aqueous 0.1% trifluoroacetic acid/acetonitrile (1:1) containing 50 mM ammonium hydrogen citrate or 50 mM ammonium sulfate. An alternative to this procedure is the incubation with 5–10 ammonium acetate activated ion-exchange beads (AG 50W-X8 Resin, 100–200 mesh, Bio-Rad, München, Germany) for 30 min at 37°C before the introduction of the sample into the mass spectrometer.
12. In general the mass of the total complex and of the crosslinked peptide plus 2-iminothiolane are detectable after crosslinking with 2-iminothiolane, whereas the crosslinked peptide mass after UV crosslinking is not detected or detectable only to a much lower extent. This indicates that UV crosslinks are more stable than 2-iminothiolane crosslinks.
13. The mass difference between the total complex and the peptide (plus 101 Daltons for 2-iminothiolane crosslinks) enables one to determine the composition of the oligoribonucleotide content. The precision of this analysis depends on the mass accuracy of the mass spectrometer used. Nevertheless, even mass spectrometers with a low resolution of about $m/\Delta m$ 200 allow the determination of the oligoribonucleotide composition with a mass accuracy of 1 Dalton for complexes with a mass of <5000 Daltons. The mass of the crosslinked peptide can be calculated from the N-terminal sequence analysis and can also be used to check the mass calibration.
14. The masses of the oligoribonucleotides are : G = 363.2 Daltons, A = 347.2 Daltons, U = 324.2 Daltons and C = 323.2 Daltons. After hydrolysis, mass differences of these ribonucleotides minus 18 Daltons (water loss) are observed, whereas the 3'-G cleavage shows a mass difference of 363.2 Daltons. The differentiation of U and C depends on the mass accuracy of the mass spectrometer used and this is complicated by adduct ions as mentioned above.
15. Cleavage of ribonucleotides from both the 5' and 3' ends occurs after alkaline hydrolysis and must be considered in the interpretation of the mass spectra. Examples of detailed interpretation of mass spectra are shown in **refs. *3* and *4***.

References

1. Urlaub, H., Kruft, V., Bischof, O., Müller, E.-C., and Wittmann-Liebold, B. (1995) Protein-rRNA binding features and their structural and functional implications in ribosomes as determined by cross-linking studies. *EMBO J.* **14,** 4578–4588.
2. Urlaub, H., Kruft, V., and Wittmann-Liebold, B. (1995) Purification scheme for isolation and identification of peptides cross-linked to the rRNA in ribosomes, in *Methods in Protein Structure Analysis* (Atassi, M. Z. and Apella, E., eds.), Plenum Press, New York and London, pp. 275–282.
3. Urlaub, H., Thiede, B., Müller, E.-C., Brimacombe, R., and Wittmann-Liebold, B. (1997) Identification and sequence analysis of contact sites between ribosomal proteins and rRNA in *Escherichia coli* 30 S subunits by a new approach using

matrix-assisted laser desorption/ionization mass spectrometry combined with N-terminal microsequencing. *J. Biol. Chem.* **272,** 14,547–14,555.
4. Urlaub, H., Thiede, B., Müller, E.-C., and Wittmann-Liebold, B. (1997) Contact sites of peptide-oligoribonucleotide cross-links identified by a combination of peptide and nucleotide sequencing with MALDI-MS. *J. Prot. Chem.* **16,** 375–383.
5. Laemmli, U. K. (1970) Cleavage of structural proteins during the assembly of the head of bacteriophage T4. *Nature* **227,** 680–685.

7

RNA Footprinting and Modification Interference Analysis

Paul A. Clarke

1. Introduction

RNA plays a central role in a wide range of processes within the cell. In some organisms RNA replaces DNA as the genetic material. In all organisms RNA is essential at all stages in the translation of genetic information to protein *(1–4)*. RNA–protein interactions have a key role in almost all of these biological processes. Therefore knowledge of the features that govern RNA–protein interactions is critical to understanding the regulation of these biological functions. Two methods used to investigate these interactions are RNA footprinting and modification interference analysis.

1.1. Principle of the Procedure

To understand the process of RNA footprinting one has to understand the principles behind the sequencing and analysis of RNA structure. The original approach to RNA sequencing involved digestion of RNA using nucleases and isolation of substrate by paper electrophoresis followed by multiple rounds of thin-layer chromatography and nuclease digestion *(5)*. This was subsequently improved by using both base-specific nucleases and chemical modifying agents to generate sequencing ladders suitable for analysis by denaturing polyacrylamide electrophoresis and autoradiography *(6,7)*. These protocols were eventually adapted for the analysis of RNA structure and the interactions between RNA and proteins.

The basis of RNA sequencing, structural analysis, and footprinting of RNA–protein complexes relies on the cleavage of radioactively end-labeled RNA by base specific agents under denaturing, semidenaturing, and native conditions *(7,8)*. Denaturing conditions, in which the molecule is unfolded in a random coiled state, give information about the sequence of the molecule. Semidenaturing conditions, in which the molecule is usually folded in the absence of divalent cations, give some information about the basic secondary structure elements of the molecule, while native conditions in the presence of divalent cations provide data on the higher order tertiary structure of the RNA. Under denaturing conditions the enzymatic or chemical probes cleave specific bases without reliance on the position or conformation of the base within the RNA mole-cule. Under semidenaturing and native conditions a reduced spectrum of cleavage is detected as some sites are protected by higher order structures. Scission conditions are chosen such that cleavage occurs at a rate of approximately one cut in every 10 molecules (*see* **Note 1**). A range of probes of different sequence and structure specificity are chosen to allow cutting at every site. The lengths of the discrete RNA fragments are determined by separation using denaturing gel electrophoresis and autoradiography. Each band on the autoradiograph corresponds to a discrete fragment, allowing the sequence to be read directly from the autoradiograph. The distance of the strand scission from the terminal label can be used to locate the position of the attacked base. For a typical structural analysis one would run a sequencing reaction and a structure reaction using sequence- and structure-specific probes. To facilitate alignment with the sequencing lane a degradation ladder with one-nucleotide increments is employed. The positions of specific cuts in the RNA structure can then be identified by comparison to this ladder and to the sequencing lanes (*see*, e.g., **Fig. 1**). Having determined the cutting pattern it is necessary to interpret the data and create a structural model. This can be done either manually or by using one of a number of software packages that are now available (*see* **Note 2**) *(9)*. Once the conditions for structural analysis are established, it is possible to examine the RNA–protein interaction by footprinting. In addition to the conditions outlined previously, one would include RNA complexed to the protein of interest. Under these conditions the RNA-bound protein will sterically inhibit the action of the probe at sites of close contact between the protein and RNA. This will be manifest in a reduced spectrum of cutting sites compared to the sequencing and structure reactions. By comparison to the hydrolysis ladder, sequencing, and structure reactions the site of protection, and by inference the site of protein binding, can be identified (**Figs. 1** and **2**).

RNA modification interference of RNA–protein interactions is analogous to that described for DNA by Siebenlist and Gilbert *(10,11)*. This method allows the identification of bases that are essential to the reaction of interest. The interference method differs from the footprinting methods described previously in that RNA molecules are modified prior to protein binding. Protein-bound RNA is isolated and the scission reaction is completed and analyzed essentially as described earlier. Modifications at positions essential for the RNA–

RNA Footprinting

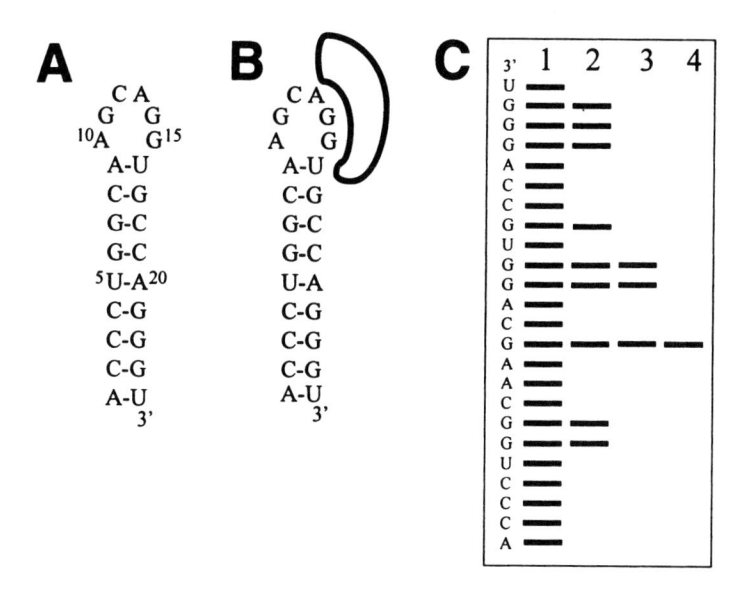

A
```
      C A
     G   G
   10A   G15
     A - U
     C - G
     G - C
     G - C
    5U - A20
     C - G
     C - G
     C - G
     A - U
        3'
```

B
```
      C A
     G   G
     A   G
     A - U
     C - G
     G - C
     G - C
     U - A
     C - G
     C - G
     C - G
     A - U
        3'
```

C

3'	1	2	3	4
U	—			
G	—	—		
G	—	—		
G	—	—		
A	—			
C	—			
C	—			
G	—	—	—	
U	—			
G	—	—	—	—
G	—	—	—	—
A	—			
C	—			
G	—	—	—	—
A	—			
A	—			
C	—			
G	—	—		
G	—	—		
U	—			
C	—			
C	—			
C	—			
A	—			

Fig. 1. RNA footprinting with single-stranded G-specific RNase T_1. (**A**) Unbound RNA. (**B**) Protein-bound RNA. (**C**) Representative autoradiograph of an RNA footprinting experiment. *Lane 1*, alkali cleavage of unbound 5' end-labeled RNA showing scission at each nucleotide. *Lane 2*, RNase T_1 digest of unbound 5'-labeled RNA carried out under denaturing conditions such that all Gs are cleaved. *Lane 3*, RNase T_1 digest of unbound 5' labeled RNA under native conditions where the stem-loop structure forms. G-11, -14, and -15 are located in a single strand loop and will be the only Gs accessible to RNase T_1. *Lane 4*, RNase T_1 footprinting of 5'-labeled RNA bound to protein. The RNA bound protein blocks cleavage of G-14 and -15; thus only scission at G-11 is detected.

protein interaction will interfere with, and block the formation of, the RNA–protein complex (*see* **Note 3**). These RNAs will be lost when the protein-complexed RNA population is separated from RNAs that are unable to bind due to the effect of the modification. In this instance gaps in the RNA sequencing ladder derived from protein-bound RNA will correspond to sites at which a modification has disrupted binding. By inference these are sites that are important to the RNA–protein interaction (**Figs. 2** and **3**).

2. Materials

1. RNase inhibitor (RNasin or equivalent).
2. RNase T_1.
3. RNase *B. cereus* (*Bc*).
4. RNase V_1.
5. RNase T_2.

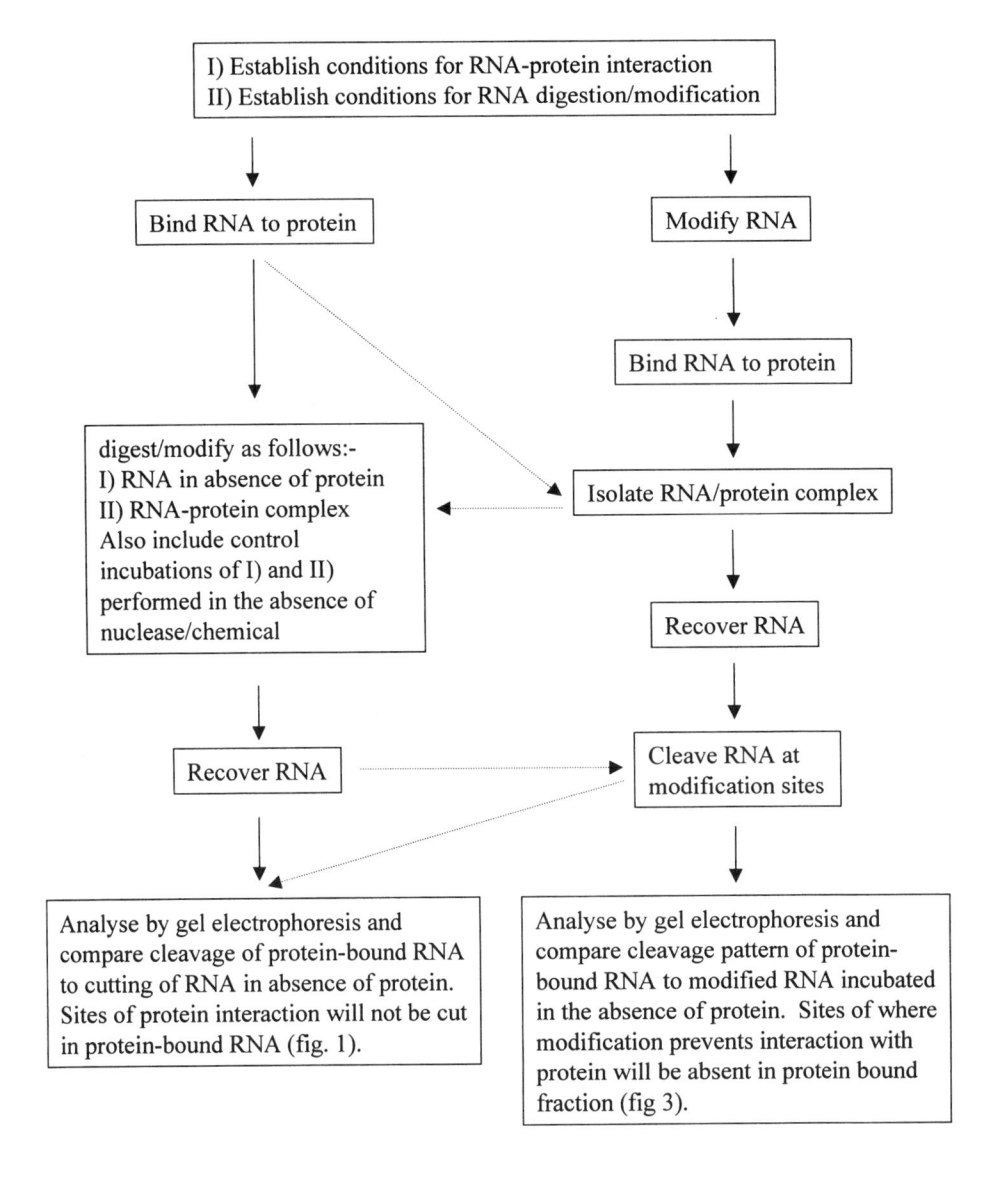

```
                I) Establish conditions for RNA-protein interaction
                II) Establish conditions for RNA digestion/modification
                            │                           │
                            ▼                           ▼
                  Bind RNA to protein              Modify RNA
                            │                           │
                            │                           ▼
                            │                  Bind RNA to protein
                            │                           │
                            ▼                           ▼
              digest/modify as follows:-  ◄──── Isolate RNA/protein complex
              I) RNA in absence of protein             │
              II) RNA-protein complex                  │
              Also include control                     ▼
              incubations of I) and II)         Recover RNA
              performed in the absence of              │
              nuclease/chemical                        │
                            │                           ▼
                            │                    Cleave RNA at
                            ▼                   modification sites
                    Recover RNA ─────────────►
                            │                           │
                            ▼                           ▼
              Analyse by gel electrophoresis and   Analyse by gel electrophoresis and
              compare cleavage of protein-bound RNA compare cleavage pattern of protein-
              to cutting of RNA in absence of protein. bound RNA to modified RNA incubated
              Sites of protein interaction will not be cut in the absence of protein. Sites of where
              in protein-bound RNA (fig. 1).       modification prevents interaction with
                                                   protein will be absent in protein bound
                                                   fraction (fig 3).

              RNASE/CHEMICAL FOOTPRINTING          MODIFICATION INTERFERENCE
```

Fig. 2. Flow chart for RNA footprinting or modification interference. For both approaches it is necessary to establish the conditions for RNA binding and conditions for digestion/modification of the RNA such that a maximum of one in 10 RNA molecules are cut, i.e., the majority (>90%) of the RNA remains intact (see **Note 1**). For footprinting the RNA–protein complex is formed and then subjected to digestion or modification. Occasionally it is necessary to isolate the RNA–protein complex prior to digestion/modification. The RNA is recovered (and cleaved at sites of modification if chemical modification protocols are used)

RNA Footprinting

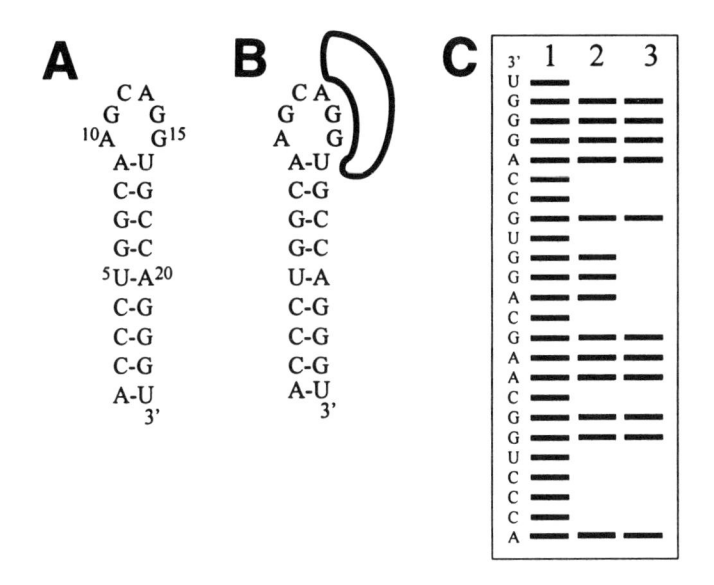

A
```
      C A
    G   G
  10A    G15
    A-U
    C-G
    G-C
    G-C
   5U-A20
    C-G
    C-G
    C-G
    A-U
     3'
```

B
```
      C A
    G   G
    A    G
    A-U
    C-G
    G-C
    G-C
    U-A
    C-G
    C-G
    C-G
    A-U
     3'
```

C

	1	2	3
3' U			
G	—	—	—
G	—	—	—
G	—	—	—
A	—	—	—
C	—		
C	—		
G	—	—	—
U	—		
G	—		
G	—	—	—
A	—	—	
C	—		
G	—	—	—
A	—	—	
A	—	—	
C	—		
G	—	—	—
G	—	—	—
U	—		
C	—		
C	—		
C	—		
A	—	—	—

Fig. 3. Modification interference analysis with A-N7 and G-N7 specific DEPC. (**A**) Unbound RNA. (**B**) Protein-bound RNA. (**C**) Representative autoradiograph of a modification interference experiment. *Lane 1*, alkali cleavage of unbound 5' end-labeled RNA showing scission at each nucleotide. *Lane 2*, DEPC-modified 5'-labeled RNA. The RNA was premodified under denaturing conditions (such that all As and Gs are modified) for use in the interference binding assay. *Lane 3*, protein- bound DEPC-modified 5'-labeled RNA separated from RNAs that could not bind. RNAs modified at positions A-13, G-14, or -15 could not bind the protein and are thus absent from the protein-bound RNA fraction. This is visualized on the autoradiograph as a gap in the sequencing "ladder."

6. Proteinase K.
7. Distilled phenol, Phenol/chloroform (1:1 v/v) and chloroform.
8. Aniline.
9. NTPαS compounds are supplied by New England Nuclear–Dupont (P.O. Box 66, Hounslow, TW5 9RT, UK).
10. Siliconized 0.5-mL and 1.5-mL tubes.
11. 0.1 M Citric acid.
12. Ethanol.

---

and analyzed by gel electrophoresis. In a parallel reaction, RNA in the absence of protein is also digested/modified. Sites of close RNA–protein interaction will prevent access of the cleavage/modification agent and thus will not be cut. Comparison of the cutting pattern of protein-bound and unbound RNA allows the identification of sites protected for digestion/modification (**Fig. 1**). Modification interference analysis requires the RNA to be modified prior to protein binding. Protein-bound RNA is compared to unbound RNA. Sites where modification prevents the binding of protein will be absent from the protein-bound RNA fraction (**Fig. 3**).

13. 0.4 M Ammonium iron (III) sulfate.
14. 2 M Sodium ascorbate.
15. 0.8 mM EDTA, pH 8.0.
16. 0.6% H_2O_2 (v/v).
17. 0.3 M NaOAc, pH 3.8.
18. 20 mM Thiourea.
19. Diethyl pyrocarbonate.
20. N-Ethyl-N-nitrosourea (handle with care, highly carcinogenic).
21. Dimethylsulfate.
22. 3 M NaOAc, pH 4.5.
23. 200 mM NaBH$_4$.
24. Anhydrous hydrazine.
25. Pure double-distilled aniline (11 M) should ideally be redistilled under nitrogen before use, and stored under nitrogen at –20°C in a light-tight bottle.
26. 1-Butanol.
27. 1% (w/v) Sodium dodecyl sulfate (SDS).
28. Diethylpyrocarbonate (DEPC)-H_2O: Add 1 mL of DEPC to 1 L of double-distilled H_2O, mix vigorously, and autoclave.
29. 10× RNA–protein binding buffer: prepare a 10× buffer that is appropriate for the RNA and protein being analyzed (*see* **Notes 4** and **5**).
30. Calf-liver tRNA: incubate for 2 h with 50 µg/mL of proteinase K at 37°C, phenol extract, and recover by ethanol precipitation (**Subheading 3.1.**). Resuspend to a final concentration of 10 mg/mL.
31. 1.5× Gel loading buffer: 10 M urea, 1.5× TBE, 0.015% (w/v) bromophenol blue, 0.015% (w/v) xylene cyanol, 1% SDS; filter sterilize and store at –20°C.
32. 10× OH buffer: 50 mM NaHCO$_3$/Na$_2$HCO$_3$, pH 9.2, 1 mM EDTA, pH 8.0; filter sterilize and store at –20°C.
33. 2× Extraction buffer: 50 mM Tris-HCl, pH 7.4, 300 mM NaCl, 0.05% Nonidet P-40 (NP40), 10 µg/mL tRNA, 1% SDS, filter sterilize and store at –20°C.
34. 2× Hydroxyl radical quenching/extraction buffer: 50 mM Tris-HCl, pH 7.4, 300 mM NaCl; 0.05% NP40, 10 µg/mL tRNA, 1% SDS, 20 mM thiourea; filter sterilize and store at –20°C.
35. A fresh saturated solution of ENU is prepared by the addition of excess ENU (handle with care, highly carcinogenic) to 100 µL of ethanol. Undissolved ENU is removed by micro-centrifugation and the supernatant is used as the modification reagent.
36. 10× N-ethyl-N-nitrosourea (ENU) modification buffer: 500 mM HEPES-KOH, pH 8.0, 10 mM EDTA; filter sterilize and store at –20°C.
37. I_2 stock solution: 10 mM I_2 in 100% ethanol.
38. 2× Dimethyl sulfate (DMS) extraction buffer: 50 mM Tris-HCl, pH 7.4, 300 mM NaCl, 0.05% NP40, 10 µg/mL tRNA, 1% SDS, 125 mM β-mercaptoethanol; filter sterilize and store at –20°C.
39. Hydrazine/0.5M NaCl and hydrazine/3M NaCl: dissolve NaCl in anhydrous hydrazine to the appropriate concentration.

**Table 1
Additional Chemical Probes and Their Substrates**[a]

Chemical	Site	Modification	Detection
Kethoxal/glyoxal	G-N1 G-N2	Adds positions across N1 and N2. Stable at low pH, unstable above pH 7.0.	Reverse transcriptase stop
CMCT	U-N3 G-N1	G-N1 slow. —	Reverse transcriptase stop
Methoxyamine	C-N4 C-C6	Adds $NOCH_3$.	—
Sodium bisulfate	C-C6	Forms 5,6-dihydrocytosine.	Mild alkali generates uridine that is detected using hydrazine.
Uranyl acetate	Ribose sugar	Minor groove modification of solvent-accessible sites.	Irradiation at 420 nM
Rb(phen)$_2$phi^{3+}	Tertiary structure	Major groove attack on ribose. Bulges, stem-loop junctions, mismatched bases, and triple base-pairs	Irradiation at 365 nM

[a]Adapted from McSwiggen *(5)* and Ehresmann et al. *(12)*.

3. Methods

The goal of this chapter is to provide a starting point for the analysis of RNA–protein complexes. The optimal reaction conditions vary considerably for different RNAs and their interaction with protein. The methods described here are intended to be provide a basis from which tailored protocols can be developed (*see* **Notes 1** and **5**). **Figure 2** outlines the steps required to successfully undertake a footprinting or modification interference study of an RNA–protein interaction. The methods outlined below require the conditions and characteristics of the particular RNA–protein interaction of interest to have been established prior to commencing the analysis. Nuclease footprinting generally gives a low-resolution view of the RNA–protein complex. The advantage of nucleases is they are generally active over a broad range of conditions, including physiological conditions, whereas the chemical modifying agents sometimes require more extreme conditions. However, RNases are relatively large and their activity can be influenced by neighboring bases (*see* **Note 1**). Chemical probes are far smaller than nucleases, often approaching solvent size, and thus offer far higher resolution of the RNA–protein interaction. For a complete study of an RNA–protein interaction the best tactic is to localize the site of interaction using nucleases and then focus on the types of interaction by using chemical probes and by modification interference analysis. **Table 1** details some additional chemical probes that are not covered in this chapter.

To undertake modification interference studies one must also have a method for reliably separating the RNA–protein complex from unbound RNA. This can be achieved by isolating the complex using gel shift analysis, filter binding, or by immunoprecipitation of the RNA–protein complex. It is essential, however, that the chosen technique does not cause release of the protein-bound RNA.

The protocols outlined below require pure, gel purified, uniquely end-labeled RNAs produced as described earlier in this volume (*see* Chapter 1). End-labeling is suitable for small RNAs less than 200 nucleotides in length or for interactions that occur within 200 nucleotides of the end of an RNA molecule. With longer RNAs or interactions beyond 200 nucleotides the cleavage sites are best identified using primer extension by reverse transcriptase *(21)* (*see* **Note 6**).

3.1. RNase footprinting

Nuclease footprinting gives a low-resolution picture of the RNA regions that interact with the protein of interest. Before attempting the footprinting reaction it is necessary to establish the concentration of nucleases required to achieve an approximate rate of a single cut once in every 10 molecules (*see* **Note 1**). This is best done in the buffer used for the RNA–protein binding (*see* **Note 5**) by using a range of RNase concentrations and a number of time points (*see* **Note 1**). There are now many RNases, with a broad range of specificities that can be used for RNA footprinting (**Table 2**). For example, for footprinting virus-associated (VA) RNA$_I$ binding to double-stranded RNA-dependent protein kinase (PKR), RNases T$_I$ (0.00012 U/5 µL reaction), T$_2$ (0.003 U/5 µL reaction), and *Bc* (0.25 U/5 µL reaction) were used to map single-stranded regions and RNase V$_I$ (0.006 U/5 µL reaction) was used to map structured regions *(14)*. RNase V$_I$ will recognize short helical regions of four to five nucleotides or longer and should always be included (*see* **Note 7**). About 50–100,000 Cerenkov cpm/reaction of 5' or 3' end-labeled RNA will be sufficient for several gels and will keep autoradiography time to a minimum.

1. For each nuclease used, perform the following reactions: a control incubation with no RNase, and a serial dilution series for each nuclease with MgCl$_2$ (native) or without MgCl$_2$ (semidenaturing), with the exception of RNase V$_1$ which absolutely requires Mg^{2+} for activity.
2. To set up each reaction add 1 µL of the appropriate 10× binding buffer, 1 µL of 10 mg/mL calf-liver tRNA (*see* **Note 8**), end-labeled RNA, 1 µL of diluted RNase, and DEPC-H$_2$O to 5 µL final volume. Incubate at 37°C for 15 min, then add 1 µL of 10 mg/mL tRNA and 10 µL of 1.5× gel loading buffer.
3. Set up an RNase T$_1$ digestion at 68°C (denaturing conditions) for 5 min using a 10× more concentrated RNase T$_1$ stock than that used for mapping. Because every G is accessible, this will generate a sequencing ladder (*see* **Note 9**).
4. Generate a hydrolysis ladder with one-nucleotide increments by alkaline lysis. Mix 1 µL of 10× OH buffer, 1 µL of 10 mg/mL calf liver tRNA, end-labeled

Table 2
Nucleases and Their Substrates

Nuclease	RNA Sub.	3' End	Comments
RNase A	ssUpN, ssCpN	3'P	Cleaves different sequences at different rates, some preference for UpA and CpA. Active at physiological pH and retains activity in EDTA.
RNase *B. cereus*	ssUpN, ssCpN	3'P	Active at physiological pH, inhibited by urea.
RNase CL3	ssCpN	3'P	—
RNase H	RNA/DNA duplex	3'P	Cuts RNA strand of RNA/DNA duplex.
RNase M_1	ssNpA, ssNpG, ssNpU	3'OH	—
RNase Phy M	ssUpN, ssApN	3'P	—
RNase T_1	ssGpN	3'P	Active at physiological pH. Retains activity in EDTA.
RNase T_2	ssRNA	3'P	Optimal activity at physiological pH, but retains activity under a wide range of conditions. Retains activity in EDTA. Inhibited by some heavy metals and calcium.
RNase U_2	ssApN	3'P	Optimal activity at pH 3.5. Some cleavage after G at physiological pH.
RNase V_1	dsRNA	3'OH	Requires divalent cations. May also cleave pseudoknots

[a]Adapted from McSwiggen *(5)*, and Ehresmann et al. *(12)*, and Knapp *(13)*.

RNA, and DEPC-H$_2$O to 10 µL and incubate for 2 min at 90°C. Terminate the reaction by adding 1 µL of 10 mg/mL calf liver tRNA, 1 µL of 0.1 *M* citric acid and 24 µL of 1.5× gel loading buffer (*see* **Note 9**).

5. Separate the products by denaturing polyacrylamide gel electrophoresis as described earlier in this volume (*see* Chapter 1). For longer RNAs it may be necessary to run several gels of different percentages for different lengths of time.
6. The labeled RNA can be detected by direct autoradiography or by phosphorimaging. For autoradiography, transfer the gel onto a piece of used X-ray film, cover with plastic wrap, and expose to X-ray film in a cassette kept at –80°C. For phosphorimaging, transfer the gel to 3M Whatman paper, cover in plastic wrap, vacuum dry, and expose to the phosphor-imaging plate at room temperature.

7. Having established the digestion conditions, where 90–95% of the input RNA remains uncleaved (*see* **Note 1**), it is then relatively straightforward to proceed to the footprinting step. In this instance cleavage under the RNase conditions established for the unbound RNA is compared to cleavage of the RNA complexed to the protein of interest (**Note 6**).
8. Prior to gel analysis the protein component of the reaction has to be removed. Dilute the reaction to 200 µL with DEPC-H_2O and reisolate the RNA by adding 200 µL of 2× extraction buffer. Add an equal volume of phenol/chloroform, vortex the mixture for a few minutes, and centrifuge in a microcentrifuge for 10 min at full speed (approx 16,000g). Recover the upper aqueous phase and reextract with an equal volume of chloroform as described earlier. Precipitate the RNA with 2.5 vol of ethanol, incubate in a dry ice/ethanol bath for 10–20 min and centrifuge at 16,000g for 15 min at 4°C. Wash the RNA pellet with ice-cold 70% ethanol and recentrifuge as described above. Resuspend the pellet in 5 µL of H_2O and 10 µL of 1.5× gel loading buffer.
9. The digestion reactions in the absence of protein should also be extracted and precipitated in parallel. To aid analysis, load approximately equal Cerenkov counts on the gel. Include an additional control of protein-bound RNA incubated in absence of nuclease and extracted as described previously. A schematic of an RNase footprinting experiment is shown in **Fig. 1** (*see* **Note 6**).

3.2. Backbone-Specific Probes for Footprinting or Modification Interference

The principle behind footprinting using chemical probes is identical to that of nuclease footprinting except the chemical probes generally give a higher resolution of the site of interaction as the probes are considerably smaller (*see* **Notes 1** and **10**). Some of these probes cut in an essentially sequence-independent manner, in which case a sequencing reaction such as an RNase T_1 digest at 68°C should be included to position the protected cut sites (*see* **Note 9** and **Subheading 3.2., step 2**).

3.2.1. Hydroxyl Radicals

The hydroxyl radical has a neutral charge and is the smallest species used for chemical footprinting. It is a short-lived species and generally cleaves in a sequence independent way at the 1' or 4' position of the ribose. The hydroxyl radical reacts equally well with both single and double-stranded RNA, although higher order structures involving the backbone will influence the cleavage rate *(14–18)*. The cleavage chemistry is mild and is compatible with a wide range of buffer, salt, temperature, and pH conditions. As before, it is necessary to establish the cleavage conditions in binding buffers before footprinting the RNA–protein complex *(15–18)* (*see* **Notes 1, 10,** and **11**).

RNA Footprinting 83

1. Prepare fresh filter-sterilized solutions of 0.4 M ammonium iron (III) sulfate; 2 M sodium ascorbate; 0.8 mM EDTA, pH 8.0; and 0.6% H_2O_2. Dilute the 0.4 M ammonium iron (III) sulfate stock to 0.4 mM and then mix with an equal volume of 0.8 mM EDTA. Dilute the sodium ascorbate to 20 mM and then mix with an equal volume of the Fe·EDTA mix and an equal volume of 0.6% H_2O_2.
2. Initially incubate the RNA in the absence of protein to identify the conditions required to give a reasonably even cleavage ladder. The cleavage efficiency is regulated by controlling the amount of iron/EDTA/H_2O_2 mixture added to the reaction (*see* **Note 1**). For VA RNA_1, the optimal conditions were an equal volume of cleavage reagent added to the binding reaction and incubation for 2 min at room temperature.
3. Quench the reaction by diluting to 200 µL final volume with DEPC-H_2O and immediately adding an equal volume of 2× extraction buffer containing 20 mM thiourea (a hydroxyl radical scavenger). Recover the RNA as described in **Subheading 3.1., step 8**.
4. Having established the cleavage conditions, the cleavage of the RNA–protein complex is compared to unbound RNA and analyzed exactly as described previously (**Subheading 3.1., steps 7** and **8**).

3.2.2. N-Ethyl-N-Nitrosourea (ENU)

ENU is a mild alkylating agent that modifies backbone phosphates by forming phosphotriester groups that can be cleaved by mild base treatment as described by Krol and Carbon *(21)*. The time of reaction and amount of ENU used should be titrated with unbound RNA to establish the modification conditions (*see* **Note 1**). Although the modification is not sequence specific the reaction may be affected at some sites by higher order structures that involve the RNA backbone. ENU can be used for both footprinting and modification interference analysis.

1. Add 2 µL of 10× binding buffer, 1 µL of 10 mg/mL calf liver tRNA, end-labeled RNA or protein-complexed RNA, 5 µL of freshly prepared saturated ENU, and H_2O to 20 µL. Incubate the mixture for 30 min at 37°C, dilute to 200 µL with H_2O and recover the RNA by extraction and precipitation as described in **Subheading 3.1., step 8**. Include a control reaction in which 5 µL of ethanol alone is used in place of the ENU solution.
2. Resuspend the RNA pellets in 200 µL of 0.3 M NaOAc, pH 3.8, and reprecipitate them with 600 µL of ethanol as described in **Subheading 3.1., step 8**.
3. To hydrolyze the phosphotriester bond, resuspend the RNA pellet in 10 µL of 100 mM Tris-HCl, pH 9.0, and incubate at 50°C for 5 min. Recover the RNA by ethanol precipitation and analyze by denaturing gel electrophoresis.
4. To premodify the RNA for interference analysis incubate as **step 1**, but use a 10× ENU modification buffer and incubate for 2 min at 80°C (*see* **Note 13** and **Subheading 3.4.**).

3.2.3. Phosphorothioate-Containing Transcripts

Sp diastereoisomers of NTPαS are substrates for bacteriophage RNA polymerases and are incorporated into RNA as efficiently as the full oxy forms of NTP. The phosphorothioate bond is sensitive to cleavage by iodine/ethanol. The reaction depends on the accessibility of the sulfur group as well as the spatial arrangement of the 2'OH that participates in backbone scission. Iodine is well suited to footprinting protocols as it is uncharged, reasonably small, reacts rapidly, and does not interfere with gel mobility *(19,20)*. This protocol can be used as an alternative to modification with ENU and offers the advantage of a faster reaction time and data that can be interpreted without the need for RNase sequencing and degradation ladders (*see* **Note 12**).

1. The phosphorothioate containing transcripts are synthesized, end-labeled and purified exactly as described earlier in this volume (*see* Chapter 1), with the exception that a separate transcription reaction is set up for each nucleotide and an additional 0.2 m*M* (5%) of the relevant NTPαS is included, i.e., for a transcript containing AαS one would include all four NTPs, but would also supplement the transcription reaction with 0.2 m*M* ATPαS.
2. To cleave at the phosphorothioate group, incubate unbound and protein-complexed RNA with 1 m*M* I$_2$ for 2 min at room temperature. An equivalent volume of ethanol alone should be used in control reactions with both the bound and unbound RNA.
3. Recover the RNA by diluting the reaction to 200 μL with DEPC-H$_2$O and adding an equal volume of 2× extraction buffer (**Subheading 3.1., step 8**). Extract, precipitate, and analyze the RNA as described in **Subheading 3.1., step 8** (*see* **Note 12**).
4. In this instance the gel is easy to read as the reaction with the unbound RNA generates a sequencing ladder that is specific to the particular NTPαS (*see* **Note 12**).

3.3. Base-Specific Probes for Footprinting or Modification Interference

The protocols described in **Subheading 3.2.** allow the mapping of protein interactions with the RNA backbone. However, other regions of the RNA molecule such as the bases can also govern the specificity of an RNA–protein interaction. A number of chemicals can be used to obtain base specific information.

3.3.1. Diethyl Pyrocarbonate (DEPC)

DEPC carboxyethylates the N-7 position of purines (A>>G) and is inhibited by both base stacking and divalent ion coordination. Again it is necessary to titrate the concentrations of probe (*see* **Note 1**). The conditions described here are essentially as descri-bed by Conway and Wickens and work well for VA RNA$_I$ *(11,14)* (*see* **Note 10**).

RNA Footprinting 85

1. Add 2 μL of 10× binding buffer, 1 μL of 10 mg/mL calf liver tRNA, end-labeled unbound or protein-complexed RNA, 2 μL of DEPC, and H$_2$O to 20 μL. Incubate the mix at 37°C for 10 min, dilute to 200 μL with H$_2$O, and recover the RNA as described in **Subheading 3.1**.
2. To generate a DEPC sequencing ladder or premodified RNA that can be used for modification interference, combine 2 μL of DEPC with end-labeled RNA, 1 μL of 10 mg/mL calf-liver tRNA, 3.3 μL of 3 *M* NaOAc, pH 4.5, and 1 μL of 200 m*M* EDTA, pH 8.0, in a final volume of 200 μL and incubate for 2 min at 90°C. Recover the RNA by the addition of 25 μL of 3 *M* NaOAc, pH 4.5, and 750 μL of ethanol and precipitate as described in **Subheading 3.1**.
3. Resuspend the RNAs recovered from **steps 1** or **2** in 200 μL of 0.3 *M* NaOAc, pH 3.8, and reprecipitate with 600 μL of ethanol as described in **Subheading 3.1**. The RNA can either be used in modification interference (*see* **Subheading 3.4.** and **Note 13**) or can be cleaved at the site of modification (**Subheading 3.3.3**.) and analyzed by denaturing gel electrophoresis (**Subheading 3.1.**).

3.3.2. Dimethylsulfate

DMS can be used to methylate base nitrogens of guanosine (N7), cytosine (N3), and adenosine (N1). Only the G and C modifications can be detected by using base cleavage as described by Krol and Carbon (21, unpublished). The G-N7 reaction is inhibited by base stacking while the C-N3 reaction is inhibited by basepairing. The conditions described here work well with VA RNA$_I$, although as discussed previously the concentration of probe required for the particular RNA being studied should be titrated (*see* **Notes 1** and **10**).

1. Add 20 μL of 10× binding buffer, end-labeled unbound or protein-complexed RNA, 1 μL of 10 mg/mL calf liver tRNA, and H$_2$O to 200 μL final volume. Add 0.5 μL of DMS and incubate at 37°C for 10 min.
2. Recover the RNA as described in **Subheading 3.1.**, using 2× DMS extraction buffer that contains 125 m*M* β-mercaptoethanol.
3. To generate a DMS sequencing ladder or premodified RNA that is suitable for modification interference, combine 1 μL of DMS with end-labeled RNA, 1 μL of 10 mg/mL calf-liver tRNA, 3.3 μL of 3 *M* NaOAc, pH 4.5, and 1 μL of 200 m*M* EDTA, pH 8.0, in a final volume of 200 μL and incubate for 1 min at 90°C.
4. Recover the RNA by the addition of 25 μL of 3 *M* NaOAc, pH 4.5, and 750 μL of ethanol and precipitate as described in **Subheading 3.1**.
5. Resuspend the RNA pellets from **steps 2** or **4** in 200 μL of 0.3 *M* NaOAc, pH 3.8, and reprecipitate with 600 μL of ethanol as described earlier. The modified RNA can be used for modification interference (*see* **Note 13**) or resuspended in 100 μL of H$_2$O, divided into two aliquots, and vacuum dried.
6. To cleave at G-N7, resuspend one aliquot in 10 μL of 1 *M* Tris-HCl, pH 8.0, plus 10 μL of freshly prepared 200 m*M* NaBH$_4$ and incubate on ice in the dark for 30 min.

7. To cleave at C-N3, resuspend the other aliquot in 10 μL of ice-cold 50% (v/v) anhydrous hydrazine and incubate on ice for 5 min.
8. Recover the RNAs from **steps 4** or **5** by the addition of 200 μL of 0.3 M NaOAc, pH 3.8, and 600 μL of ethanol and precipitate as described previously. Resuspend the RNA in 200 μL of 0.3 M NaOAc, pH 3.8, add 600 μL of ethanol, and reprecipitate. The RNA is then ready for aniline cleavage (*see* **Subheading 3.3.3.**).

3.3.3. Aniline Cleavage of Modified RNA

1. Resuspend the modified RNA in 20 μL of 1 M aniline (freshly diluted with 0.3 M NaOAc, pH 3.8) and incubate at 60°C for 20 min in the dark *(11)*.
2. Stop the reaction by adding 1.4 mL of 1-butanol, vortex briefly, and microcentrifuge at room temperature for 10 min at full speed (approx 16,000g). Carefully remove the butanol and resuspend the RNA pellet in 150 μL of 1% (w/v) SDS. Add 1.4 mL of butanol, vortex briefly, and pellet as before.
3. Rinse the pellet with 100% (v/v) ethanol and dry thoroughly. Resuspend the pellet in 4 μL of DEPC-H$_2$O and 8 μL of 1.5× gel loading buffer (*see* **Note 14**). The sample is ready for gel analysis. An aniline-treatment of the unmodified labeled RNA should also be included.

3.4. Modification Interference Analysis

As discussed earlier, modification interference analysis uses premodified RNAs to identify moieties important for the RNA–protein interaction. DEPC, ENU, and DMS modifications and phosphorothioate transcripts can all be used for interference analysis (*see* **Subheadings 3.2.2.**, **3.2.3.**, **3.3.1.**, and **3.3.2.** for protocols). In the case of the chemical modifications the RNA is modified under denaturing conditions to ensure all sites are available for modification. The modified RNA is recovered (*see* **Note 13**) and then used in a binding reaction. The RNA–protein complex is separated from RNA that cannot bind and the protein-bound RNA is extracted as described in **Subheading 3.1**. The RNAs are then cleaved at the site of modification as described in previous sections. An aliquot of the premodified RNA is run in parallel to provide a sequencing ladder. Unmodified RNA is also included as a control for nonspecific cleavage. The cleavage products are analyzed by denaturing gel electrophoresis and autoradiography as described earlier (**Figs. 2** and **3**). An additional modification procedure for analyzing pyrimidine bases is outlined in **Subheading 3.4.1**.

3.4.1. Hydrazine

Hydrazine can be used to remove pyrimidine bases for modification interference analysis. Depending on the choice of conditions the reaction is specific for U or C alone or both U and C *(11)* (*see* **Note 15**). The conditions described here were suitable for modifying VA RNA$_I$ (*see* **Note 1**) and are essentially as described by Conway and Wickens *(11,14)*.

1. Precipitate end-labeled RNA and 1 µL of 10 mg/mL calf liver tRNA as described in **Subheading 3.1**.
2. For the U-specific reaction, resuspend the RNA pellet in 10 µL of H$_2$O and mix with 10 µL of anhydrous hydrazine. Incubate the mixture for 10 min on ice.
3. Perform the U- and C-specific reaction exactly as described previously, except resuspend the RNA pellet in 20 µL of freshly prepared anhydrous hydrazine/0.5 M NaCl and incubate on ice for 30 min.
4. For the C only reaction, resuspend the RNA pellet in 20 µL of freshly prepared anhydrous hydrazine/3 M NaCl and incubate on ice for 30 min.
5. Precipitate the RNAs from **steps 2**, **3**, or **4** with 200 µL of 0.3 M NaOAc, pH 3.8, and 600 µL of ethanol as described earlier. Resuspend the RNA in 200 µL of 0.3 M NaOAc, pH 3.8, reprecipitate with 600 µL of ethanol, and recover the RNA for interference binding studies and aniline cleavage as described in **Subheading 3.3.3**.

4. Notes

1. It is necessary to keep the rate of cutting low (approx one cut in 10 RNA molecules or >90% of the input RNA remains intact) to prevent the generation of secondary cuts. Initial cuts can sometimes alter RNA structure such that regions previously inaccessible to a single-stranded probe become accessible to secondary cutting. Under conditions of rapid cutting in which RNA molecules will undergo multiple cuts, some of these sites would be falsely identified as *bona fide* cut sites. Secondary cuts are generally detected only with one of the two end-labeled species, whereas primary cuts will only be seen with both. Another related problem is the existence of very accessible sites close to the labeled terminus; cleavage of these sites results in an accumulation of short products that, in extreme situations, rapidly exhausts the pool of full-length RNA such that only the short products are detected. This can be a problem with RNase Bc (single-stranded U/C-specific) and 3' labeled RNA polymerase III transcripts that have single-stranded (U)$_n$ 3' termini. The required concentration of RNase can vary over orders of magnitude for similarly sized molecules. Factors that influence the rate of cutting include both base composition and degree of structure of the molecule. For example, a base-stacked single-stranded nucleotide in a very accessible region may be cut at the same rate as an unstacked single-stranded base in a less accessible region. For the enzymatic and chemical reagents described here, a range of concentrations from 50× above (in some protocols 50× above will not be possible) to 50× below, tested over a number of time points, is a good starting point. In the majority of cases this will allow one to identify the optimal cleavage or modification conditions that result in a cleavage rate of one cut per 10 RNA molecules or lower (although there may be exceptions that require a greater range of concentrations to be tested).
2. A number of software packages for predicting RNA structure are available. We use the FOLDRNA program from the Genetics Computer Group software package or the RNAFOLD program in the Intelligenetics suite. Most software has the

facility to prevent basepairing of bases identified as single-stranded and to force regions identified as double-stranded to basepair.
3. Modification interference analysis will detect only bases that are absolutely essential for the RNA–protein interaction. If there are multiple interactions of equal importance the modification of a single base may not be sufficient to interfere with binding. This method will also fail if the site of modification is not involved in the interaction; e.g., phosphate backbone modifications may not influence interactions involving the ribose moiety.
4. The rate of RNA scission is deliberately kept low; the slightest RNase contamination will cause background that will complicate the analysis of *bona fide* cut sites. Therefore, all stock solutions or reaction buffers should be made with DEPC-treated double-distilled H_2O and sterilized by autoclaving or filtration.
5. It is essential that the RNA structure analysis and footprinting are performed under identical buffer conditions and that the buffer conditions reflect what is known about the RNA–protein interaction. Differences in monovalent and divalent ion concentrations can have profound effects on RNA structure that can lead to the misassignment of protected regions.
6. To get a clean footprint it is essential that there is no unbound RNA in the cleavage reaction. Generally this is the case, as the concentration of gel-purified RNA in the initial binding reaction is low and the protein will be in excess. However, if the background is unacceptable it may be necessary to separate unbound RNA from the protein-bound RNA. This can be done by gel shift, filter binding, or immunoprecipitation of the complex, provided the purification step does not cause release of the protein-bound RNA (*see* **Chaps. 9, 10, and 20**). One drawback is the loss of sample during purification. At best it is usually only possible to recover 30–50% of the input RNA, and therefore it is necessary to scale up the initial binding reaction.
7. The terminal group, either phosphate or hydroxyl, affects the relative migration of the digestion products. One must be careful when comparing reactions that produce different termini, e.g., RNase T_I and V_I (**Table 2**).
8. Since the labeled RNA is used at a very low concentration, unlabeled carrier tRNA is added to standardize the RNA concentration in each reaction. The source and purity of the calf liver tRNA can affect the RNA–protein interaction and the quality of the footprinting reaction. As a precaution the tRNA stock is treated for 2 h with 50 µg/mL of proteinase K at 37°C and then extracted and precipitated (as described in **Subheading 3.1.**) before use.
9. Sequencing and degradation ladders can be prepared in advance and used for up to 10 d provided they are stored in urea loading buffer at –80°C.
10. It is crucial to ensure that the chemical probes used do not cause the release of the protein-complexed RNA. For example, DEPC can modify histidine residues, DMS can alkylate proteins, and the H_2O_2 component of the hydroxyl radical reaction can oxidize sulfhydryl groups *(16)*. In most cases this is not a major problem as the protein component of the reaction is usually in excess and this RNA-free protein will dilute out most protein-modification effects.

11. The production of hydroxyl radicals is inhibited by the presence of radical scavengers. These include glycerol at concentrations greater than 0.5%, while other common buffer components such dithiothreitol (DTT), Tris, HEPES, and bovine serum albumin (BSA) will, to a far lesser extent, reduce hydroxyl radical production. This can be avoided by increasing the amount of cleavage reagent. However, the nucleic acid binding abilities of some proteins are inhibited by H_2O_2 *(16)* in which case it is better to increase the Fe·EDTA and ascorbate concentrations and leave the H_2O_2 concentrations low. The Fe·EDTA solution should have a light green tint; a rusty color indicates the formation of Fe(II) in which instance a fresh Fe·EDTA mix should be prepared.
12. Care has to be taken when interpreting footprinting data using phosphorothioate RNAs. These RNAs are in effect premodified. Therefore on analysis, gaps in the protein-bound fraction may be caused by probe exclusion (footprinting) or by binding interference of the sulfur (modification interference). However, both explanations still infer a site of close interaction. With low probe concentrations extra protections have been reported; these disappear as the iodine concentration increases and probably represent weak interactions *(19,20)*.
13. The production of premodified RNAs for modification interference requires more extreme conditions that can sometimes result in background RNA breakdown. In this case it is sometimes necessary to repurify the end-labeled modified RNA before further use. It is prudent to analyze an aliquot of the modified RNA by gel electrophoresis prior to use in RNA–protein binding experiments.
14. Care must be taken to ensure complete drying of the RNA pellet as residual ethanol will affect gel electrophoresis. When analyzing aniline-cleaved RNA a urea-based loading buffer is preferable to a formamide-based loading buffer *(11)*.
15. The U+C and C reactions with 5' end-labeled RNA are poorly resolved on gel analysis, owing to hydrazine attacking the ribose moiety at the 3' end of the labeled RNA *(11,12)*.

Further Reading

Until recently, mutagenic studies have been the only way to translate footprinting data to the in vivo situation. However, with the advent of linker ligation coupled with reverse transcription-polymerase chain reaction (PCR) it is now possible to compare in vitro footprints to in vivo footprints of RNA–protein interactions *(22)*.

Acknowledgments

These protocols were developed while the author was working in the laboratory of Dr. M. B. Mathews. His support is gratefully acknowledged. This work was funded by the Human Frontiers Science Programme. The author is currently supported by the Cancer Research Campaign.

References

1. Wimberly, B., Varani, G., and Tinoco, I. (1991) Structural determinants of RNA function. *Curr. Opin. Struct. Biol.* **1,** 405–409.
2. Nagai, K. (1992) RNA-protein interactions. *Curr. Opin. Struct. Biol.* **2,** 131–137.
3. Pleij, C. W. A. (1990) Pseudoknots: a new motif in the RNA game. *Trends Biochem. Soc.* **15,** 143–147.
4. Steitz, T. A. (1990) Structural studies of protein-nucleic acid interaction: the sources of sequence-specific binding. *Q. Rev. Biophys.* **23,** 205–280.
5. McSwiggen, J. A. (1990) RNA sequencing. *Editorial Comments (United States Biochemical Corp.).* **17,** 1–7.
6. Donis-Keller, H., Maxam, A. M., and Gilbert, W. (1977) Mapping adenines, guanines and pyrimidines in RNA. *Nucleic Acids Res.* **4,** 2527–2538.
7. Peattie, D. A. (1979) Direct chemical methods for sequencing RNA. *Proc. Natl. Acad. Sci. USA* **76,** 1760–1764.
8. Peattie, D. A. and Gilbert, W. (1980) Chemical probes for higher-order structure in RNA. *Proc. Natl. Acad. Sci. USA* **77,** 4679–4682.
9. Jaeger, J. A., Turner, D. H., and Zuker, M. (1990) Predicting optimal and suboptimal secondary structure for RNA. *Methods Enzymol.* **183,** 281–306.
10. Siebenlist, U. and Gilbert, W. (1980) Contacts between *Escherichia coli* RNA polymerase and an early promoter of phage T7. *Proc. Natl. Acad. Sci. USA* **77,** 122–129.
11. Conway, L. and Wickens, M. (1989) Modification interference analysis of reactions using RNA substrates. *Methods Enzymol.* **180,** 369–379.
12. Ehresmann, C., Baudin, F., Mouget, M., Romby, P., Ebel, J.-P., and Ehresmann, B. (1987) Probing the structure of RNAs in solution. *Nucleic Acids Res.* **15,** 9109–9128.
13. Knapp, G. (1989) Enzymatic approaches to probing of RNA secondary and tertiary structure. *Methods Enzymol.* **180,** 192–212.
14. Clarke, P. A. and Mathews, M. B. (1995) Interactions between the double-stranded RNA binding motif and RNA: definition of the binding site for the interferon-induced protein kinase (PKR) on adenovirus VA RNA. *RNA* **1,** 7–20.
15. Wang, X. and Padgett, R. A. (1989) Hydroxyl radical "footprinting" of RNA: application to pre-mRNA splicing complexes. *Proc. Natl. Acad. Sci. USA* **86,** 7795–7799.
16. Dixon, W., Hayes, J. J., Levin, J. R., Weidner, M. F., Dombroski, B. A., and Tullius, T. D. (1991) Hydroxyl radical footprinting. *Methods Enzymol.* **208,** 380–413.
17. Tulius, T. D. (1987) Chemical "snapshots" of DNA: using the hydroxyl radical to study the structure of DNA and DNA-protein complexes. *Trends Biochem. Soc.* **12,** 297–300.
18. Lathem, J. A. and Cech, T. R. (1982) Defining the inside and outside of a catalytic RNA molecule. *Science* **245,** 276–282.
19. Schatz, D., Leberman, R., and Eckstein, F. (1991) Interaction of *Escherichia coli* tRNA[ser] with its cognate aminoacyl-tRNA synthetase as determined by footprinting with phosphorothioate-containing transcripts. *Proc. Natl. Acad. Sci. USA* **86,** 6132–6136.

20. Rudinger, J., Puglisi, J. I., Putz, J., Schatz, D., Eckstein, F., Florentz, C., and Giege, R. (1992) Determinant nucleotides of yeast tRNA[asp] interact directly with aspartyl-tRNA synthetase. *Proc. Natl. Acad. Sci.* **89,** 5882–5886.
21. Krol, A. and Carbon, P. (1989) A guide for probing native small nuclear RNA and ribonucleoprotein structures. *Methods Enzymol.* **180,** 212–227.
22. Grange, T., Bertrand, E., Espinas, M. L., Froment-Racine, M., Rigaud, G., Roux, J. and Pictet, R. (1997) In vivo footprinting of the interaction of proteins with DNA and RNA. *Methods.* **11,** 151–163.

8

Oligonucleotide-Targeted RNase H Protection Analysis of RNA–Protein Complexes

Arthur Günzl and Albrecht Bindereif

1. Introduction

This chapter focuses on the analysis of RNA–protein complexes (RNPs, ribonucleoprotein particles) by oligonucleotide-targeted RNase H digestion, a powerful approach to probe the domain structure of an RNP, both in crude extracts and in purified preparations. RNase H requires DNA-RNA hybrids of at least four basepairs for cleaving the RNA strand *(1)*. Therefore, RNase H-mediated cleavage can be directed to a short, specific RNA sequence by hybridization of a complementary DNA oligonucleotide. Oligonucleotide-targeted RNase H cleavage was first used to analyze RNA structures (*see*, e.g., the analysis of U1 snRNA secondary structure by Lazar and Jacob *[2]* and Rinke et al. *[3]*, the characterization of RNA lariat structures *[4]*, and the structural probing of a specific mRNA by Hwang et al. *[5]*). The approach has been adopted to RNA–protein complexes, and an important early example is the study of the basepairing potential of the 5' end of U1 snRNA in the U1 snRNP *(6)*. Subsequently, oligonucleotide-targeted RNase H cleavage has been used in various systems to disrupt RNPs for functional studies and in protection assays to map RNP protein-binding domains. Examples include in vitro studies that established snRNA sequences essential for spliceosome assembly and pre-mRNA splicing in the mammalian system and in yeast (*see*, e.g., **refs. 7–12**), studies on the assembly of the yeast and mammalian spliceosome *(13,14)*, investigations on the U7 snRNA function in histone 3' end processing (e.g., *see* **ref. 15**), and the analysis of snRNA requirements for *trans* splicing in trypanosome cells *(16)*.

For RNase H cleavage the RNA target sequence has to be accessible to oligonucleotide annealing. Target sequences that participate in stable double-

From: *Methods in Molecular Biology, Vol. 118: RNA-Protein Interaction Protocols*
Edited by: S. Haynes © Humana Press Inc., Totowa, NJ

stranded structures or that are tightly bound by protein can usually not be cleaved by RNase H. The goal of the oligonucleotide-targeted RNase H protection analysis of RNA–protein complexes is to discriminate between these two possibilities and to identify RNA regions that are involved in protein binding (for a schematic representation of the approach, *see* **Fig. 1**). It is therefore essential to compare RNase H digestion patterns obtained from an RNP preparation and from a control containing an equivalent quantity of deproteinized RNA prepared from the same RNP source. The protocols described here are derived from our work on snRNPs in the trypanosome *trans*-splicing system *(17–19)*, but should be applicable to characterize domain structures of any RNP. In addition to analyzing the RNA fragments after RNase H digestion we have also characterized the RNP fragments after RNase H digestion; this can yield valuable information on mapping protein components of the RNP.

2. Materials

All buffers and solutions should be RNase-free.

2.1. RNP Preparation and DEAE Chromatography

1. Buffer D: 20 mM HEPES, pH 8.0, 100 mM KCl, 0.2 mM EDTA, 20% glycerol. Autoclave. Just before use add dithiothreitol (DTT) to 1 mM and phenylmethylsulfonyl fluoride (PMSF) to 0.1 mM.
2. DEAE-Sepharose CL-6B (Pharmacia, Piscataway, NJ, USA).
3. Buffer D$_{100}$: 20 mM HEPES, pH 8.0, 100 mM KCl, 10 mM MgCl$_2$, 20% glycerol. Autoclave. Just before use add DTT to 1 mM and PMSF to 0.1 mM.
4. Buffer D$_{400}$: same as buffer D$_{100}$, but with 400 mM KCl.

2.2. RNA Preparation

1. Proteinase K (PK). Prepare a 10 mg/mL stock solution in dH$_2$O. Store at –20°C.
2. 2× PK buffer: 200 mM Tris-HCl, pH 7.5, 300 mM NaCl, 25 mM EDTA, 2% (w/v) SDS.
3. DNase I, RNase-free (10 U/μL, Boehringer Mannheim, Indianapolis, IN, USA).
4. 10× DNase I buffer: 100 mM Tris-HCl, pH 7.5, 50 mM MgCl$_2$.
5. Buffered phenol/chloroform solution: mix equal volumes of chloroform and phenol equilibrated first with 100 mM Tris-HCl, pH 7.5, and then with TE buffer (10 mM Tris-HCl, pH 7.5, 1 mM EDTA).
6. 3 M Sodium acetate, pH 7.0. Autoclave.

2.3. Oligonucleotide-Targeted RNase H Protection Assay

1. DNA oligonucleotides complementary to their RNA target sequences and 12–20 nt long.
2. RNasin (40 U/μL, Promega, Madison, WI, USA).
3. RNase H from *E. coli* (1 U/μL, Boehringer-Mannheim).

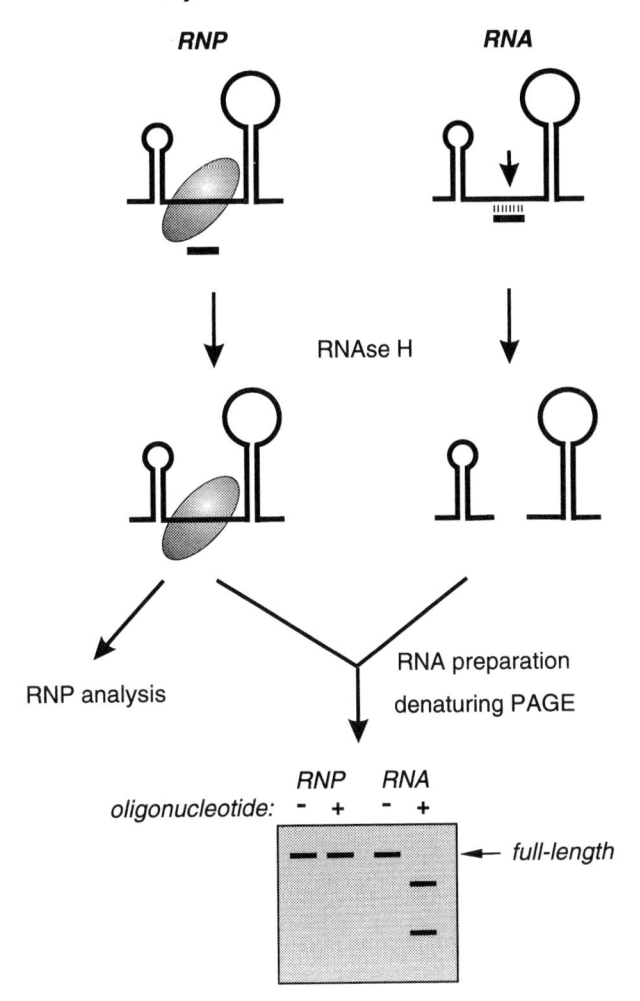

Fig. 1. Schematic representation of the oligonucleotide-targeted RNase H protection assay of RNA–protein complexes. An RNA–protein complex (RNP) and, as a control, the corresponding RNAs are treated with a specific DNA oligonucleotide and RNase H. RNA is prepared from both reactions (RNP, RNA; + *lanes*) as well as from control reactions done in the absence of oligonucleotide (– *lanes*) and analyzed by denaturing PAGE. The RNA analysis indicates that the region targeted by the oligonucleotide is protected by protein, as RNase cleavage occurs with the RNA, but not with the RNP. In addition, the RNP can be characterized further after the RNase H cleavage reaction (*see* **Subheading 3.5.**).

2.4. Northern Blot and Denaturing Polyacrylamide Gel Electrophoresis (PAGE)

1. TBE buffer: 90 mM Tris-borate, 2 mM EDTA.

2. Acrylamide/urea stock solutions for denaturing polyacrylamide gel electrophoresis (PAGE): Prepare two stock solutions with 0% and 20% acrylamide (acrylamide:*bis*-acrylamide 20:1), respectively, each containing 50% urea and 1× TBE buffer. Filter solutions and keep in dark bottles at room temperature.
3. 10% Ammonium persulfate (APS) in dH_2O.
4. *N,N,N',N'*,-Tetramethylethylenediamine (TEMED).
5. Urea gel-loading buffer: 50% Urea (w/v), 1× TBE buffer, 0.05% bromophenol blue (w/v), 0.05% xylene cyanol (w/v).
6. Transfer buffer: 10 m*M* Tris-acetate, pH 7.8, 5 m*M* NaOAc, 0.5 m*M* EDTA.
7. Electroblotting equipment (Trans-Blot Cell from Biorad, Hercules, CA, USA).
8. Nylon membrane: We have obtained good results with either uncharged (Hybond N, Amersham, Little Chalfont, Buckinghamshire, UK) or positively charged nylon membranes (Boehringer Mannheim).
9. Chromatography paper 3MM (Whatman, Clifton, NJ, USA).
10. 20× SSPE buffer: 3.6 *M* NaCl, 200 m*M* Na_2HPO_4/NaH_2PO_4, pH 7.7, 2 m*M* EDTA.
11. Denhardt's solution (100×): 2% (w/v) bovine serum albumin (fraction V), 2% (w/v) Ficoll 400, 2% (w/v) polyvinylpyrrolidone.
12. 10% SDS (w/v) stock solution.
13. Aqueous (pre)hybridization solution: 5× SSPE buffer, 5× Denhardt's, 0.1% SDS, 50 µg/mL tRNA.
14. ^{32}P-labeled oligonucleotide (1 ng/µL; specific activity of $1–5 \times 10^7$ cpm/µg DNA). Use T4 polynucleotide kinase and [γ-^{32}P]ATP for 5' end-labeling.

2.5. Primer Extension

1. AMV reverse transcriptase (5–10 U/µL, Promega, *see* **Note 1**), including 5× reaction buffer.
2. Nucleotide solution containing 10 m*M* each of dATP, dCTP, dGTP, and dTTP.
3. ^{32}P-labeled DNA oligonucleotide (*see* **Subheading 2.4.**, **step 14**).

2.6. Native RNP Gel Electrophoresis

1. Agarose.
2. 20% Acrylamide solution (acrylamide:*bis*-acrylamide 80:1).
3. RNP gel loading buffer: 0.3× TBE buffer, 80% glycerol, 0.05% xylene cyanol (w/v), 0.05% bromophenol blue (w/v).
4. 1.5-mm Spacers and comb for a small vertical gel (17 cm × 18 cm).

3. Methods

3.1. RNP Preparation

Depending on the cellular localization of the RNP, either whole cell, cytoplasmic, or nuclear extracts can be prepared (for protocols and further references, *see*, e.g., **refs. 20** and **21,** and Chapters 20–24, 26, and 32). Although RNase H cleavage assays have been done successfully in crude extracts, these results may be inconclusive, as the same RNA often occurs in different RNPs or in different conformations. In addition, degradation of RNPs and of RNase

H cleavage products may be a serious problem when working in crude extracts. These potential problems can be avoided by partially purifying RNPs, e.g., by DEAE chromatography, which has been used successfully in many cases to enrich for spliceosomal snRNPs (*see* below, this section). However, this procedure imposes some stringency on the RNA–protein interaction, and during the high-salt elution step many RNPs are reduced to stable core complexes.

All the steps of DEAE chromatography described in the following should be carried out at 4°C.

1. Prepare a 5-mL DEAE-Sepharose column for the fractionation of 50 mL of crude extract, which had been dialyzed against buffer D. Rinse column with water and ethanol, and just before use with a few milliliters of buffer D_{100}. Then slowly pack column with a 1:1 suspension of DEAE-Sepharose in D_{100}. This is best done with a peristaltic pump. Rinse with 5 mL buffer D_{100}.
2. Adjust $MgCl_2$ concentration of extract to 10 mM. Load extract slowly to column. Do not let the column dry out. Wash column with 50 mL of buffer D_{100}.
3. For elution of snRNPs, change to buffer D_{400} and collect 15 1-mL fractions.
4. Assay fractions for snRNP peak by determining the protein peak (usually between fractions 5 and 10). Pool appropriate fractions, freeze in liquid nitrogen, and store at –70°C.

3.2. RNA Preparation

To control for whether RNase H protection reflects protection by protein or an RNA conformation inaccessible to the enzyme it is essential to carry out control RNase H cleavage reactions under comparable ionic conditions with deproteinized RNA. The RNA concentration in the control reaction and in the reactions with extract should be equivalent. The following protocol can be scaled up or down as required.

1. Mix 100 µL of extract or RNP preparation with 100 µL of 2× proteinase K buffer (protein-rich extracts may become cloudy at this step). Add proteinase K to a final concentration of 0.2 mg/mL and incubate the reaction for 1 h at 50°C.
2. Phenol/chloroform extract the reaction: Add an equal volume of buffered phenol/chloroform, vortex for 10 s, and centrifuge for 2 min at 16,000g in a minifuge. Transfer the upper, aqueous phase to a new reaction tube. If a substantial white protein interphase is visible, it is necessary to repeat this extraction step.
3. Precipitate nucleic acids with 3 vol of 100% ethanol, wash with 70% ethanol, and dry the pellet (*see* **Note 2**).
4. Resuspend the pellet in 89 µL of dH_2O, add 10 µL of 10× DNase I buffer and 1 µL of DNase I, and incubate at 37°C for 30 min.
5. Extract with an equal volume of phenol/chloroform (*see* **step 2**). Precipitate the RNA by adding 1/10 vol of 3 M NaOAc, pH 7.0, and 3 vol of 100% ethanol; wash with 70% ethanol and dry the pellet.
6. Resuspend the RNA in 100 µL of dH_2O and store at –20°C.

3.3. Oligonucleotide-Targeted RNase H Protection Assay

Some knowledge of the potential RNA secondary structure is important for choosing the target sequences of DNA oligonucleotides, which should reside in single-stranded or loop regions. Although four basepairs of annealed oligonucleotide are sufficient for RNase H cleavage *(1)*, a minimum length of approx 12 basepairs is usually necessary for efficient and specific annealing. Therefore it may be best to compare and evaluate empirically several DNA oligonucleotides covering the RNA of interest. For selecting and testing oligonucleotides directed to a specific RNA, efficiencies in primer extension may be used as a guideline (*see* **Subheading 3.4.2.**). When working in crude extracts the presence of RNA annealing/helicase activities may play an additional role (*see* **Note 3**).

1. A standard RNase protection reaction is carried out in 25 µL and contains 15 µL of extract or RNP preparation (*see* **Note 4**), 40 µg/mL of DNA oligonucleotide, 12 m*M* HEPES, pH 8.0, 60 m*M* KCl, 3 m*M* MgCl$_2$, 1 m*M* DTT, 20 U of RNasin, and 1 U of *E. coli* RNase H (*see* **Notes 5–7**). For control purposes, set up a reaction with the corresponding amount of deproteinized RNA from the same extract or RNP preparation using the same ionic conditions as in the extract. Incubate the reactions at 30°C for 60 min.
2. For analysis of RNA fragments, terminate the reaction by adding 75 µL of dH$_2$O, 100 µL of 2× PK buffer, and 4 µL of proteinase K (10 mg/mL), and prepare RNA as described previously (*see* **Subheading 3.2., steps 1–5**).
3. For analysis of RNP fragments, the reactions can be used directly (*see* **Subheading 3.5.**).

3.4. Analysis of RNase H Cleavage Products: RNA Fragments

After RNase H cleavage the analysis usually focuses on the RNA fragments. They can be detected either by Northern blot analysis (**Subheading 3.4.1.**) or by primer extension (**Subheading 3.4.2.**).

3.4.1. Northern Blot

In Northern blot analysis of RNA fragments the probes of choice are complementary DNA oligonucleotides because they allow specific detection of even very small RNA fragments. Thus, in the following protocol, the particular hybridization and washing conditions refer to oligonucleotide probes.

1. Gel preparation for denaturing PAGE: Assemble glass plates for a small vertical gel (17 cm × 18 cm) using 0.4-mm spacers. For an 8% acrylamide gel, mix 10 mL of 20% acrylamide/50% urea and 15 mL of 0% acrylamide/50% urea stock solutions. Then add 200 µL of 10% APS and 20 µL of TEMED, mix, and pour the gel immediately. After a polymerization period of at least 30 min, prerun the gel in 1× TBE buffer at 800 V for 30 min.

RNase H Protection Analysis

2. Resuspend RNA pellets in 5 μL of urea loading buffer. Just before loading heat the sample to 90°C for 1 min, collect all material at the bottom of the reaction tube by a short centrifugation in a minifuge, and load sample onto the gel. Run the gel in TBE buffer at 800 V until the bromophenol blue reaches the bottom of the gel.
3. Transfer of RNA from polyacrylamide gel to nylon membrane is done by electroblotting. A sandwich has to be assembled in which a tight contact of the gel and the membrane is established. Soak two foam pads and four sheets of 3MM chromatography paper, and equilibrate the gel and nylon membrane in transfer buffer. Use an appropriate sandwich holder and assemble in the following order avoiding any air bubbles trapped between individual layers: foam pad, two sheets of 3MM paper, gel, nylon membrane, 2 sheets of 3MM paper, foam pad. Place sandwich in electroblot apparatus filled with transfer buffer chilled to 4°C. Electroblot the RNA at 50 V for 2 h at 4°C.
4. Disassemble the sandwich and let the membrane air dry for 10 min. To crosslink the RNA to the membrane, place the membrane on a UV transilluminator (312 nm) in such a way that the RNA faces the light source, and UV-irradiate membrane at maximum power for 10 min. Let the membrane air dry completely.
5. Prehybridize the membrane in aqueous hybridization solution at 37°C for at least 2 h. Then add 200 ng of ^{32}P-labeled oligonucleotide probe and hybridize at 37°C overnight.
6. Wash the membrane twice in 5× SSPE/0.1% SDS solution and then twice in 2× SSPE/0.1% SDS solution for 15 min at 37°C for each wash (*see* **Note 8**).
7. Completely enclose the wet membrane in plastic wrap and expose it to X-ray film.

3.4.2. Primer Extension

1. Annealing reaction: Resuspend the RNA pellet in 13.5 μL of dH$_2$O, add 4 μL of 5× reverse transcription buffer and 1 μL of ^{32}P-labeled oligonucleotide (1 ng/μL). Incubate at 70°C for 5 min and then on ice for 5 min.
2. Reverse transcription reaction: Add 1 μL of 10 m*M* dNTPs, and 0.5 μL of reverse transcriptase to the 18.5-μL annealing reaction (*see* **Note 1**). Incubate at 42°C for 45 min.
3. Precipitate the nucleic acids by adding 2 μL of 3 *M* NaOAc, pH 7.0, and 60 μL of 100% ethanol. Dry and resuspend the pellet in urea loading buffer and separate by denaturing PAGE (*see* **Subheading 3.4.1.**, **steps 1** and **2**).
4. Dry the gel and expose it to X-ray film.

3.5. Analysis of RNase H Cleavage Products: RNP Fragments

If site-specific RNase H cleavage occurs, it is often very informative to characterize the RNP fragments after RNase H digestion. This may yield insight into the domain structure of the RNP, i.e., information on which protein components are bound to which RNP subdomains. Experimental approaches to separate various RNP fragments from each other and from RNA include glycerol gradient sedimentation, isopycnic CsCl density gradient centrifugation

(22), and native RNP gel electrophoresis (*see* below, this section, and, alternatively, **ref. 23**). In each case, the RNA component of RNP fragments can be detected by Northern blot analysis or by primer extension, as described previously (*see* **Subheadings 3.4.1.** and **3.4.2.**). The following protocol describes the steps carried out during native gel electrophoresis.

1. Assemble glass plates for a small vertical gel using 1.5-mm thick spacers.
2. Prepare the acrylamide solution by mixing 12.6 mL of 20% acrylamide (acrylamide:*bis*-acrylamide 80:1), 14.4 mL of 50% glycerol, 2.2 mL of 10× TBE buffer, 5.8 mL of dH$_2$O, and 0.96 mL of 10% APS.
3. Prepare the agarose solution by dissolving 0.36 g of agarose in 36 mL of dH$_2$O in a microwave oven. Allow the agarose solution to cool until it reaches approx 60°C. Then add the agarose solution to the acrylamide solution.
4. Mix in 35 µL of TEMED and immediately pour the gel. Let the gel polymerize for at least 1 h.
5. Standard oligonucleotide-targeted RNase H cleavage reactions combined with 5 µL of RNP loading buffer can be directly used as electrophoresis samples (*see* **Note 9**). For comparison dilute one reaction with 75 µL of dH$_2$O and prepare RNA as described previously (*see* **Subheading 3.2., steps 1–3**). Resuspend RNA in 10 µL of dH$_2$O and add 15 µL of buffer D (*see* **Subheading 2.1.**) as well as 5 µL of RNP loading buffer.
6. Prerun the gel for 30 min at 60 V in 0.3× TBE at room temperature (*see* **Note 10**).
7. Load samples on the gel and run it for at least 4–5 h at 175 V (17 V/cm) until the bromophenol blue dye reaches the bottom of the gel. Depending on the RNP fragment, longer gel runs may be required (*see* **Note 11**). The temperature of the gel should remain constant, because a temperature increase may result in denaturation of RNPs.
8. Transfer the RNA components of RNPs to a neutral nylon membrane by electroblotting (*see* **Subheading 3.4.1., step 3**). Transfer for at least 5 h at a constant voltage of 60 V at 4°C. Make sure that the buffer does not heat up during transfer; otherwise use a cooling coil. After transfer, the membrane is handled as described previously (*see* **Subheading 3.4.1., steps 4–7**).

4. Notes

1. In our original experiments we used AMV reverse transcriptase for primer extension analysis. When signals are very weak, the primer-extension efficiencies may be improved by using Superscript™ (GIBCO BRL, Rockville, MD, USA) or Expand™ (Boehringer Mannheim) reverse transcriptases.
2. When very small quantities of RNA have to be ethanol precipitated, 10 µg of glycogen (RNase-free, Boehringer Mannheim) should be included as a carrier before the addition of ethanol.
3. The fate of the DNA oligonucleotide and of the cleavage products after the RNase H reaction depends on their stabilities, the extract source, preparation, and incu-

bation conditions. For example, in yeast extracts DNA oligonucleotides are degraded completely by endogenous nucleases, following an additional incubation after the RNase H reaction *(24)*.

4. Reconstituted RNPs are an alternative substrate for oligonucleotide-targeted RNase H protection experiments. An RNP is formed by incubating a ^{32}P-labeled RNA in a suitable extract or with purified RNP proteins. Because the RNA component is labeled, RNA fragments after RNase H cleavage can be visualized directly by gel electrophoresis and autoradiography. Accessibility to RNase H cleavage can therefore be used as a criterion for correct RNP formation (e.g., *see* refs. 17–19).

5. A standard oligonucleotide-targeted RNase H cleavage reaction is done with extract or RNP preparation that has been dialyzed against buffer D. However, we have also performed cleavage reactions in the presence of up to 400 mM KCl without apparent effects on the RNase H activity; we have observed that at this ionic strength the activity of other RNases causing RNA degradation is reduced.

6. Our standard reaction is carried out in the absence of ATP. However, some RNPs may change dramatically depending on the presence or absence of ATP; such conformational changes may result in differences in accessibility of an RNP target region (*see*, e.g., the effect of ATP on mammalian snRNPs *[7,8,25]*).

7. Although crude extracts often contain endogenous RNase H activity, we and others *(8)* have found that cleavage reactions are more consistent when *E. coli* RNase H is added. This may be due to variations in RNase H activity in different extract preparations. Moreover, the addition of *E. coli* RNase H to standard reactions ensures that control reactions with deproteinized RNA can be done under comparable conditions.

8. Hybridization and washing conditions of the Northern blot protocol are derived from our experience with DNA oligonucleotides of about 20 nt in length. We do not see nonspecific background under these conditions. However, when shorter oligonucleotides are used as probes (we have successfully used 12 mers), the stringency for hybridization and washes may have to be reduced, e.g., by lowering the incubation temperature.

9. The stability of RNP fragments under conditions of increased stringency can be further characterized by an additional incubation step prior to electrophoresis, e.g., in the presence of higher salt concentrations or heparin *(18)*.

10. RNP gel electrophoresis may have to be carried out at 4°C, because some RNPs may not be stable during electrophoresis at room temperature. In our experience gel runs in the cold have generated sharper bands in some cases.

11. The migration rate of an RNP during native gel electrophoresis depends on its overall charge, size, and conformation. Therefore, the optimal time of electrophoresis has to be determined empirically for each RNP. RNPs usually travel toward the anode during electrophoresis owing in part to their negatively charged RNA component. However, after RNase H cleavage the RNA component can be significantly shortened and, if the bound proteins carry a positive net charge, the RNP fragment may run in the opposite direction.

References

1. Donis-Keller, H. (1979) Site specific enzymatic cleavage of RNA. *Nucleic Acids Res.* **7**, 179–192.
2. Lazar, E. and Jacob, M. (1982) Accessibility of U1 RNA to base pairing with a single-stranded DNA fragment mimicking the intron extremities at the splice junction. *Nucleic Acids Res.* **10**, 1193–1201.
3. Rinke, J., Appel, B., Blöcker, H., Frank, R., and Lührmann, R. (1984) The 5'-terminal sequence of U1 RNA complementary to the consensus 5' splice site of hnRNA is single-stranded in intact U1 snRNP particles. *Nucleic Acids Res.* **12**, 4111–4126.
4. Ruskin, B., Krainer, A. R., Maniatis, T., and Green, M. R. (1984) Excision of an intact intron as a novel lariat structure during pre-mRNA splicing *in vitro*. *Cell* **38**, 317–331.
5. Hwang, S. P., Eisenberg, M., Binder, R., Shelness G. S., and Williams, D. L. (1989) Predicted structures of apolipoprotein II mRNA constrained by nuclease and dimethyl sulfate reactivity: stable secondary structures occur predominantly in local domains via intraexonic base pairing. *J. Biol. Chem.* **264**, 8410–8418.
6. Krämer, A., Keller, W., Appel, B., and Lührmann, R. (1984) The 5' terminus of the RNA moiety of U1 small nuclear ribonucleoprotein particles is required for the splicing of messenger RNA precursors. *Cell* **38**, 299–307.
7. Krainer, A. R. and Maniatis, T. (1985) Multiple factors including the small nuclear ribonucleoproteins U1 and U2 are necessary for pre-mRNA splicing *in vitro*. *Cell* **42**, 725–736.
8. Black, D. L., Chabot, B., and Steitz, J. A. (1985) U2 as well as U1 small nuclear ribonucleoproteins are involved in premessenger RNA splicing. *Cell* **42**, 737–750.
9. Black, D. L. and Steitz, J. A. (1986) Pre-mRNA splicing *in vitro* requires intact U4/U6 small ribonucleoprotein particles. *Cell* **46**, 697–704.
10. Krämer, A. (1987) Analysis of RNAse-A-resistant regions of adenovirus 2 major late precursor-mRNA in splicing extracts reveals an ordered interaction of nuclear components with the substrate RNA. *J. Mol. Biol.* **196**, 559–573.
11. Frendewey, D., Krämer, A., and Keller, W. (1987) Different small nuclear ribonucleoprotein particles are involved in different steps of splicing complex formation. *Cold Spring Harb. Symp. Quant. Biol.* **52**, 287–298.
12. Krämer, A. (1990) Site-specific degradation of RNA of small nuclear ribonucleoprotein particles with complementary oligodeoxynucleotides and RNAse H. *Methods Enzymol.* **181**, 284–292.
13. Rymond, B. C. and Rosbash, M. (1986) Differential nuclease sensitivity identifies tight contacts between yeast pre-mRNA and spliceosomes. *EMBO J.* **5**, 3517–3523.
14. Bindereif, A., Ruskin, B., and Green, M. R. (1987) An analysis of specific RNA-factor interactions during pre-mRNA splicing *in vitro*. *UCLA Symposia on Molecular and Cellular Biology* (Granner, D. K., Rosenfeld, G., and Chang, S., eds.), new series, Vol. **52**, pp. 181–194.

15. Cotten, M., Gick, O., Vasserot, A., Schaffner, G., and Birnstiel, M. L. (1988) Specific contacts between mammalian U7 snRNA and histone precursor RNA are indispensable for the *in vitro* 3' RNA processing reaction. *EMBO J.* **7,** 801–808.
16. Tschudi, C. and Ullu, E. (1990) Destruction of U2, U4, or U6 small nuclear RNA blocks *trans* splicing in trypanosome cells. *Cell* **61**, 459–466.
17. Cross, M., Günzl, A., Palfi, Z., and Bindereif, A. (1991) Analysis of small ribonucleoproteins (RNPs) in *Trypanosoma brucei* — structural organization and protein-components of the spliced leader RNP. *Mol. Cell. Biol.* **11**, 5516–5526.
18. Günzl, A., Cross, M., Palfi, Z., and Bindereif, A. (1992) Domain structure of U2 and U4/6 snRNPs from *Trypanosoma brucei:* identification of trans-spliceosomal specific RNA-protein interactions. *Mol. Cell. Biol.* **12**, 468–479.
19. Günzl, A., Cross, M., Palfi, Z., and Bindereif, A. (1993) Assembly of the U2 small nuclear ribonucleoprotein from *Trypanosoma brucei. J. Biol. Chem.* **268**, 13336–13343.
20. Krämer, A. and Keller, W. (1990) Preparation and fractionation of mammalian extracts active in pre-mRNA splicing. *Methods Enzymol.* **181**, 3–19.
21. Cheng, S.-C., Newman, A., Lin, R.-J., McFarland, G. D., and Abelson, J. N. (1990) Preparation and fractionation of yeast splicing extract. *Methods Enzymol.* **181**, 89–96.
22. Brunel, C. and Cathala, G. (1990). Purification and characterization of U small nuclear ribonucleoproteins in cesium chloride gradients. *Methods Enzymol.* **181**, 264–273.
23. Konarska, M. M. (1989) Analysis of splicing complexes and small nuclear ribonucleoprotein particles by native gel electrophoresis. *Methods Enzymol.* **180**, 442–453.
24. Fabrizio, P., McPheeters, D. S., and Abelson, J. (1989) *In vitro* assembly of yeast U6 snRNP: a functional assay. *Genes Dev.* **3**, 2137–2150.
25. Black, D. L. and Pinto, A. L. (1989) U5 small nuclear ribonucleoprotein: RNA structure analysis and ATP-dependent interaction with U4/U6. *Mol. Cell. Biol.* **9**, 3350–3359.

9

Nitrocellulose Filter Binding for Determination of Dissociation Constants

Kathleen B. Hall and James K. Kranz

1. Introduction

Nitrocellulose filter binding has been used to measure the association of proteins to both DNA and RNA for many years. Yarus and Berg *(1)* first described the application of the method for characterization of aminoacyl-tRNA synthetase:tRNA association; other RNA applications include the R17 coat protein:RNA hairpin interaction *(2)*, ribosomal protein:RNA binding *(3)*, and TFIIIA:5S *(4)*. The method has also been extensively applied to DNA:protein systems.

The basis of the methodology is the observation that (most) proteins bind to nitrocellulose. If a protein is associated with a nucleic acid, then the complex can also be retained on a nitrocellulose filter. However, the association must be tight enough to survive the filtration, and the protein must be able to retain the bound nucleic acid when it is in turn bound to the filter (there are some reports of proteins denaturing upon adhering to the nitrocellulose). For some systems only a percentage of the complexes is retained by the filter: perhaps the RNA is too long and wraps around the protein so that it is not trapped by the filter; perhaps long RNA pulls associated proteins through the filter; perhaps there are several binding modes, only one of which is stable enough to be trapped during filtration. As Draper et al. *(3)* point out, some rRNA:rprotein complexes are not retained by nitrocellulose at all. Thus, not only is the success of this method dependent on the system in an unpredictable way, it is also important to remember that nitrocellulose filter binding is not a true equilibrium method.

The typical experiment is performed with a constant trace amount of [^{32}P]-RNA that is titrated with increasing amounts of protein. The opposite configuration (using protein as a reporter) is not possible since the protein alone sticks

From: *Methods in Molecular Biology, Vol. 118: RNA-Protein Interaction Protocols*
Edited by: S. Haynes © Humana Press Inc., Totowa, NJ

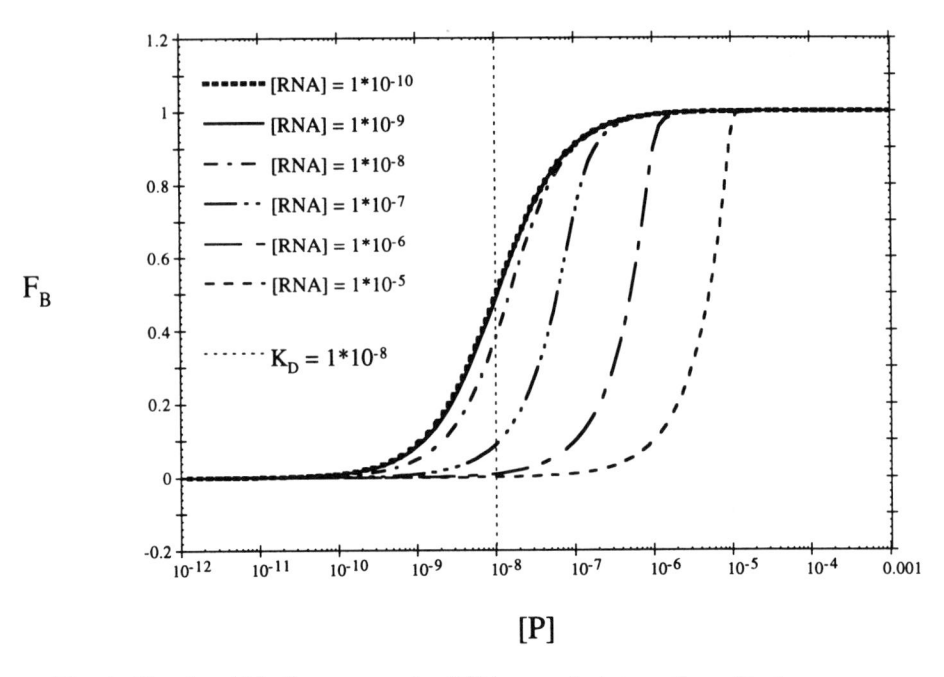

Fig. 1. Simulated binding curves for RNA–protein interactions. Each curve represents a different, constant RNA concentration, with the fraction of RNA bound, F_B, plotted as a function of log[protein]. All curves were simulated assuming a 1:1 stoichiometry, with a $K_d = 1 \times 10^{-8}$ M. Six different concentrations of the RNA are shown. At low RNA concentrations (<1 n*M*), all binding isotherms are identical in shape, and provide a useful estimate of the dissociation constant. When the RNA concentration is equivalent to the K_d, the profile of the isotherm begins to be misshapen. At RNA concentrations much higher than the K_d, the shape of the isotherm no longer provides accurate information on the strength of the RNA–protein interaction.

to the nitrocellulose. The important points to remember are: (1) To accurately measure a dissociation constant (K_d), one component must be present at an invariant concentration below the K_d. When it is the RNA concentration that is invariant, then if the RNA is ^{32}P-labeled, its specific activity must be adjusted to obtain sufficient signal. At concentrations of components at or above the K_d, the experiment measures the stoichiometry of association. As an illustration of the consequence of this mistake for determination of a binding constant, isotherms are shown in **Fig. 1** where the RNA concentration is increased to exceed the K_d. The curve becomes progressively steeper as the concentration of RNA increases; when the concentration of RNA becomes equal to that of protein, the experiment becomes one that determines binding stoichiometry (*see* **Note 1**).

Nitrocellulose Filter Binding

(2) At high concentrations of protein necessary for measuring weak binding, the protein may aggregate (or dimerize, etc.), which will change its binding characteristics. Likewise, when the RNA is at high concentrations, a hairpin may form a dimer, or it may self-associate. Thus binding at high concentrations of either component can be complicated by the state of the molecules. In addition, the capacity of the nitrocellulose is limited, and especially at high concentrations of protein, retention of the complex can be very inefficient. For example, with Schleicher and Schuell membranes, a protein concentration of 10^{-5} *M* saturates the binding capacity; thus no meaningful data can be obtained at higher protein concentrations.

2. Materials

All buffers should be prepared with MilliQ H_2O to remove ionic, organic, and biological contaminants. Nucleases are a common problem, so each solution is filtered through a nitrocellulose-based filter that retains the nucleases; we use a Nalgene® disposable filterpack and store solutions in the attached sterile bottle (*see* **Note 2**). Solutions are transferred by pipet or with plastic pipet tips from stacked racks (not from bags). Only high-purity reagents should be used to avoid both nuclease contaminations and impurities that could alter the dissociation constant.

For use with the dot blot apparatus (*see* **Notes 4** and **5**):

1. 4 in. × 5 in. Schleicher and Schuell (Keene, NH, USA) 0.2 µm supported nitrocellulose filter (BAS).
   or: 4 in. × 5 in. DEAE membranes and 4 in. × 5 in. Schleicher and Schuell 0.45 µm nitrocellulose filters (BA45).
2. 96-Well polypropylene microtiter trays.
3. Modified dot-blot apparatus.
4. Eight-channel pipettor.

For single-filter applications:

1. Nitrocellulose filters: Schleicher and Schuell BA85 0.45- or 0.2-µm filters.
2. Millipore glass-fritted 25-mm filter apparatus.

Store the nitrocellulose filters at 4°C in plastic bags to keep them from drying out. Old filters are subject to uneven wetting and also become fragile when dehydrated, making them unreliable. For use without the DEAE membrane, the supported filters are more robust and do not rupture when vacuum is applied. Use the unsupported variety with DEAE. Keep in mind that nitrocellulose is explosive when it is heated too high or hit with a hammer, aged filters being most unstable.

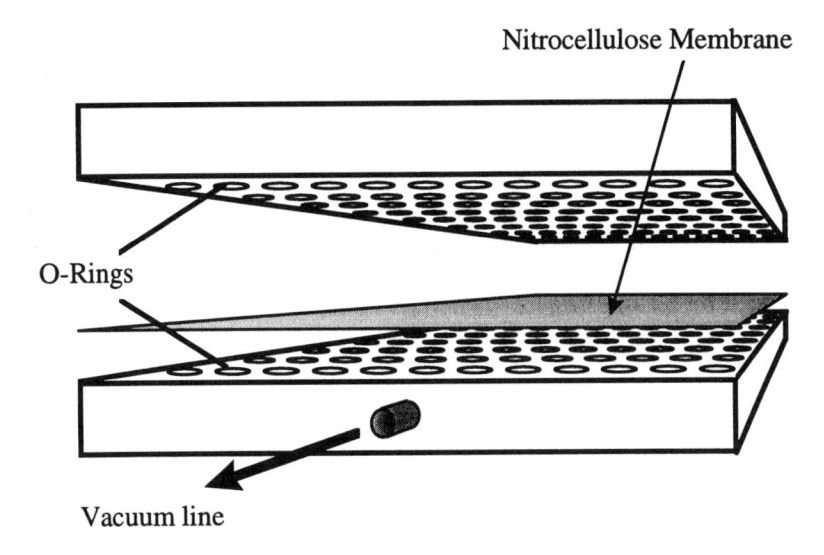

Fig. 2. Schematic of the modified Minifold dot-blot apparatus.

The polypropylene trays can be acid washed for 30 min in 1 N HCl to remove any trace RNases and to remove adhered protein. This allows them to be reused many times. Polycarbonate and polystyrene trays are not acid-resistant, and are more prone to trap protein on the surface.

The modified dot-blot apparatus was designed by Wong and Lohman (5), based on the original Schleicher and Schuell Minifold I device. It is illustrated in **Fig. 2**; the filter(s) are clamped between two sample well plates in which each well is encircled by a silicon O-ring. The tight seal afforded by this design eliminates seepage of the sample beyond the well. Construction of an apparatus requires the standard Minifold I (Delrin) system with an additional sample well plate; two holes are machined in the additional plate for alignment with the existing pins.

An alternative scheme is to use individual filters for each sample. The disadvantage of this method is the increase in the error due to filter variability.

3. Methods
1. Presoak the nitrocellulose filter in the appropriate binding buffer, at the appropriate temperature (*see* **Notes 1** and **3**), for at least 30 min prior to use, making sure that all surfaces are in contact with buffer. Any filter that does not completely saturate with buffer (i.e., is resistant to wetting by the solution) should be discarded. Avoid contact with fingers, as they shed nucleases.
2. Prepare the samples, which are of constant volume, in a polypropylene microtiter plate. Assuming a constant amount of [^{32}P]-RNA is being titrated with an increas-

Nitrocellulose Filter Binding

ing amount of protein, a necessary control is a well containing RNA only, to measure the nonspecific binding of RNA to membrane. A typical titration curve will span three or four orders of magnitude in protein concentration, with at least five points (more are recommended to define the midpoint accurately) per decade, with the caveat that both upper (saturated) and lower (unsaturated) baselines are adequately sampled. Most experiments are done in duplicate, and the values averaged.

3. Prepare the filter by lightly blotting it with a Kimwipe after soaking to remove the excess buffer. Without such blotting, the samples applied tend to diffuse over the surface.

4. Insert the membrane into the apparatus just prior to filtering. Turn on the vacuum and immediately begin the transfer of samples from tray to filter. Start pipetting from the least concentrated protein samples and progress to the highest; the same pipet tips can be used for a single titration before ejecting the row. Prolonged suction dries the filter and leads to unreliable retention, so this step should be done swiftly; hence the use of the same tips for a single titration.

   Bubbles may form between the filter and the solution, blocking filtration; these must be dislodged, but without poking a hole in the membrane or disrupting the equilibrium of the solution. Tapping the apparatus gently on the bench is usually sufficient. Very concentrated protein solutions may foam upon suction, which is a sign that they are denaturing; the result will be unreliable retention of RNA (*see* **Notes 5** and **6**).

5. Remove the filter and gently blot the underside again to remove excess buffer that will diffuse across the surface and obscure the individual dots (*see* **Note 5**).

6. For quantitation of bound RNA using a phosphorimager, the filter can either be dried or damp; if damp, wrap it in Saran wrap. If the DEAE filter is used, then it provides a direct count of the total amount of material applied in each well; in its absence, single filter(s) with an amount of [^{32}P]-RNA equal to that added to each sample well must be included to provide an estimate of total RNA added. For these references, the RNA is applied directly to the filter and allowed to dry. Bound radiolabeled RNA is quantified, and the retained counts (B) are normalized to the total RNA (T) present. The RNA bound to the filter in the absence of protein is designated as the background (O), and this value is subtracted from each data point, to give (B – O)/T = F_B (fraction bound). Plotting F_B vs log[protein] gives the binding isotherm, as illustrated in **Fig. 1**.

7. Analysis of binding assays. Complex formation is assumed to be described by a bimolecular association; stoichiometry of binding is 1:1 (*see* **Note 7**). The data are fit to a Langmuir isotherm to determine the dissociation constant, following the formalism of Lin and Riggs *(6)*.

   The association is bimolecular:

   $$[RNA] + [P] \rightleftharpoons [R:P].$$

   At equilibrium, the concentrations of [RNA] and [P] change by the amount [R:P], and the equilibrium constant can be written

   $$K_{eq} = K_d = ([RNA] - [R:P])*([P] - [R:P])/[R:P].$$

Multiplying out this expression leads to the quadratic equation in [R:P] that is normalized by dividing the expression by [RNA]. The expression is solved for [R:P] at each value of [P], with K_{eq} as the variable. Fitting is done using nonlinear regression. The final form of the equation is expressed in terms of the fraction bound, F_B:

$$F_B = [R:P]/[RNA]$$
$$= (C/2[RNA]) * ((K_d + [RNA] + [P]) - ((K_d + [RNA] + [P])^2 - 4[RNA][P])^{1/2})$$

The constant C is the fit for the upper baseline (in cases when retention efficiency is less than 100%; *see* **Note 8**), which is then used to normalize for varying retention efficiency. Only after an individual data set has been normalized can it be analyzed with replicate data sets. Thus, obtain a value for C from the initial fit, then compute $F_{B,norm}$ by dividing each F_B by the fitting parameter C, giving $F_{B,norm}$ as a function of protein concentration. Fit again, using $F_{B,norm}$ in place of F_B, with C omitted (or set to 1.00). For analysis of individual data sets, a plotting program such as KaleidaGraph™ (Synergy Software) for a Macintosh or PC is adequate; other software packages that can handle more sophisticated analysis (e.g., global fitting) include Scientist™ (MicroMath Scientific Software, Inc.).

4. Notes

1. The minimum protein concentration range needed for accurate measurement of the dissociation constant is at least three log units (e.g., 1×10^{-9} M to 1×10^{-6} M) centered around the dissociation constant. Therefore, the dissociation constant must be known approximately prior to its accurate measurement using this method. Preliminary experiments may be needed to locate the protein concentration range needed for a successful K_d determination, using only a limited number of points over a wide range of protein concentrations (e.g., four or five different points per log unit, over five or six log units), taking care to ensure that the RNA concentration is limiting at all protein concentrations. Alternatively, the dissociation constant may be estimated by an alternative method, such as in a gel shift assay (*see* Chapter 10). The gel shift method is similar in principle to filter binding, but does not reliably measure the true dissociation constant for every system; because the gel shift assay is a nonequilibrium technique, the measured dissociation constant does not always agree with other true equilibrium methods of measuring binding affinities *(7)*.

    Typically, the value of the dissociation constant in nucleic acid:protein interactions varies with conditions. Changing salt concentrations, pH, or temperature often can raise or lower the K_d dramatically. Therefore, the choice of conditions dictates the protein concentration range needed for K_d measurement. At very high salt concentrations, the binding affinity is generally weakened which can bring the specific dissociation constant into the range where nonspecific RNA:protein binding can interfere with its measurement; the concentration of carrier tRNA and BSA can often be adjusted to block nonspecific binding (*see* **Note 2**). At very low salt concentrations, problems can arise in retention efficiency as well as nonspecific association of the RNA to the nitrocellulose membrane. Extremes of pH should be avoided as both acid- and base-catalyzed hydrolysis of the protein and RNA can occur; nitrocellulose dissolves in highly basic conditions (pH > 13).

Measuring the temperature-dependence of the dissociation constant using the 96-well dot-blot apparatus can be cumbersome, as it is necessary to equilibrate both the samples and the filter unit at the same temperature for accurate determination of the K_d. In cases in which this is not possible, the single-filter method should be used as a means of minimizing changes in temperature during filtration. Buffer selection is most important when the temperature-dependence of the interaction is being measured; a buffer with a low heat of ionization should be used such that the pH remains invariant over a wide range of temperatures.

2. The buffers used in the binding experiment typically contain bovine serum albumin (BSA) and might also contain tRNA, in addition to the protein and RNA of interest. Because proteins often stick to glass and plastic, the BSA is present to coat any surfaces, leaving the protein of interest free in solution. Also, when the protein is present at low concentrations, the BSA will help to stabilize it. The (acetylated, nuclease-free) BSA is typically added to nuclease-free buffers to a working concentration of 20–50 µg/mL (0.1–1 µM), although the optimum concentration should be determined for each system. The tRNA is present to decrease nonspecific association of the protein to RNA, and also to reduce nonspecific [^{32}P]-RNA binding to the nitrocellulose. tRNA is prepared from bulk yeast tRNA (Boehringer Mannheim, Indianapolis, IN, USA) that is treated with Proteinase K and phenol extracted, ethanol precipitated, and resuspended in MilliQ H_2O. It is stored in small aliquots at 1 mg/mL. In binding reactions, a tRNA concentration might be 10 µg/mL; it should be calculated to compete effectively with nonspecific binding under conditions in which that is observed. This is a variable that needs to be adjusted with the system. If tRNA creates a problem, poly(A) is a common alternative.

3. Several filters can be soaked together, and the soaking solution can be reused. However, the filters release fine material that accumulates in the solution. Because this material could clog the filters, the solution should be changed frequently. Some filters also release some foamy compounds and their accumulation in the soaking solution is not desirable. Note that if experiments are to be done at several salt concentrations, then the filters for these experiments need to be soaked independently. The soaking solutions contain salts and buffer, but nothing else.

4. With this method, a complete titration in the 96-well apparatus uses a single filter to collect all the points, eliminating the variability in individual nitrocellulose filters. The variability among filters is perhaps the single greatest source of error in this method: filters can wet to different extents, which alters their retention efficiency; different lots of filters have different properties (never mix lots in an experiment) and filters from Schleicher and Schuell and Millipore behave differently. This method can accommodate several binding isotherms on a single filter, so that an experiment can be done in duplicate or triplicate to increase the accuracy of the data.

5. A modification of this method is the inclusion of a DEAE filter beneath the nitrocellulose that traps unbound RNA. This allows very accurate quantitation of the free RNA added to each sample, rather than having to use an external representative sample. However, the DEAE membrane may significantly perturb the bound complex, for in some systems the associated RNA is pulled out of the nitrocellulose and into the DEAE. This effect is system dependent, and must be experimentally tested. Preparation of the optional DEAE is given in Wong and Lohman *(5)*.
6. Note that no wash of the filter after filtration of sample is included here. Although washing with buffer can reduce the level of nonspecific association of labeled RNA to the filter, with some systems, washing can also strip off the associated RNA. This must be assessed for each system; when washing removes the RNA, it may also be true that retention is dependent on the volume of the sample applied to the filter (larger volumes mean longer suction times and so more loss of RNA). Also, the vacuum is the house vacuum line (12–15 in. Hg), and there is a trap between the apparatus and the house line intake.
7. Assumptions in the data analysis are that all protein added is active and that the stoichiometry of complex formation is 1:1. The filter binding assay can also be used to assess these assumptions, with some caveats. To determine the stoichiometry, comparable concentrations of RNA and protein are used; usually the [^{32}P]-RNA has a lower specific activity for these measurements. For these experiments, it is best to keep the RNA concentration above the K_d, and titrate the protein to achieve saturation. The break point of the titration gives the [P]/[RNA] ratio of the complex, as illustrated in **Fig. 3**. Also notable in that figure is the consequence of using different concentrations of components that approach the value of the dissociation constant. It is important to note that this method will not distinguish between 1:1 stoichiometry and two binding sites on the RNA (2:1) where only 50% of the protein is active. However, when the system is particularly simple (e.g., an RNA hairpin and a small protein) where it might be assumed that the stoichiometry is in fact 1:1, then this method could indeed report on the percent active protein.

    If there are multiple binding sites on the RNA, then this method will be unable to distinguish their individual isotherms. When [^{32}P]-RNA is the reporter of complex formation, the first RNA:protein association that binds to the filter will be observed; any subsequent protein association to that RNA will not be detected, as that event produces no additional signal. Proteins bound to different RNA sites could also have different retention efficiencies, making analysis very difficult. In addition to this practical problem, the expressions for binding isotherms cannot be solved combinatorially for the case of two concurrent binding events *(8)*.
8. At complete saturation of RNA by protein, there is typically significantly less than 100% of the input RNA retained by the filter. The percentage of input RNA bound at saturation is the "retention efficiency" of the system, and may vary from 25% to 90% (i.e., the constant C ranging from 0.25 to 0.9 in the fit). For a given preparation of RNA, in given solution conditions, retention efficiency

Nitrocellulose Filter Binding

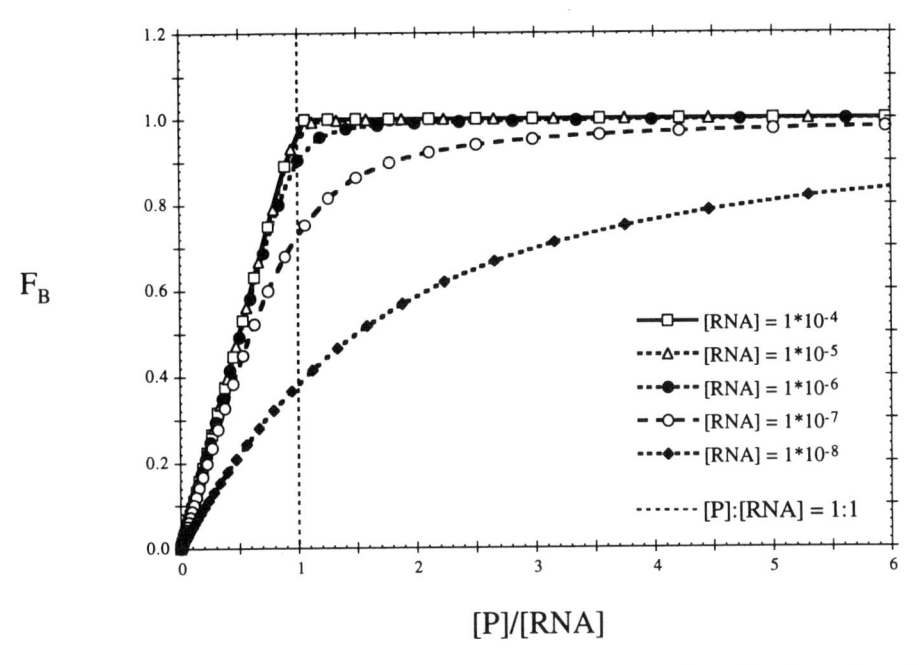

[Figure: Plot of F_B vs [P]/[RNA] with curves for:
- [RNA] = 1*10⁻⁴ (open squares, solid line)
- [RNA] = 1*10⁻⁵ (open triangles, dashed)
- [RNA] = 1*10⁻⁶ (filled circles, dashed)
- [RNA] = 1*10⁻⁷ (open circles, dashed)
- [RNA] = 1*10⁻⁸ (filled diamonds, dotted)
- [P]:[RNA] = 1:1 (vertical dashed line)]

Fig. 3. Simulated curves for RNA–protein interactions. Data from **Fig. 1** are plotted as a linear function of protein concentration, normalized by the RNA concentration used to simulate the curve. These [P]/[RNA] plots are useful for measuring the stoichiometry of the interaction, determined from the break point in the curve. Based on the $K_d = 1 \times 10^{-8}$ M, only those simulations with a high concentration of RNA allow for accurate measurement of the stoichiometry.

should be constant; variation is typically observed when different RNA preparations are compared, although the value of the dissociation constant should be unchanged. When the apparent retention efficiency of a given RNA preparation becomes progressively lower during the course of several titrations, the usual cause is a contaminating nuclease. The only recourse is to either repurify the RNA or synthesize a new batch; it may be prudent to make new solutions at this point to safeguard against a potential nuclease contamination.

Retention efficiency depends not only on the specific RNA and protein, but also on the salt concentration, pH, and temperature, as well as the concentration of added BSA. At higher salt concentrations and at higher temperatures, retention efficiency is reduced; at higher concentration of BSA, the filter becomes saturated. For proper analysis of the binding data, the retention is normalized before comparison of K_d values.

References

1. Yarus, M. and Berg, P. (1970) On the properties and utility of a membrane filter assay in the study of isoleucyl-tRNA synthetase. *Anal. Biochem.* **35**, 450–465.

2. Carey, J., Cameron, V., deHaseth, P., and Uhlenbeck, O. C. (1983) Sequence-specific interaction of R17 coat protein with its ribonucleic acid binding site. *Biochemistry* **22**, 2601–2610.
3. Draper, D. E., Deckman, I. C., and Vartikar, J. V. (1988) Physical studies of ribosomal protein-RNA interactions. *Methods Enzymol.* **164**, 203–220.
4. Romanuik, P. J. (1985) Characterization of the RNA binding properties of transcription factor IIIA of *Xenopus laevis* oocytes. *Nucleic Acids Res.* **13**, 5369–5387.
5. Wong, I. and Lohman, T. M. (1993) A double-filter method for nitrocellulose-filter binding: application to protein–nucleic acid interactions. *Proc. Natl. Acad. Sci. USA* **90**, 5428–5432.
6. Lin, S. Y. and Riggs, A. D. (1972) lac repressor binding to nonoperator DNA: detailed studies and a comparison of equilibrium and rate competition methods. *J. Mol. Biol.* **72**, 671–690.
7. Hall, K. B. and Stump, W. T. (1992) Interaction of N-terminal domain of U1A protein with an RNA stem/loop. *Nucleic Acids Res.* **20**, 4283–4290.
8. Ackers, G. K., Shea, M. A., and Smith, F. R. (1983) Free energy coupling with macromolecules: the chemical work of ligand binding at the individual sites in cooperative systems. *J. Mol. Biol.* **170**, 223–242.

10

Measuring Equilibrium and Kinetic Constants Using Gel Retardation Assays

David R. Setzer

1. Introduction
1.1. Outline of the Problem

When analyzing ribonucleoprotein complexes, it is important to keep in mind that any such complex in solution must coexist with some concentration of the free (uncomplexed) RNA and protein components of the RNP. This can be formalized by the equilibrium equation:

$$RP \rightleftharpoons R + P$$

where R represents the free RNA, P the free protein, and RP the RNA–protein complex for the simplest possible case of a 1:1 molar complex of RNA and protein. Thus, at equilibrium, the distribution of components between free and bound forms is determined by the concentrations of RNA and protein and by the equilibrium binding constant (K_d, or its reciprocal, K_a) which describes the interaction. Formally, for a simple binary complex, $K_d = [R][P]/[RP]$. Of course, a more complex analysis, involving multiple equilibrium constants, is necessary if the RNP contains more than two components. The equilibrium binding constant can be treated as the ratio of the rate constants describing RNP formation and dissociation ($K_d = k_{off}/k_{on}$). It is often the case that the association rate constant for an RNA–protein interaction is determined by the diffusion rates of the RNA and protein reactants, and therefore exhibits relatively little variability. When this is true, differences in equilibrium binding constants are directly reflected in corresponding differences in dissociation rate constants. There are certainly exceptions to this general observation, however, and it is therefore useful to distinguish between affinity (defined by the equilibrium binding constant) and stability (defined by the dissociation rate con-

From: *Methods in Molecular Biology, Vol. 118: RNA-Protein Interaction Protocols*
Edited by: S. Haynes © Humana Press Inc., Totowa, NJ

stant). Any complete description of an RNA–protein interaction must include measurements of these constants which describe the kinetics and thermodynamics of the formation and dissociation of the complex. Even when less quantitative approaches are used to assess the relative affinities of various RNA–protein complexes, it is important to have at least a rough estimate of the equilibrium and rate constants that define the interaction. In the absence of information concerning equilibrium binding constants, it is possible (and distressingly common) that the concentrations of RNA and/or protein used in the binding studies may make the experimental results relatively insensitive to differences in binding affinity, leading to serious misinterpretations. In other cases, variations in rate constants could result in failure to measure "affinity" differences at equilibrium (*see* **Note 3**).

1.2. Choice of Methods

To measure equilibrium binding and dissociation rate constants, it is necessary to have a method that distinguishes bound and free forms of either the protein or, more commonly, the RNA. One such method that is simple, informative, and enjoys widespread use is the gel retardation assay. This technique makes use of nondenaturing polyacrylamide gels to resolve the free RNA from the RNA–protein complex, which has lower mobility and is therefore "retarded" in the gel. Other methods that may be used to distinguish free and bound forms of the RNA include nitrocellulose filter-binding assays (*see* Chapter 9) and, less commonly, fluorescence-based assays. Gel retardation assays are sensitive and directly provide information concerning the stoichiometry of binding, which must generally be inferred indirectly by other methods. Disadvantages of gel retardation assays include the fact that very unstable complexes may be hard to detect and that rapid kinetics (on a time scale of less that 0.25–1 min) cannot generally be analyzed. Although gel retardation assays require no specialized equipment and are quite straighforward, they are nonetheless considerably more time-consuming than either filter-binding or fluorescence-based assays and therefore limit the number of experimental points that can be reasonably analyzed. On balance, however, gel retardation is a good choice for many investigators who wish to analyze quantitatively an RNA–protein interaction.

1.3. Experimental Design for Determining Dissociation Rate Constants (k_{off})

The dissociation of simple binary complexes consisting of one molecule of RNA and one molecule of protein can generally be treated as a simple first-order reaction:

$$RP \rightarrow R + P, \text{ so that } [RP]t = [RP]_0 e^{-kt} + C$$

where [RP] is the concentration of RNA–protein complex at a particular time (t), [RP]$_0$ + C is the concentration of RNA–protein complex at $t = 0$, k is the dissociation rate constant (k_{off}), and C represents the amount of RNA–protein complex remaining at infinite time (normally zero). Experimental determination of k_{off} is achieved by incubating a mixture of radiolabeled RNA with protein until equilibrium is reached. The time course is then initiated by either diluting the sample sufficiently to prevent free protein and RNA from subsequently associating (which is not generally feasible given the constraints on sample volume imposed by the gel retardation assay) or, more commonly, by adding a large molar excess of unlabeled RNA to the reaction so that any free protein that remains or that is generated via dissociation of complexes binds to the unlabeled rather than to the labeled RNA. Aliquots of the reaction mixture are loaded onto a running nondenaturing polyacrylamide gel at various time points and subsequent analysis of the amount of labeled RNA–protein complex at these time points is used to estimate the k_{off} using the equation given previously (*see* **Note 6**).

1.4. Experimental Design for Determining Equilibrium Binding Constants

As noted previously, the equilibrium binding constant K_d = [R][P]/[RP]; K_d is expressed in molar (*M*) units (or more typically, n*M* or μ*M*). Equilibrium constants can also be expressed as a K_a = [RP]/[R][P], with units of M^{-1}; thus, $K_a = 1/K_d$. A common method for measuring equilibrium binding constants involves holding the concentration of radiolabeled RNA constant in a series of binding reactions containing variable concentrations of protein, and measuring the concentration of bound and free RNA after equilibrium has been reached in each case. Because K_d = [P] when [RP] = [R], an estimate of the K_d can be obtained by determining the free protein concentration at which 50% of the RNA is in bound form. Since the concentration of free protein cannot generally be easily measured, such an experiment should be performed at a concentration of RNA that is low relative to the K_d, so that the concentration of free protein is approximately equal to the concentration of total protein (i.e., [P] ~ [P] + [RP]). This approach can be used to obtain a "quick-and-dirty" estimate of the equilibrium binding constant, but is not recommended for precise determinations. First, it ignores potential data points over most of the binding titration curve (from 0% bound to 100% bound) and instead focuses only on points in the neighborhood of 50% bound. Second, it requires one to approximate the concentration of free protein as equal to the total protein concentration, and this is true only when the RNA concentration is much below the K_d. Finally, simple curve-fitting programs are generally available and can be used to extract a better estimate of the K_d from the entire range of data.

A limitation of using protein titration at a fixed RNA concentration to estimate the K_d as described previously is that one must know the active protein concentration at each point in the titration. This is not always possible, either because the protein is impure or because only a fraction of the protein is active in RNA binding. This latter point is often not fully appreciated and may lead to underestimates of affinity, particularly when recombinant protein that has been subjected to denaturation during purification is used. An alternative approach that avoids this problem is to perform titrations with variable concentrations of RNA at a fixed (though unknown) concentration of protein. When the concentrations of free and bound RNA are then determined at each input RNA concentration, the K_d can be estimated using a traditional Scatchard analysis *(1)*, which makes use of the following rearrangement of the equilibrium binding equation:

$$B/F = (-1/K_d)B + (P_T/K_d)$$

where B is the concentration of RNA in bound form, F is the concentration of free RNA, and P_T is the total concentration of active protein in the reaction mixture, assuming a binding stoichiometry of 1:1. By plotting B/F vs B, one can obtain the K_d from the slope $(-1/K_d)$ of the best-fit line and the active protein concentration from the *x*-intercept (P_T). More complex versions of this approach must be used if the binding stoichiometry is not unitary. While numerous limitations of the Scatchard approach have been noted (e.g., *2–4*), most can be avoided and accurate and precise estimates of the K_d can be obtained by carefully designing the binding titration so that one covers that part of the saturation curve corresponding to 20–80% of the protein bound *(5)*. An alternative computational approach that is available to most investigators makes use of nonlinear curve-fitting methods. In this case, any of a number of rearrangements of the equilibrium binding equation can be used to obtain a best-fit curve for the data when values for both K_d and P_T are allowed to vary simultaneously. One such equation is:

$$[RP] = \{(P_T + R_T + K_d) - [(-P_T - R_T - K_d)^2 - (4P_T R_T)]^{1/2}\}/2$$

where P_T is the unknown total protein concentration, R_T is the total RNA concentration for each binding reaction (the independent variable), and $[RP]$ is the concentration of RNA–protein complex in each reaction mixture (the dependent variable). Inexpensive nonlinear curve-fitting algorithms will determine values of K_d and P_T that provide the best fit to the experimental data (*see* **Note 6**).

1.5. Source and Preparation of RNA and Protein

This protocol makes the assumption that the relevant radiolabeled RNA can be prepared and that this RNA is not excessively heterogeneous in size and/or structure. In particular, it is important that the RNA migrate as a discrete, unique species on the nondenaturing gel used to separate free RNA from RNA–protein complexes. Even though it may be possible to distinguish several different forms of RNA and RNA–protein complexes after electrophoresis, the computational analysis suggested here makes the assumption that all the RNA in the reaction mixture can be partitioned into two forms, bound and free. It is possible in some cases to use more sophisticated methods of data analysis to obtain binding constants or rate constants for multiple different RNA structures and/or RNA–protein complexes in a single reaction mixture, but those methods are beyond the scope of this simplified description (*see* **Note 5**).

Similarly, we make the assumption that a source of the RNA-binding protein in question is available to the investigator. As described previously, it is not absolutely necessary that the protein be purified to obtain equilibrium or rate constants, but more purified preparations are almost always preferable and will generate fewer complicating artifacts. Again, it is important that the RNA–protein complex detected after electrophoresis be discrete. If multiple bands are observed, one cannot simply ignore some bands while analyzing others; as noted previously, the computational analysis suggested here assumes that all of the RNA in the reaction mixture exists in one of two discrete states. As already noted, more sophisticated methods may be in order if multiple products of the binding reaction are observed.

2. Materials

1. Radiolabeled RNA to be analyzed as well as identical RNA in unlabeled form (*see* Chapter 1). Chemical concentration in molar units must be known with high accuracy. In most cases, this is best achieved by preparing a sufficiently large stock solution to permit spectrophotometric determination of the concentration. If the extinction coefficient of the specific RNA is unknown, as is generally the case, use an estimate of 1 OD_{260} = 40 µg/mL.
2. Preparation of protein to be analyzed. An estimate of the active protein concentration is useful but not absolutely necessary. Purified protein should be used whenever possible, although equilibrium binding constants and dissociation rate constants can be measured in most cases even if contaminating proteins are present.
3. Binding buffer: 20 mM Tris-HCl, pH 7.5, 50 mM KCl, 5 mM $MgCl_2$, 1 mM dithiothreitol, 10% glycerol, 100 µg/mL bovine serum albumin. Alternatively, a 2× stock of the same buffer can be used.
4. Standard electrophoresis power supply, box, plates, spacers, and comb for polyacrylamide gel electrophoresis.
5. Electrophoresis stock solutions and reagents (solutions are prepared as separate stocks): 40% acrylamide; 2% *bis*-acrylamide; 80% glycerol; 4× stock of Tris-glycine buffer (1× buffer is 25 mM Tris base, 0.2 M glycine); ammonium persulfate; N,N,N',N'-tetramethylethyltenediamine (TEMED).

6. If at all possible, either a phosphor imager or a direct beta particle detection system (such as those available from Ambis/Scanalytics, BetaScope, or Packard). Alternatively, but less desirably, X-ray film and either a liquid scintillation counter or a scanning densitometer.
7. Software for linear regression analysis and/or nonlinear curve-fitting (*see* **Note 7**).

3. Methods
3.1. Binding Reactions for Determination of Equilibrium Binding Constants

1. Make a reasonable guess for the values of the K_d and the active protein concentration (*see* **Note 8**), and use these values to calculate a series of RNA concentrations that will result in equally spaced points on a Scatchard plot; the end points should correspond to 20% and 80% of the protein bound. I recommend a total of 8–12 different data points. The protein concentration should be invariant in the various binding reactions and should be such that a maximal variation in the percentage of RNA bound will occur over the range of the titration. At protein concentrations equal to about two times the K_d, the percentage of RNA bound will vary from about 30% to about 65% when the percentage of protein bound varies from 20% to 80%. *See* **Note 7** for help with the calculations.
2. Using the values just calculated, prepare the set of binding reactions on ice. This can be achieved efficiently by using a constant, low concentration of radiolabeled RNA and supplementing with various amounts of the same RNA in unlabeled form. Both RNA and protein stocks can be diluted into 1× binding buffer and mixed to achieve the desired final concentrations; alternatively, the protein can be diluted into 2× binding buffer and the RNA prepared in dilute buffer without other salts (e.g., 10 m*M* Tris-HCl, pH 7.5). In this case, the protein should be added in 50% of the final reaction volume to the premixed RNA solution, also present in 50% of the final reaction volume. It will be necessary to optimize the binding buffer for the particular interaction you are studying (*see* **Note 1**). Monovalent salt and Mg^{2+} concentrations are particularly crucial. For a starting buffer, I recommend 20 m*M* Tris-HCl, pH 7.5, 50 m*M* KCl, 5 m*M* $MgCl_2$, 1 m*M* dithiothreitol, 100 µg/mL bovine serum albumin, and 10% glycerol. The final reaction volume should be 20 µL (*see* **Note 1**).
3. Incubate the reaction mixes at a physiological temperature for 30 min. Temperature and reaction time may need to be empirically determined. For most interactions, 30 min will be more than adequate to permit equilibrium to be reached, but this should be confirmed directly.
4. Load the reaction mixes directly onto the running nondenaturing gel (*see* **Subheading 3.3.**).

3.2. Binding Reactions for Determining Dissociation Rate Constants

1. As a preliminary experiment, set up a series of binding reactions similar to those described in **Subheading 3.1.**, but containing a single, low concentration of radio-

labeled RNA. The concentration of RNA should be chosen so that as little as 5% of the RNA can be readily detected and quantified in a protein–RNA complex after electrophoresis. Add a variable amount of protein to each tube. After incubation and gel-loading (**Subheading 3.1.**) and electrophoresis and quantification (**Subheadings 3.3.** and **3.4.**), analyze the results to determine the protein concentration at which 60–80% of the RNA is bound.

2. In a second preliminary experiment, use the amount of protein and RNA determined in **step 1** above in setting up another series of binding reactions. In this case, the concentrations of radiolabeled RNA and protein are held constant, but variable amounts of unlabeled RNA are added. Titrate the concentration of unlabeled RNA so that it is in large excess relative to the labeled RNA when at the highest concentrations. Run these reaction mixes on another nondenaturing gel and analyze to determine the concentration of unlabeled RNA that is required to reduce the percentage of labeled RNA found in the RNA–protein complex to less than 3–5% of the percentage bound in the absence of unlabeled competitor.

3. To determine the dissociation rate constant, set up two reaction mixtures as described in this and the following steps. In tube 1 (20 µL total), incubate radiolabeled RNA and protein in 1× binding buffer at the concentrations determined in **step 1** above in the presence of the concentration of unlabeled competitor determined in **step 2**. This will serve as a control to demonstrate that the concentration of competitor used is sufficient to effectively prevent binding of free or dissociated radiolabeled RNA to protein during the dissociation time course.

4. In tube 2, prepare a reaction mixture equal in volume to $10(n + 2)$ µL, where n is the number of different time points to be taken during the dissociation time course. I recommend $n = 8$–10. This mixture should contain concentrations of RNA and protein equal to those determined in **step 1** above, and should also contain 1× binding buffer.

5. In tube 3, prepare a solution of unlabeled competitor RNA at a concentration equal to 2× that determined in **step 2** (total volume about 10 µL greater than that in tube 2). This RNA solution should be prepared in 1× binding buffer, identical to that used in tubes 1 and 2.

6. Incubate all three tubes at physiological temperature for 30 min or until equilibrium is reached.

7. At this point, load all of the reaction mixture in tube 1 onto a running nondenaturing gel. Also load 10 µL from tube 2 in a separate well. This will serve as a control to determine the percentage of RNA bound at equilibrium, prior to addition of competitor.

8. Quickly mix a volume from tube 3 with the remaining mixture in tube 2. The volume used from tube 3 should be equal to the volume remaining in tube 2. Immediately load a 20-µL aliquot of this onto the running gel and note the time. This time is defined as $t = 0$ for the dissociation time course.

9. At subsequent intervals calculated based on the expected rate of dissociation, load 20-µL aliquots onto sequential lanes of the running gel. For a starting point, try time points at 3-min intervals.

10. Run the gel for the standard electrophoresis time (*see* **Subheading 3.3.**), measured from the time of final sample loading.

3.3. The Nondenaturing Polyacrylamide Gel

1. The gel can be prepared well in advance.
2. Prepare plates by cleaning thoroughly with water and ethanol. Use spacers of about 0.7 mm thickness and a comb with 12 teeth of dimensions 0.8 cm × 1.5 cm. The gel plates are 23 cm long and 18 cm wide (*see* **Note 2**).
3. Mix from appropriate stock solutions to obtain 40 mL solution containing 6% acrylamide, 0.12% *bis*-acrylamide, 25 mM Tris base, 0.2 M glycine, and 5% glycerol. Add 15 mg of ammonium persulfate and 30 µL of TEMED. Pour between the prepared glass plates and insert the comb to a depth of about 1–1.2 cm. Allow the gel to polymerize completely and to equilibrate at the final running temperature (*see* **Note 2**).
4. Remove the bottom spacer (if used) and mount the gel in the electrophoresis box. Fill both upper and lower chambers with running buffer (25 mM Tris base and 0.2 M glycine). Blow out the wells with a drawn Pasteur pipet and make sure all air bubbles are removed both from the wells and from any space at the bottom of the gel. Prerun the gel at 300 V for at least 30 min prior to loading samples (*see* **Note 2**).
5. Load the samples directly on the running gel. Continue electrophoresis for 3 h at 300 V (*see* **Note 2**).
6. Empty the electrophoresis chambers and remove the gel. For gels of the specific dimensions run at the specific voltage recommended here, all the radioactivity, including any unincorporated nucleoside triphosphates remaining from the original RNA-labeling reaction, will remain in the gel and the lower reservoir buffer should not be radioactive. If any of the experimental parameters are changed, however, one should directly monitor the lower reservoir buffer to determine if it is radioactive (a good practice in any case). Remove one of the gel plates and, depending upon the subsequent method of analysis, mount the gel on Whatman filter paper and dry it on a standard gel dryer, fix (12% methanol, 10% acetic acid) the gel and then mount and dry it, or wrap the wet gel in plastic wrap.

3.4. Quantification of Results

1. It is now necessary to determine the quantities of radioactive material in bands corresponding to free and bound RNA. The method for doing so will depend upon the equipment available. At least four options are commonly available.
    a. Autoradiography and densitometry. The wet or dried gel can be exposed to X-ray film, preferably without an intensifying screen, and the resulting autoradiograph can be analyzed with a scanning densitometer. The primary difficulty with this approach is the inherently small linear response range of X-ray film, which can lead to substantial errors in quantification. If this approach is used, you should construct a standard curve to convert densitometric units to units

of radioactivity in order to minimize problems with nonlinear responses of the film. In general, I do not recommend this method.
   b. Autoradiography, followed by cutting and counting. The wet gel can be directly exposed to X-ray film and can subsequently be used to identify the locations of bands on the gel. These bands can then be excised and counted directly in a liquid scintillation counter (measuring Cerenkov radiation to avoid the use of scintillation fluid). Although this approach avoids problems of nonlinearity associated with densitometry of exposed X-ray film, it is tedious and subject to errors in excision of bands.
   c. Phosphorimaging. The dried gel can be exposed to a phosphor screen that can be subsequently processed and analyzed using commercial phosphorimaging equipment and analysis software. This method is much more precise and is preferable to either alternative described previously.
   d. Direct detection of beta emissions. A number of instruments for direct detection of beta emissions from gels and/or filters are commercially available (e.g., Ambis/Scanalytics, BetaScope, and Packard). These instruments generally lack the high resolution available with phosphorimagers, but are more than adequate for analysis of gel retardation data. Generally, only short scans (<1 h) are required to detect and quantify the quantities of radiolabeled RNA used in the kinds of experiments described here. This is my preferred method of analysis.

2. Whatever method of quantification is used, the end result should be values representing the quantity of radioactivity in free and bound RNA. The units may be arbitrary ones, so long as they are consistent within an experiment. It is possible that some of the RNA–protein complexes will dissociate during electrophoresis, resulting in a smear of radioactivity running just ahead of the low mobility RNA–protein complex (*see* **Fig. 1**). This material should be included in the quantification of the amount of bound RNA, as the goal is to establish the distribution of RNA in bound and free forms at the time of gel loading, not at later points during electrophoresis.

3.5. Data Analysis: Equilibrium Binding Constants

1. Calculate the concentration (in molar or nanomolar units) of RNA in bound and free form in each binding reaction mixture at equilibrium. This is done by determining the fraction of RNA bound or free, and multiplying this fraction by the total input RNA concentration in that reaction mixture. Perform this calculation for each reaction mixture.
2. To produce a Scatchard plot, graph B/F vs B, where B is the concentration of bound RNA and F is the concentration of free RNA. The K_a is the negative slope of the best-fit line calculated by regression analysis. The standard error of the K_a is the standard error of the regression coefficient, as defined in standard least-squares linear regression. The concentration of active protein present in the binding reactions is given by the x-intercept, assuming a binding stoichiometry of 1 (*see* **Notes 4** and **7**).

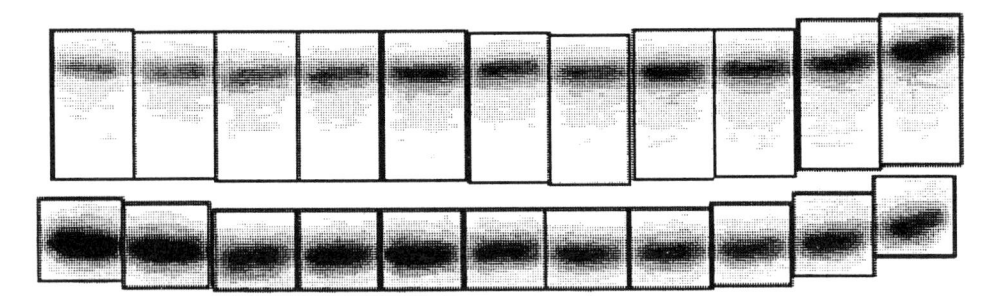

Fig. 1. Gel retardation analysis of a set of equilibrium binding reactions containing *Xenopus* transcription factor IIIA (TFIIIA) and variable concentrations of 5S rRNA. In this particular example, TFIIIA was present at a concentration of 2.6 nM, uniformly ^{32}P-labeled synthetic 5S rRNA was present at a concentration of 0.027 nM, and unlabeled 5S rRNA was present at concentrations ranging from 0.9 nM to 9.6 nM. After electrophoresis, the gel was dried and scanned on an Ambis Radioanalytic Scanner. Bound and free 5S rRNA were quantified in each case using the rectangles shown to define the resolved complexes and free 5S rRNA. Analysis of the results indicated that the percentage of RNA bound varied from 20% to 59%, and that the percentage of protein bound varied from 21% to 75%. The K_d calculated from the slope of a Scatchard plot was 1.5 nM with a standard error of 15% for this single determination.

3. Results from an initial determination (K_d and active protein concentration) can be used to calculate a new set of conditions for another binding titration as described in **Subheading 3.1., step 1**. By this iterative method, relatively high precision in the determination of binding constants can be achieved *(6)*.

3.6. Data Analysis: Dissociation Rate Constants

1. Determine the fraction of radiolabeled RNA in bound form at each time point. The fraction bound at $t = 0$ may be lower than the fraction bound at equilibrium because of the time required to load the sample on the gel and for the RNA–protein complex to enter the gel and become resistant to dissociation.
2. Plot the natural logarithm of the fraction bound vs time. The slope of the best-fit line is equal to the dissociation rate constant. As described in **Subheading 3.5., step 2**, the standard error of the rate constant is the standard error of the regression coefficient when the best-fit line is calculated by least-squares linear regression (*see* **Notes 4** and **7**).

3.7. Combining Data from Multiple Determinations

1. As an arbitrary standard, I recommend discarding results in which the standard error of the regression coefficient is >30% of the regression coefficient itself. This is rarely a problem in the determination of dissociation rate constants, but is

Gel Retardation Assays

often the case in determinations of equilibrium binding constants. As a second arbitrary standard, I recommend continuing to collect data until at least three independent determinations with <30% error are made in the case of equilibrium binding constants, and at least two independent determinations for rate constants.
2. A common practice for arriving at a final value for equilibrium binding or dissociation rate constants is to simply calculate the arithmetic average of multiple determinations. Because this weights all determinations equally, even though some may have greater inherent error than others, the simple arithmetic average may not be as precise an estimate of the relevant constant as can be obtained. An alternative approach is to calculate a weighted average based upon the relative standard errors of the regression coefficients obtained in each case. Perhaps a more rigorous approach is to make use of a well-established statistical method called analysis of covariance. Although rarely used in molecular biology and biochemistry, this computational method is designed to obtain a single best-estimate of a common slope from multiple regression plots, along with associated measurements of goodness-of-fit. This is well-suited for arriving at a final, single estimate of equilibrium binding constants or dissociation rate constants measured as described above. The details of analysis of covariance can be found in more advanced textbooks of statistics or regression analysis (*see also* **Note 7**).

4. Notes

1. The conditions given for the binding reaction are intended only as a starting point and should be optimized for a particular RNA–protein interaction. Monovalent salt and Mg^{2+} concentrations may be most important. Low concentrations of Zn^{2+} (10 μM–100 μM) may also be advisable if there is any chance the protein being analyzed contains zinc fingers or other zinc-stabilized domains. Some glycerol is required in the binding buffer so that the sample will be of higher density than the gel running buffer and will thus settle to the bottom of the electrophoresis well. Bovine serum albumin (BSA) is recommended as a stabilizing agent and to minimize losses of dilute proteins as a result of adsorption to surfaces; it may be dispensable, however. In some cases, it may be advisable to include some nonspecific RNA or DNA carrier as well to minimize nonspecific losses of RNA when present at very low concentrations. We have often used 10 μg/mL poly (dI-dC) for this purpose. It is important to confirm, however, that this "nonspecific" nucleic acid does not compete for specific binding sites on the protein at the concentrations used. I recommend carrying out binding reactions at physiological temperatures, but it may be necessary to use lower temperatures to detect weak interactions or to minimize problems with contaminating nucleases or proteases.
2. It may be necessary to manipulate gel and electrophoresis conditions to detect RNA–protein complexes or to maximize the resolution of complexes from free RNA. Acrylamide and *bis*-acrylamide concentrations can be altered to improve performance. The gel length recommended is somewhat longer than is typical and shorter gels will suffice in many cases. The voltage gradient described here (13 volts/cm) can be increased substantially without introducing problems with heat dissipation or, alternatively, can be reduced. Running the gel at 4° may re-

sult in "tighter" bands, presumably because of a reduced rate of diffusion, and may also aid in detection of low-affinity or unstable complexes. If gels are run at low temperatures, however, it is important that the gel and running buffer be equilibrated at the lower temperature prior to sample loading. The Tris-glycine buffer recommended has worked well for us, but Tris-borate and Tris-borate-EDTA buffers are often used as well. In general, EDTA-containing buffers seem ill-advised on theoretical grounds because of the importance of divalent cations (Mg^{2+} and Zn^{2+}, in particular) in RNA–protein interactions. Nonetheless, many reports of successful gel retardation analysis of nucleic acid–protein complexes using EDTA-containing electrophoresis buffers have appeared in the literature. The sample wells described will hold up to about 50 μL, but samples of lower volume are preferred to minimize contamination between wells and to decrease the time required for samples to enter the gel.
3. The methods described here are designed to measure affinity and stability, not specificity. Specificity can readily be assessed, however, by analyzing equilibrium binding reactions containing variable concentrations of unlabeled "specific" and "nonspecific" competitors.
4. Data analysis as described here makes use of various transformations of the relevant equilibrium and kinetic equations to permit graphical interpretation of the data when presented as linear plots. An attractive alternative is to make use of nonlinear curve-fitting algorithms to fit the data to nonlinear equations, as described in **Subheadings 1.3.** and **1.4.** These nonlinear methods have some advantages when compared to the use of least-squares regression analysis to analyze the more traditional linear transformations of the same equations *(7)* and should be seriously considered. A disadvantage is that analysis of covariance methods cannot be used with nonlinear representations to combine data from multiple independent determinations.
5. Numerous problems can result in the appearance of multiple bands in a gel-retardation analysis. In most cases, these extra bands must be eliminated before the quantitative methods described here can be applied. As noted previously, one potential source of multiple bands is the occurrence of different conformational isomers or structural forms of the RNA. Denaturation/renaturation protocols with synthetic RNAs can sometimes reduce the structural complexity, as can renaturation under different conditions (again, Mg^{2+} concentrations can be particularly crucial). Choosing an RNA fragment of minimal size is also generally helpful. Contaminating RNA-binding proteins or proteolytic fragments of the protein being analyzed can also lead to the appearance of spurious bands; additional purification of the protein of interest and/or taking measures to minimize proteolysis is recommended. If the additional bands represent nonspecific binding by contaminating proteins, the judicious use of nonspecific competitors at an empirically determined concentration can be helpful. One must confirm, however, that the "nonspecific" competitor does not affect the availability of the protein of interest for specific binding. The analysis described here also assumes equimolar stoichiometry of binding. Higher binding stoichiometries can also give rise to additional low-mobility bands. If the molecular identities of these bands can be

established, it may sometimes be possible to quantify all the bands present and deconvolute the data in a more sophisticated analysis to arrive at estimates of binding and/or dissociation constants. A more pragmatic approach is to empirically establish conditions under which unique complexes are formed.

6. It is sometimes more useful to characterize a binding equilibrium by defining the ΔG for the binding reaction rather than the equilibrium binding constant. The conversion from K_a to ΔG is straightforward:

$$\Delta G = -RT[\ln(K_a)]$$

where ΔG is units of cal/mol, R is the universal gas constant equal to 1.99 (cal)(K)$^{-1}$(L)$^{-1}$, and K_a is units of M^{-1}. Similarly, it is often easier to think in terms of half-lives than rate constants. The half-life ($t_{1/2}$) of an RNA–protein complex is related to the dissociation rate constant (k_{off}) by the equation

$$t_{1/2} = (\ln 2)/k_{off}$$

7. A variety of Microsoft Excel® 5.0 spreadsheets have been prepared to assist in the planning and analysis of experiments of the sort described here. These spreadsheets are available for downloading at no cost from the World Wide Web at http://www.cwru.edu/med/microbio/setzlab.htm. They are particularly helpful in designing binding titrations with constant protein and variable RNA so as to generate data in the optimal range for K_d determination by Scatchard analysis, for analysis of the data generated in such an experiment using Scatchard plots and linear regression analysis, and for use of analysis of covariance to combine data from multiple independent determinations of the K_d or k_{off}.

8. The approach outlined in **Subheading 3.1.** requires one to make an estimate of the K_d and active protein concentration in the design of the initial titration experiment used to make an experimental determination of the K_d. An estimate of the active protein concentration can be made based on the percent purity of the protein and the total protein concentration, determined using standard methods (Bradford assays *[8]*, e.g., using reagents marketed by Bio-Rad), assuming all of the RNA-binding protein in question is active. Of course, this value can be reduced if there is good reason to believe a fraction of the total RNA-binding protein is inactive. The K_d can then be roughly approximated in an initial experiment in which a fixed, low concentration of labeled RNA (e.g., <1 n*M*) is titrated with increasing concentrations of protein and the results analyzed by gel retardation. As noted in **Subheading 1.4.**, the K_d for the RNA–protein interaction will be approximately equal to the concentration of protein at which 50% of the RNA is bound. Alternatively, the same initial experiment can be used to estimate the active protein concentration if a reasonable guess of the K_d can be made—the active protein concentration is equal to the K_d when 50% of the RNA is bound, so long as the RNA concentration is low relative to the K_d. Although the equilibrium binding constants for RNA–protein interactions can vary over several orders of magnitude, in the absence of any other information a reasonable guess for the K_d of a sequence-specific interaction might be about 5 n*M*.

References

1. Scatchard, G. (1949) The attractions of proteins for small molecules and ions. *Ann. NY Acad. Sci.* **51,** 660–672.
2. Weder, H. G., Schildknecht, J., Lutz, R. A., and Kesselring, P. (1974) Determination of binding parameters from Scatchard plots. Theoretical and practical considerations. *Eur. J. Biochem.* **42,** 475–481.
3. Zierler, K. (1989) Misuse of nonlinear Scatchard plots. *TIBS* **14,** 314–317.
4. Norby, J. G., Ottolenghi, P., and Jense, J. (1980) Scatchard plot: common misinterpretation of binding experiments. Anal. Biochem., 102, 318–320.
5. Deranleau, D. A. (1969) Theory of the measurement of weak molecular complexes. I. General considerations. *J. Am. Chem. Soc.* **91,** 4044–4049.
6. Setzer, D. R., Menezes, S. R., Del Rio, S., Hung, V. S., and Subramanyan, G. (1996) Functional interactions between the zinc fingers of *Xenopus* transcription factor IIIA during 5S rRNA binding. *RNA* **2,** 1254–1269.
7. Johnson, M. L. (1992) Why, when, and how biochemists should use least squares. *Anal. Biochem.* **206,** 215–225.
8. Bradford, M. M. (1976) A rapid and sensitive method for the quantitation of microgram quantities of protein utilizing the principle of protein-dye binding. *Anal. Biochem.* **72,** 248–254.

11

PACE Analysis of RNA–Peptide Interactions

Christopher D. Cilley and James R. Williamson

1. Introduction

The PACE assay is a relatively new addition to the arsenal of techniques used to examine quantitatively the interactions of proteins and peptides with DNA and RNA *(1)*. Polyacrylamide coelectrophoresis (PACE) involves electrophoresis of a labeled nucleic acid through a gel medium that contains the target peptide or protein ligand. In this way, the nucleic acid is maintained in a constant concentration of the ligand throughout the electrophoresis, and the conditions for binding equilibrium are maintained throughout the experiment. This avoids the requirement for formation of complexes that are kinetically stable under the nonequilibrium conditions typical of a gel mobility shift experiment. A particularly powerful aspect of the PACE experiment is the ability to probe interactions that are too weak to be observed in other binding assays such as gel shift or filter binding (*see* Chapters 9 and 10).

The PACE assay is straightforward. Polyacrylamide plugs containing different peptide concentrations are sequentially poured into a gel perpendicular to the orientation during electrophoresis. This results in a discrete step gradient of peptide concentration in the gel from left to right that can span a wide range of peptide concentrations from picomolar to micromolar. The PACE gel can be cast such that there are multiple wells for each gel segment at a given peptide concentration, providing for the simultaneous measurement of the binding interaction of a wild-type RNA sequence and several mutant RNAs. Radiolabeled RNA samples are subjected to electrophoresis through the gel, and their mobilities are analyzed as a function of peptide concentration. In general, for complexes that rapidly exchange between free and bound forms (i.e., weak interactions) the mobility of the RNA is directly proportional to the fraction of the time spent bound to the peptide during the electrophoresis. As a result, the mobility of a given RNA will depend on the peptide concentration.

From: *Methods in Molecular Biology, Vol. 118: RNA-Protein Interaction Protocols*
Edited by: S. Haynes © Humana Press Inc., Totowa, NJ

In the standard gel mobility shift assay, unstable complexes can dissociate just prior to or shortly after entering the polyacrylamide gel *(2,3)*. As a consequence, there may be no clear difference in the mobilities of the bound and unbound RNA species if the binding is sufficiently weak. In the PACE experiment, the presence of the peptide ligand in the gel at a uniform concentration ensures that the complex remains at equilibrium during electrophoresis, and the degree to which complex formation occurs at any given peptide concentration is reflected in the decreasing mobility of the nucleic acid. The PACE assay is very robust, and a variety of factors that affect the binding interaction can be surveyed, including salt, temperature, pH, and detergent.

The PACE assay is related to affinity electrophoresis *(4)*, which involves covalent linkage of one member of a ligand pair to a stationary matrix. The rate of electrophoresis of the complementary mobile component through the affinity matrix is retarded by its interaction with the immobilized component. In an extension of this methodology (affinity *co*electrophoresis *[5]*, or ACE), labeled nucleic acid substrates are electrophoresed through an agarose gel containing different concentrations of protein in each gel slab. To increase resolution, we developed a version of the ACE assay using polyacrylamide in place of agarose. The smaller pore sizes obtainable with polyacrylamide extend this methodology to include small peptide–ligand complexes.

During PACE electrophoresis, the RNA is assumed to exist in two distinct states: an unbound state that has a mobility equal to that of the RNA in the absence of peptide, and a bound state that has a mobility that is different from the free state. The presence of multiple conformations of the RNA–peptide complex (of differing mobilities), or multiple complexes of different stoichiometries will complicate analysis of the PACE experiment, as well as other binding experiments. It is also assumed that the free and bound states are in rapid equilibrium with an exchange rate that is fast relative to the electrophoresis time. A model exploring the effects of binding kinetics on affinity electrophoresis suggests that slow kinetics result in a spreading of the bands at peptide concentrations close to the apparent dissociation constant, $K_{d,app}$ *(6)*. It is important to note that the acrylamide gel matrix and the mechanics of pouring a PACE gel may have some effect on the binding interaction. However, the utility of PACE is its ability to test the relative affinities of multiple RNA species under these same conditions.

2. Materials

2.1. Equipment

1. Standard equipment for polyacrylamide gel electrophoresis including power supply, gel box, clamps, and heat sink (typically a 1/8 in. thick aluminum plate).
2. Wide, short gels are preferable for the PACE assay because a wider gel allows for more steps in the concentration gradient, and a shorter gel helps reduce the

Polyacrylamide Coelectrophoresis 131

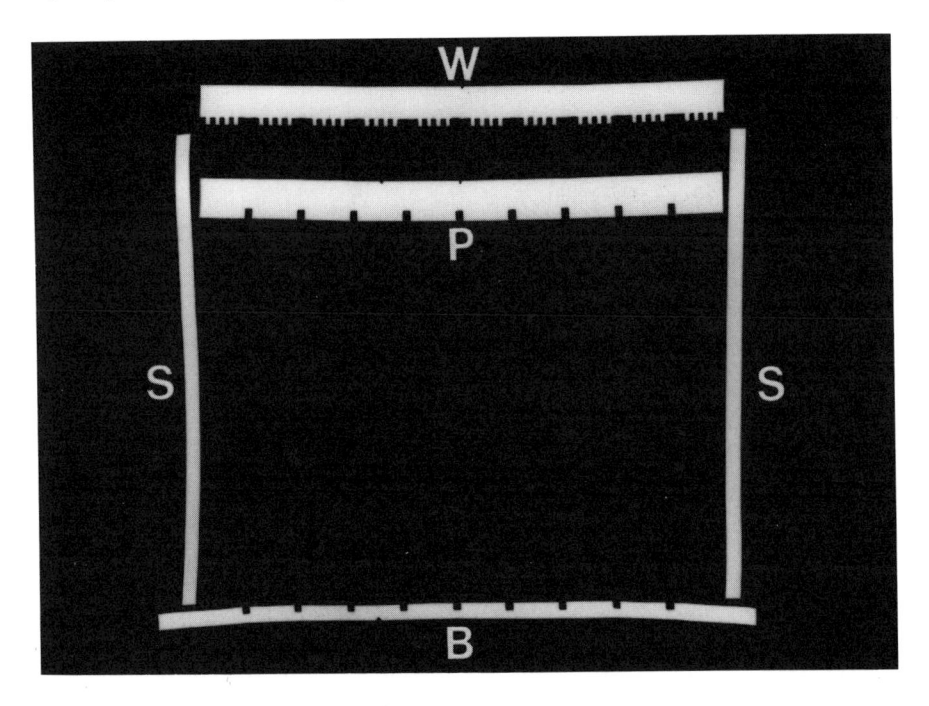

Fig. 1. Combs and spacers for the PACE gel are cut from 1/32 in. thick Teflon sheets. The side spacers (S) are 29.5 × 1 cm. The bottom spacer (B) is 38.5 × 1 cm with 5 × 5 mm notches spaced 3 cm apart and 5.5 cm of space from the ends. The pouring comb (P) is 34.5 × 2.5 cm with 5 mm wide and 6 mm deep notches cut with 3 cm spacing between each other and the ends (*see* **Note 12**). The well comb (W) is cut from a 34.5 × 2.5 cm rectangle of Teflon. The teeth are 3 mm wide and 5 mm long. They are spaced 3 mm apart with 14 mm between each grouping of four teeth. There is 4.5 mm of space from the first and last teeth to the ends of the comb. *See* **Note 13** for details on obtaining a guide for cutting the combs and spacers. Reprinted from **ref.** *1* with permission from Cambridge University Press.

amount of peptide needed for each polyacrylamide plug. Our gel plates were 24 × 36 cm and 26 × 36 cm ($h \times w$). The actual size of the PACE gel is not critical, and the procedure can be readily implemented for gels of other sizes, depending upon the particular application.
3. Combs and spacers custom cut for PACE gels. The PACE assay as described here utilizes combs and spacers for a 10-step gradient of peptide concentration across a single gel. The combs and spacers are cut as illustrated and described in **Fig. 1**.
4. 10-mL Luer syringes, 25-gage needles, gloves, and parafilm.
5. Light-emitting labels.
6. Chromatography paper, plastic wrap, and a gel dryer.

7. Autoradiography cassettes, X-ray film, and developer for the autoradiograph.
8. Light box and ruler for quantifying the PACE data.

2.2 Reagents

1. A stock of 5× TBE for use as the gel buffer and the electrophoresis running buffer: 450 mM Tris base, 450 mM boric acid, and 10 mM EDTA (*see* **Note 1**). To a final volume of 1 L add 54 g of Tris base, 27.5 g of boric acid, 20 mL of 0.5 M EDTA, pH 8.0, and double-distilled water.
2. 40% Acrylamide stock solution (29:1 acrylamide:N',N'-methylene *bis*-acrylamide). See **Note 2** for comments on working with acrylamide.
3. 10% Ammonium persulfate (APS) (w/v) in water, 99% N',N',N',N'-tetramethyl-ethylene diamine (TEMED).
4. ^{32}P-radiolabeled substrate RNAs. The PACE binding experiment can be performed with either 3'-, 5'-, or internally labeled RNA. Standard procedures can be used for preparation of RNA samples (*7*; *see* Chapter 1). A high specific activity (>200 dpm/fmol) is desired as only trace concentrations of RNA are loaded onto the PACE gel (*see* **Notes 3** and **4**).
5. Peptide and proteins are prepared and purified. Because the PACE assay is very sensitive, high purity and an accurate measurement of peptide concentration are essential *(8)*. Reverse-phase high-performance liquid chromatography (RP-HPLC) or a similar high-resolution purification step is recommended as part of the purification. Peptide or protein stock solutions should be prepared by serial dilutions into double-distilled H$_2$O or appropriate storage buffer. These stocks will be diluted 100-fold into the gel mix to give the final concentration for each lane in the PACE gel (*see* **Notes 5** and **6**).
6. RNA samples diluted in 6× Type III loading dye: 0.25% bromophenol blue, 0.25% xylene cyanol FF, and 30% glycerol in water *(9)*.
7. Coomassie protein staining: 0.25% Coomassie brilliant blue, 45% methanol, and 10% glacial acetic acid in water *(10)*. This is used for calibrating the run time of the PACE gel.

3. Methods

3.1. Assay

1. Carefully clean each gel plate with nonabrasive soap and water, followed by ethanol. Ideally, siliconize both plates to ensure even spreading of gel mix when poured. Assemble the gel sandwich with the pouring comb and spacers arranged (albeit touching) as in **Fig. 1**. Turn the final gel sandwich 90° and slide the "upper" side spacer out to create a pouring gap as illustrated in **Fig. 2**. Draw lines between the "tops" of the notches as shown to help determine when sufficient gel solution has been added and label each lane to keep track of the appropriate peptide stock to use.
2. Prepare 200 mL of gel mix: 0.5× TBE, 15% (29:1) acrylamide, and 0.02% ammonium persulfate (20 mL of 5× TBE stock, 75 mL of 40% acrylamide stock, 400 µL of 10% APS, 104.6 mL of ddH$_2$O). Any additional salts (Na$^+$, K$^+$, Mg^{2+}, etc.), detergents, or other buffer components for the particular application should

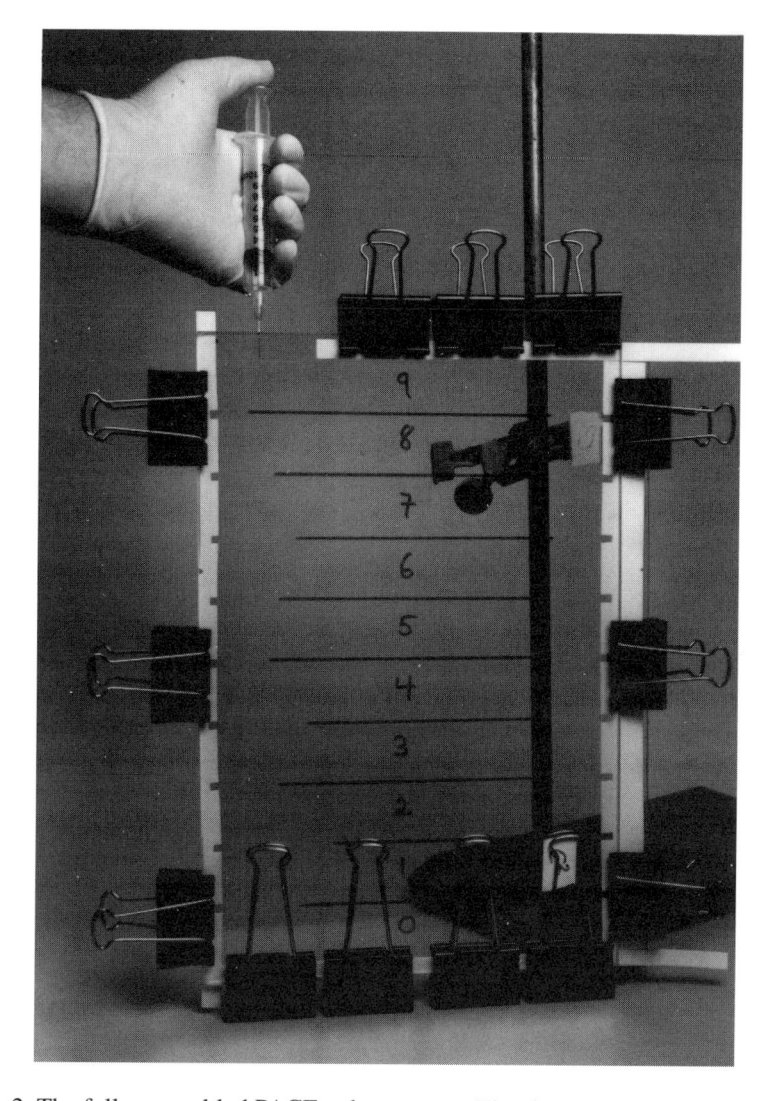

Fig. 2. The fully assembled PACE gel apparatus. The ring stand with a clamp in the background is used to support the gel plate sandwich, which is tilted back slightly to lean against the tube clamp. There are typically five clamps along the long sides of the gel sandwich (where the bottom spacer and pouring comb are). Only three clamps are shown here to better show the sandwich assembly.

be added to the appropriate final concentration (*see* **Notes 7** and **8**). To facilitate rapid and complete polymerization, the gel mix is degassed under aspirator vacuum until vigorous bubbling ceases. If detergent is being used, it should be added **after** degassing. The final volume of the gel mix should be 200 mL.

3. Remove the plunger from a 10-mL syringe. Hold a piece of parafilm over the small hole at the bottom of the syringe barrel while holding it upright. Add 70.7 µL of 100× peptide stock solution (or buffer for the "0" peptide lane). Add 7 mL of gel mix. Add 14 µL of TEMED. Tightly cover the end of the syringe with your finger while gently putting the plunger back into the syringe barrel until it just seats. Invert the syringe so that the nozzle points up and the trapped air bubble rises to the top. Remove your finger and the parafilm, pointing the syringe away from you because a small amount of gel mix will likely squirt out. Push the plunger until it is securely in the barrel. Again cover the end of the syringe and invert the syringe several times to mix the solutions. Remove your finger and push the plunger to get the air out of the syringe until the gel mix is just short of coming out.
4. Attach a 25-gage needle. Insert the needle between the plates at the top of your gel sandwich (as in **Fig. 2**). Slowly squirt the gel mix in until the level of the gel just reaches the line drawn between notches.
5. Wait 10 min for the gel plug to polymerize (*see* **Note 9**). While waiting, remove the needle and dispose of any remaining gel solution into a proper receptacle. Rinse out the syringe and needle with distilled water in order to re-use them for the next lane and be sure to shake them to remove excess water.
6. After the gel has set, use a strip of Whatman paper just narrow enough to insert in the pouring gap and slide it between the plates to wick out the shallow layer of unpolymerized gel solution. Tilt the gel sandwich towards the filter paper to soak all of the unpolymerized gel mix into the filter paper.
7. Repeat from **step 4** with each successive peptide concentration until the second to the last lane is completed.
8. Remove the clips from the "top" of the gel and slide the side spacer until the pouring gap is very narrow. This will reduce the amount of gel mix exposed to air for the last plug. Replace the clamps.
9. Pour the last lane. It is helpful to tip the gel sandwich so that the gel solution runs down toward the pouring comb and the air pocket stays up near the needle as long as possible. Wait 10 min for the gel to polymerize.
10. Lay the gel sandwich flat and then slightly raise (about 2 in.) the end with the pouring comb. Remove the pouring comb and use Whatman paper to wick out any unpolymerized gel solution that remains (*see* **Fig. 3**).
11. Make up 7 mL of gel solution containing no peptide. Add 14 µL of TEMED, then fill the gel sandwich until the gel mix beads up over the edge of the top plate. Add about 1 mL of TEMED to a tissue and use that to wipe down the well comb. Insert the well comb and clamp into place and allow the gel to polymerize for 15 min.
12. Remove the bottom spacer from the gel sandwich. Remove the well comb. Put the gel sandwich into an electrophoresis box and be certain to attach a metal plate (or appropriate heat sink) to the outside.
13. Fill the buffer chambers with 0.5× TBE (containing any additional salts or detergents being used) and rinse out the wells and the gap at the bottom of the gel where the bottom spacer was. Do not prerun the gel. Load the RNA samples into

Polyacrylamide Coelectrophoresis

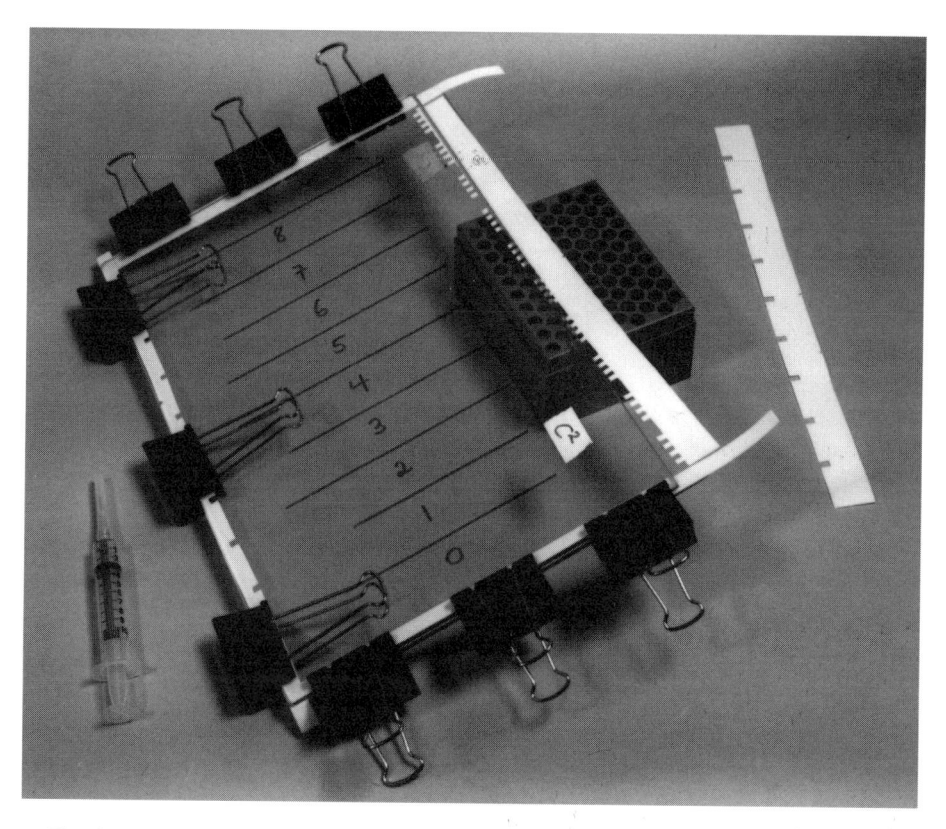

Fig. 3. Forming the wells of the PACE gel. A pouring and well comb combination is necessary because such small wells are difficult to form the way the PACE gel is poured. After the lanes are poured, the gel is laid almost flat (an empty tip rack is used here for propping up the top of the gel sandwich) before the well comb is used.

the appropriate wells. The wells are small so only about 2 µL of sample can be loaded in each well. *See* **Note 4** for comments on how many dpm of RNA to load for a good signal.

14. The gel should be run at a constant low power to eliminate or reduce heating. Our experience has been that 3 W for a 24 × 36 cm gel that is 0.8 mm thick (regardless of the salt concentration) is effective. Run the gel just long enough to keep the RNA and peptide migration fronts from passing (*see* **Note 10**).
15. Take down the gel sandwich. Remove one of the plates (*see* **Note 11**). Cover the gel with plastic wrap. Be sure to mark on the plastic wrap which lanes correspond to what peptide concentration. Peel the gel off the other plate and onto the plastic wrap, and lay a piece of Whatman paper onto the gel.
16. Dry the gel on a gel dryer for about 1 h. Attach the light emitting labels in a couple of spots for reference markers. Expose the gel to film for an appropriate amount of time (*see* **Note 4**).

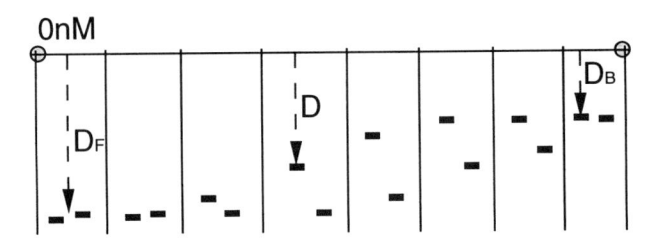

Fig. 4. Idealized PACE autoradiogram. Migration distances are measured directly from the film.

17. Develop the film and allow the autoradiogram to dry. Align the film to the gel using the light emitting labels and mark the interface between the peptide-containing region and the well-forming region. Measure the distance from that line to the center of each band, or to the lowest point of a band (sometimes necessary if the lanes are streaked or cross a boundary between lanes). It is important to use a consistent measuring scheme.

3.2. Analysis

Ideally, the PACE gel autoradiogram will exhibit a single band for each lane, and the mobility usually decreases with increasing peptide concentration. The total migration distance (D) for each RNA at each peptide concentration is measured directly from the autoradiogram (illustrated in **Fig. 4**). The migration distance of the free RNA (D_F) is also measured as a reference, to allow results from different gels to be compared. In the analysis of a PACE gel, a simple binding equilibrium is assumed:

$$R + P \underset{}{\overset{K_d}{\rightleftharpoons}} RP \quad K_d = [R][P]/[RP] \quad (1)$$

Where [R] and [P] represent the free, equilibrium RNA and peptide concentrations, respectively. The fraction of RNA bound, Θ, is given by:

$$\Theta = [R]_{Bound} / [R]_{Total} = [RP] / ([RP] + [R]) \quad (2)$$

Under the conditions used in the PACE experiment, $[P]_{Total} \gg [R]_{Total}$, so that $[P] = [P]_{Total}$. Combination of Eqs. 1 and 2 and rearrangement yields an equation for Θ:

$$\Theta = [P]_{Total} / (K_{d,app} + [P]_{Total}) \quad (3)$$

Where $K_{d,app}$ is the apparent dissociation constant for the complex under the particular PACE gel conditions.

When D, the migration distance, equals D_F, there is no significant interaction between the RNA and peptide. At sufficiently high peptide concentrations, the RNA is completely bound by peptide and is maximally retarded, and the migration distance of the fully formed complex is given by D_B. The migration distance D_B is a characteristic of the particular system and may be different for different proteins and peptides (compare **Figs. 5A** and **B**). The fraction of RNA bound at any given peptide concentration can be determined by:

$$\Theta = (D - D_F) / (D_B - D_F) \qquad (4)$$

Rearrangement of Eq. 4 and substitution from Eq. 3 yields:

$$D = ([P]_{Total} (D_B - D_F)) / (K_{d,app} + [P]_{Total}) + D_F \qquad (5)$$

Fitting the PACE derived distance data (D) as a function of $[P]_{Total}$ to Eq. 5 yields values for the $K_{d,app}$, D_B, and D_F. After fitting, the data can be normalized using the fit parameters D_B and D_F to allow plotting of Θ vs log $[P]$ for the direct comparison of different experiments.

Figure 5 illustrates actual PACE data and analysis. A wild-type and mutant peptide derived from the λ bacteriophage antitermination protein N are compared for binding to the wild-type RNA hairpin target of N protein and three RNA mutants (*see* **ref. 1** for details). In this experiment (**Figs. 5A** and **B**), each peptide concentration spans 6 orders of magnitude (50 pM to 20 µM). Notice that the concentration gradient is spread over two gels and that each gel contains a "0" peptide lane that allows for calibrating the migration distances between the two gels. **Figures 5C** and **D** show the results of plotting and fitting the data to Eq. 5. The PACE experiment readily differentiates binding affinities that differ by as little as a factor of 2 and as much as several orders of magnitude.

4. Notes

1. The pH of the 0.5× TBE gel buffer is approx 8.4. The use of other gel buffering systems with different pK_as is recommended to examine the effect of pH on the interaction.
2. All work with acrylamide solutions should be performed wearing latex gloves. Acrylamide is a potent, cumulative neurotoxin and there is ample opportunity for inadvertent contact.
3. It is important to keep the total RNA concentration low to ensure that it is always present in trace quantities with respect to the peptide concentration. A good guideline should be to use an RNA sample that is at least 10-fold lower than the expected $K_{d,app}$. Your first PACE experiment will roughly determine the $K_{d,app}$. With this rough $K_{d,app}$ you can design your subsequent PACE experiments to bracket the appropriate peptide concentrations to accurately determine the $K_{d,app}$. An important control to do early on is to show that the PACE results do not change if the RNA concentration is increased 5- to 10-fold.
4. To see a clear band on an autoradiogram exposed overnight, 2000 dpm of [^{32}P]RNA should to be applied for each lane. You should determine how many

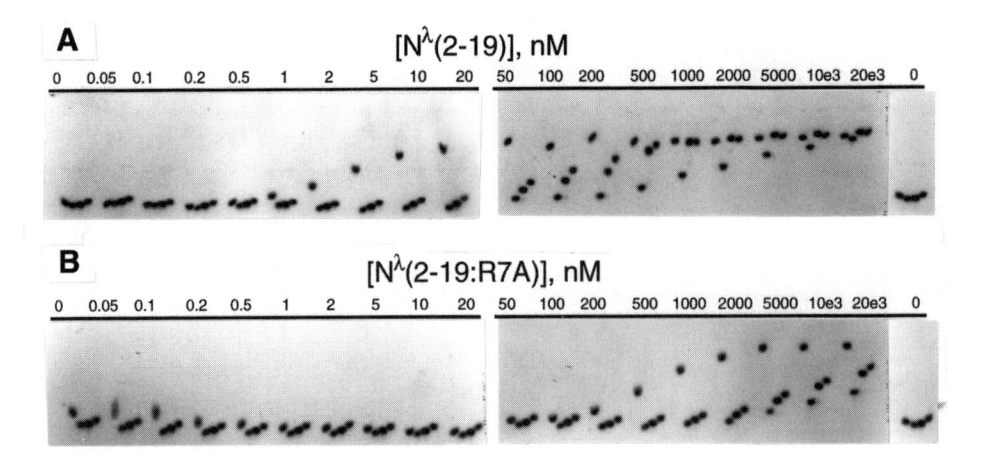

Fig. 5. PACE data and analysis. (**A**) PACE experiment using a peptide corresponding to the 2nd through 19th residues of the λ N protein. The RNAs loaded onto the gel are the wild-type *boxB* RNA hairpin in the first lane of each segment and three mutant RNAs. (**B**) The same experiment as in **A**, but using the R7A mutant peptide. The actual height of the gel slab pictured is 10 cm. *(continued opposite page)*

   moles of your lowest activity RNA gives 2000 total dpm and make the appropriate dilutions into water and Type III loading dye such that 2 µL equals 2000 dpm. If the labeled RNA has a much lower specific activity, fewer dpm can be used, and the exposure time should be proportionally increased. It is important to load the same number of dpm for each lane to ensure that no one lane overexposes the film and masks the other bands.
5. The amount of protein or peptide required to run a PACE experiment, as presented here, is fairly large compared to other methods. For the concentration range of 50 p*M* to 20 µ*M* in the experiments in **Fig. 5**, ≈275 nmol of peptide were used.
6. The oxidizing environment that exists during the polymerization of the PACE gel could adversely affect any free cysteines in the peptide of interest.
7. Ionic and nonionic detergents can influence the apparent affinity and specificity of peptide–RNA complexes. Detergent can be added directly to the gel mix after degassing. Ionic detergent should be added to both the gel mix and the running buffer since an ionic detergent will migrate during electrophoresis. Sometimes the addition of detergent can cause the gel to be so slippery that it will actually slide out from between the glass plates during the run. This problem is avoided if one or both of the gel plates is **not** siliconized.
8. The presence of mono- and divalent salts will affect the binding interaction. Salts will also dramatically alter the current passing through the gel during electrophoresis. This additional current can cause heating of the gel. To avoid this, use a lower wattage, or a better heat sink that can compensate for additional heating (i.e., a water bath surrounding the gel).

Polyacrylamide Coelectrophoresis

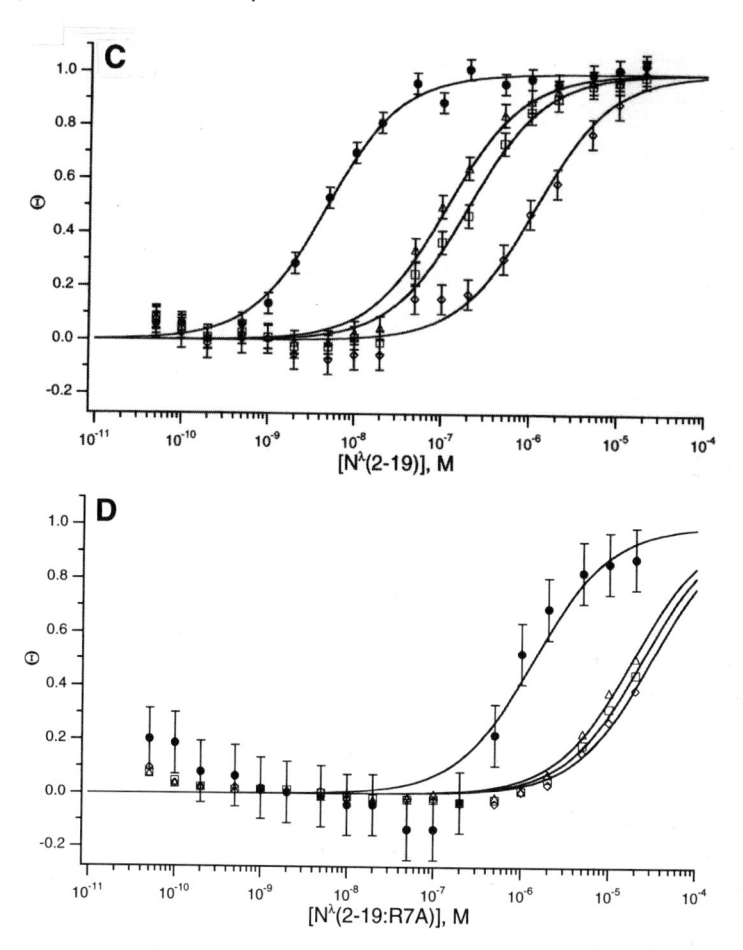

Fig. 5 (cont.). (**C**) and (**D**) are the normalized data derived by measuring migration distances on the gels shown in **A** and **B**. The solid line represents the curve fit using the derived values of K_d, D_F, and D_B. The closed circles (•) are for the wild-type RNA date and the open symbols are for the mutant RNAs. The error bars are estimates from the χ^2 of the least-squares fit to the data. Reprinted from **ref. 1** with permission from Cambridge University Press.

9. If you have problems with the gel mix not polymerizing near the spacers or comb, which results in channels, 20 µL of TEMED can be pipetted along the outside edge of the spacer or comb just at the level where the next plug is going to be poured. The TEMED will wick along the spacer and help to completely polymerize the gel mix.
10. Before the first PACE assay is run, it is important to calibrate the migration rates for both the peptide and the RNA. The RNA will migrate downward through the gel, while RNA-binding proteins and peptides are generally basic and will mi-

grate upward through the gel. It is important to ensure that the migration fronts do not pass one another during the PACE experiment, or the shape of the binding curve may be affected. Pour a PACE gel as described, but add peptide to only one lane at a final concentration of about 1 μM, which is sufficient to visualize by Coomassie blue staining. Load a single sample of labeled RNAs in a lane containing no peptide. Run the PACE gel at 3 W for about 3 h (for a 24 × 36 × 0.1 cm gel). Cut the gel in half vertically. Coomassie stain the half with the peptide lane, and dry down and expose to X-ray film the half with the labeled RNAs. Basic peptides will migrate through the gel bottom-to-top and the RNAs will migrate top-to-bottom. The bottom of the peptide lane will not be stained owing to the upward migration of the peptide. Measure the distance migrated by the peptide and RNAs during the 3-h run and determine their migration rates in mm/min. The PACE gel should be run just long enough for the peptide and RNA fronts to be about 2 cm apart, given by:

$$\text{Time, min} = \frac{\text{Gel length, mm - 20 mm}}{\text{Peptide migration rate, mm / min + RNA migration rate, mm / min}}$$

It is important to recalibrate the gel running time whenever you adjust the assay conditions by changing the salt, detergent, pH, or change the peptide or RNA sequence. This is also the time to check, with either a strip thermometer or by touch, if the electrophoresis conditions result in any heating. If the gel is warm at the end of the 3 h, reduce the wattage and recalibrate.

11. If you have trouble with the gel sticking to both plates after the run, the gel sandwich can be placed on a warmed gel dryer for a few minutes to heat up one plate, which will cause the gel to release more readily.
12. The notches that are cut in the bottom spacer and the pouring comb help to prevent channeling that results from incomplete polymerization along the edge of the Teflon.
13. The authors have set up a web page to foster use of the PACE technique. We have made available a FAQ (Frequently Asked Questions), postscript templates for cutting combs and spacers for download, and a facility for getting technical help. It is accessible at http://williamson.scripps.edu/PACE.

References

1. Cilley, C. and Williamson, J. R. (1997) Analysis of bacteriophage N protein and peptide binding to *boxB* RNA using polyacrylamide gel coelectrophoresis (PACE). *RNA* **3**, 57–67.
2. Cann, J. R. (1989) Phenomenological theory of gel electrophoresis of protein–nucleic acid complexes. *J. Biol. Chem.* **264**, 17,032–17,040.
3. Carey, J. (1991) Gel retardation. *Methods Enzymol.* **208**, 103–116.
4. Horejsí, V. (1981) Affinity electrophoresis. *Anal. Biochem.* **112**, 1–8.
5. Lim, W. A., Sauer, R. T., and Lander, A. D. (1991) Analysis of DNA-protein interactions by affinity coelectrophoresis. *Methods Enzymol.* **208**, 196–210.

6. Matousek, V. and Horejsí, V. (1982) Affinity electrophoresis: a theoretical study of the effect of the kinetics of protein-ligand complex formation and dissociation reactions. *J. Chromatogr.* **245**, 271–291.
7. Martin, G. And Keller, W. (1998) Tailing and 3'-end labeling of RNA with yeast poly(A) polymerase and various nucleotides. *RNA* **4**, 226–230.
8. Gill, S. C. and von Hippel, P. H. (1989) Calculation of protein extinction coefficients from amino acid sequence data. *Anal. Biochem.* **182**, 319–326.
9. Sambrook, J., Fritsch, E. F., and Maniatis, T. (1989), in *Molecular Cloning: A Laboratory Manual*, Cold Spring Harbor Press, Cold Spring Harbor, NY, pp. 6.12.
10. Sambrook, J., Fritsch, E. F., and Maniatis, T. (1989) Staining SDS-polyacrylamide gels with Coomassie brilliant blue, in *Molecular Cloning: A Laboratory Manual*, Cold Spring Harbor Press, Cold Spring Harbor, NY, pp. 18.55.

12

Detection of Nucleic Acid Interactions Using Surface Plasmon Resonance

Robert J. Crouch, Makoto Wakasa, and Mitsuru Haruki

1. Introduction

Several different methods can be used for detection of interactions between RNA and other molecules (e.g., proteins and other nucleic acids). Among these are: (1) enzymatic assays in which there is a measurable change in the property of the substrate such as an increase or decrease in size (mobility in gels or solubility in trichloroacetic acid), (2) gel-shift assays in which the interaction can be detected by altered migration of a complex in gel-electrophoretic analysis, (3) Northwestern assays and Northern analysis in which either proteins or RNA are transferred to a membrane after gel electrophoresis and RNAs are added in solution to interact with the protein or anneal with the RNA on the membrane and others. Each of these techniques has advantages and limitations.

Many interesting biochemical interactions do not necessarily result in an alteration of either component but are simply the result of an interaction that may be transitory in nature. Several methods, ranging from immunoprecipitation to analytical ultracentrifugation, are available for assessing association between molecules based solely on an interaction. The recent development of biosensors employing surface plasmon resonance (SPR) measurements has provided a useful method for obtaining data about real-time interactions. Basically, one interacting component is immobilized to the gold surface of a chip on which a thin film of gold has been deposited on glass. The biosensor measures the change in refractive index that occurs when there is an interaction between the immobilized molecules and those being passed over the surface of the chip. SPR offers several useful features and can be performed with relatively small amounts of materials. Interactions are observed in real time permitting the determination of both rates of association and dissociation. The

From: *Methods in Molecular Biology, Vol. 118: RNA-Protein Interaction Protocols*
Edited by: S. Haynes © Humana Press Inc., Totowa, NJ

methods and results presented here are those related to the BIACORE® instrument manufactured by BIACORE AB, Uppsala, Sweden (http://www.biacore.com). Other companies have surface plasmon detection instruments with varying degrees of sophistication and cost. Their Internet sites are Affinity Sensors (http://www.affinity-sensors.com), Texas Instruments (http://www.ti.com/research/docs/spr/spr.html), and Quantech (http://www.biosensor.com).

1. What Information Can Be Obtained Using SPR?

The questions asked using the SPR fall into two classes. First: Does substance A interact with substance B? Second: What are the association and dissociation rates for an interaction?

$$A + B \rightleftharpoons AB$$

For the question of A interacting with B, there is the simple question of does A interact with B? Once it is known that A interacts with B, one can determine if there is any B in the solution being tested and what its concentration is. Also, given an interaction between two large molecules, the region required for interaction can be determined (epitope mapping [1,2]) by using different portions of A and B. One need not be limited to A and B, but C and D may also be examined.

$$A + B + C + D \rightleftharpoons ABCD$$

A might interact with B and also C. One can ask if binding of B to A inhibits binding of C to A or, in a more complex assembly process, maybe D interacts with B bound to A but with neither A nor B alone. Thus, SPR can be useful in determining the order of addition of proteins and nucleic acids to build a rather large complex. Finally, one can attempt to reveal the identity of substance B. Using SPR interactions as an assay, column fractions produced from classic biochemical procedures can be followed for the presence of something interacting with the material immobilized on the chip. Eventually, the unknown substance can be obtained in quantities sufficient for determining its identity.

Although SPR is not unique in obtaining kinetic data, the ease of use and high throughput are useful properties for obtaining such information. In theory, kinetic data can be obtained when only one of the two materials is pure, namely that bound to the surface. It should be possible to determine the concentration of an interacting substance in an impure sample and then measure the association rates based on that information. In practice, hardly anyone uses an impure solution for collecting kinetic data.

Table 1 lists some of the limits of the BIACORE instrument (from the *BIACORE Handbook*). Reaction rates outside those listed in **Table 1** can be determined under special circumstances dependent upon the molecules being examined. It is also possible to use molecules with molecular masses less than

**Table 1
Range of On and Off Rates and Other System Parameters for the BIACORE Instrument**

Equilibrium constant, KD	10^{-4}–$10^{-12} M^a$
Association constant, k_a	10^3–$10^7 /M \cdot s^a$
Dissociation constant, k_d	10^{-6}–$10^{-1}/s^a$
Minimum molecular weight	200 Daltons[b]
Temperature range	4–40°C
Flow rate	1–100 µL/min

[a]May be extended under favorable circumstances.
[b]May be lower if used as bound material or in competition experiments.

200 Daltons by either immobilizing the small molecule on the surface and having the larger molecule flow over the surface or by attaching the small molecule to a larger molecule that does not interfere with the interaction *(3)*. If an interaction can be detected by SPR, changes in the interaction can be monitored upon addition of a small molecular weight substance *(4–7)*. For example, *E. coli* RNase HI binds an RNA–DNA hybrid in the absence of a divalent metal ion. If the hybrid is immobilized and we add $MgCl_2$ in the buffer with RNase HI, the SPR signal given by the hybrid will decrease due to hydrolysis *(8)*.

One rather powerful use of SPR is to determine stoichiometry. Because the signal measured is directly related to the mass of the two interacting molecules, it is possible to determine if two Bs are binding to one A at saturating levels of B. Finally, SPR can be used in conjunction with other methods to identify the substance interacting with the bound material.

2. A Short Description of the Detection System

The sensor chip is composed of a plastic holder and a thin glass surface uniformly coated with a thin layer of gold. One substance of interest is attached to the gold surface (usually to dextran bound to the gold). The second material is passed over the surface in an aqueous buffer. When the chip is in the BIACORE, "flow cells" are formed on the surface of the chip by interaction with an integrated µ-fluidic cartridge (IFC) that provides for delivery of sample and buffers to the materials immobilized on the chip (**Fig. 1**). The light source and detection unit are on the same surface and opposite that of the immobilized substance and its interacting partner. Light that is monochromatic and *p*-polarized is focused on the glass–gold interface of the sensor chip. When two transparent media are interfaced (in this case, glass and water, with a thin film of gold separating the two) light coming from the side of higher refractive index (glass) is either reflected or refracted. When the incident light is below a certain angle, light is totally reflected. However, a component of the electromag-

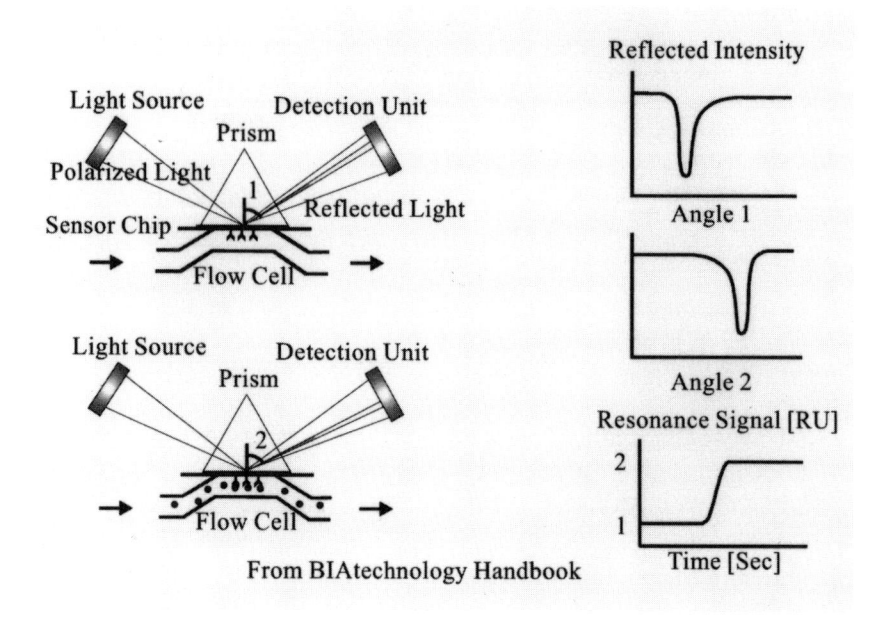

Fig. 1. Detection of changes in surface plasmon resonance resulting from interaction of two components. The **left panel** shows a flow cell and a sensor chip docked to each other. Polarized light passes from the source through a prism and impinges on the glass surface. Light is reflected at angle 1 resulting in the reflected intensity signal seen in the **top right panel**. As shown, there are three molecules attached to the Sensor chip. In the **bottom left panel** small spherical molecules are shown passing through and interacting with the immobilized molecules. The increase in mass near the gold surface produces a change in the reflected intensity signal as shown in the middle of the right panel which can in turn yield the change in resonance signal vs time (**lower right panel**). Reproduced with permission from *BIACORE Handbook*.

netic field (termed an evanescent wave) penetrates about one wavelength into the medium of lower refractive index (water). At a specific angle, the intensity of the reflected light is significantly reduced. The angle at which this decrease occurs depends on, among other things, the refractive index of the medium into which the evanescent wave penetrates. Binding of molecules near the surface interface changes the refractive index, and thereby the angle of minimal reflected light (**Fig. 1**). The reflected light is monitored by a diode array detector, and data are collected at intervals set by the software.

By definition, 1000 RUs (resonance units) corresponds to 1 ng protein/mm^2 (*BIACORE Handbook*). This value seems to hold for many different proteins containing normal amino acids. The value may be different for glyco- or lipoproteins, which will have lower SPR values. Nucleic acids give a slightly higher signal (1000 RUs corresponds to 0.8 ng of nucleic acid/mm^2) (*9,10*).

3. The Process

One of the molecular species is immobilized to the surface of a chip that can be placed in the instrument of choice and is the ligand. A solution containing the second material (analyte) is passed over the surface containing the ligand, and data on the interaction detected are collected by the software. The association phase is measured from changes in the SPR signal as the first part of the sample comes in contact with the surface of the chip. A plateau may be reached where association and dissociation are in equilibrium, followed by drop in SPR signal when no more analyte is present but buffer is now passing across the surface of the chip. Rates of dissociation are determined from the data collected during this phase. Fast dissociation rates lead to a rapid decline in SPR signal whereas slow rates are characterized by a slow decrease in signal. Data are analyzed using various software programs. To complete the cycle, it is necessary to return the surface of the chip to the state present at the beginning of injection of the analyte.

4. Choice of Chip

Sensor chips available from Biacore have an inert linker joining the gold to a matrix of carboxymethyl dextran. This surface is the most commonly used matrix, with many methods available for derivatizing dextran for covalent linking of substances such as proteins or modified nucleic acids. However, it is possible to attach substances directly to the gold surface *(11)* or to other, less well-characterized surfaces. Recently, Biacore has introduced such chips known collectively as "Pioneer" chips. This latter approach is suggested for those willing to work in the mode of explorers. There may be some differences in signal when using no dextran linkage or when the dextran is shorter. Such differences could arise due to changes in signal intensity, a function of the distance between the glass–gold interface and the molecular interaction site. This may be particularly important as one builds a complex thereby increasing the signal response for the most distant component.

5. Immobilization

5.1. Considerations for Choice of Method for Immobilization

Ideally, each component being studied should be immobilized for one set of experiments and in solution in a second set of studies to demonstrate that immobilization does not alter the interactions. For experiments in which one is asking if A interacts with B, it is not necessary to have A (or B) attached to the surface of the chip in a unique manner. In fact, random attachment would provide the greatest chance for interaction when the points of contact between the two molecules are unknown.

5.2. Various Coupling Methods

Details of coupling are not presented here. Each instrument may require variations in the procedures for attaching molecules to the chip surface. Here, we discuss some of the commonly used methods of attachment that are of general use and may be important for studying protein–nucleic acid interactions.

Initially, one usually wants to know if there is an interaction detectable by SPR. For protein immobilization, amine coupling is probably the method of choice since all that is required is a free amino group (e.g., N-terminal amino acid or an internal lysine). Moreover, the protein will likely be linked in several different orientations exposing different surfaces for interaction, thereby providing a high probability that the surface with which the second material is to interact will be available even though the nature (location, epitope, etc.) may be initially unknown. However, for kinetic studies a unique, homogeneous ligand is required *(12)*. Thiol coupling is also a possibility. Cysteine residues occur at a relatively low frequency in proteins and can provide linking to the dextran on the chip in a limited number of orientations. Some proteins may have no available cysteines in which case thiols be introduced with various reagents. Attachment of aldehydes created by hydrazine can be used for glycoproteins, polysaccharides, or other substances with potential aldehyde formation. Studies involving membranes are also possible using a hydrophobic chip, allowing membrane-bound receptors to be immobilized in a more native environment. Biotinylated ligands (either protein or nucleic acids) can be attached to the surface of a chip containing streptavidin.

5.3. Antibody Capture

Fusion proteins expressed from a wide variety of plasmid vectors can be immobilized using specific binding reagents. Attaching antibodies directed against His-tagged *(13)*, Flag-tagged (Application Note 104, BIACORE), or GST-tagged (Application Note 104, BIACORE) proteins to the surface via amine linkage generates surfaces capable of interacting with the correspondingly tagged fusion protein. One can use antibodies directed against any substance provided the interaction is sufficiently stable *(14,15)*. An antibody can be useful if its interaction is stable relative to the duration of the experiment. For anti-GST antibodies a dissociation rate of less than 0.03 RU/s (2 RU/min) was reported to be acceptable when the total RU of the GST-luciferase bound was about 300–400 (Application Note 104, from BIACORE). The GST system may suffer from dimerization of the GST domains generating spurious results *(16)*.

5.4. Biotinylated Proteins

It is possible to produce fusion proteins containing a sequence that is biotinylated in vivo at a unique site *(17,18)*. Such proteins can be purified on monovalent streptavidin columns from which they can be eluted with biotin. Removal of the free biotin can be performed in several ways producing protein suitable for immobilization to a streptavidin chip *(17)*.

5.5. Nucleic Acid Immobilization

The most common method of attachment of nucleic acids is via a biotin added to the nucleic acid, usually added to the 5'- or 3'- terminal residue. Other modifications such as thiol containing adducts can in theory be used. What if duplex nucleic acids are being studied? Typically, one of the two strands of duplex nucleic acid bears a biotin residue as orientation specificity is desirable. Annealing of the complementary strand to an immobilized nucleic acid is possible. In fact, several reports have measured the rates of association/dissociation using surface plasmon resonance. There are other routinely used methods for measuring such associations but if one needs to measure these rates for many samples, an automated device such as the high-end BIACORE instrument could be useful. We have found that preannealing to form duplex nucleic acid is easy and provides us the opportunity to use an excess of the nonderivatized (cheaper) nucleic acid to drive the annealing reaction *(8)*. The unannealed nucleic acid will not bind to the surface during attachment of the duplex nucleic acid.

One rather innovative procedure, developed in Buc's laboratory *(10)*, is to attach a biotinylated oligonucleotide to the chip and anneal the oligo- or polynucleotide to be used in the experiments to the bound oligonucleotide (**Fig. 2**). Regeneration of the surface can be accomplished by denaturing the double-stranded region of bound nucleic acid. In essence, a new type of chip has been created whose surface is now a nucleic acid surface to which complementary strands of RNA or DNA can be annealed. Of course, it will be necessary to demonstrate that any results obtained are not related to the duplex structure formed. Also, the fraction of surface–nucleic acid that anneals with the mobile phase-nucleic acid may be less than 100% thereby generating a nonhomogenous surface.

For kinetic experiments, it is important that the nucleic acid be of a homogeneous size and sequence. If the nucleic acid is short and has been synthesized by chemical means, it must be purified by gel electrophoresis. For longer RNAs the procedures described by Ferre-D'Amare and Doudna *(19)*, by Price et al. *(20)*, and by Schwienhorst and Lindemann *(21)* may be useful. Normal precautions are used to protect RNA from degradation by RNases. Most chemical syntheses produce more than enough oligomers for many experiments, unless they are used as competitors. We have used streptavidin chips made in our laboratory from CM5 chips from Biacore as well as SA chips purchased from Biacore. It is important to remove any unbound streptavidin by washing with 1 M NaCl plus 50 mM NaOH for three consecutive 1-min injections. Often, we find a slow decrease in the RU value that seems to result from loss of streptavidin

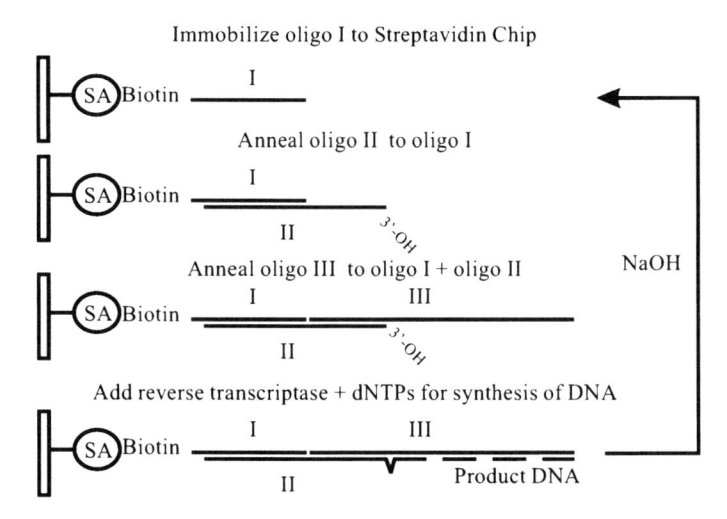

Fig. 2. Regeneration of streptavidin–biotin–nucleic acid surface. Streptavidin chip interacts with biotinylated nucleic acid. Once the streptavidin is reacted with a 5'-biotinylated nucleic acid and unoccupied streptavidin is blocked with biotin, the chip is converted to a nucleic acid chip. Annealing oligonucleotide II creates a duplex nucleic acid with a region of 3'-single-stranded extension. Annealing oligonucleotide III to the immobilized duplex generates a substrate (primer-template) for DNA polymerases, including reverse transcriptases. Synthesis is initiated by addition of dNTPs and the appropriate polymerase. If oligonucleotide I is DNA, injection of NaOH would melt the nucleic acid, thereby restoring the surface to one with only the 5'-biotinylated nucleic acid bound.

from the surface of the chip. When using low levels of nucleic acid on the surface, this decrease in RU caused by loss of streptavidin can be unacceptable.

6. How much to immobilize?

One thousand RUs is by definition the increase in signal when 1 ng of protein is present on 1 mm^2 of surface (*BIACORE Handbook*). For nucleic acid, the number is 0.8 ng *(9,10)*. When performing kinetic analysis, it is best to work with 100 RUs or less. 100 RUs (0.08 ng) of a 30-basepair RNA–DNA hybrid (MW ≈ 20,000) is about 4 fmol/flowcell. It is also necessary to repeat the experiments with slightly different amounts of nucleic acid bound to the chip for elimination of any variables due to association and reassociation. We use a solution of RNA–DNA hybrid that is 50–100 n*M* (moles of hybrid) in 10 m*M* Tris-HCl, 50 m*M* NaCl, 1 m*M* EDTA, 1 m*M* β-mercaptoethanol, 0.005% Tween P20, pH 8.0, for immobilization. Twenty microliters injected over a streptavidin surface at 5 µL/min gives an increase between 50 and 100 RU. The rate of injection, the concentration of RNA–DNA hybrid, the ionic strength, and the pH all can be varied to affect the amount of material immobilized.

7. Orientation — Does It Make a Difference?

Immobilization can have significant effects on the accessibility of the material bound. One of the first experiments we performed was to immobilize *E. coli* RNase HI to a CM5 chip via amine coupling (*see* **Subheading 5.2.**) or thiol linkage. For amine coupled RNase HI, we could readily detect binding of RNase HI to poly(rA)-poly(dT). However, because the thiol coupling involved a single Cys residue on the surface of the protein that is known to interact with the RNA–DNA hybrid, we were unable to observe binding of the nucleic acid. Demonstration that RNase HI was immobilized was accomplished by use of an anti-RNase HI antiserum (unpublished data). It is useful to confirm kinetic parameters when one substance is immobilized and when the same material is in the mobile phase.

8. The Interaction Phase

This phase can be divided into (1) injection, (2) association, and (3) dissociation. Injection is accomplished in the BIACORE by a robotic device that injects the sample. The concentration of the sample to be analyzed should be varied over a wide range and with various ionic strengths, pH, and any other variable that may be known for the interaction under study. Association rates that are very fast create considerable problems in measuring the dissociation rates due to rebinding once dissociation occurs. Two procedures can be used to aid the study of high rates of association. First, flow rates can be increased and/or a competitor molecule can be injected immediately after the sample is applied. Second, the density of the immobilized material can be kept to a minimum. Immobilization of molecules in the range of 10–150 RUs gives sufficient response signals for data analysis. Fast association rates also can be problematic due to mass transport limited reactions *(15,22)*. Molecules passing over the surface can be depleted from the solution at or near the surface, and the rate of interaction then becomes limited by the rate of mass transport of the soluble molecule. For our experiments on *E. coli* RNase HI, surface densities of about 100 RUs were employed and we used poly(rA)-poly(dT) injected immediately in the mobile phase to act as a competitor for the RNase HI leaving the RNA–DNA hybrid bound to the chip surface.

9. Data Analysis

Software from Biacore is useful for data analysis. Dr. David Myszka also has a web site for data analysis (http://www.hci.utah.edu/cores/biacore). Myszka's web site provides tutorials for his program CLAMP to take advantage of global fitting. The program can be downloaded to PCs running Windows® 3.1, 95 or NT. Myszka's site is very informative, describing numerous possibilities for problems and how to obtain the best data possible. Much of this is also found in his article in *Current Opinion in Biotechnology* *(22)*. Other useful articles concerning data analysis are available *(3,12,23–27)*.

10. Regeneration

Most of the methods for immobilizing ligands convert the original chip to one dedicated to measuring a specific interaction. For example, immobilization via amine coupling of an antibody converts a CM5 chip to an antibody chip. Reversing this process to return the chip to a CM5 surface has been the subject of one report *(28)* but there is little in the literature to indicate that this is a widely used step. Regeneration of an nickel nitrilotriacetic acid (NTA) surface by addition of EDTA or imidazole, and a thiol surface by reducing agents returns the surface to that of the chip as first made (i.e., no ligand remains on the surface). Although this property of these chips means that they can be used to bind a totally different ligand, it is often important to regenerate the chip to the condition in which the ligand remains bound.

A 96-well microtiter dish can be placed in the BIACORE so that many samples can be injected over the surface, provided the surface can be regenerated after each prior injection. Ideally, treatment with some agent such as NaCl at high concentrations will destroy the interaction between analyte and ligand, returning the surface to its condition after the original immobilization of the ligand. A number of different chemicals have been successfully used for surface regeneration, including glycine at pH 2.5–3.0 for antigen–antibody interactions and SDS for general protein–nucleic acid interactions. Surprisingly, 100 mM HCl can be used to remove RNA–DNA bound to *E. coli* RNase HI without altering binding in a second injection. Once the surface SPR signal returns to its starting point, a second (and subsequent) injection of samples with different concentrations of analyte, different materials altogether, or any other parameters such as flow rate can be performed.

11. Stoichiometry

A number of experiments can be performed using SPR that provide useful information, but determining stoichiometry of interactants is a most valuable tool. A change in RU value is directly related to the mass of the substance binding to the surface. In many instances, the molecular weight of the interacting protein or nucleic acid is known and from the increase in RU value, the amount of material binding can be accurately determined. For example, we wished to know how many basepairs of RNA–DNA hybrid is bound by a single molecule of *E. coli* RNase HI. To measure the stoichiometry, we immobilized the RNA–DNA hybrids depicted in **Fig. 3** and injected *E. coli* RNase HI over the surfaces. By increasing the amount of RNase HI, we were able to saturate the available sites on each hybrid and could determine the amount of enzyme bound to each RNA–DNA molecule. Because we did not know the exact number of basepairs to which *E. coli* RNase HI would bind, we used various lengths of hybrid molecules. From these data, we could determine the amount of *E.*

Surface Plasmon Resonance

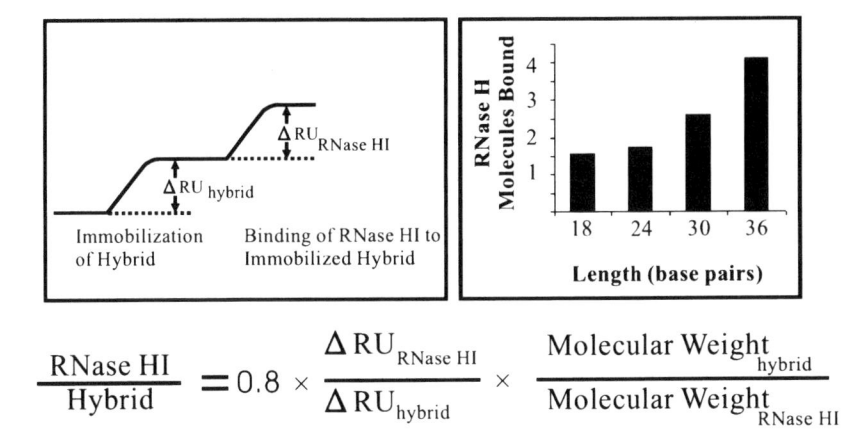

$$\frac{\text{RNase HI}}{\text{Hybrid}} = 0.8 \times \frac{\Delta RU_{\text{RNase HI}}}{\Delta RU_{\text{hybrid}}} \times \frac{\text{Molecular Weight}_{\text{hybrid}}}{\text{Molecular Weight}_{\text{RNase HI}}}$$

Fig. 3. Stoichiometry determinations. Immobilization of RNA–DNA of known composition yields an increase in the RU value from that of the streptavidin surface to some value, indicated as $\Delta RU_{\text{hybrid}}$ on the **left panel**. Injection of various concentrations of RNase HI will produce sensorgrams that will have different slopes. At saturation of the nucleic acid bound to the surface, the maximum plateau value is related to the amount of RNase HI bound to the hybrid. For the example shown here, a single injection of RNase HI is shown and the value $\Delta RU_{\text{RNase HI}}$ is achieved. We used the equation indicated to calculate the RNase HI molecules bound to each RNA–DNA immobilized. To determine the number of basepairs bound per RNase HI molecule, hybrids of various lengths were immobilized. The number of protein molecules bound for each hybrid tested is indicated in the **right panel**. Our estimate is that each RNase HI binds to 9–10 basepairs.

coli RNase HI bound on the various hybrids and could deduce that about 9–10 basepairs were bound by a single molecule of E. coli RNase HI (8).

Similar studies permitted McHenry's group (17,29,30) to determine the composition of E. coli DNA polymerase III and gather information about subunit interactions. DNA polymerase III contains 10 subunits that can be assembled in vitro to form an active complex. Two subunits ($\delta\delta'$) interact with one another. These in turn interact with another complex $\chi\psi$ and either τ or γ to form $\tau_4\delta_1\delta'_1\chi_1\psi_1$ or $\gamma_4\delta_1\delta'_1\chi_1\psi_1$. δ immobilized via amine coupling interacted with the $\delta' + \chi + \psi + \tau$ or $\delta' + \chi + \psi + \gamma$ but the complex did not appear to have the τ or γ subunit in the expected ratio of 4:1. As mentioned previously, the random linkage of δ probably created a surface on which not all of the δ could interact with δ'. Thus, the stoichiometry of this complex could not be determined. However, the complex that did form was relatively stable (half-life of about 1.4 h) and could be used as a surface for measuring the interaction between the $\tau_4\delta_1\delta'_1\chi_1\psi_1$ or $\gamma_4\delta_1\delta'_1\chi_1\psi_1$ complexes and the core subunits (α-ϵ-Θ). Assuming that the $\tau_4\delta_1\delta'_1\chi_1\psi_1$ structure formed, it bound two core complexes, with the new complex having a half-life of 1.5 h. Interestingly, the $\gamma_4\delta_1\delta'_1\chi_1\psi_1$ failed to interact with core. Injection of the β subunit over the surface with core plus $\tau_4\delta_1\delta'_1\chi_1\psi_1$ resulted in a complex containing all of the DNA polymerase III subunits except γ. The DNA polymerase III (γ-less) was capable of forming a stable complex (half-life 1.5 h) with a DNA coated with single-stranded binding protein.

12. Kinetics of Enzymatic Reactions on a Chip

Measuring elongation rates of MoMuLV reverse transcriptase (RT) has been accomplished by immobilization of a nucleic acid primer-template to the surface of a CM5-streptavidin chip, and subsequent addition of MuLV RT and dNTPs. One primer-template employed was poly(rA)–(dT)$_{20}$ with the (dT)$_{20}$ having a biotin residue at its 5'-terminus. Synthesis of poly(dT) converts the primer-template to a completed RNA–DNA duplex. Regeneration of the surface is not possible because there is now no original primer-template left. To circumvent this problem, Buckle et al. *(10)* devised a reusable primer-template surface. **Figure 2** depicts linking a biotinylated DNA oligomer (I) to a streptavidin surface. Annealing a complementary DNA oligomer (II) to the immobilized oligomer (I) generates a duplex DNA with a 3'-single-stranded region. Annealing a third DNA oligomer (III) to oligomer (II) forms a primer-template that when extended produces a duplex DNA the length of which can be regulated by choice of the sequence of the single-stranded template. Buckle et al. *(10)* had 25 G followed by 25 T residues such that they could extend DNA with dCTP for 25 nucleotides and then add dATP for synthesis of additional 25 bases. Upon completion, injection of NaOH over the surface would denature the DNA returning the surface to its status of having the biotinylated DNA oligomer attached to streptavidin. The cycle can be repeated and conditions varied in an automated fashion. Buckle *et al.* *(10)* report an efficiency of annealing of oligo (I) to oligo (II) of ~ 75% and annealing of oligo(III) to oligo(I) + oligo(II) of ~ 70%, resulting in approx 50% of oligo(I) immobilized to the surface participating in interaction with MoMuLV reverse transcriptase.

13. What About RNase Degradation of Immobilized RNA?

Various extracts have been injected into our BIACORE over the past several years yet we are still able to put RNA samples through the machine without noticeable RNA degradation. The manufacturer provides two routines for cleaning the complete system. "Desorb" removes materials (e.g., proteins) from the autosampler and the cartridge through which the fluid is directed to the flow cells on the surface of the chip. This buffer contains SDS. "Sanitize" is used for more vigorous cleaning and follows "Desorb." "Sanitize" includes a treatment with a disinfectant and a nonionic detergent. We include an injection of a commercial RNase remover such as "RNase Zap" from Ambion (Austin, TX, USA; 20 µL at 20 µL/min followed by a water wash; this can be repeated for several times if desired).

14. Why Would One Want to Do Experiments on an SPR Instrument that Can Be Performed in Other Ways?

Several studies have been performed on the SPR instruments that can also be carried out using conventional techniques. Nucleic acid annealing *(31–37)*, nucleic acid synthesis *(10,32)*, ligation *(32)*, phage display *(38,39)*, and other such techniques have been carried out on SPR instruments. Why have the experiments been done? First, because they can be done. Second, the ability of the more sophisticated instruments to perform repeated analyzes with little or no intervention required by the operator permits rather high throughput analyses. Finally, sensor chips with small quantities of materials deposited on the surface of the sensor chip may be useful as small reaction vessels. Recently, Nelson's group *(40)* has shown that an antibody attached to the surface of a chip can capture its antigen from a complex mixture (e.g., extracts of *E. coli* to which small amounts of the antigen have been added). Placing the chip removed from a BIACORE instrument into a mass spectrometer permits identification of the molecular mass of the bound antigen. In principle, it should be possible to learn a great deal about the material interacting with immobilized molecules. For instance, if one knows that the material binding is a protein from *S. cerevisiae* its identity could be known or at least limited to a small set of proteins by searching a database for proteins of that molecular weight. If one were to capture enough material for enzymatic digestion for peptide analysis, the sizes of the peptides could also limit the protein to one or a few candidates.

15. Nucleic Acid Interactions

Nucleic acid interactions can be divided into three classes: (1) annealing, (2) protein–nucleic acids, and 3) other. Articles describing nucleic acid annealing experiments are in **refs.** *31–37*. Protein–nucleic acid interactions are studied in **refs.** *7–10,29,32*, and *41–46*. In the other category is a study of a DNA binding agent, chromomycin, with DNA *(47)*. It may be helpful to examine what methods each of these articles reports.

16. How to Do an SPR Experiment

1. Find someone who has an SPR instrument.
2. Become informed about the operation of the instrument.
   a. Discuss the problem to be studied with an experienced operator, or
   b. Follow directions in the manual or on a training CD. Initially, it may take 3 wk to become sufficiently comfortable with operation and analysis.

3. Perform an experiment that is known to work.
4. Immobilize your material following standard procedures. Make certain that the material is stable to the conditions used. Take advantage of what is known about the interaction.
5. Test second interactant in aqueous solution for binding to the surface prepared in **step 4**.
6. Regenerate the surface. There are several methods used for regeneration that vary depending on the nature of the materials being studied. You can assess the efficacy of the regeneration by observing a return to the RU value seen prior to binding and obtaining the same signal when the aqueous material is injected over the same surface a second time.
7. Repeat the experiment with different levels of immobilization, flow rates, salt concentrations, pH, and other changes related to the information you may have about the molecules under study. For example, you may know from previous electrophoretic mobility shift studies that Tris buffer at pH 7.2 is useful but the studies with SPR suggest HEPES buffer. Try both conditions.
8. Perform the same interaction studies switching the immobilized and aqueous materials. The data should be identical regardless of which interactant is immobilized.
9. Analyze the data.

17. A Problem in Progress

The example given below is used to point out that each set of experiments may reveal its own problems, and what worked for others or previously in your laboratory cannot be applied universally.

17.1. His-Tagged Proteins

We have several proteins that are expressed as fusions to a His-tag, and we wish to use them in SPR analysis by binding to an Ni^{2+}–NTA surface. Two chips can be used. First, we can generate an Ni^{2+}–NTA surface as described by Gershon and Khilko (48), the O'Shannessy group (49), Sigal et al. (11) or we can purchase an Ni^{2+}–NTA chip. Are they equivalent? The answer is "Maybe yes and maybe no." Do all His-tagged proteins bind with sufficient half-lives to be useful? The answer is No. Several examples of a monomeric proteins with six histidines that do bind are *E. coli* RNase HI (8,41) and some chimeric versions of the same protein (41) as well as vaccinia virus VP55 and VP39 (48) and the human La protein (48). However, some others are reported to bind with off rates too high to be of any use in measuring interactions (50). To circumvent this anomaly, multiple copies of the six-His-tag can be attached to the protein either in a cluster or at different parts of the protein (C-terminal and N-terminal or on different subunits if the protein has more than one subunit [50]). However, the necessity of recloning makes this option less desirable. It should also be noted that the NTA surface was the commercial version for the experiments of Nieba et al. (50).

An alternative to the Ni^{2+}–NTA surface is to employ anti-His antibodies. We have recently used this method with some degree of success. Antibodies from Qiagen can be immobilized to a CM5 chip via amine coupling, and these chips can be used to immobilize His-tagged proteins that, in our hands, do not form stable surfaces with Ni^{2+}–NTA chips. Previously, we showed that an RNA–DNA hybrid employed as ligand reacted with analyte RNase HI with the same kinetic properties as when the protein was ligand and the analyte contained the RNA–DNA hybrid *(8)*. Unfortunately, the binding constants are different for *E. coli* RNase HI when it is bound to the surface via the anti-His antibody than when it is linked to the Ni^{2+}–NTA chips (Uyeda and Crouch, unpublished). This means we need to search for other methods on linking, and points out the fact that each experiment may involve considerable trial and error.

References

1. D'Ettorre, C., DeChiara, G., Casadei, R., Boraschi, D., and Tagliabue, A. (1997) Functional epitope mapping of human interleukin–1 beta by surface plasmon resonance. *Eur. Cytokine Network* **8,** 161–171.
2. Laune, D., Molina, F., Ferrieres, G., Mani, J. C., Cohen, P., Simon, D., Bernardi, T., Piechaczyk, M., Pau, B., and Granier, C. (1997) Systematic exploration of the antigen binding-activity of synthetic peptides isolated from the variable regions of immunoglobulins. *J. Biol. Chem.* **272,** 30937–30944.
3. Karlsson, R. and Stahlberg, R. (1995) Surface-plasmon resonance detection and multi-spot sensing for direct monitoring of interactions involving low-molecular-weight analytes and for determination of low affinities. *Anal. Biochem.* **228,** 274–280.
4. Floer, M., Blobel, G., and Rexach, M. (1997) Disassembly of RanGTP-karyopherin beta complex, an intermediate in nuclear protein import. *J. Biol. Chem.* **272,** 19538–19546.
5. Kuhlmann, J., Macara, I., and Wittinghofer, A. (1997) Dynamic and equilibrium studies on the interaction of Ran with its effector, RanBP1. *Biochemistry* **36,** 12027–12035.
6. Heierhorst, J., Mann, R. J., and Kemp, B. E. (1997) Interaction of the recombinant S100A1 protein with twitchin kinase, and comparison with other Ca^{2+}-binding proteins. *Eur. J. Biochem.* **249,** 127–133.
7. Pond, C. D., Holden, J. A., Schnabel, P. C., and Barrows, L. R. (1997) Surface plasmon resonance analysis of topoisomerase I-DNA binding: effect of Mg^{2+} and DNA sequence. *Anti-Cancer Drugs* **8,** 336–344.
8. Haruki, M., Noguchi, E., Kanaya, S., and Crouch, R. J. (1997) Kinetic and stoichiometric analysis for the binding of *Escherichia coli* ribonuclease HI to RNA–DNA hybrids using surface plasmon resonances. *J. Biol. Chem.* **272,** 22015–22022.
9. Fisher, R. J., Fivash, M., Casasfinet, J., Erickson, J. W., Kondoh, A., Bladen, S. V., Fisher, C., Watson, D. K., and Papas, T. (1994) Real-time DNA-binding measurements of the ETS1 recombinant oncoproteins reveal significant kinetic differences between the p42 and p51 isoforms. *Protein Sci.* **3,** 257–266.

10. Buckle, M., Williams, R. M., Negroni, M., and Buc, H. (1996) Real time measurements of elongation by a reverse transcriptase using surface plasmon resonance. *Proc. Natl. Acad. Sci. USA* **93,** 889–894.
11. Sigal, G. B., Bamdad, C., Barberis, A., Strominger, J., and Whitesides, G. M. (1996) A self-assembled monolayer for the binding and study of histidine tagged proteins by surface plasmon resonance. *Anal. Chem.* **68,** 490–497.
12. Kortt, A. A., Oddie, G. W., Iliades, P., Gruen, L. C., and Hudson, P. J. (1997) Nonspecific amine immobilization of ligand can be a potential source of error in BIACORE binding experiments and may reduce binding affinities. *Anal. Biochem.* **253,** 103–111.
13. Lindner, P., Bauer, K., Kremmer, E., Krebber, C., Honegger, A., Klinger, B., Mocikat, R., and Pluckthun, A. (1997) Specific detection of his-tagged proteins with recombinant anti-his tag scFv-phosphatase or scFv-phage fusions. *Biotechniques* **22,** 140–149.
14. Karlsson, R., Michaelsson, A., and Mattsson, L. (1991) Kinetic analysis of monoclonal antibody–antigen interactions with a new biosensor based analytical system. *J. Immunol. Methods* **145,** 229–240.
15. Myszka, D. G., Morton, T. A., Doyle, M. L., and Chaiken, I. M. (1997) Kinetic analysis of a protein antigen–antibody interaction limited by mass transport on an optical biosensor. *Biophys. Chem.* **64,** 127–137.
16. Ladbury, J. E., Lemmon, M. A., Zhou, M., Green, J., Botfield, M. C., and Schlessinger, J. (1995) Measurement of the binding of tyrosyl phosphopeptides to SH2 domains —a reappraisal. *Proc. Natl. Acad. Sci. USA* **92,** 3199–3203.
17. Kim, D. R. and McHenry, C. S. (1996) Biotin tagging deletion analysis of domain limits involved in protein–macromolecular interactions — mapping the tau binding domain of the DNA polymerase III alpha subunit. *J. Biol. Chem.* **271,** 20,690–20,698.
18. Smith, P. A., Tripp, B. C., DiBlasio-Smith, E. A., Lu, Z. J., LaVallie, E. R., and McCoy, J. M. (1998) A plasmid expression system for quantitative in vivo biotinylation of thioredoxin fusion proteins in *Escherichia coli*. *Nucleic Acids Res.* **26,** 1414–1420.
19. Ferre-D'Amare, A. R. and Doudna, J. A. (1996) Use of *cis*- and *trans*-ribozymes to remove 5' and 3' heterogeneities from milligrams of in vitro transcribed RNA. *Nucleic Acids Res.* **24,** 977–978.
20. Price, S. R., Ito, N., Oubridge, C., Avis, J. M., and Nagai, K. (1995) Crystallization of RNA–protein complexes. 1. Methods for the large-scale preparation of RNA suitable for crystalographic studies. *J. Mol. Biol.* **249,** 398–408.
21. Schwienhorst, A. and Lindemann, B. F. (1998) Novel vector for generating RNAs with defined 3' ends and its use in antiviral strategies. *Biotechniques* **24,** 116–124.
22. Myszka, D. G. (1997) Kinetic analysis of macromolecular interactions using surface plasmon resonance biosensors. *Curr. Opin. Biotechnol.* **8,** 50–57.
23. Bowles, M. R., Hall, D. R., Pond, S. M., and Winzor, D. J. (1997) Studies of protein interactions by biosensor technology: an alternative approach to the analysis of sensorgrams deviating from pseudo-first-order kinetic behavior. *Anal. Biochem.* **244,** 133–143.

24. Christensen, L. H. (1997) Theoretical analysis of protein concentration determination using biosensor technology under conditions of partial mass transport limitation. *Anal. Biochem.* **249,** 153–164.
25. Hall, D. R., Cann, J. R., and Winzor, D. J. (1996) Demonstration of an upper limit to the range of association rate constants amenable to study by biosensor technology based on surface plasmon resonance. *Anal. Biochem.* **235,** 175–184.
26. Schuck, P. (1997) Reliable determination of binding affinity and kinetics using surface plasmon resonance biosensors. *Curr. Opin. Biotechnol.* **8,** 498–502.
27. Oddie, G. W., Gruen, L. C., Odgers, G. A., King, L. G., and Kortt, A. A. (1997) Identification and minimization of nonideal binding effects in BIACORE analysis: ferritin/anti-ferritin Fab' interaction as a model system. *Anal. Biochem.* **244,** 301–311.
28. Chatelier, R. C., Gengenbach, T. R., Griesser, H. J., Brighamburke, M., and O'Shannessy, D. J. (1995) A general-method to recondition and reuse BIACORE sensor chips fouled with covalently immobilized protein peptide. *Anal. Biochem.* **229,** 112–118.
29. Kim, D. R. and McHenry, C. S. (1996) In vivo assembly of overproduced DNA polymerase III — overproduction, purification, and characterization of the alpha, alpha-epsilon, and alpha-epsilon-theta subunits. *J. Biol. Chem.* **271,** 20681–20689.
30. Dallmann, H. G. and McHenry, C. S. (1995) DNAX complex of *Escherichia-coli* DNA-polymerase-III holoenzyme — physical characterization of the DNAX subunits and complexes. *J. Biol. Chem.* **270,** 29,563–29,569.
31. Nilsson, P., Persson, B., Larsson, A., Uhlen, M., and Nygren, P. A. (1997) Detection of mutations in PCR products from clinical samples by surface plasmon resonance. *J. Mol. Recog.* **10,** 7–17.
32. Nilsson, P., Persson, B., Uhlen, M., and Nygren, P. A. (1995) Real-time monitoring of DNA manipulations using biosensor technology. *Anal. Biochem.* **224,** 400–408.
33. O'Meara, D., Nilsson, P., Nygren, P. A., Uhlen, M., and Lundeberg, J. (1998) Capture of single-stranded DNA assisted by oligonucleotide modules. *Anal. Biochem.* **255,** 195–203.
34. Persson, B., Stenhag, K., Nilsson, P., Larsson, A., Uhlen, M., and Nygren, P. A. (1997) Analysis of oligonucleotide probe affinities using surface plasmon resonance: a means for mutational scanning. *Anal. Biochem.* **246,** 34–44.
35. Watts, H. J., Yeung, D., and Parkes, H. (1995) Real-time detection and quantification of DNA hybridization by an optical biosensor. *Anal. Chem.* **67,** 4283–4289.
36. Woods, S. J. (1993) DNA DNA hybridization in real-time using BIACORE. *Microchem. J.* **47,** 330–337.
37. Jensen, K. K., Orum, H., Nielsen, P. E., and Norden, B. (1997) Kinetics for hybridization of peptide nucleic acids (PNA) with DNA and RNA studied with the BIAcore technique. *Biochemistry* **36,** 5072–5077.
38. Malmborg, A. C., Duenas, M., Ohlin, M., Soderlind, E., and Borrebaeck, C. K. (1996) Selection of binders from phage displayed antibody libraries using the BIACORE™ biosensor. *J. Immunol. Methods* **198,** 51–57.

39. Sibille, P., Ternynck, T., Nato, F., Buttin, G., Strosberg, D., and Avrameas, A. (1997) Mimotopes of polyreactive anti-DNA antibodies identified using phage-display peptide libraries. *Eur. J. Immunol.* **27,** 1221–1228.
40. Krone, J. R., Nelson, R. W., Dogruel, D., Williams, P., and Granzow, R. (1997) BIA/MS: interfacing biomolecular interaction analysis with mass spectrometry. *Anal. Biochem.* **244,** 124–132.
41. Zhan, X. Y. and Crouch, R. J. (1997) The isolated RNase H domain of murine leukemia virus reverse transcriptase — retention of activity with concomitant loss of specificity. *J. Biol. Chem.* **272,** 22,023–22,029.
42. Dutreix, M. (1997) (GT)(n) repetitive tracts affect several stages of RecA-promoted recombination. *J. Mol. Biol.* **273,** 105–113.
43. Fisher, R. J., Fivash, M., Casas-finet, J., Bladen, S., and McNitt, K. L. (1994) Real-time BIACORE measurements of *Escherichia coli* single-stranded-DNA binding protein to polydeoxythymidylic acid reveal single state kinetics with steric cooperativity. *Methods: A Companion to Methods Immunol.* **6,** 121–133.
44. Fisher, R. J., Rein, A., Fivash, M., Urbaneja, M. A., Casas-Finet, J. R., Medaglia, M., and Henderson, L. E. (1998) Sequence-specific binding of human immunodeficiency virus type 1 nucleocapsid protein to short oligonucleotides. *J. Virol.* **72,** 1902–1909.
45. Rutigliano, C., Bianchi, N., Tomassetti, M., Pippo, L., Mischiati, C., Feriotto, G., and Gambari, R. (1998) Surface plasmon resonance for real-time monitoring of molecular interactions between a triple helix forming oligonucleotide and the Sp1 binding sites of human Ha-ras promoter: effects of the DNA-binding drug chromomycin. *Int. J. Oncol.* **12,** 337–343.
46. Tanchou, V., DeLaunay, T., Derocquigny, H., Bodeus, M., Darlix, J. L., Roques, B., and Benarous, R. (1994) Monoclonal antibody-mediated inhibition of RNA-binding and annealing activities of HIV type-1 nucleocapsid protein. *AIDS Res. Hum. Retroviruses* **10,** 983–993.
47. Gambari, R., Bianchi, N., Rutigliano, C., Borsetti, E., Tomassetti, M., Feriotto, G., and Zorzato, F. (1997) Surface plasmon resonance for real-time detection of molecular interactions between chromomycin and target DNA sequences. *Int. J. Oncol.* **11,** 145–149.
48. Gershon, P. D. and Khilko, S. (1995) Stable chelating linkage for reversible immobilization of oligohistidine tagged proteins in the BIACORE surface-plasmon resonance detector. *J. Immunol. Methods* **183,** 65–76.
49. O'Shannessy, D. J., O'Donnell, K. C., Martin, J., and Brighamburke, M. (1995) Detection and quantitation of hexa-histidine-tagged recombinant proteins on western blots and by a surface-plasmon resonance biosensor technique. *Anal. Biochem.* **229,** 119–124.
50. Nieba, L., Nieba-Axmann, S. E., Persson, A., Hamalainen, M., Edebratt, F., Hansson, A., Lidholm, J., Magnusson, K., Karlsson, A. F., and Pluckthun, A. (1997) BIACORE analysis of histidine-tagged proteins using a chelating NTA sensor chip. *Anal. Biochem.* **252,** 217–228.

13

An *Escherichia coli*-Based Genetic Strategy for Characterizing RNA Binding Proteins

Chaitanya Jain

1. Introduction

In recent years, interest in RNA-binding proteins (RNA-BPs) and RNA–protein interactions has grown considerably. An important reason for this interest is the increased recognition that RNA-BPs are involved in a wide variety of critical viral and cellular processes including transcription, translation, RNA processing, RNA transport, gene regulation, and splicing. A second reason lies in the technical advances and the generation of new methods that have simplified the analysis of RNA–protein interactions. Nevertheless, a thorough analysis of an RNA-BP is not a trivial task, and, at present, there are only a few RNA-BPs that have been characterized in detail.

It was with a view toward expediting such analyses that a genetic method was developed for the characterization of RNA-BPs *(1)*. The method is based on a strategy in which binding of a heterologous RNA-BP to its RNA target in *E. coli* results in translation repression. This method has been successfully applied to a variety of different RNA-binding proteins, including the human immunodeficiency virus-1 (HIV-1) Rev protein, the *Bacillus Stearothermophilus* ribosomal protein S15, and the human proteins U1A and nucleolin *(1,2*, unpublished data).

In *E. coli*, translation is initiated by the recruitment of ribosomes to a region of the mRNA that spans the initiation codon and an element called the Shine–Dalgarno (S/D) region, usually located a few nucleotides upstream of the initiation codon. Upon binding mRNAs, ribosomes cover a region of about 40 nucleotides, roughly centered about the initiation codon *(3,4)*. Studies have shown that this region must be kept free of impediments such as strong secondary structures and associated proteins for translation to proceed efficiently *(4,5)*. It was reasoned that modifications to a reporter mRNA that introduce the

From: *Methods in Molecular Biology, Vol. 118: RNA-Protein Interaction Protocols*
Edited by: S. Haynes © Humana Press Inc., Totowa, NJ

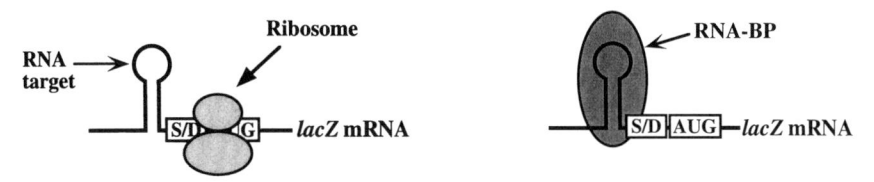

Fig. 1. Schematic representation of the translational repression strategy by heterologous RNA-BPs. The left hand side of the figure depicts the fate of *lacZ* mRNAs modified to contain a high-affinity RNA target for a specific RNA-BP (shown here as a hairpin loop structure) upstream of the *lacZ* S/D sequence. These transcripts are actively translated by ribosomes, resulting in the production of substantial levels of β-galactosidase. The right side of the figure depicts the effect of RNA-BP binding to its RNA-target. Because the bound RNA-BP is located close to the *lacZ* initiation codon (AUG), it sterically hinders ribosome binding. As a result, a significant reduction in *lacZ* translation and β-galactosidase production takes place.

RNA target for a heterologous RNA-BP upstream of the S/D sequence, but overlapping the ribosome binding site, should not affect translation greatly *per se*. However, expression and binding of the RNA-BP to its RNA target in *E. coli* could sterically hinder ribosome binding and reduce expression of the reporter. A schematic representation of this strategy is illustrated in **Fig. 1**.

The practical application of the genetic method involves the construction and use of two plasmids that are transformed into an appropriate *E. coli* strain. One of the plasmids expresses a *lacZ* reporter mRNA, suitably modified to contain a high-affinity RNA target upstream of the *lacZ* S/D region. The second plasmid expresses the RNA-BP that recognizes the RNA target in *E. coli*. Transformation of these two plasmids into a *lacZ*– strain should lead to expression of β-galactosidase activity by the reporter plasmid, to be substantially reduced due to binding by the expressed RNA-BP to the RNA target. The β-galactosidase activity of these cells can be qualitatively assessed by phenotypic screening of the transformed colonies on indicator plates. The most straightforward application of this genetic system is to introduce mutations into the RNA-BP or the RNA-target, and to identify mutants defective for binding based on their inability to repress *lacZ* expression. These and other applications of this genetic strategy are discussed in **Subheading 3.4.**

The reporter plasmid encodes a modified *lacZ* gene. To reduce β-galactosidase expression to a range suitable for phenotypic screening, a small segment of the weakly expressed IS10 transposase gene was fused to a plasmid-borne *lacZ* gene. As a result, the transcriptional and translational signals of the IS10 transposase-β-galactosidase fusion protein are derived from the IS10 segment.

Further modifications were made by introduction of unique restriction sites to facilitate cloning of RNA targets upstream of the IS10 S/D sequence. Finally, signals for translation and transcription termination were introduced upstream of IS10 translation and transcription signals, respectively, to reduce background expression. The sequence of the expression region of the final plasmid, pLacZ-Rep, is shown in **Fig. 2A**.

The second plasmid is one that allows high-level but regulated expression of RNA-BPs. A plasmid that is suitable for cloning and expression of RNA-BPs is pREV1 *(1)*, which encodes the HIV-1 Rev protein. Regulated transcription from this plasmid may be achieved in strains expressing Lac repressor, due to the presence of Lac repressor binding sites overlapping transcriptional signals in this plasmid. The transcriptional and translational signals responsible for high-level expression as well as relevant restriction sites are shown in **Fig. 2B**.

In principle, this genetic method is applicable to any cloned RNA binding protein, once its RNA target has been adequately defined. The first part of the chapter describes how to construct new plasmids required for applying this method to a specific RNA-BP and how to assay for translational repression. In the latter part of the chapter, potential uses of this genetic system are discussed. These include the use of the system to identify variants of the RNA-BP or RNA target that affect RNA–protein interactions, identification of suppressor RNA-BP variants that can bind to mutant RNA targets, and a rapid spectrophotometric assay method for estimating the binding affinity of RNA or protein variants. The use of the techniques described in this chapter should greatly expedite the analysis of RNA-BPs in the future.

2. Materials

1. Plasmids: pREV1, pLacZ-Rep, and pACYC184 (available from the author).
2. Strains: WM1 and WM1/F' (available from the author).
3. X-Gal stock: 30 mg/mL of 5-bromo 4-chloro 3-indoyl-β-D-galactoside (X-Gal) in dimethylformamide. Store at –20°C.
4. IPTG stock: 1 M isopropyl-β-D-thiogalactopyranoside (IPTG) in dH$_2$O. Store at –20°C.
5. *lacZ* buffer: 60 mM Na$_2$HPO$_4$, 40 mM NaH$_2$PO$_4$, 10 mM KCl, 1 mM MgSO$_4$, pH 7.0.
6. ONPG solution: 4 mg/mL *o*-nitrophenylgalactoside in distilled water. Store at 4°C. Discard if the solution turns yellow with age.
7. 1 M Sodium carbonate solution.
8. 2-Mercaptoethanol.
9. Sodium dodecyl sulfate (SDS).
10. Chloroform.
11. Indicator plates: LB-agar plates containing ampicillin (100 µg/mL), chloramphenicol (35 µg/mL), X-Gal (60 µg/mL), and/or IPTG (*see* **Subheading 3.3.1.**).
12. Growth medium: LB containing ampicillin (100 µg/mL), chloramphenicol (35 µg/mL), and/or IPTG.

A

```
              -35                        -10                    ┌──► IS10 transcription start
AGTTAAGGTAGATACACATCTTGTCATATGATCAAATGGTTTCGCTAGACTAGTCTAGCGAACCGC
                                                    translational terminators
```

```
                         ┌──► T7 RNA polymerase start
ACTTAATACGACTCACTATAGGTACCAATCCATTGCACTCCGGATTGAATTCAGACAACAAGATG
                       Kpn I                         EcoR I  S/D        Initiation
                                                                          codon
```

B

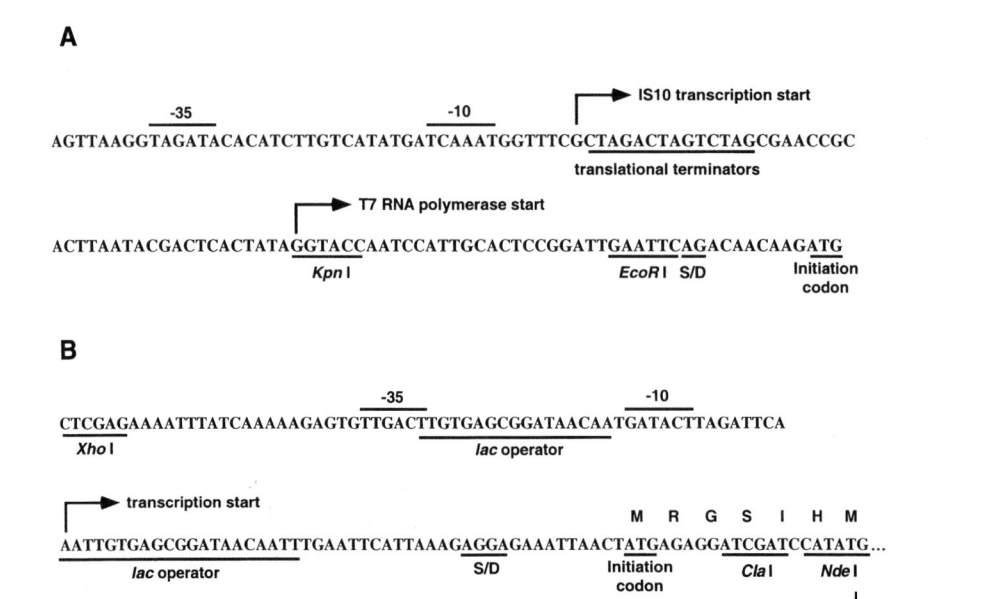

```
                          -35                      -10
CTCGAGAAAATTTATCAAAAAGAGTGTTGACTTGTGAGCGGATAACAATGATACTTAGATTCA
Xho I                              lac operator
```

```
      ┌──► transcription start
                                                 M  R  G  S  I  H  M
AATTGTGAGCGGATAACAATTTGAATTCATTAAAGAGGAGAAATTAACTATGAGAGGATCGATCCATATG...
     lac operator                   S/D       Initiation       Cla I  Nde I
                                                codon
                                                                        └──► Rev
```

Fig. 2. Sequences and motifs within the expression region of the *lacZ* reporter plasmid and the RNA-BP plasmid. **(A)** Reporter plasmid: The reporter plasmid, pLacZ-Rep, is calculated to be 5752 basepairs (bps) in size. In addition to the expression region shown, the plasmid also contains an IS10-*lacZ* gene fusion that runs clockwise from bps 186–3440, a *col*E1 origin of replication, and an ampicillin-resistance gene that runs counterclockwise from bps 5367 to 4527. A unique *Hin*dIII site at bp 372 is present at the junction of IS10 and *lacZ* sequences. The positions of other unique restriction sites are *Spe*I (106), *Kpn*I/*Acc*65I (144), *Bsp*EI (162), *Eco*RI (170), *Bsu*36I (603), *Eco*RV (1491), *Dra*III (1565), *Bss*HII (1878), *Sac*I/*Ecl*136II (2316), *Acc*I (3143), *Bst*1107I (3143), *Bsi*WI (3151), *Blp*I (3392), *Sap*I (3555), *Ahd*I (4566), *Bsa*I (4638), *Sca*I (5049), and *Eco*O109I (5546). The transcription and translational signals of the *lacZ* expression plasmid are indicated in the figure. The –35 and –10 regions of the IS10 promoter are overlined. The minimal S/D sequence and the initiation codon are underlined. Other features include a T7 RNA polymerase promoter (TTAATACGACTCACTATAGG), which can be used to produce RNA in vitro, and translational terminators in all three frames, that may help reduce background levels of β-galactosidase. Upstream of the –35 region (not shown) are situated transcriptional terminators derived from the *E. coli rrnB* operon that may also help reduce background β-galactosidase expression. **(B)** RNA-BP expression plasmid: The expression plasmid, pREV1, is calculated to be 3647 bps in size. This plasmid, derived from pACYC184 *(6)*, contains a chloramphenicol-resistance gene and a P15A origin of replication. There is also an f1 origin of replication in the plasmid which can be used to produce single stranded DNA corresponding to the sense strand in *E. coli*. The Rev gene spans bps 2627–2974, upstream of which are unique *Cla*I and *Nde*I sites at positions 2617 and 2624, respectively (underlined). Downstream of the Rev gene are a unique *Bsu*36I restriction site at position 2989, and a *Sal*I

3. Methods
3.1. Reporter Plasmid (pLacZ-Rep)

New RNA targets can be conveniently introduced into pLacZ-Rep by polymerase chain reaction (PCR) cloning. The RNA target should be present as close as possible to the S/D region, to maximize the potential for translational repression when the RNA-BP is bound to its target. Thus, an oligonucleotide corresponding to the sequence 5' AAAGGATCC[minimal RNA target sequence]AGACAACAAGATGTGCGAAC 3' used in conjunction with a primer complementary to the *lacZ* gene (e.g., 5' CGACGGGATCGATCCCCC 3') can be used for PCR, using pLacZ-Rep as a template. The purified ≥ 0.25 kb PCR product is then digested with *Kpn*I and *Hin*dIII, and cloned between the unique *Kpn*I and *Hin*dIII sites of pLacZ-Rep. The resulting plasmid construct places the RNA target immediately upstream of the IS10 S/D sequence. Alternatively, the RNA target can be cloned directly between the unique *Kpn*I and *Eco*RI sites in pLacZ-Rep.

3.2. RNA–BP Expression Plasmid (pREV1)

Replacement of the Rev gene with a DNA segment encoding a different RNA-BP can be achieved by excising the Rev gene using the unique restriction sites *Cla*I or *Nde*I preceding the Rev gene, and the restriction sites *Sal*I or *Bsu*36I downstream of the Rev gene, and by substituting with a DNA segment containing the new RNA-BP gene. Care should be taken to ensure that the replacement allows the RNA-BP to be translated in frame with respect to the short open reading frame preceding the Rev gene (*see* **Fig. 2B**). Thus, precise cloning of a heterologous RNA-BP into pREV1 may require prior introduction of restriction sites overlapping the initiation codon. Alternatively, a set of primers that overlap the beginning and end of the RNA-BP coding region and also introduce appropriate restriction sites can be used for PCR amplification of the RNA-BP gene, followed by cloning into pREV1. If the RNA-BP gene contains internal restriction sites for the enzymes mentioned previously, *Cla*I compatible restriction sites (e.g., GGCGCC, AACGTT, and TTCGAA which are

---

restriction site at position 2984 (another SalI site is present at bp 2751 within the Rev gene). The positions of some other unique restriction sites are *Nhe*I (593), *Ase*I (1405), *Not*I (1436), *Eco*RV (1450), *Dra*III (1728), *Xho*I (2497), *Spe*I (2932), *Sca*I (3233), and *Nco*I (3347). The transcription and translation signals of the expression plasmid are indicated. The –35 and –10 regions of the strong P_{A1} promoter are overlined. The presence of dual *lac* operator sites allows expression of this promoter to be regulated by Lac repressor. The S/D sequence and the initiation codon are underlined. This plasmid expresses the wild-type HIV-1 Rev protein preceded by a short peptide leader corresponding to the amino acids MRGSIH.

digested by the restriction enzymes *Nar*I, *Psp*1406I, and *Bst*BI, respectively), an *Nde*I compatible restriction site (ATTATT, which is a substrate for *Ase*I) or a *Sal*I compatible restriction site (CTCGAG, a substrate for *Xho*I) may be used instead (*see* **Note 1**).

3.3. Testing for Translational Repression

Once the new reporter and expression plasmids have been made, it is a relatively straightforward matter to determine whether the heterologous RNA-BP can repress expression of the modified *lacZ* reporter when expressed in *E. coli*. Experimentally, this is done by co-transforming the two plasmids into the *lacZ*⁻ *E. coli* strain WM1/F', plating on X-gal indicator plates, and incubating overnight at 37°C. Expression of β-galactosidase and cleavage of X-gal by this enzyme yields a blue product imparting color to colonies expressing this enzyme. The colony color intensity is directly related to the amount of β-galactosidase activity produced by the cell, and it is often possible to distinguish between colonies that have a twofold difference in β-galactosidase activity. Accordingly, the colony color phenotype of individual colonies containing the reporter plasmid and the RNA-BP expression plasmid is compared with that of colonies containing the reporter plasmid and a compatible control plasmid, such as pACYC184 *(6)* to determine whether *lacZ* expression is impaired because of the presence of the RNA-BP in the cell. A significant reduction in the blue color of colonies containing the expression plasmid, as compared to control colonies, is an indication that translational repression has been achieved.

In practice, determining conditions for translational repression may require finding optimal conditions for expression of the RNA-BP. The signals for transcription and translation of RNA-BPs in pREV1 are strong, and expression of some RNA-BPs at high levels may be toxic for *E. coli*. However, the promoter expressing the RNA-binding protein in pREV1 is regulated by Lac repressor and transcription can be reduced substantially in strains that express elevated levels of Lac repressor, such as WM1/F'. In these strains, intermediate levels of transcription can be achieved by the addition of the *lac* inducer, IPTG, to the growth medium at micromolar concentrations.

3.3.1. Visualizing Translation Repression

1. Cotransform the reporter plasmid with either a control plasmid (e.g., pACYC184) or the RNA-BP expressing plasmid using competent WM1/F' cells (*see* **Note 2**).
2. Plate aliquots of both transformation mixes on LB-agar plates containing ampicillin, chloramphenicol, X-gal, and 0, 1, 2, 5, 10, 20, 50, or 1000 μM IPTG. Incubate plates at 37°C overnight.
3. The next day, examine the colonies and compare the colony color phenotype of isolated transformants containing the expression plasmid with transformants con-

taining the control plasmid. Parallel comparisons should be made for colonies grown on plates containing the same level of IPTG. Determine the lowest concentration of IPTG on plates that allows expression of the RNA-BP to cause a marked reduction in the intensity of the blue color relative to control transformants. The β-galactosidase activity of these transformants can then be determined spectrophotometrically by assaying cell extracts.

3.3.2. β-Galactosidase Assays

1. From plates containing the lowest concentration of IPTG necessary to visualize a color difference, disperse four to six colonies corresponding to each transformation into separate tubes containing 1 mL of LB supplemented with ampicillin, chloramphenicol, and IPTG at the same concentration present in the plates. Colonies from plates containing higher concentrations of IPTG may also be inoculated into growth media containing the higher concentrations of IPTG. Grow overnight with shaking at 37°C.
2. On the next day, inoculate tubes that each contain 1.5 mL of LB + antibiotics + IPTG with 45 µL of the overnight culture. Grow at 37°C for 1.5–2.0 h or until the cells reach mid-log phase (OD_{600} ~ 0.5).
3. Place the tubes on ice for 20 min.
4. Pour the cultures into polystyrene cuvets. After wiping away any condensation on the outside of the cuvets, measure OD_{600} using a cuvet containing the growth medium as a blank.
5. Prepare several glass culture tubes, each containing 0.5 mL of *lacZ* buffer supplemented with 0.01% SDS and 50 m*M* β-mercaptoethanol. Dispense two drops of chloroform into each tube. Add 0.5 mL of cell culture to each tube. Also prepare a blank tube to which 0.5 mL LB medium has been added. Vortex each tube vigorously for 15 s to permeabilize the cells, and incubate for 5 min in a 28°C water bath.
6. Add 200 µL of *o*-nitrophenyl-β-D-galactopyranoside (ONPG) solution to the tubes, noting the time at which the solution was added to each tube. Allow the ONPG enzymatic hydrolysis reaction to proceed at 28°C until the color in tubes becomes yellow. To stop a reaction, add 0.5 mL of 1 *M* sodium carbonate solution and vortex the tube briefly, noting the time at which each reaction was stopped. If no significant color develops within 3–5 h, stop the reaction anyway. Transfer the solutions to microfuge tubes, centrifuge in a microfuge for 5 min at 16,000*g* (14,000 rpm), and pour the supernatants into polystyrene cuvets, taking care to avoid the transfer of chloroform droplets.
7. Measure the OD_{420}, and then subtract the reading obtained for the LB blank reaction.
8. The β-galactosidase activity of each culture (in Miller units) = $(2000 \times OD_{420})/(\Delta T \times OD_{600})$, where ΔT is the time interval of the ONPG hydrolysis reaction in minutes *(7)*.

Assuming minimal experimental error, some positive indications that the expression of the RNA-BP is having no adverse effects on cell growth are that independent cultures grow well upon subculturing and that there is little sample-to-sample variation in the level of β-galactosidase activity (i.e., the fractional standard deviation of β-galactosidase activity from independent cultures expressing the RNA-BP should not be significantly greater than that of cultures containing the control plasmid). In that case, the amount of repression by the RNA-BP can be quantitated. Thus, at any IPTG concentration, if the mean β-galactosidase activity in strains containing the reporter plasmid and the control plasmid is X, and the mean β-galactosidase activity in strains containing the reporter plasmid and the RNA-BP expression plasmid is Y, then a quantity defined as the repression ratio (R) can be calculated ($R = X/Y$). The repression ratio is a measure of the degree of translational repression due to binding of the RNA-BP to its RNA target in *E. coli*. For the genetic strategies described below to be useful, conditions under which the RNA-BP represses expression of the reporter by a factor of five or greater are generally necessary. If $R \leq 5$, then the cultures should be grown and assayed in a medium containing higher levels of IPTG, to allow more RNA-BP to be produced. On the other hand, if there is significant sample-to-sample variation in the β-galactosidase activity of independently derived cultures expressing the RNA-BP at any IPTG concentration, this may be an indication that expression of the RNA-BP is having a toxic effect on *E. coli*. In that case, cells should be grown using lower levels of IPTG in the growth medium (*see* **Note 3**).

Once repression has been demonstrated, it may be desirable to confirm that it is due to specific binding of the RNA-BP to its RNA target. The simplest way to obtain evidence of specific binding is to introduce known disruptive mutations into the RNA target or into the RNA-BP expression plasmid. Introduction of mutations that have previously been shown to impair RNA–protein interactions in vitro should lead to loss of translational repression, providing evidence that the repression observed with the wild-type plasmids is due to specific binding of the RNA-BP to its RNA target in *E. coli*.

3.4. Applications of the Genetic System

3.4.1. Identification of Mutants that Affect Binding Affinity

Once conditions for translational repression have been established, the genetic system can be used to screen for RNA-BP mutants that bind to the RNA target with altered affinity. The identification of such variants is based on the expected correlation between the affinity of the RNA-BP for its RNA target and the degree of translational repression observed in *E. coli* (**Subheading 3.4.2.**). Thus, variants of the RNA-BP that increase binding in vivo should repress *lacZ* synthesis to a greater extent than the wild-type protein, whereas mutants that are impaired for binding should repress *lacZ* expression less effectively. Using indicator plates, it should be possible to identify such variants from a pool of RNA-BP mutants by observing the colony color phenotype of transformed colonies.

There are several variations on this theme that can be implemented using the genetic system. In the most general case, it might be of interest to the investigator to generate a large number of mutants of the RNA-BP, and to identify the subset of mutants that affect RNA binding and translational repression. A comprehensive set of mutants will allow RNA-binding surfaces of the RNA-BP to be delineated (an especially detailed experimental analysis for the MS2 coat RNA-BP used a similar genetic method for identifying defective protein variants *[8]*). If little is known about the RNA binding determinants of the RNA-BP, it is probably best to randomly mutagenize the entire protein and screen for variants that affect *lacZ* expression, followed by sequencing to determine which amino acid changes are responsible for the altered repression phenotype. In situations where RNA-BP regions important for target recognition have been identified, detailed information can be obtained by localized mutagenesis and screening for variants that affect *lacZ* repression. Finally, the contribution of single amino acids can be tested by mutating or randomizing a specific codon and identifying variants that affect *lacZ* activity.

From a practical standpoint, there are many different ways of mutagenizing DNA. In general, it is usually desirable that the overall mutagenic frequency be kept at a level corresponding to an average of about one amino acid change per RNA-BP gene or slightly less. Significantly lower rates of mutagenesis will increase the amount of screening necessary to find interesting mutants, whereas higher rates of mutagenesis may complicate the analysis of multiply mutated variants that have been identified by genetic screening.

When small regions of a protein are to be mutagenized, a convenient way to do so is by using a "doped" oligonucleotide that corresponds to the sequence of the region to be mutagenized, but is spiked with nonwild-type nucleotides during synthesis. For example, if a region encoding 10 amino acids is being mutagenized, then by adjusting the fraction of nonwild-type bases incorporated at each of the 30 nucleotide positions to $\frac{1}{30}$ (1.1% of each of the three nonwild-type nucleotides), a significant proportion of the oligonucleotides will contain just one altered codon. Subsequently, the oligonucleotide can be incorporated into the coding region of the RNA-BP by site-directed mutagenesis. To facilitate this process, pREV1 contains an f1 origin of replication, which is useful for the production of single-stranded DNA, an intermediate in some oligonucleotide mutagenesis protocols *(9,10)*.

When mutagenesis of larger coding regions (> 50–100 basepairs) is required, oligonucleotide-based mutagenesis can become impractical. Although there is no perfect method for evenly mutagenizing large regions of DNA, a useful method currently available is by error-prone PCR *(11)*. This method relies upon the generation of mutagenized DNA by amplification of DNA under conditions that increase the rate of misincorporation of nucleotides by *Taq* polymerase. Subsequently, the mutagenized DNA is recloned into the protein expression vector. A more detailed discussion of the considerations involved in PCR-based mutagenesis is provided in **ref. *12*.**

After a pool of plasmid mutants has been generated, it can be directly transformed into competent cells containing the reporter plasmid (*see* **Note 4**). Alternatively, the mutant library may be first transformed into high-efficiency electrocompetent cells and amplified. An aliquot of the amplified DNA is then transformed into competent cells containing the reporter plasmid.

After transformation, colonies are plated on indicator plates containing ampicillin, chloramphenicol, X-Gal, and if appropriate, IPTG. In parallel, control transformants containing the reporter plasmid and the wild-type RNA-BP expression plasmid should be plated as well. After overnight incubation at 37°C, individual colonies can be picked according to colony color phenotype (enhanced or diminished blue color as compared to transformants expressing the wild-type RNA-BP, as desired).

Many of these transformants will be expected to express variants of the RNA-BP that bind to the RNA target with altered affinity. However, a sizable fraction of these variants may affect the *lacZ* phenotype of the transformed colonies not because they change the binding affinity, but because they alter the concentration of the RNA-BP in *E. coli*. A simple way to identify and discard this class of mutants is by preparing protein lysates from mid-log cultures of cells containing the wild-type RNA-BP or candidate RNA-BP mutants (grown in LB supplemented with chloramphenicol, and IPTG, if appropriate). Following polyacrylamide gel electrophoresis and dye staining or Western-blot analysis using antibodies specific to the RNA-BP (or to an epitope tag engineered at one end, *see* **Note 1**), the amount of RNA-BP expressed in each case is evaluated and transformants that express significantly altered amounts of protein can be discarded. Thereafter, attention may be focused upon variants that affect *lacZ* expression without significantly perturbing the concentration of the RNA-BP in *E. coli*. In situations where it is not possible to visualize RNA-BP levels directly, an indirect strategy that measures growth inhibition may be helpful to distinguish between variants that are genuinely defective for RNA-binding and variants that are differentially expressed in *E. coli*. This strategy is implemented by growing overnight cultures of candidate mutants, as well as the wild-type RNA-BP containing cells in LB-chloramphenicol medium lacking IPTG. Dilute the saturated overnight cultures 1:100 into LB-chloramphenicol medium containing IPTG at levels that are somewhat inhibitory for growth of the wild-type RNA-BP. After growth for ~ 2–3 h at 37°C, the optical density of cell cultures is measured at 600 nm. Cultures that grow to the same optical density as wild-type RNA-BP transformants are likely to express RNA-BP variants at similar levels in the cell, whereas cultures that express RNA-BP variants at increased or decreased levels usually grow to lower or higher densities, respectively and may be discarded.

After a set of interesting candidates has been identified through screening, the *lacZ* phenotype should be confirmed by assaying for β-galactosidase activity (**Subheading 3.3.2.**). The phenotype of potential mutants should be confirmed prior to detailed in vitro analysis by isolating plasmid DNA, retransforming into the reporter strain, and performing a second set of β-galactosidase activity assays. In this manner it should be possible to identify amino acids involved in RNA binding.

There are two other similar applications that can provide useful information about RNA–protein interactions. First, the RNA target can be mutagenized in a manner analogous to the RNA-BP. Thereafter, the mutagenized reporter library is transformed into strains containing the wild-type RNA-BP, followed by screening for RNA mutants that are differentially repressed by the RNA-BP. In this way, regions of the RNA target involved in RNA–protein interaction can be identified. The mutants can be characterized further by measuring the β-galactosidase activity produced from the mutant reporters in strains containing the wild-type RNA-BP plasmid or pACYC184, and by calculating repression ratios.

A second variation involves the use of the genetic system to identify altered specificity mutants of the RNA-BP. Altered specificity mutants are characterized by their ability to restore high affinity binding to specific defective RNA targets. A major benefit of identifying altered specificity mutants is that such protein variants often provide evidence of direct contacts between specific amino acids in the protein and nucleotides in the RNA, knowledge of which can be used to provide constraints in building structural models of RNA–protein complexes. References *(1,2)* describe how the identification of altered specificity mutants allowed RNA–protein complexes to be modeled.

3.4.2. Assessment of the Binding Affinity of RNA or RNA-BP Mutants

The genetic system, as developed, allows expression of the *lacZ* reporter at low levels and the RNA-BP at significantly higher levels in *E. coli*. A simple model for translational repression predicts that there should be a correlation between the degree of translational repression by the RNA-BP in *E. coli* and the binding affinity of the protein for its RNA target *(12)*. Specifically,

$$R - 1 \propto 1/K_d \tag{1}$$

The validity of this relationship was tested by comparing the repression ratios for variants of the RNA binding proteins Rev and nucleolin with the disso-

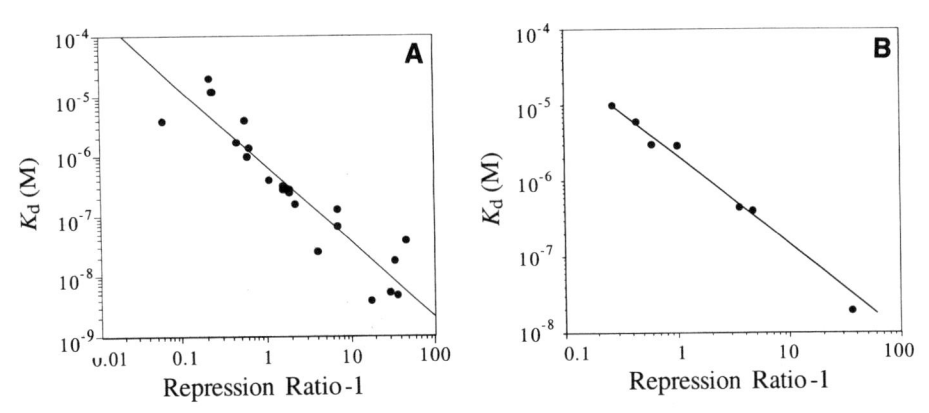

Fig. 3. Correlation between dissociation constants measured in vitro and repression ratios measured in vivo. A number of mutants of Rev and nucleolin were isolated and purified *(11,12)*. These proteins were used for gel-shift RNA-binding assays in vitro to determine the dissociation constant for complexes with wild-type and mutant RNA targets. The same protein/RNA combinations were also used to determine repression ratios (R) in *E. coli*. The dissociation constants measured in vitro were plotted on a log-log scale vs $R - 1$ observed in vivo. Best-fit straight lines have been drawn through the data points. **(A)** Rev and its mutants **(B)** nucleolin and its mutants.

ciation constants measured in vitro using purified RNA and protein. The relationship between the two parameters plotted on a log-log scale is shown in **Fig. 3** for both proteins. In both cases, there is a linear correlation, suggesting that linear relationships between $R - 1$ measured in vivo and K_d measured in vitro may be expected for other RNA-BPs.

The possibility of obtaining reliable estimates of the binding affinity of RNA-BP variants using a simple spectrophotometric assay to measure repression ratios should greatly expedite the characterization of RNA-BPs. Once a large number of variants have been identified using genetic screening methods described in the previous section, a few (six to eight) selected variants may be purified and tested for binding in vitro to determine dissociation constants (*see* Chapters 9 and 10). A standard graph can be drawn, similar to the ones shown in **Fig. 3**, using the dissociation constants and repression ratios measured for the mutants. With the help of this graph, the dissociation constants for the remaining mutants can be estimated by interpolation after measuring the repression ratios for these mutants in *E. coli*. Alternatively, dissociation constants can be inferred directly using Eq. 1. For example, if the repression ratio with the wild-type protein is 20 and for a particular protein mutant it is 1.5, then it can be inferred that the mutant protein binds to the RNA target more weakly by a factor of 38 [(20 – 1)/(1.5 – 1)]. Thus, if the dissociation constant of the wild-type protein–RNA complex is 10 nM, then the mutant protein will be expected to bind to the RNA target with a K_d of 380 nM. Of the two methods for estimating dissociation constants, the first method, which involves the generation of a standard graph, is preferable, as the relationship between R-1 and K_d may not be exactly linear [with Rev, the slope of the best-fit line for log (K_d) plotted against log (R – 1) is –1.25, indicating that $1/K_d \propto (R – 1)^{1.25}$. With nucleolin, the slope is –1.1].

4. Notes

1. Notes on construction of the RNA-BP plasmid: (a) One caveat to digesting pREV1 with *Cla*I is that its site in pREV1 overlaps the sequence GATC (**Fig. 2B**), which is methylated by the *E. coli* enzyme Dam methylase. As a result the DNA becomes refractory to *Cla*I digestion. To effect cleavage by *Cla*I, it is therefore necessary to extract pREV1 DNA from a *dam*– strain. (b) The sequence of the degenerate *Bsu*36I site in pREV1 is 5' CCTCAGG 3'. (c) During the cloning stage, it may also be desirable to engineer a small antibody epitope tag, particularly if mutants of the RNA-BP will be screened subsequently, and antibodies reacting with the RNA-BP are unavailable (*see* **Subheading 3.4.1.**). (d) The ligation reactions for construction of new RNA-BP plasmids should be transformed into WM1/F' or any other *lac*I^q strain to avoid possible toxic effects of high-level RNA-BP expression.

2. Notes on *E. coli* transformation: WM1 or WM1/F' competent cells can be made using a modified CCMB method *(12)*, although other methods also work well. For cotransformation of plasmids, in general 10–100 ng of each plasmid should be adequate to obtain several transformants containing both plasmids. After transformation and expression of antibiotic resistance, the transformation mix should be spread over one half of a dry indicator plate, and using a sterile wire loop or glass rod, some of the mix should be streaked out on the other half. In this manner, it will usually be possible to get individual, well isolated colonies to appear following overnight incubation.

3. Notes on optimizing RNA-BP mediated translational repression: The initial experiments for finding optimal conditions for translational repression should be directed toward determining conditions of RNA-BP expression that confer a significant degree of *lacZ* translational repression without seriously affecting cell growth. In the course of these initial experiments, it may become apparent that expression can be induced at maximal levels (by adding 1 mM IPTG to the growth medium) without causing toxicity. In that case, further experiments can be conducted in the strain WM1, which is isogenic with WM1/F' but lacks the F' episome containing the Lac repressor gene. In this strain, the RNA-BP is constitutively expressed at the maximal level, and further addition of IPTG to plates or liquid media becomes unnecessary.

4. Notes on transforming mutagenized DNA and screening for mutants: In situations where it is desirable to screen a large number of mutants, as may be the case if a large region of the RNA-BP coding region is being mutagenized, it may be useful to consider the following suggestions: First, prior to transformation, an estimate of the number of transformants to be screened should be made. For example, if the size of the region being mutagenized is 500 basepairs, and the goal is to introduce point mutants over this region, then the number of possible single base mutants is 1500. For a library containing an average of 0.5 mutations per plasmid, screening of 3000 transformants will be necessary for reasonable coverage of possible mutants. In practice, one may desire to screen two or three times that number to be reasonably confident that most of the interesting mutants have been screened. Once the desired number of transformants to be screened is calculated, one half of the mutagenized library can be transformed into competent cells of WM1/F' containing the reporter plasmid (or WM1/F' containing the RNA-BP plasmid, if it is the target RNA on the reporter plasmid that is being mutagenized). After expression of antibiotic resistance, a small aliquot of the transformation mix should be diluted in LB medium and plated on LB plates containing ampicillin and chloramphenicol followed by overnight incubation at 37°C, whereas the bulk of the transformation mix should be made up to 20% in glycerol and stored at –70°C. The next day, the number of colonies are counted, providing an estimate of the total number of transformants in the frozen sample. If that number is significantly lower than the number of transformants desired, then it would be advisable to transform the other half of the mutagenesis mix into very highly competent cells (e.g., electrocompetent DH12S cells obtained from BRL) and to use amplified DNA for genetic screening, or to redo the mutagenesis on a larger scale. On the other hand, if the number of transformants exceeds the calculated number, then the frozen culture may be thawed, diluted appropriately, and evenly plated on a number of indicator plates at a density of 250–500 transformants per 8.5 cm diameter plate (higher plating densities may result in the colony color phenotype not being clearly distinguishable).

Conclusions

The genetic system for RNA-binding proteins described in this chapter can potentially be applied to any cloned RNA-BP whose RNA target has been identified. As such, it offers a simplified method for identifying and studying mutants that affect RNA–protein interactions, and may develop into a standard technique for studying RNA-BPs in the future. Already, this method has been tried for several RNA-BPs, and in most cases, a significant degree of translational repression was achieved. Therefore, it appears likely that the method may be applicable for a variety of RNA-binding proteins. However, there may be some circumstances in which translational repression does not take place. Some of the possible reasons for such difficulties and troubleshooting advice have been discussed *(12)*.

Acknowledgments

I am grateful to Dr. Joel Belasco for providing encouragement to write this chapter. I would also like to thank Drs. Joel Belasco, Deborah Lu, and Martina Rimmele for critical reading of the manuscript and for providing constructive suggestions. I am grateful to Academic Press for giving permission to use **Figs. 1** and **3A**, which are similar to figures in **ref. *12***.

References

1. Jain, C. and Belasco, J. G. (1996) A structural model for the HIV-1 Rev-RRE complex deduced from the altered specificity of Rev variants isolated by a rapid genetic screening strategy. *Cell* **87**, 115–125.
2. Bouvet, P., Jain, C., Belasco, J. G., Almaric, F., and Erard, M. (1997) RNA recognition by the joint action of two nucleolin RNA-binding domains: genetic analysis and structural modeling. *EMBO J.* **16**, 5235–5246.
3. Dreyfus, M. (1988) What constitutes the signal for the initiation of protein synthesis on *E. coli* mRNAs? *J. Mol. Biol.* **204**, 79–94.
4. Gold, L. (1988) Posttranscriptional regulatory mechanisms in *Escherichia coli*. *Annu. Rev. Biochem.* **57**, 199–233.
5. de Smit, M. H. and van Duin, J. (1990) Secondary structure of the ribosome binding site determines translational efficiency: a quantitative analysis. *Proc. Natl. Acad. Sci. USA* **87**, 7668–7672.
6. Rose, R. E. (1988) The nucleotide sequence of pACYC184. *Nucleic Acids. Res.* **16**, 355.
7. Miller, J. H. (1972) in *Experiments in Molecular Genetics*, Cold Spring Harbor Laboratory Press, Cold Spring Harbor, NY.
8. Peabody, D. (1993) The RNA binding site of bacteriophage MS2 coat protein. *EMBO J.* **12**, 595–600.
9. Kunkel, T. A., Bebenek, K., and McClary, J. (1991) Efficient site-directed mutagenesis using uracil-containing DNA. *Methods Enzymol.* **204**, 125–139.
10. Olsen, D. B., Sayers, J. R., and Eckstein, F. (1993) Site-directed mutagenesis of single-stranded and double-stranded DNA by phosphorothioate approach. *Methods Enzymol.* **217**, 189–217.
11. Cadwell, R. C. and Joyce, G. F. (1992) Randomization of genes by PCR mutagenesis. *PCR Methods Appl.* **2**, 28–33.
12. Jain, C. and Belasco, J. G. (1997) A genetic method for the study of RNA-binding proteins, in *mRNA Formation and Function* (Richter, J., ed.), Academic Press, New York, NY, pp. 263–284.

14

Screening RNA-Binding Libraries Using a Bacterial Transcription Antitermination Assay

Kazuo Harada and Alan D. Frankel

1. Introduction

A number of genetic assays have recently been developed for detecting RNA–protein interactions, several of which have been applied to screen cDNA or combinatorial libraries for RNA-binding peptides and proteins (*1–5*) (*see* Chapters 13 and 15). These methods are useful for understanding the mechanisms of RNA–peptide and RNA–protein interactions, and may provide tools for designing compounds of therapeutic interest.

Here we describe the use of a bacterial genetic assay for detecting polypeptide–RNA interactions, which was used to screen combinatorial libraries for novel arginine-rich peptides that bind to the Rev-response element (RRE) of human immunodeficiency virus (HIV) (*1*). The method utilizes the transcription antitermination activity of the bacteriophage λ N protein, which causes RNA polymerase to read through transcription termination sites by forming an antitermination complex (**Fig. 1A**). Formation of this complex is mediated partly through the interaction of the N protein with an RNA element in the nut site of the nascent RNA transcript called box B. A bacterial two-plasmid reporter system has been devised that accurately monitors λ N-mediated antitermination (*6*), and this system has been modified to study heterologous polypeptide–RNA interactions (*1*).

In the two-plasmid system, the N protein is expressed from one plasmid and binds to the nut site RNA expressed from a second plasmid, resulting in transcription antitermination of a reporter gene. By replacing the RNA-binding domain of N with a heterologous binding domain and replacing the nut site with a corresponding target RNA, the RNA–protein interaction can be monitored by measuring reporter gene expression. In the system, N is expressed

From: *Methods in Molecular Biology, Vol. 118: RNA-Protein Interaction Protocols*
Edited by: S. Haynes © Humana Press Inc., Totowa, NJ

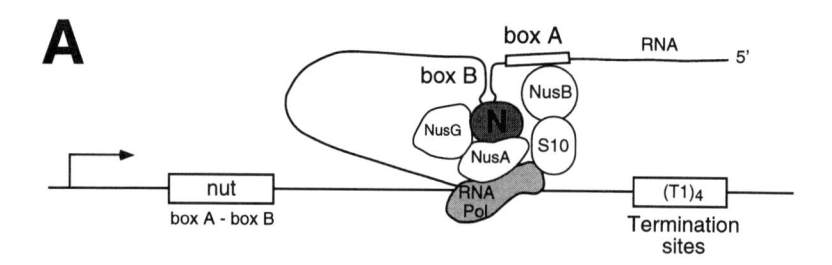

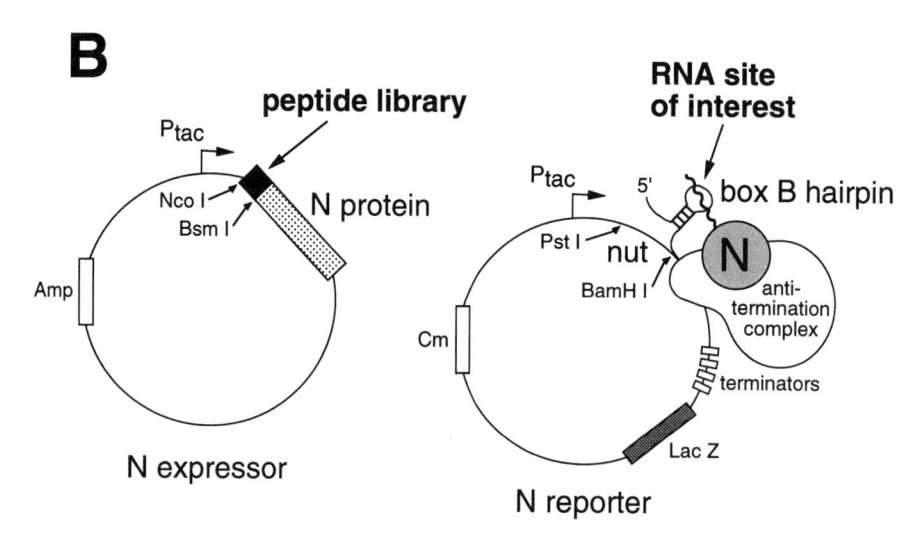

Fig. 1. Antitermination by the λ N protein and the two-plasmid system for detecting polypeptide–RNA interactions. (A) The λ N antitermination complex. *E. coli* proteins known to participate in antitermination and the approximate arrangement of the complex are shown *(18)*. (B) The two-plasmid system for measuring transcription antitermination by the λ N protein *(6)*.

under the control of the *tac* promoter on a pBR322-derived plasmid. The reporter gene, *lacZ*, is expressed on a compatible pACYC plasmid, also under the control of a *tac* promoter, and has a nut site and termination sites upstream so that binding of N to box B is required to assemble the antitermination complex, resulting in the expression of β-galactosidase *(6)* (**Fig. 1B**). The RNA-binding domain of the N protein, which is a short arginine-rich domain located at the N-terminus of the protein, can be replaced by other RNA-binding polypeptides by cloning synthetic oligonucleotide cassettes into the unique *Nco*I and *Bsm*I sites of the N-expressor plasmid, thereby creating fusion proteins at

A

```
          M  D  A  Q  T  R  R  R  E  R  R  A  E  K  Q  A  Q  W  N  A
    5'  CATGGATGCACAAACACGCCGCCGCGAACGTCGCGCAGAGAAACAGGCTCAATGGAATGCA  3'
    3'      CTACGTGTTTGTGCGGCGGCGCTTGCAGCGCGTCTCTTTGTCCGAGTTACCTTAC    5'
        NcoI                                                   BsmI
```

B

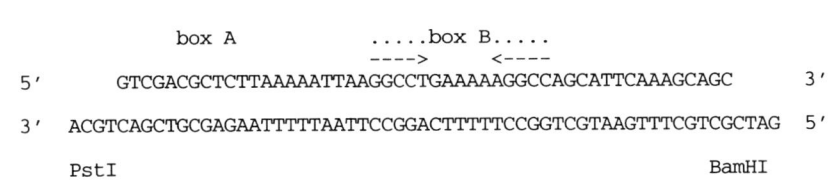

```
              box A             .....box B.....
                     ---->         <----
    5'      GTCGACGCTCTTAAAAATTAAGGCCTGAAAAAGGCCAGCATTCAAAGCAGC         3'
    3'  ACGTCAGCTGCGAGAATTTTTAATTCCGGACTTTTTCCGGTCGTAAGTTTCGTCGCTAG     5'
        PstI                                                  BamHI
```

Fig. 2. Regions of the N-expressor and N-reporter plasmids used for cloning. (**A**) Sequence of the DNA cassette encoding the RNA-binding domain (amino acids 1–20) of the N protein (the amino acid sequence is shown above) in the N-expressor plasmid. Expressor plasmids containing heterologous RNA-binding polypeptides are prepared by replacing the RNA-binding segment of N with *Nco*I–*Bsm*I fragments encoding the polypeptide or combinatorial polypeptide library of choice. (**B**) The sequence of the nut (box A–box B) region of the N-reporter plasmid. Reporter plasmids containing the heterologous RNA targets are prepared by replacing the box A–box B segment of pACnut⁻TAT13 *(6)* with a synthetic *Pst*I–*Bam*HI fragment where the 15 nt box B stem-loop (indicated by the dotted line) is replaced by the target RNA sequence.

amino acid 19 of the N protein (**Fig. 2A**). To prepare N-reporter plasmids with heterologous RNA sites, oligonucleotides containing box A of the nut site and the heterologous RNA site in place of box B are cloned into the unique *Pst*I and *Bam*HI sites of pACnut⁻TAT13 (**Fig. 2B**).

This chapter focuses on using the antitermination system for identifying novel polypeptide–RNA interactions from combinatorial libraries. Although most of our studies have used short arginine-rich sequences as frameworks for generating libraries, the protocols may also be adapted to studying larger RNA-binding protein frameworks *(7)*. The system is described in four parts:

1. The β-galactosidase colony color assay described in **Subheading 3.1.** is used for semiquantitative assessment of antitermination activities and for library screening. N-expressor and N-reporter plasmids are transformed into *E. coli* N567 host cells and blue colony color is scored on plates using a relative scale of zero to

five plusses, with wild-type nut reporter cells in the absence of N scoring zero (or –) and the nut reporter in the presence of the wild-type N scoring +++++ *(1)*.
2. More quantitative assessment of β-galactosidase is carried out using a solution assay as described in **Subheading 3.2**. In general we have observed that colony color intensity on plates and the level of β-galactosidase activity measured in solution correlate well with in vitro binding affinities. At low activities, however, the colony color assay appears to be considerably more sensitive than the solution assay, presumably because colony color reflects β-galactosidase accumulated during up to 48 h of growth while activity in solution reflects enzyme accumulation after 1 h of isopropyl-β-d-thiogalactopyranoside (IPTG) induction. The high sensitivity of the colony color assay makes it ideal for use in the initial library screens.
3. Preparation of the combinatorial oligonucleotide cassette and ligation of the library into the library-N-expressor plasmid is described in **Subheading 3.3**. We show an example (**Fig. 3**) of a degenerate oligonucleotide encoding a 14-residue library consisting of four amino acids and enriched in arginines, that we used to identify peptides that bind to the HIV RRE site *(1)*. The library was designed to consist only of arginine, serine, aspargine, and histidine (RSNH library). The degenerate oligonucleotide possesses a 3'-flanking sequence to which a primer will bind, allowing synthesis of the second strand, and restriction sites at both the 5' and 3' ends to allow ligation of the double stranded cassette into the *Nco*I and *Bsm*I sites of the N-expressor plasmid (**Fig. 3**).

The strategy used to design combinatorial peptide libraries is critical because the number of bacterial colonies that can be conveniently screened using the colony color assay is limited to approx 10^6–10^7. Thus, the investigator must devise ways to limit the complexity while still ensuring an efficient search of sequence space. We have used two protocols for generating combinatorial libraries encoded by synthetic oligonucleotides. We refer to the first protocol as "randomization," which is carried out either by completely randomizing a small number of amino acid positions using all 20 amino acids, or a larger number of positions using a small subset of the 20 amino acids *(1)*. The second protocol involves "doping" of a prototype sequence by codon-based mutagenesis *(8,9)*. Nucleotide-based methods such as mutagenic polymerase chain reaction (PCR) are advantageous for low-level "doping" of large DNA fragments *(10,11)*, but are biased in the amino acid changes that arise because each codon position will only have one nucleotide change, resulting in an amino acid substitution to an average of only 5.7 of the 19 other amino acids *(12)*. The strategy for library construction depends on the particular application, and general considerations for library selection experiments have been discussed in detail *(13,14)*.
4. Finally, we describe the selection of novel RNA-binding polypeptides from large combinatorial libraries (**Subheading 3.4.**). The antitermination system allows the identification of tight RNA-binding peptides from mixtures of peptides having a wide range of affinities toward a particular RNA target *(1)*. It is possible to identify strong binders from 10^6 to 10^7 sequences, the practical limit for colony

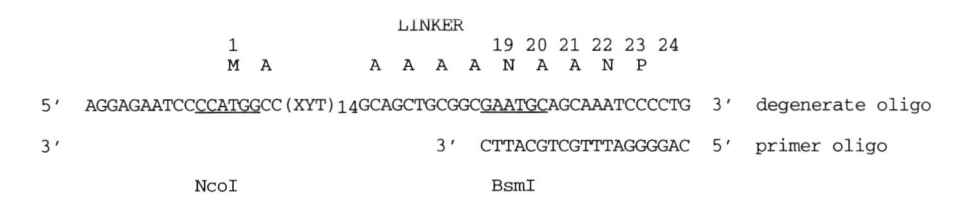

```
                       LINKER
            1                  19 20 21 22 23 24
            M  A       A A A A  N  A  A  N  P
5'  AGGAGAATCCCCATGGCC(XYT)14GCAGCTGCGGCGAATGCAGCAAATCCCCTG  3'   degenerate oligo
3'                                 3'  CTTACGTCGTTTAGGGAC   5'   primer oligo
              NcoI                          BsmI
```

Fig. 3. Example of a library oligonucleotide used to screen for RRE-binding peptides. A synthetic oligonucleotide encoding a randomized 14-mer peptide library consisting of arginine, serine, aspargine, and histidine (RSNH library) and a C-terminal alanine linker is shown *(1)*. The amino acid sequence encoded is indicated above, and the numbering is that of wild-type N. The randomized codon (XYT), where *X* is a C:A mixture at a 3:1 ratio and Y is an A:G mixture at a 1:3 ratio, encodes R:S,H:N at a ratio of 56.25%:18.75%:6.25%, biased toward arginines. Thymine (T) was chosen at the wobble position to optimize codon usage in *E. coli*. The nature of the linker between the polypeptide and N may be important, particularly for short peptides. We have found, e.g., that for α-helical peptides, an alanine linker is preferred over a glycine linker, whereas a glycine linker is preferred for nonhelical peptides *(14)*.

screening, but a number of problems are encountered: First, the antitermination system displays a relatively high rate of false positives in the initial screen, and approx 0.2–0.5% of colonies show blue color that appears to result from spontaneous mutation of the reporter plasmid. Second, some of the positive clones that survive a secondary screen are nonspecific in that they antiterminate with multiple reporters. Third, false-positives may result from mutations in the N-expressor plasmid outside of the library region, though this has not yet been observed. Each of these problems is eliminated by the sequential screens described in **Subheading 3.4**.

2. Materials

All solutions are prepared in deionized or distilled water unless indicated. The protocols assume a working knowledge of basic recombinant DNA techniques and use standard cloning procedures except where noted.

2.1. Detecting Polypeptide–RNA Interactions Using a β-Galactosidase Colony Color Assay

1. Stock solutions (can be stored at −20°C for several months): ampicillin (100 mg/mL); chloramphenicol (40 mg/mL in methanol); 5-bromo-4-chloro-3-indolyl-β-D-galactopyranoside (X-gal, 40 mg/mL in dimethylformamide); isopropyl-β-D-thiogalactopyranoside (IPTG, 100 m*M*) (*see* **Note 1**).
2. Tryptone broth: 10 g of Bactotryptone and 5 g of sodium chloride to 1 L, autoclaved.

3. Tryptone plates: 10 g of Bactotryptone, 5 g of sodium chloride, and 15 g of Bactoagar to 1 L, autoclaved, containing 0.05 mg/mL of ampicillin, 0.015 mg/mL of chloramphenicol, 0.08 mg/mL of X-gal, and 50 µM IPTG and poured into 100 mm or 150 mm diameter plates (*see* **Note 2**).
4. N-expressor (pBR322-derived, ampicillin resistant) and N-reporter plasmids (pACYC184-derived, chloramphenicol resistant) (*see* **Note 3**).
5. Competent N567 cells containing the appropriate N-reporter or N-expressor plasmid (prepared using the standard $CaCl_2$ method) (*see* **Note 4**).

2.2. Quantitating Polypeptide–RNA Interactions Using a β-Galactosidase Solution Assay

1. Stock solutions: 1 M glucose; 1 mg/mL vitamin B1; 1 M Mg_2SO_4; each sterile filtered through 0.2 µm filters; *o*-nitrophenyl-β-D-galactopyranoside (ONPG, 4 mg/mL in A medium and stored at 4°C); 1 M sodium carbonate; 0.1% sodium dodecyl sulfate (SDS); chloroform.
2. A medium: 1.5 g of K_2HPO_4, 4.5 g of KH_2PO_4, 1.0 g of $(NH_4)_2SO_4$, 0.5 g of sodium citrate dihydrate, H_2O to 1 L; autoclaved.
3. Z buffer: 0.06 M Na_2HPO_4, 0.04 M NaH_2PO_4, 0.01 M KCl, 0.001 M Mg_2SO_4, 0.05 M β-mercaptoethanol.

2.3. Preparing Combinatorial Libraries

1. Stock solutions: 200 mM dithiothreitol (DTT; stored at –20°C); a mixture of 2.5 mM of each deoxynucleotide triphosphate (dNTP, prepared from 100 mM dNTP solutions from Pharmacia, Piscataway, NJ, USA, stored at –20°C); 0.5 M EDTA-Na, pH 8.0; phenol/chloroform/isoamylalcohol solution (1 M Tris-HCl [pH 7.5]-saturated phenol:chloroform:isoamyl alcohol [25:24:1]); 3 M sodium acetate, pH 5.2; 50× TAE (242 g of Tris base, 57.1 mL of glacial acetic acid, and 100 mL of 0.5 M EDTA, H_2O to 1 L); NuSieve 3:1 agarose (FMC Bioproducts, Rockland, ME, USA).
2. 20 µM Degenerate oligonucleotide consisting of the randomized DNA sequence encoding the library, flanked by fixed sequences with the *Nco*I and *Bsm*I sites thereby creating an N-terminal library sequence fused at position 19 of N (*see* **Note 5** and **Fig. 3**).
3. 20 µM Primer oligonucleotide complementary to the fixed sequence at the extreme 3'-end of the degenerate oligonucleotide (*see* **Fig. 3**).
3. Enzymes and accompanying commercial buffers: *Nco*I (10 U/µL; NEB, Beverly, MA, USA); *Bsm*I (5 U/µL; NEB); Sequenase 2.0 (13 U/µL; USB, Cleveland, OH, USA); T4 DNA ligase (1 U/µL; Gibco-BRL, Gaithersburg, MD, USA).

2.4. Screening for Novel RNA-Binding Polypeptides from Large Randomized Libraries

1. N-expressor plasmid combinatorial library (from **Subheading 3.3.**).
2. Competent *E. coli* N567 cells containing the appropriate reporter plasmids.

3. Tryptone broth, tryptone plates, and antibiotics (*see* **Subheading 2.1.**).
4. Sterile multiple-well plates (typically 96-well plates) with lids that allow aeration of cultures.

3. Methods

3.1. Detecting Polypeptide–RNA Interactions Using a β-Galactosidase Colony Color Assay

1. Transformation of the pBR N-expressor plasmid into N567/pAC reporter cells: Mix 10 ng of the pBR plasmid in 1–10 µL volume with 50–100 µL of competent reporter cells and incubate on ice for 10 min in 10-mL round-bottom tubes. Incubate the mixture for 2 min at 37°C and immediately place on ice. Add 0.5–1 mL of tryptone broth and allow the cells to recover from heat shock by shaking at 37°C for 1 h.
2. Spread one fifth to one tenth of the transformation mixture onto plates containing antibiotics, X-gal, and IPTG (to induce the *tac* promoters; *see* **Subheading 2.1.**) and incubate for up to 48 h at 34°C.
3. Score colony color by comparison with the appropriate positive and negative controls to evaluate antitermination activity (*see* **Note 6**).

3.2. Quantitating Polypeptide–RNA Interactions Using a β-Galactosidase Solution Assay

1. Grow overnight cultures from single representative blue colonies at 37°C in tryptone broth containing 0.05 mg/mL of ampicillin and 0.015 mg/mL of chloramphenicol.
2. Dilute the tryptone cultures 50-fold into 3–5 mL of A medium containing 22.4 mM glucose, 1 µg/mL vitamin B1, 1 mM Mg$_2$SO$_4$, and both antibiotics. Grow the bacteria at 37°C (*see* **Note 7**) to an OD$_{600}$ of approx 0.2 (3–6 h). Add IPTG to a concentration of 0.5 mM and grow the cells for an additional hour to an OD$_{600}$ of 0.4–0.5. Record the OD$_{600}$ of the culture for subsequent calculation of β-galactosidase activity.
3. Mix 0.1 mL of the cell culture, 0.9 mL of Z buffer, 25 µL of chloroform, and 12 µL of 0.1% SDS, and vortex vigorously for 10 s. The solution will be cloudy, and cells will be permeabilized.
4. Add 0.2 mL of ONPG solution to the permeabilized cells at 10- to 15-s intervals, and incubate at 28°C (*see* **Note 8**).
5. When yellow color develops, add 0.4 mL of 1 M sodium carbonate to stop the reaction. Record the times at which the ONPG was added and the reactions were stopped.
6. Read the OD$_{420}$ and OD$_{550}$ of the reactions after centrifuging for 1 min at 11,000g to pellet cell debris. Calculate the β-galactosidase activity using the following equation:

$$\text{Units activity} = 1000 \times (\text{OD}_{420} - 1.6 \times \text{OD}_{550})/(t \times 0.1 \times \text{OD}_{600})$$

where t = reaction time in minutes and OD$_{600}$ is from **step 2**.

3.3. Preparing Combinatorial Libraries

1. Mix 20 µL of primer oligonucleotide, 15 µL of degenerate oligonucleotide, 100 µL of Sequenase buffer, and 272.3 µL of H$_2$O. Anneal primer and degenerate oligonucleotide by heating to 65°C and slow cooling to room temperature.
2. Add 25 µL of DTT, 60 µL of the dNTP solution, and 7.7 µL (100 U) of Sequenase to the solution containing the annealed oligonucleotides, and incubate 20 min at 37°C. Quench the reaction by adding 4 µL of 0.5 *M* EDTA. Retain a small aliquot of the solution prior to adding Sequenase for comparison with the extended product by electrophoresis on 4% agarose gels.
3. Phenol extract the reaction mixture by adding 500 µL of phenol/chloroform/isoamylalcohol solution, vortexing vigorously, and centrifuging at 11,000*g* for 1 min. Transfer the aqueous phase (upper layer) into a fresh tube. Add 500 µL of chloroform and repeat the above step. Add one tenth the volume of 3 *M* sodium acetate and precipitate DNA with 2.5 vol of ethanol.
4. Redissolve the precipitated DNA in 290 µL of H$_2$O, 50 µL of NEB buffer 2, and 40 µL of DTT. Retain a small aliquot for comparison with the digested products by electrophoresis on 4% agarose gels. Add 40 µL of *Nco*I and incubate at 37°C for 2 h. Retain a small aliquot. Add 80 µL of *Bsm*I and incubate at 65°C for 2 h. Retain a small aliquot. Phenol extract and ethanol precipitate as described previously. The extent of digestion is usually examined on gels prior to DNA purification (**step 5**).
5. Isolate the digested DNA on a 12%, 1.6 mm polyacrylamide gel (19:1 acrylamide:*bis*-acrylamide). Visualize the DNA band by UV shadowing (using a precoated silica gel TLC plate [EM Science, Gibbstown, NJ, USA] and a handheld 254 nm UV lamp), and cut out the gel slice. Crush the gel and elute the DNA into 300 m*M* sodium acetate and precipitate with 2.5 vol of ethanol.
6. Perform a test 20-µL ligation reaction containing 50 ng of *Nco*I–*Bsm*I-digested plasmid, 5 ng of insert DNA from **step 5**, 1× ligase buffer, and 2 µL of T4 DNA ligase, and transform the mixture into 100 µL of N567 cells to determine the number of transformants. A density of no more than 2000–3000 colonies per 100 mm diameter plate gives appropriate colony sizes for scoring blue color. Scale the ligation reaction proportionally, depending on the number of transformants desired.

3.4. Screening for Novel RNA-Binding Polypeptides from Large Randomized Libraries

1. Primary screen. Transform the ligation reaction containing the combinatorial library plasmid (from **Subheading 3.3.**) in 200-µL aliquots into 1.5 mL of competent reporter cells and spread onto plates (150 mm diameter) containing antibiotics, IPTG, and X-gal (*see* **Note 9**). Pick desired blue colonies (even the lightest blue colonies are usually chosen in an initial screen) and suspend individual colonies into 96-well plates containing 100 µL of tryptone broth and antibiotics (*see* **Note 10**). Grow bacteria overnight to saturation to ensure that all clones are represented by an equal number of cells. Pool cultures and isolate

Selection of RNA-Binding Polypeptides

the pBR plasmids using a standard alkaline lysis procedure followed by purification on agarose gels.

2. Secondary screen to eliminate reporter-related false positives. Retransform plasmid DNA from the pooled positives into fresh reporter cells, and spread onto tryptone plates to give the desired number of colonies. This time, pick blue colonies and inoculate into 3 mL of tryptone broth containing ampicillin, grow overnight to saturation, and isolate plasmid DNAs from individual clones.
3. Tertiary screen to eliminate nonspecific positives. Retransform each clone into reporter cells containing the specific RNA reporter or nonspecific reporters and eliminate those that score positive on both. The frequency that one may expect to observe nonspecific positives depends on the nature of the combinatorial library. We have observed frequencies of 50% and 80% (of the clones obtained from **step 2**) for two randomized libraries that we have tested.
4. Quarternary screen to eliminate false positives due to changes outside of the library. Sequence the library portion of the selected positives from **step 3**. Resynthesize the selected sequence, reclone into pBR, and test for antitermination activity.

4. Notes

1. Antibiotic solutions should be stored in small aliquots (1 mL) at –20°C to avoid repetitive freezing and thawing.
2. Plates should be prepared just prior to use. Plates left for long periods of time at room temperature or at 4°C often show weak colony color intensities (possibly due to X-gal precipitation). Plates containing antibiotics and IPTG, but not X-gal, can be prepared beforehand, and X-gal can be spread onto plates prior to use (at the same concentration as if added prior to pouring the plates).
3. N-expressor and reporter plasmids have been generously provided by N. Franklin, Department of Biology, University of Utah and may be requested from Dr. Franklin.
4. Competent cells are prepared and transformed using standard $CaCl_2$ and heat shock procedures, which produce a sufficient number of colonies (~10^6/µg plasmid DNA) for β-galactosidase screening. *E. coli* strain N567 is derived from the C600 strain and is highly transformable. Requests for this strain may be directed to Dr. Franklin (*see* **Note 3**). It is possible that other *laqI*q *rho*– strains may be used, but alternatives have not been tested. The antitermination activity of the polypeptide and RNA binding site of interest can be tested by cotransformation of the two plasmids into N567 cells. However, the efficiency of cotransformation is relatively low, and it is difficult to obtain a uniform number of colonies from plate to plate. Since colony color appears to increase with increasing colony size, the scoring of blue color becomes difficult. We therefore recommend preparing competent cells containing either the N-expressor or N-reporter plasmid. We typically transform N-expressor plasmids into competent N-reporter cells to test antitermination activity. The specificity of the antitermination activity is confirmed using reporters containing different RNA targets or mutant RNAs that are known to have low binding activity.

5. In designing the degenerate oligonucleotide, it is important to search for sequences within the randomized region that may fortuitously form restriction sites used for cloning (in this case *Nco*I and *Bsm*I), as these sequences would be cleaved upon cloning and would not be represented in the library. Such sequences can often be avoided by choosing different nucleotides at the degenerate third codon position that still retain the same amino acid representation.
6. In general we have observed that substitution of N-peptide and box B hairpin with heterologous polypeptide–RNA of comparable binding affinities results in weaker antitermination (about 10-fold) as determined by β-galactosidase solution assays (*see* **Subheading 3.2.**). This may be due to the absence of an additional interaction of box B with *Nus*A *(15)* and/or other RNA–protein or protein–protein interactions *(16)*, all of which appear to contribute to the overall stability of the antitermination complex *(17)*. Several attempts have been made to overcome this decrease in activity by adding back the box B hairpin in different orientations relative to the target RNA, but no significant increases were observed. Nevertheless, the antitermination activities generated by the heterologous interactions are readily detectable and show excellent correlations to in vitro binding affinities *(1)*.
7. Comparison of β-galactosidase levels when bacterial cultures were grown and induced at 37°C, 34°C, and 30°C gave similar values.
8. Reactions should ideally be stopped at an OD_{420} of approx 0.5, and not higher than 1.0, which is above the linear range of detection. Typical incubation times are 5–60 min.
9. We typically obtain $2–3 \times 10^5$ colonies per microgram of digested plasmid when the ligation reaction is carried out with a plasmid to insert ratio of 1:3.
10. When searching for rare peptides that occur with frequencies similar to the background level of blue colonies (0.2–0.5%), all visibly blue colonies should be picked to avoid losing positive clones. In one screen for RRE-positives using the library shown in **Fig. 3**, 600,000 colonies were screened and 1920 blue colonies were picked.

References

1. Harada, K., Martin, S. S., and Frankel, A. D. (1996) Selection of RNA-binding peptides in vivo. *Nature* **380**, 175–179.
2. Jain, C. and Belasco, J. G. (1996) A structural model for the HIV-1 Rev-RRE complex deduced from altered-specificity rev variants isolated by a rapid genetic strategy. *Cell* **87**, 115–125.
3. Laird-Offringa, I. A. and Belasco, J. G. (1995) Analysis of RNA-binding proteins by in vitro genetic selection: identification of an amino acid residue important for locking U1A onto its RNA target. *Proc. Natl. Acad. Sci. USA* **92**, 11,859–11,863.
4. Martin, F., Schaller, A., Eglite, S., Schumperli, D., and Muller, B. (1997) The gene for histone RNA hairpin binding protein is located on human chromosome 4 and encodes a novel type of RNA binding protein. *EMBO J.* **16**, 769–778.

5. Wang, Z. F., Whitfield, M. L., Ingledue, T. C. R., Dominski, Z., and Marzluff, W. F. (1996) The protein that binds the 3' end of histone mRNA: a novel RNA-binding protein required for histone pre-mRNA processing. *Genes Dev.* **10**, 3028–3040.
6. Franklin, N. C. (1993) Clustered arginine residues of bacteriophage lambda N protein are essential to antitermination of transcription, but their locale cannot compensate for boxB loop defects. *J. Mol. Biol.* **231**, 343–360.
7. Wilhelm, J. E. and Vale, R. D. (1996) A one-hybrid system for detecting RNA–protein interactions. *Genes to Cells* **1**, 317–323.
8. Glaser, S. M., Yelton, D. E., and Huse, W. D. (1992) Antibody engineering by codon-based mutagenesis in a filamentous phage vector system. *J. Immunol.* **149**, 3903–3913.
9. Cormack, B. P. and Struhl, K. (1993) Regional codon randomization: defining a TATA- binding protein surface required for RNA polymerase III transcription. *Science* **262**, 244–248.
10. Cadwell, R. C. and Joyce, G. F. (1992) Randomization of genes by PCR mutagenesis. *PCR Methods Appl.* **2**, 28–33.
11. Vartanian, J.-P., Henry, M., and Wain-Hobson, S. (1996) Hypermutagenic PCR involving all four transitions and a sizeable proportion of transversions. *Nucleic Acids Res.* **24**, 2627–2631.
12. Hermes, J. D., Blacklow, S. C., and Knowles, J. R. (1990) Searching sequence space by definably random mutagenesis: improving the catalytic potency of an enzyme. *Proc. Natl. Acad. Sci. USA* **87**, 696–700.
13. Clackson, T. and Wells, J. A. (1994) In vitro selection from protein and peptide libraries. *Trends Biotechnol.* **12**, 173–184.
14. Harada, K. and Frankel, A. D. (1998) In vivo selection of specific RNA-binding polypeptides using a transcription antitermination reporter assay, in *RNA–Protein Interactions: A Practical Approach* (Smith, C., ed.), Oxford Univesity Press, Oxford, pp. 217–236.
15. Chattopadhyay, S., Garcia-Mena, J., DeVito, J., Wolska, K., and Das, A. (1995) Bipartite function of a small RNA hairpin in transcription antitermination in bacteriophage lambda. *Proc. Natl. Acad. Sci. USA* **92**, 4061–4065.
16. Mogridge, J., Mah, T.-F., and Greenblatt, J. (1995) A protein–RNA interaction network facilitates the template-independent cooperative assembly on RNA polymerase of a stable antitermination complex containing the λ N protein. *Genes Dev.* **9**, 2831–2844.
17. Rees, W. A., Weitzel, S. E., Yager, T. D., Das, A., and von Hippel, P. H. (1996) Bacteriophage lambda N protein alone can induce transcription antitermination in vitro. *Proc. Natl. Acad. Sci. USA* **93**, 342–346.
18. Greenblatt, J., Nodwell, J. R., and Mason, S. W. (1993) Transcriptional antitermination. *Nature* **364**, 401–406.

15

In Vitro Genetic Analysis of RNA-Binding Proteins Using Phage Display

Ite A. Laird-Offringa

1. Introduction

RNA-binding proteins (RNA-BPs) play an essential role in key processes in all living organisms, from viruses to mammals. They are involved in RNA packaging, pre-mRNA splicing, translation, and RNA localization, to name a few. For these proteins to function properly it is imperative that they recognize their correct RNA target sequences or structures. As yet, relatively little is known about the mechanisms by which RNA-BPs specifically recognize RNA. To gain more insight into RNA–protein interaction, several genetic systems have been developed in the last few years *(1–4)* (*see* Chapters 13 and 14). The system described in this chapter is based on in vitro genetics. This means that the selective step (in this case, the binding of an RNA-BP to RNA) occurs outside of a living organism, in the test tube.

1.1. Phage Display

In vitro genetics has proven to be a very powerful tool to study the characteristics of nucleic acids that allow them to bind specifically to proteins (the SELEX system) *(5,6)*. Using SELEX, one can rapidly select from a pool of random sequences, those that bind to a particular protein (*see* Chapter 16). The system described here is aimed at isolating *protein* variants that can bind to a particular RNA, from a pool of mutated proteins. This can be very useful to identify the determinants that allow certain regions of a protein to bind specifically to a particular RNA target. To do this, the RNA target that is used must be attached to a solid support. The pool of mutated proteins is then added, and molecules that can bind are retained, while molecules that do not bind are washed away (**Fig. 1**). The technical problem that is then encountered, is how to determine the identity of bound protein molecules. To solve this problem,

From: *Methods in Molecular Biology, Vol. 118: RNA-Protein Interaction Protocols*
Edited by: S. Haynes © Humana Press Inc., Totowa, NJ

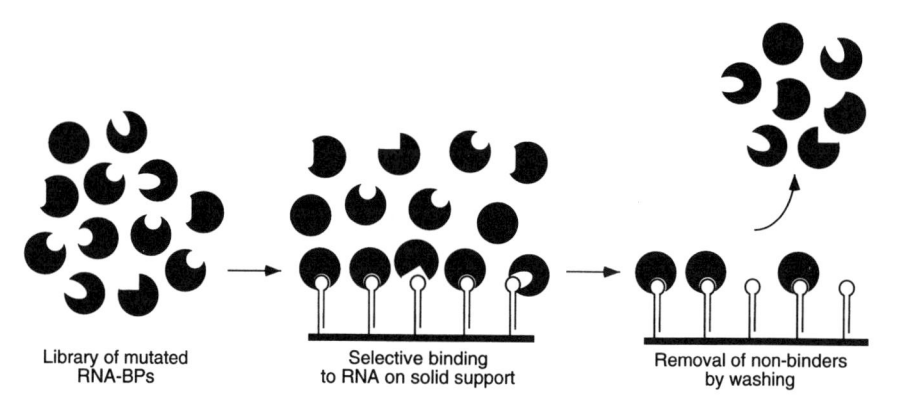

Fig. 1. The principle of in vitro selection of RNA-BPs. A pool of mutated RNA-BPs is incubated with an RNA target attached to a solid support. Unbound and weakly binding proteins are then washed away, leaving variants that can bind to the target sequence.

the RNA-BP molecules are each physically coupled to the gene by which they are encoded. This is accomplished by "phage display," a technique developed in 1985 by George Smith *(7)*. Phage display is the genetic fusion of a foreign gene to a coat protein gene of a bacteriophage in such a way that in the resulting phage particle, the foreign protein is "displayed" on the outside of the phage particle, while the fusion gene is present on the inside (**Fig. 2**). Using this system to display a library of RNA-BP mutants, one can simultaneously capture the protein variants that bind to an RNA, and the genes by which they are encoded.

The bacteriophage that are commonly used to display proteins are of the filamentous type (M13, fd, f1). Foreign proteins are fused to either of two phage proteins *(8,9)*. One is the major coat protein encoded by phage gene VIII, present in about 2700 copies per virion. The second is the attachment protein encoded by gene III, which is present in five copies at one end of the phage and allows the phage to bind to the host. Although gene VIII fusions may be useful in some cases, drawbacks are the importance of the integrity of this protein for particle assembly and the high copy number per phage, which could result in multivalent particles displaying more than one copy of the RNA-BP. Therefore, we use gene III fusions. To avoid problems of multivalency, and to ensure optimal infectivity, a wild-type copy of gene III (provided by the helper phage; *see* **Subheading 3.2.**) is also expressed during display phage production (*see* **Note 1** and **Fig. 2**).

Since the conception of phage display, proteins have been displayed on a variety of organisms, including various bacteriophage and even *E. coli* *(10,11)*. The type of organism chosen for display will affect the kinds of assays one can perform. The drawback of using filamentous phage for dis-

In Vitro Genetics of RNA-BPs

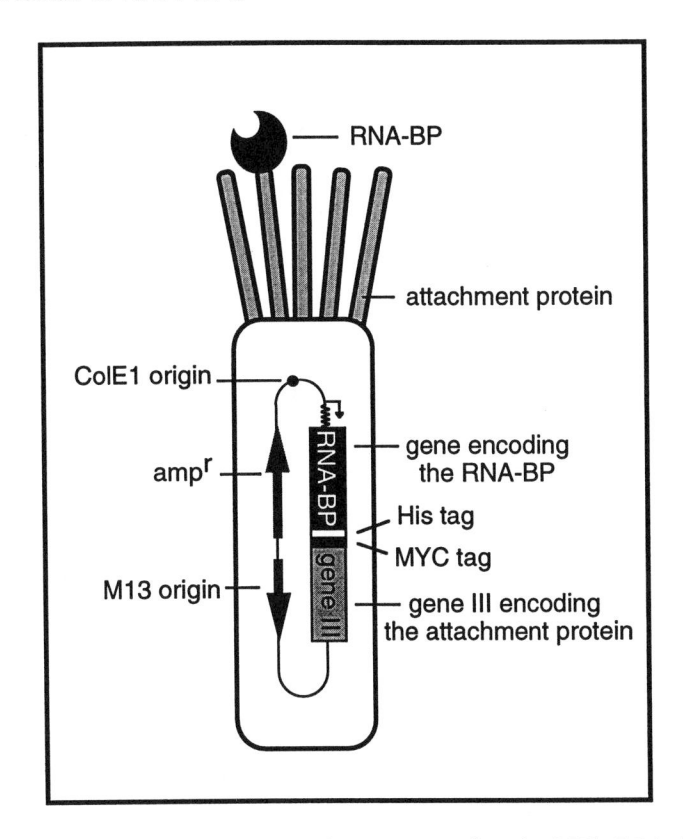

Fig. 2. Display of RNA-BPs on phage. The gene encoding the RNA-BP is fused to a hexahistidine tag, a c-*myc* epitope tag and the bacteriophage gene III in such a way that the RNA-BP is displayed on the outside of the mature phage particle. The fusion gene is present inside the phage particle, so that the RNA-BP and the gene by which it is encoded are physically linked. In this case, the fusion gene has been inserted into a phagemid, or plasmid carrying a phage origin of replication. The phage origin allows the plasmid to replicate and be packaged in the presence of helper phage. The helper phage also provides wild-type copies of the attachment protein, ensuring monovalent display and optimal infectivity.

play is that to be displayed, proteins must be successfully secreted through the inner bacterial membrane. Filamentous phage do not kill (lyse) their host, and reproduce by extruding phage particles through the host membranes. While this can complicate display of certain proteins that are poorly exported, it offers a huge advantage. Because the host cells are intact, it is very easy to purify clean preparations of phage particles that are relatively free of RNases. This is important, because during the genetic selection, the displayed proteins are retained on beads by binding to RNA. Degradation of the RNA target would result in loss of bound protein.

Once the population of phage displaying RNA-BPs that bind to the RNA target has been selected, the bound phage is used to infect *E. coli*, and to generate a new population of phage, which is now enriched for "binders." This new population can be used to repeat the selection procedure. Thus, as with SELEX, iterative cycles of binding, selection, and amplification will rapidly yield mutants that can bind very tightly to the target. At any moment, the identity of the individual mutants or the composition of the pool can be determined by sequencing the single-stranded genetic material inside the particles.

1.2. A Simple Method to Determine Whether a Protein can be Displayed

As mentioned previously, a prerequisite for the display of a foreign protein on filamentous phage is the ability of the protein to be successfully secreted. Because RNA-BPs are usually nuclear or cytoplasmic proteins that have not been evolutionarily selected to be competent for export, this may pose a problem. Our phage display vector contains a signal sequence upstream of the region into which the gene of interest is to be inserted, which will direct the protein to the inner bacterial membrane *(12)*. However, certain properties of the foreign protein can interfere with secretion. These are the presence of many charged residues close to the signal peptide cleavage site (in particular positive charge can be a problem *[13,14]*), and the rapid folding of the protein into a complex structure before it is secreted. The effect of charged residues can be avoided by choosing the foreign protein fragment carefully, or by inserting a sequence encoding an uncharged linker between the signal sequence and the gene of interest. Rapid folding of the foreign protein can be minimized by growing the cells at lower temperatures. Since it is often unknown whether charge or folding will interfere with the display of a protein, we developed a display vector that can first be used to measure ability of the foreign protein to be secreted *(15)*. Export of the foreign protein can be visualized on indicator plates, which will show blue colonies upon proper secretion. Subsequently, a simple modification of the vector allows the protein to be displayed on phage. This vector is called pDISPLAYblue-B (**Fig. 3**).

The blue color screen is accomplished by linking the gene encoding the foreign protein at its 3' end to the *E. coli* alkaline phosphatase gene, such that a fusion protein is produced. Alkaline phosphatase is active only in the bacterial periplasm, where disulfide bonds essential for its activity are formed, and can be used to measure protein export *(16)*. If the foreign protein prevents the alkaline phosphatase from reaching the periplasm, the bacteria will be white. In this case, the construct or the growth temperature of the cells can be optimized. Following successful secretion, the alkaline phosphatase gene is removed by a simple digestion and religation, which results in fusion of the foreign gene to

In Vitro Genetics of RNA-BPs

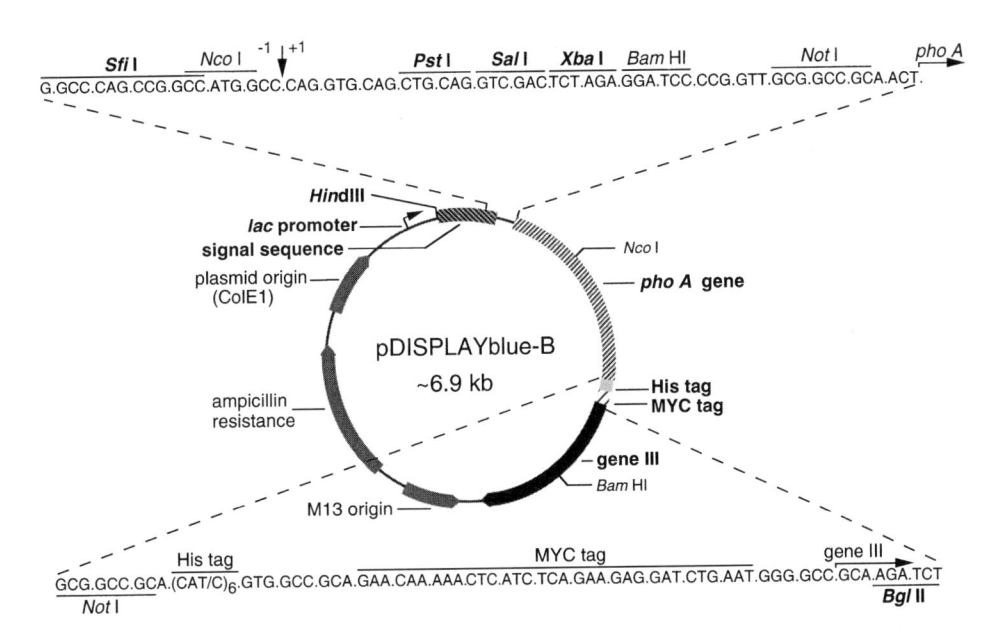

```
      Sfi I         Nco I   -1 +1          Pst I      Sal I    Xba I    Bam HI              Not I            pho A
G.GCC.CAG.CCG.GCC.ATG.GCC.CAG.GTG.CAG.CTG.CAG.GTC.GAC.TCT.AGA.GGA.TCC.CCG.GTT.GCG.GCC.GCA.ACT.
```

Fig. 3. The pDISPLAYblue-B vector. Relevant regions of the plasmid are indicated. The sequence of the polylinker region, into which foreign gene fragments are inserted, is given at the *top*. The appropriate reading frame is marked by *dots*. The signal peptide cleavage site is marked by an *arrow*. The sequence just downstream of the *phoA* cassette is given at the *bottom*. Restriction sites of interest are marked, with unique sites represented in bold. A related vector lacking the *Not*I *phoA* cassette is called pDISPLAY-B.

the phage coat protein. Thus, using the pDISPLAYblue vector, one can easily determine if phage display is a feasible approach to study the protein of interest. In our work with various RNA-binding proteins carrying multiple globular RNA-binding domains of the RRM type (RNA Recognition Motif, also referred to as RNP motif *[17,18]*), reduction of the bacterial growth temperature to 30°C was sufficient to allow secretion, even of large protein domains carrying multiple RRMs *(15)*. The largest fragment displayed by us to date was from the yeast poly(A) binding protein (~ 50 kDa), and contained four RRMs motifs.

1.3. Using In Vitro Genetics to Study RNA-Binding Proteins

In vitro genetics using phage display is a particularly powerful approach to study RNA–protein interaction. First, the iterative nature of the process allows the isolation of even very rare mutants from a large population of RNA-BP variants. Second, the fact that the selective step occurs in vitro means that the binding conditions can be completely controlled by the researcher. Thus, the optimal salt and buffer conditions to study the RNA–protein combination can be chosen. Third, it offers the possibility of studying complex interactions between multiple proteins and RNA. In a multiprotein complex, one or more of the proteins could be displayed on phage. The other proteins (as well as modifiers of proteins such as protein kinases) could be added separately to the reaction. Finally, it is a rapid and inexpensive procedure, with each selection cycle taking just a few days.

This chapter describes the different steps that are involved in setting up the in vitro genetic analysis for an RNA-BP of interest. The methods section is divided into seven parts: (1) preparation of display constructs and analysis on indicator plates, (2) preparation of helper phage stock, (3) preparation of display phage, (4) detection of the displayed protein by immunoblotting, (5) determination of percentage phage displaying the RNA-BP, (6) performing cycles of in vitro genetic selection, and (7) interpretation of genetic data. The first five sections describe the basic steps involved in display of RNA-BPs, while the last two sections deal with actual genetic analysis of libraries of RNA-BP mutants.

2. Materials

Distilled deionized water is used for all applications.

2.1. The pDISPLAYblue-B Vector

The vector developed for the display of RNA-BPs on filamentous bacteriophage is called pDISPLAYblue-B. It can be obtained from the author. Its composition is diagramatically represented in **Fig. 3**. It is based on the phagemid pHEN1 *(19)*. The plasmid backbone is derived from the pUC 119 phagemid (for a sequence see accession number U07650). A phagemid is a plasmid carrying a filamentous phage origin of replication, which allows it to replicate and be packaged by an *E. coli* host in the presence of helper phage. Helper phage is required to supply the phage proteins needed for replication and packaging. Relevant sections of the pUC119 part of pDISPLAYblue-B are the plasmid origin of replication, the phage origin of replication, the ampicillin resistance gene, and the *lac* promoter flanking the polylinker sequence. This inducible promoter can be used to drive the expression of inserted genes. All sequences relevant to display are located between the *Hin*dIII site just downstream of the *lac* promoter and the *Eco*RI site of the original pUC119 plasmid. These are, starting from the *Hin*dIII site, a signal sequence, a polylinker region, an *E. coli* alkaline phosphatase gene (*phoA*) cassette, a hexahistidine tag, a c-*myc* epitope tag, and the filamentous fd phage gene III. The histidine tag can be used later for the purification of interesting RNA-BP mutants that are found in the genetic screen. To this end, the fragment containing the RNA-BP and the tags is subcloned into a bacterial expression vector (e.g., pET [20]) and the protein is induced and purified using Ni^{2+} affinity beads (Qiagen, Valencia, CA, USA). The MYC tag allows many different fusion protein products to be recognized using a single antibody (21). The sequence of the polylinker region into which the foreign gene fragment is inserted, is shown at the top in **Fig. 3**. For the foreign protein to be fused to the signal peptide, it must conform to the indicated reading frame. The signal peptide cleavage site is marked by an arrow. Although the *Sfi*I and *Nco*I sites can be used to insert gene fragments, care must be taken to restore the identity of the amino acids in the signal peptide to retain optimal signal peptide cleavage. Note that the *phoA* gene cassette can be removed by a *Not*I digestion and religation, resulting in an in-frame fusion of the foreign gene and gene III. pDISPLAYblue-B from which the *phoA* gene has been removed is called pDISPLAY-B, and can be used to directly display foreign gene fragments.

2.2. Bacterial Strains and Helper Phage

Bacterial strains to be used for phage display must have the following characteristics: (1) They must carry an F factor, which allows them to be infected by filamentous phage (*see* **Note 2**). (2) They must carry the *lacIq* gene, which represses the *lac* promoter driving the fusion gene. This is necessary because the expression of the fusion gene can often be toxic to the host cell. (3) Ideally, the strain should also be recombination deficient, ensuring the stability of plasmids. *E. coli* strain TG1 (K12, Δ[*lac-pro*], *supE, thi, hsdD5*/F'*traD36, proA+B+, lacIq. lacZ*ΔM15) meets the first two criteria (19). We inserted a chloramphenicol resistance gene into the *recA* gene (*recA::cat*) of TG1 to make strain IAL-3, which also meets the third requirement. IAL-3 can be obtained from the author. This is a very pleasant strain to work with. It can be easily made competent by the calcium/manganese-based (CCMB) procedure (22) and produces good quality miniprep DNA for sequencing. Very recently, we have worked with strain DH12S (ϕ80d*lacZ*ΔM15, *mrcA*, Δ[*mrr-hsdRMS-mcrBC*], *araD139,* Δ[*ara, leu*]7697, Δ*lacX74, galU, galK, rpsL, deoR, nupG, recA1*/ F'*proAB+, lacIq* ZΔM15). This strain has two characteristics that may be advantageous when screening large libraries of RNA-BP mutants. First, it cannot suppress amber codons, which can survive and sometimes even be selected for in an amber suppressing strain (1). Second, highly electrocompetent cells can be purchased (Gibco-BRL, Gaithersburg, MD, USA), so that very large display libraries can be made. A drawback of this strain is that it seems to display proteins up to fivefold less efficiently than IAL-3.

As a helper phage we use M13K07, a phage that has been tagged with a kanamycin resistance gene and that replicates less efficiently than wild-type phage (23). It can be purchased from Gibco-BRL.

2.3. Phage Display and Genetic Selection

2.3.1. Antibiotics

Filter sterilize and store at –20°C.

1. Ampicillin: 1000× stock solution of 100 mg/mL in water.
2. Kanamycin: 1000× stock solution of 70 mg/mL water.
3. Methicillin: 1000× stock solution of 200 mg/mL in water (we used "Staphcillin" (Bristol-Myers Squibb, Princeton, NJ, USA), bought from a pharmacy).

2.3.2. Media

1. LB: 10 g of Bacto-tryptone, 5 g of Bacto-yeast extract, and 10 g of NaCl per liter of water, sterilized. For plates add 15 g of Bacto-agar per liter medium and autoclave.
2. Top agar: LB containing 7 g of Bacto-agar per liter of medium, autoclaved.
3. 2xYT: 16 g of Bacto-tryptone, 10 g of Bacto-yeast extract, 5 g of NaCl per liter of water, sterilized.

2.3.3. Media Additives

1. BCIP: 5-bromo-4-chloro-3-indolyl phosphate (Sigma, St. Louis, MO, USA), stored at –20°C as a 40 mg/mL (1000× concentrated) stock solution in N,N-dimethyl formamide.
2. Glucose stock solution: 20% w/v glucose, dissolved in water and filter sterilized.
3. X-gal: 5-bromo-4-chloro-3-indolyl-β-D-galactopyranoside, 40 mg/mL stock solution (1000× concentrated) in *N,N*-dimethyl formamide, stored at –20°C.

2.3.4. Other Reagents

1. Bead-washing solutions: 0.1 M NaOH, 0.1 M NaCl for the first wash and 0.1 M NaCl for the second wash.
2. Paramagnetic beads and a magnet: Dynabeads M-280 Streptavidin were obtained from Dynal, Lake Success, NY, USA, and a magnet from CPG, Lincoln Park, NJ, USA.
3. PEG/ammonium acetate: 20% polyethylene glycol 8000, 3.5 M ammonium acetate.
4. Protease inhibitors: phenylmethylsulfonyl fluoride (PMSF), 100 mM stock in ethanol, stored at –20°C, used at 500 µM; pepstatin, 1 mg/mL stock in methanol, stored at 4°C, used at 0.7 µg/mL; leupeptin, 1 mg/mL stock in water, stored at –20°C, used at 0.5 µg/mL.
5. RNase inhibitor: RNasin (Promega, Madison, WI, USA).
6. TE: 10 mM Tris-HCl, pH 8.0, 1 mM EDTA, autoclaved.
7. TENT binding buffer: We use 1× TENT (10 mM Tris-HCl, pH 8.0, 1 mM EDTA, 250 mM NaCl, 0.5% (v/v) Triton X-100), but any buffer suitable for studying the interaction between the RNA-BP of interest and RNA can be used (*see* **Note 3**).
8. tRNA: *E. coli* tRNA, 20 mg/mL stock solution in water.

2.4. Immunoblotting

1. Antibodies: Primary antibody 9E10 recognizes the MYC epitope tag in the vector and is available from multiple vendors. Antibodies that recognize histidine tags are also becoming available (e.g., from Qiagen, Valencia, CA, USA). The secondary antibody was goat antimouse IgG, alkaline phosphatase-conjugated.
2. 10× Blot buffer: 250 mM Tris base, 1.9 M glycine and 1% (w/v) SDS. For blotting, mix 100 mL of the 10× blot buffer, 700 mL of water, and 200 mL of methanol per liter blot buffer.
3. Bovine serum albumin (BSA): Fraction V (Amersham/USB, Arlington Heights, IL, USA).
4. Coomassie (de)staining solution: Staining solution: 0.2% (w/v) Coomassie brilliant blue R (Sigma, St. Louis, MO, USA) in destaining solution (45% (v/v) methanol, 9% (v/v) acetic acid in water).
5. Gel drying frame and cellophane: Research Products International, Mt. Prospect, IL, USA.
6. Gel preparation solutions: You will need the following individual stock solutions: (a) 2 M Tris-HCl, pH 8.8; (b) 1 M Tris-HCl, pH 6.8; (c) 10% (w/v) sodium dodecyl sulfate (SDS); (d) 30% (w/v) acrylamide solution containing *bis*-acrylamide and acrylamide in a 0.8:29.2 ratio; (e) 10% (w/v) ammonium persulfate (APS) in water (prepared freshly); (f) *N,N,N',N'*-tetramethylethylenediamine (TEMED); (g) water-saturated butanol (*see* **Note 4**).
7. 2× Layer mix: 125 mM Tris-HCl, pH 6.8, 10% SDS, 20% glycerol, 0.02% bromphenol blue. For each milliliter of 2× layer mix, mix 900 µL with 100 µL of β-mercaptoethanol just prior to use.
8. Minigel electrophoresis and blotting system: Mini Protean system (Bio-Rad Laboratories, Hercules, CA, USA).
9. Molecular weight markers: We use Kaleidoscope prestained markers (Bio-Rad Laboratories).
10. NBT/BCIP mix: Add 66 µL of NBT (nitro blue tetrazolium [Sigma, St. Louis, MO, USA], 50 mg/mL in 70% (v/v) dimethyl formamide) and 41 µL of BCIP (5-bromo-4-chloro-3-indolyl phosphate [Sigma], 40 mg/mL in dimethylformamide) to 10 mL of AP buffer (100 mM Tris-HCl, pH 9.5, 100 mM NaCl, 5 mM MgCl$_2$).
11. Nitrocellulose membrane: Micron Separations, Westborough, MA, USA.
12. 10× Running buffer: 250 mM Tris base, 1.2 M glycine, and 1% (w/v) SDS. (The pH of this solution should automatically be about 8.3.)
13. TBST: 10 mM Tris-HCl, pH 8.0, 150 mM NaCl, 0.05% (v/v) Tween 20.

2.5. RNA Target Preparation

1. Biotinylated oligonucleotide: The 5' biotinylated oligonucleotide complementary to the 3' end of the RNA was purchased from Integrated DNA Technologies (Coralville, IA, USA) (*see* **Note 5**).
2. Chloroform: A mixture of chloroform and isoamyl alcohol in a 49:1 ratio.
3. Electrophoresis equipment: V16 system (Gibco-BRL, Gaithersburg, MD, USA).

4. Ethanol.
5. Formamide: Molecular biology grade formamide, aliquotted and stored at –20°C.
6. Gel preparation solutions: (a) 10× TBE: 0.89 M Tris base, 0.89 M boric acid, 0.02 M EDTA; (b) 30% (w/v) acrylamide solution containing *bis*-acrylamide and acrylamide in a 1:19 ratio; (c) 10% (w/v) APS in water (prepared freshly); (d) TEMED; (e) urea (powder); *see* **Note 4**.
7. In vitro transcription kits: For T7 polymerase: MEGAshortscript kit; for SP6 polymerase: MEGAscript kit (both from Ambion, Austin, TX, USA).
8. Molecular weight marker: ϕX174 DNA cut with *Hae*III.
9. 3 M NaOAc, pH 5.2.
10. Phenol/chloroform: A mixture of phenol, chloroform and isoamyl alcohol in a 50:49:1 ratio.
11. Sephadex G50: Powder (Pharmacia Biotech, Piscataway, NJ, USA), hydrated in TE and autoclaved *(24)*.
12. TENT binding buffer: *See* **Subheading 2.3**.

3. Methods

To use in vitro genetics to analyze an RNA-BP, it is important to first determine whether the RNA-BP can be displayed on phage (**Subheadings 3.1.–3.4.**), and whether it is functional (**Subheading 3.5.**). Once this has been confirmed, libraries of RNA-BP mutants can be displayed, and variants that are able to bind to a particular RNA target can be selected (**Subheadings 3.6.–3.7.**).

3.1. Preparation of Display Constructs and Analysis on Indicator Plates

First one must choose the area of the gene that is to be inserted into the display vector. The whole coding region could be used, or segments thereof containing an RNA-binding domain. Obviously, the stop codon must not be included. Considerations in choosing the area are the avoidance of many charged residues close to the signal sequence cleavage site, the availability of restriction endonuclease sites that are convenient, and the maintenance of the reading frame. Unique sites in the polylinker region of the DISPLAY vectors are *Sfi*I, *Pst*I, *Sal*I, and *Xba*I (**Fig. 3**). When *Sfi*I is used, the last segment of the signal sequence should be restored to its original sequence. Because ideal restriction sites are often absent, we generally amplify the fragment of interest by high-fidelity PCR (polymerase chain reaction), creating restriction sites on either end.

In spite of the presence of the *lacI*q gene in the host, there is still a low level of transcription from the *lac* promoter. This can be minimized by the addition of glucose to the medium to a final concentration of 2%. Omission of glucose could lead to selection against the construct *(19)*. Therefore, glucose should be added to the culture medium and plates *at all times*, except when display phage is made.

Once the constructs are obtained, the level of secretion of the *phoA*-fusion protein can be measured on indicator LB plates containing 100 µg/mL of ampicillin, 40 µg/mL of BCIP and glucose, incubated at 37°C overnight. For comparison, cells containing pDISPLAYblue-B, and a vector lacking the *phoA* gene (pDISPLAY-B; *phoA* removed by a *Not*I deletion) should be compared. Ideally, cells expressing the different constructs are streaked out on separate areas of the same plate. The "empty" vector (e.g., pDISPLAYblue-B) should give dark blue colonies, while the cells containing the negative control plasmid pDISPLAY-B will be white to very pale blue. The activity of the endogenous (chromosomal) *phoA* gene is too low to significantly affect colony color. The colonies containing the new fusion constructs will have an intermediate level of blueness, depending on their level of secretion. If a more accurate quantitation of the *phoA* activity is desired, a liquid assay using cell extracts can be performed *(25)*. We use the *phoA* assay only as an indication that our fusion protein is secreted. Even colonies only slightly darker than the negative control indicate that the protein can be exported. In one case, the absence of color alerted us to the fact that an unexpected stop codon was present in 3' end of the foreign gene fragment. If poor export is seen, and the effect of reduced temperatures on secretion needs to be measured, the indicator plates can be incubated overnight at the desired temperature, and then incubated at 37°C to detect the *phoA* activity in each colony (*see* **Note 6**).

After the constructs pass the *phoA* secretion test, the *phoA* gene cassette is removed by a *Not*I digestion followed by religation. This results in fusion of the foreign gene to gene III. The construct is now ready for display. Although this rarely occurs, the presence of a *Not*I site within the gene segment will complicate the removal of the *phoA* cassette, necessitating a partial digestion. When no secretion problems are anticipated, foreign gene fragments can be cloned directly into pDISPLAY-B. Due to the absence of the *phoA* fragment, this vector has unique *Nco*I and *Not*I sites in addition to the unique restriction sites marked in bold in **Fig. 3**.

3.2. Preparation of Helper Phage Stock

To generate display phage, the bacteria containing the pDISPLAY construct must be infected with helper phage. We make our own helper phage stocks, as described below. Aerosol blocking pipet tips are used when working with phage or phagemid stocks to prevent cross-contamination.

Day 1
1. Inoculate the *E. coli* host strain from a fresh plate into 2 mL of LB and shake overnight at 37°C, 250 rpm.

Day 2
2a. Dilute the culture 1:50 into 10 mL of 2×YT and incubate with shaking (250 rpm) for 30 min at 37°C.
2b. Also make a 1:500 dilution of host cells in LB and incubate for ~6 h at 37°C with shaking. These cells will be used to determine the helper phage titer.
3. Add helper phage M13K07 (from a previously obtained stock) to the culture grown in 2×YT, to a final concentration of 10^8 plaque-forming units (pfu)/mL and shake the culture for 30 min at 37°C.
4. Add kanamycin to 70 µg/mL, and shake the culture at 37°C, 250 rpm for 4 h. This selects for cells infected by the helper phage, which carries a kanamycin resistance gene.
5. Pipet several 1.3-mL aliquots into 1.5-mL microfuge tubes and centrifuge for 5 min at maximal speed (~15,000g).
6. Pipet 1-mL aliquots of supernatant into clean tubes (leaving the bacterial pellet untouched) and centrifuge again.
7. Take 900 µL of supernatant from each tube, deposit into a clean tube, and heat at 65°C for 10 min to kill any remaining cells. The helper phage stock is stable for months at 4°C.
8. Determine the titer by making four serial 1:100 dilutions (5 µL plus 495 µL) in LB.
9. Take 100 µL of the 10^{-6} and 10^{-8} dilutions and add 200 µL of host cells grown in LB to each (*see* **step 2b** above).
10. Pipet each cell/phage mix into 3 mL of warm top agar (50°C), quickly mix, and pour onto warm (37°C) LB plates. Include plates with host cells alone and phage alone as controls.
11. After the top agar has solidified, incubate the plates overnight at 37°C. Plaques will be visible as small circular areas of retarded growth on a lawn of cells.
12. Determine the titer (in pfu/mL) by counting the number of plaques on a plate, multiplying it by the dilution factor used, and multiplying by 10, since 100 µL was plated while the titer is expressed in milliliters. For example, 20 plaques on the 10^{-8} means $20 \times 10^8 \times 10 = 2 \times 10^{10}$ pfu/mL. Host cells alone should show no plaques, while phage alone should show no bacterial growth. Typical helper phage titers are $1-2 \times 10^{11}$ pfu/mL.

If desired, a more concentrated helper phage stock can be prepared by scaling up the volume of the culture and performing a PEG precipitation as described in **Subheading 3.3.**, **steps 7–9** (no need to add protease inhibitors).

3.3. Preparation of Display Phage

This procedure is similar to the one described above except that the host contains the pDISPLAY construct, the phage is grown overnight, and colonies instead of plaques are titered.

Day 1
1. Inoculate a fresh colony of host cells containing the display construct of interest into 2 mL of 2×YT containing 100 µg/mL of ampicillin and 2% glucose. Shake overnight at 37°C, 250 rpm.

Day 2

2. Dilute the culture 1:200 into 20 mL of 2×YT containing 100 μg/mL of ampicillin, and 2% glucose. Shake at 37°C, 300 rpm, until an OD_{600} of ~0.5 is reached (about 3–5 h).
3. Pipet 15 mL of the culture into a 50-mL Falcon centrifuge tube and add helper phage to a final concentration of 6×10^{10} pfu/mL. Let stand at 37°C for 30 min.
4. Centrifuge for 10 min at room temperature at 1500g (3500 rpm in a Sorvall SS34 rotor).
5a. Remove as much supernatant as possible. Resuspend the bacterial pellet in 30 mL of 2×YT, 100 μg/mL of ampicillin, 70 μg/mL of kanamycin and shake overnight at 30°C, 300 rpm.
5b. Also inoculate a host culture in LB for growth overnight; this will be used for titer determination.

Day 3

6a. Dilute the host stock (for titration) 1:500 in LB and grow for 6 h at 37°C with shaking.
6b. Centrifuge the phage culture for 20 min at 4°C at 3000g (5000 rpm in a Sorvall SS34 rotor).
7. Transfer 16 mL of supernatant to a clean tube and add 4 mL of polyethylene glycol (PEG)/ammonium acetate. Mix by inversion and set on ice for 1 h.
8. Centrifuge the solution for 15 min at 4°C at 15,000g (12,000 rpm in a Sorvall SS34 rotor).
9. Discard the supernatant and resuspend the pellet in 1 mL of TE containing protease inhibitors.
10. Transfer the sample to a microfuge tube and centrifuge for 2 min at 15,000g.
11. Transfer the supernatant to a clean microtube, leaving behind any remaining bacteria.
12. Add 250 μL of PEG/ammonium acetate and incubate the mixture on ice for 1 h.
13. Centrifuge the tubes for 5 min at 15,000g and discard the supernatant.
14. Resuspend the pellet in 100 μL of TE containing protease inhibitors. The display phage stocks are stable for months at 4°C, as measured by titer determination and immunoblotting.
15. Determine the titer by making six serial dilutions in LB (5 μL plus 495 μL LB for each 100-fold dilution). Take 100 μL of the 10^{-8}, 10^{-10}, and 10^{-12} dilutions and add 100 μL of host culture to each (**step 6a**; *see* **Note 2**). Incubate at 37°C for 30 min. As controls, use 100 μL of host bacteria alone, and 100 μL of phage alone (10^{-2} dilution).
16. Plate the mixtures on LB plates containing 100 μg/mL of ampicillin and 2% glucose. Each phagemid particle will give rise to an ampicillin-resistant colony. Note that this is different from the titering of the helper phage, where each particle gives rise to a plaque on a lawn of cells. Determine the titer as described in **Subheading 3.2., step 12**. Phagemid alone and host alone should give no ampicillin resistant colonies (*see* **Note 7**). Determining the titer provides no information about whether the RNA-BP is displayed or not. This must be assessed by immunoblotting (*see* **Subheading 3.4.**).

3.4. Detection of the Displayed Protein by Immunoblotting

The analysis of phage preparations by immunoblotting is a quick way to determine whether the foreign protein is displayed. When analyzing a new display phage preparation, it is convenient to have a stock of display phage without insert for comparison, e.g., a prep of pDISPLAY-B phage. The "empty" vector produces phage that displays an attachment protein carrying the hexahistidine and Myc tags. This protein will generate a band migrating at about 65 kDa in an SDS protein gel (the actual molecular weight is about 45 kDa). Insertion of a foreign gene fragment into the DISPLAY vector will cause this band to migrate more slowly. Thus, insertion of a single RRM motif will result in a band that has increased in size by about 11 kDa. We usually run two 8% mini SDS-polyacrylamide gels side by side using duplicate samples. One gel is stained with Coomassie blue to visualize the amount of protein loaded, the other gel is transferred to nitrocellulose for immunoblot analysis.

3.4.1. Preparation of Gels and Transfer to Nitrocellulose

Two 8% mini gels are prepared, which requires about 20 mL of gel mix.

1. Mix 3.75 mL of 2 M Tris-HCl, pH 8.8, 5.33 mL of 30% acrylamide solution, 10.6 mL of water, 0.2 mL of 10% SDS, 0.1 mL of freshly made 10% APS, and 25 µL of TEMED.
2. Pour gels as directed by the manufacturer, leaving a 3 cm distance from the top of the gel. Overlay with about 200 µL of water-saturated butanol and let polymerize for about 30 min.
3. Prepare the stacking gel mix: 0.625 mL of 1 M Tris-HCl, pH 6.8, 0.835 mL of 30% acrylamide, 3.5 mL of water, 50 µL of 10% SDS, and 25 µL of 10% APS.
4. Pour off the butanol and rinse with water. Remove the last of the water with a piece of filter paper.
5. Add 7.5 µL of TEMED to the stacking mix and pour it onto the top of the gels (we use a syringe). Insert the combs and let the gel polymerize for 30 min.
6. For each duplicate sample, add 10 µL of phage to 10 µL of 2× layer mix, heat at 100°C for 5 min, and load (one sample on each gel). Also load a molecular weight marker according to the manufacturer's recommendations.
7. Run the gels at 150 V until the blue dye is at the bottom (about 60 min).
8a. Stain one gel with Coomassie solution for 15–60 min.
8b. Transfer the proteins from the second gel to nitrocellulose in an electroblotting device as described by the manufacturer. Blot for 1 h at 100 V/500 mA (use high current power supply such as Gibco-BRL 250EX).
9. Destain the first gel until the bands are clearly visible, then mount it and dry it in a drying frame between two sheets of cellophane wet with water.
10. Mark the side of the nitrocellulose filter carrying the proteins.

3.4.2. Detection of the Displayed Protein

1. Block the filter at least 2 h (or overnight) with TBST containing 4% BSA. All steps are performed at room temperature and with gentle rocking. A 10-mL solution should be sufficient for a blot from one small gel, when a small container is used (such as a 96-well plate lid).
2. Add the primary antibody (diluted as specified by manufacturer) in TBST containing 4% BSA and incubate for 2 h.
3. Wash the blot 3× for 10 min each with TBST containing 0.3% BSA.
4. Add the secondary antibody (diluted according to manufacturer's specifications) in TBST containing 4% BSA, for 1–2 h.
5. Wash as in **step 3**.
6. Drain the liquid from the blot and develop with NBT/BCIP solution.
7. When sufficient color has developed, stop the reaction with water. Store the filter away from light.

The control "empty" vector should show a thick band migrating at about 65 kDa. This is the tagged attachment protein without a foreign gene insert. Display phage carrying a foreign protein should show bands that have increased in size by the length of the inserted protein fragment. For an example, *see* **Fig. 4**. The immunoblot result will show whether the foreign protein fragment is displayed and what the relative amounts of displayed protein are in different phage preparations. It will also provide an indication of the amount of proteolysis. Some degradation of the displayed protein is often seen (we suspect this occurs during phage assembly). This is not usually a problem as long as intact copies are also visible. The immunoblot analysis will not give any indication about the functionality of the displayed protein. For example, the protein could be displayed but folded incorrectly. To assess functionality, a binding study must be done (*see* **Subheading 3.5.5.**).

3.5. Determination of Percentage Phage Displaying the RNA-BP

The RNA-binding activity of the displayed protein is assessed by determining the percentage of phage that can bind to RNA targets attached to paramagnetic beads. The first step is to prepare beads carrying the RNA target. The RNA targets are generated by in vitro transcription from linearized plasmid templates. They are attached to streptavidin-coated paramagnetic beads, which can be drawn to the side of a tube using a powerful magnet. To be attached to the beads the RNA target is annealed to a biotinylated oligonucleotide complementary to the nucleotides at the 3' end of the RNA. These complementary nucleotides are not in the region of the RNA that is bound by the RNA-BP, but are derived from downstream transcribed polylinker sequences. Annealing the RNA to a biotinylated oligo is better than incorporating biotin into the RNA because: (1) the biotin is outside of the target area and thus will not interfere with bind-

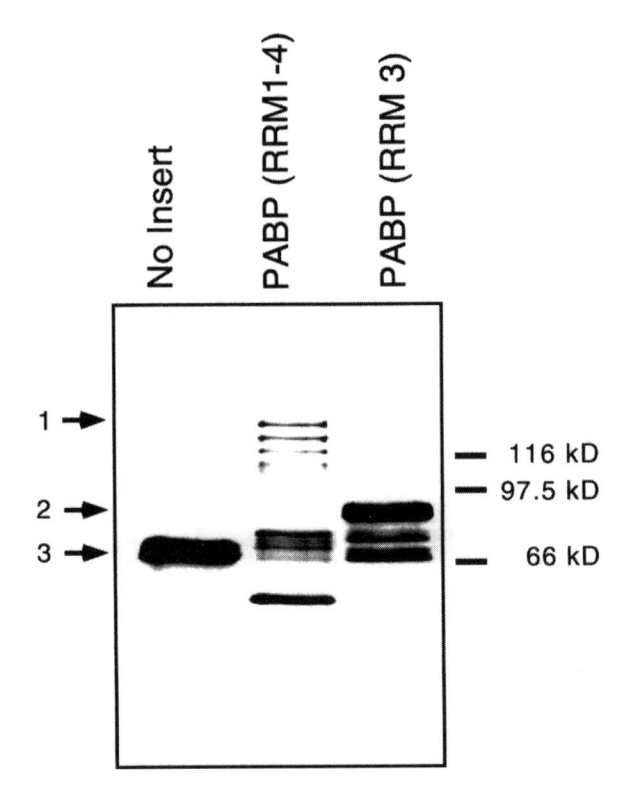

Fig. 4. Immunoblot analysis of display phage preparations. pDISPLAY-B phagemids containing no insert, RRMs 1–4 of poly(A)-binding protein (PABP), or RRM 3 of PABP were used to produce display phage. Equivalent aliquots of each were loaded on an SDS polyacrylamide gel. The blot was probed with an antibody against the MYC epitope tag. The sizes of the molecular weight markers are indicated at the *right*. *Arrows* mark the full-length fusion proteins: (1) RRM 1–4 fused to the attachment protein, (2) RRM 3 fused to the attachment protein, (3) no foreign protein fragment displayed (except for the tags). The bands below the full-length protein bands are proteolytic fragments. Note that proteolysis seems to occur between the different RRMs (*middle lane*), an indication that they are folded into the proper domain structure.

ing, (2) high yields of the in vitro transcribed RNA can be obtained, and (3) a single biotinylated oligonucleotide can be used to link many different target RNAs to the beads.

3.5.1. Preparation of the RNA Target

The target RNA is transcribed from a linearized plasmid containing the sequences encoding the RNA under control of a T7 or SP6 promoter. We use pGEM-derived plasmids (Promega, Madison, WI, USA), and insert the RNA-encoding sequences (two annealed oligonucleotides) between the *Hin*dIII and *Pst*I sites. Thus far, we have used only small RNA targets of about 70 nucleotides. Prior to transcription, the plasmids are linearized with *Ava*I (which lies about 45 nucleotides downstream) and purified. The biotinylated oligonucleotide (5' biotin-ACCGGGGATCCTCTAGAGTC-3') hybridizes to the sequences between the *Pst*I and the *Ava*I sites (*see* **Note 8**). We obtain high yields of short RNA using Mega(short)script kits (Ambion, Austin, TX, USA). Synthesizing the RNA in the presence of a small amount of labeled NTP (according to manufacturer's instructions) will facilitate quantitation and purification of the RNA, and will allow the process of annealing the RNA to the biotinylated oligonucleotide and the attachment to the beads to be followed (*see* Chapters 1 and 22). However, if labeled RNA is not required, we prepare and purify RNA as outlined below. Use autoclaved solutions and RNase free reagent stocks, and wear gloves.

1. Transcribe the RNA as directed by the manufacturer. Generally we use about 8 µg of template DNA per 20 µL reaction, and allow transcription to proceed overnight.
2. Pour a 5–8% denaturing acrylamide gel. We use an 8% gel *(24)* for RNA targets <100 nucleotides. Mix 6 mL of 10× TBE, 16 mL of 30% acrylamide, 0.5 mL of 10% APS, 25 g of urea, and water to a volume of 60 mL (*see* **Note 4**). Add 50 µL of TEMED to initiate polymerization.
3. Purify the RNA from the transcription reaction by phenol:chloroform extraction and ethanol precipitation.
4. Prerun the gel for 30 min at 275 V in 1× TBE at room temperature to warm the gel to ensure that the RNA remains denatured (*see* **Note 9**).
5. Dissolve the RNA pellet in 4 µL of TE. Add 4 µL of formamide and 2 µL of layer mix. As a molecular weight marker we use 4 µL of φX174 (~1 µg) cut with *Hae*III plus 8 µL of formamide and 2 µL of layer mix.
6. Heat the samples for 5 min at 85°C and load them immediately on the prewarmed gel.
7. Run the gel for 1–2 h at 250 V.
8. Disassemble the gel, and cut off lanes that were not used. Stain the gel for 15 min in a rocking container with 1× TBE, 10 µg/mL ethidium bromide. Destain the gel twice for 10 min in 1× TBE. Even though these procedures are not really RNase-free, the RNA seems to be protected against degradation, probably due to the urea in the gel.
9. Wrap the gel in Saran wrap and photograph if desired. Cut out the RNA band using a razor blade.
10. Place the gel slice inside a 0.5-mL microtube (punctured at the bottom with a 25-gage needle) sitting in a 1.5 mL microtube. Shred the slice by centrifuging for 2 min in a microfuge.
11. Add an equal volume of TE, 0.2% SDS to the gel fragments. Elute the RNA by mixing for 1 h, or mix and leave at 4°C overnight.

12. Pipet the slurry onto a sterile Sephadex G-50 column *(24)* (*see* **Note 10**).
13. Centrifuge at 150*g* in a tabletop centrifuge for 2 min.
14. Centrifuge an additional 50 µL of TE/SDS over the column, into the same microfuge tube.
15. Extract the sample once with phenol:chloroform, once with chloroform, and precipitate by the addition of 1/10 vol of 3 *M* NaOAc and 2.5 vol of ethanol. Wash the pellet with 70% ethanol and dissolve in 50 µL of TE.
16. Measure the absorbance of a 1:100 dilution at 260 nm to determine the concentration (A_{260} 1= 40 µg/mL).
17. Analyze 40 pmol of RNA on a gel as described in **steps 5–9** to confirm the recovery and integrity of the RNA.

Although this approach may seem somewhat laborious, the result is clean, unlabeled RNA of a defined size, which can be stored indefinitely, and can be used for many bead-binding experiments. It is most efficient to prepare multiple RNA targets at a time. If desired, the RNA can be radioactively labeled later using polynucleotide kinase (*see* Chapter 1).

3.5.2. Binding of RNA to Beads

When binding the RNA to beads for the first time we suggest monitoring the annealing step by annealing a constant amount of radiolabeled RNA with increasing amounts of DNA oligo and analyzing the samples on a nondenaturing polyacrylamide TBE gel *(24)*. Conversion of free to the annealed RNA should be visible by a change in mobility. Radiolabeled annealed RNA can also be used to follow the binding to the beads (*see* **Note 11**). The procedure for binding RNA to the beads is described in the next paragraph, using a single binding reaction as an example. Be sure to resuspend beads when pipeting because they tend to sink to the bottom of the tube. When generating beads for multiple binding reactions with the same RNA target, the amounts should be scaled up.

The binding capacity of the beads is 2–5 pmol annealed RNA per 10 µL of beads. This equals > 10^{12} binding sites (if there is one site per RNA). Thus, 10 µL of beads per binding reaction should be sufficient for many kinds of applications.

1. For each phage binding reaction, add 5 pmol of RNA (1 µL) to 5 pmol of oligo (1.5 µL) and heat the solution for 3 min at 85°C.
2. Cool the samples at room temperature for 3 min.
3. Add one volume of 2× TENT buffer (other buffers may be used).
4. Prepare beads (10 µL per binding reaction) by washing twice with at least 1 vol of 0.1 *M* NaOH, 0.1 *M* NaCl, and twice with at least 1 vol of 0.1 *M* NaCl.
5. Resuspend the beads in 1 vol of 1× TENT.
6. Remove the buffer from 10 µL of beads and add 5 µL of DNA/RNA duplex (5 pmol) and 5 µL of tRNA (40 µg). Rotate the tube (along lengthwise axis or beads will get trapped along the lid of the microfuge tube) for 30 min at room temperature.

7. Wash the beads twice with 1× TENT to remove unbound RNA.
8. Resuspend in 10 µL of 1× TENT.

3.5.3. Binding of Display Phage to Beads

To determine how many of the phage in a preparation display an active RNA-binding domain, the titer of the preparation is first measured. This tells us how many phagemid particles are present. Subsequently, a sample of this phage stock is bound to beads carrying an excess of RNA target. After washing, the number of bound phage is determined. The ratio of bound phage divided by the number in the original preparation multiplied by 100 is the percentage "binders." In this type of experiment it can be informative to include a control phage that does not display the RNA-BP, and that is distinguishable from the display phage. To this end, we use a *lacZ*-tagged phagemid *(15)*, that gives rise to blue colonies when infected cells are plated on X-gal-containing plates. The number of blue colonies before and after selection will provide an indication of the amount of background binding. If no *lacZ*-tagged phage is included, the level of background binding must be determined from a parallel experiment using beads that carry no or an irrelevant RNA target. The selective binding procedure is as follows:

1. Assemble phage binding reactions (30 µL) by combining ~1×10^{10} cfu of the display phage preparation, 40 µg of tRNA, and 40 U of RNasin (Promega, Madison, WI, USA) in 1× TENT. A fivefold excess of *lacZ*-tagged phage can be included in this mix. Prepare more of this mix than is needed for the binding reaction, so that part of the mix can be used to determine the number of phage in the original ("input") mix.
2. Add the phage mix to the RNA-carrying beads, from which the buffer has just been removed. Gently rotate the binding reaction for 30 min at room temperature.
3. Wash the beads twice quickly with 1 mL of 1× TENT, and resuspend them in 100 µL of 1× TENT.
4. Make serial dilutions of these 100 µL for titering as described in **Subheading 3.3., step 15**. There is no need to release the phage from the beads; however, care should be taken to mix the bead solution well by vortexing during each dilution step. Also make serial dilutions of the "input" mix.
5. Add 100 µL of each dilution to 100 µL of host cells (diluted and grown for 6 h the same day), incubate at 37°C for 30 min, and plate on LB agar plates containing 100 µg/mL ampicillin, 2% glucose, and 40 µg/mL of X-gal (if the *lacZ*-tagged phage is included).
6. Incubate the plates at 37°C overnight, and count the number of colonies the next day. If the *lacZ*-tagged phage is included, the percentage of blue colonies remaining after binding selection should provide a good indication of the level of background binding. The ratio of blue to white colonies should decrease greatly after selective binding. Generally, 2–5% of our display phage can bind to beads

carrying the target. Background binding is usually less than 0.005% (as measured by retention of *lacZ*-tagged phage on RNA beads, or display phage on "bare" beads). Thus, a single round of binding selection can result in a purification of three to four orders of magnitude. Also *see* **Note 12**.

3.6. Performing Cycles of in Vitro Genetic Selection

3.6.1. Preliminary Considerations

Once the RNA-BP of interest has been shown to be displayed in active form, in vitro genetic analysis can begin. The genetic system described in this chapter is best suited for the detailed study of small regions of a protein that are known to be involved in RNA binding. Such regions can be selectively mutated to several possible amino acid identities, or randomized to all possible amino acids. The choice of a mutagenic strategy will influence the number of amino acid positions that can be studied simultaneously. Owing to the limitations of bacterial transformation (10^{10} at best), a library of a complexity of more than 10^9–10^{10} different clones cannot be fully sampled. If each randomized amino acid position can have 21 possible identities (20 amino acids or a stop codon), then a library containing maximally seven randomized amino acids could still be fully represented. However, if a series of amino acids is changed to only two possible identities, a region of about 30 amino acids could be scanned. This does not mean that libraries larger than 10^{10} cannot be used; it simply means that when analyzing libraries with such complexities, not all clones can be sampled at once. Which approach should be taken depends on the questions to be answered. For example, in our study of the spliceosomal protein U1A, a nine amino acid region was mutated to two possible identities at each position *(1)*. This yielded a small library of 512 clones, which was still large enough to answer questions about specificity determinants. Currently, we are working with libraries that are orders of magnitude larger than what can be sampled, to find proteins with new binding specificities. In this case, it seems more important to create more diversity than to be able to sample all the clones at one time. This is based on the assumption that there may be many possible amino acid configurations that will allow proteins to bind to a particular RNA, and as long as any of those is represented in our library, binders will be selected. Thus, the first consideration in making a library of mutants is how large the mutated region will be, and what the mutations at those positions will be.

3.6.2. Generating the Phagemid Library

Once a genetic strategy has been decided upon, the technical approach to generate the library can be chosen. There are two ways to randomize areas of the RNA-BP. The first one is to replace a section of the gene by synthetic oligonucleotides that contain one or more randomized regions. Smaller oligos that hybridize to constant regions in the randomized oligonucleotides are used to prime the synthesis of the complementary strand. The synthetic fragment is then inserted into the RNA-BP encoding gene, replacing the original sequence. This requires the presence of restriction sites flanking the mutated region. If these are absent they can first be introduced by site-directed mutagenesis. We have used this approach to generate an RNA-BP library with a complexity exceeding 10^{10} different clones.

The second approach is to use a partially randomized oligonucleotide as a primer for PCR. Such an oligo should contain a 3' priming region identical to the template (the gene to be mutated), a 5' randomized region, and 5' of that a restriction site that is used to insert the PCR fragment into the original gene. The second ("reverse") primer is an oligonucleotide identical to the target sequence which also contains a unique restriction site that can be used to insert the PCR fragment. In this scenario, only one restriction site close to the mutagenized region is required, the other one can lie quite far away. This strategy was used to generate the U1A library mentioned above *(1)*. In this approach, the template must be carefully chosen because it may influence the composition of the generated library due to differential priming.

The phagemid library generated by either of the methods outlined above is transformed into highly competent *E. coli* host cells and plated on a series of large plates (25 × 25 cm) containing 100 µg/mL of ampicillin and 2% glucose (maximally ~100,000 colonies per plate). If the library is larger than several million, it may be grown in liquid culture for a few hours in the presence of ampicillin, and 200 µg/mL of methicillin, which will inhibit the growth of "satellite" (non-ampicillin-resistant) cells. However, in liquid culture there is a risk that certain cells with a growth advantage may take over the population. The colonies on plates are pooled by gently scraping in 2×YT (use as little as possible, add 1 mL at a time), and adding glycerol to 16%. Bacteria from a liquid culture are concentrated by centrifugation and resuspended in a smaller volume of 2×YT, 16% glycerol. The glycerol stock, representing the initial or "Round-0" (R-0) library of phagemid in host cells, is stored at –80°C. Cells from this stock are used to generate the display library. They can also be used for the analysis of the initial library by sequencing individual clones (from colonies on a plate) or a multiclone sample from the pool. The latter is a quick and simple way to obtain an indication of the success of the mutagenic strategy.

3.6.3. Generating the Display Library

To ensure adequate representation of the library, it is useful to determine the number of host cells per microliter in the bacterial library stock (make dilutions and plate on LB plates containing 100 µg/mL of ampicillin and 2% glucose). The resulting display phage library (R-0 display phage library) is titered so that the volume of phage stock needed to represent the library can be determined (the estimated percentage display as described in **Subheading 3.5.** should be taken into account). To generate the initial display phage library, a sample of the R-0 bacterial library is taken from the glycerol stock and grown for several hours in 30 mL of 2×YT containing 100 µg/mL of ampicillin and 2% glucose until an OD_{600} of about 0.5 is reached. Then the culture is infected with helper phage and display phage is prepared as described in **Subheading 3.3., steps 3–14**.

3.6.4. In Vitro Selection by Binding to an RNA Target

The method of binding selection depends on the complexity of the displayed library. If a very large library with many different clones exhibiting many different binding affinities is studied, the procedure as described in **Subheading 3.5.** is followed. In that case, an excess of RNA that has already been bound to beads is used. If a library of proteins with similar binding affinities is studied, more stringent selection conditions may be required. This is achieved by reducing the RNA concentration. In this case, the phage are first incubated with the free RNA–DNA duplex, and are subsequently trapped on beads. This way, the RNA concentration during binding can be accurately controlled, and diffusion of the RNA is not limited by the attachment to beads.

The procedure described below uses plates for bacterial propagation, and utilizes limiting amounts of target RNA. It can be adapted for the study of very complex libraries by growing the cells in liquid culture and using an excess of RNA target. It is recommended to do parallel experiments with a negative (non-binding) and a positive (binding) phage population, if these are available. It is also useful to have a fresh overnight culture of host cells available every day for titer determinations.

Day 1

1a. Dilute an overnight culture of host cells 1:500 and grow for 6 h at 37°C (for titer determination).
1b. Incubate an appropriate number of display phage (adequately representing the library complexity) with RNA–DNA duplex (prepared as described in **Subheading 3.5.2.**). First anneal RNA and DNA, then dilute to a concentration that will provide the desired selectivity (e.g., RNA concentrations close to or below the K_d). The binding solution should contain the appropriate optimal buffer and salt concentration, 2 µg/µL of tRNA, and 2 U/µL of RNasin.
2. After allowing the reaction to equilibrate, add the binding solution to beads (prewashed as in **Subheading 3.5.2., steps 4–5**) from which the supernatant has been just removed. Gently tumble the beads for 30 min at room temperature.
3. Quickly wash twice with 1 mL of binding buffer to remove unbound phage (*see* **Note 13**). Resuspend the beads in 100 µL of binding buffer. We call this the "Round-1 (R-1) bound phage" population. It is stored at 4°C.

4. Make serial dilutions of the bead suspension in the ranges of $10^{-4}-10^{-8}$, to determine the titer. Initially, it is wise to plate a wide range of concentrations because it may not be clear how many colonies to expect. When RNA concentrations at or below the K_d are used, expect a reduction in the number of colonies. Also titer the "input" phage mix at the same time to get an accurate comparison.

Day 2
5. Dilute an overnight grown host culture as in **step 1a** above.
6. Count the colonies on titer plates to determine the titer of the input and R-1 phage, and the percentage of phage that bound to the beads during selection.
7. Infect the diluted host cells (from 5) with an aliquot (or dilution) of R-1 bound phage that will adequately represent the complexity of the library. Use an excess of host cells to ensure uptake of all phage. (A 1:500 dilution grown for 6 h contains approx 10^9 cells/mL.) Incubate at 37°C for 30 min.
8. Plate the infected cells on large LB plates containing 100 µg/mL of ampicillin, 2% glucose. Grow the cells overnight at 37°C.

Day 3
9. Pool the colonies as described in **Subheading 3.6.2**. This library, representing the R-1 phagemid library in host cells, is stored at –80°C and will be used to give rise to a new display library (the amplified R-1 phage library) by infecting with helper phage as in **Subheading 3.6.3**. Each set of RNA binding, host infection, and phage production represents a cycle of the in vitro genetic selection. The procedure is summarized in **Fig. 5**. If host cells and phage stocks are always titered, this takes five days, because one needs to grow the titer plates overnight to count the colony numbers. However, if the titers can be estimated (e.g., when an experiment is repeated) the selection could theoretically be shortened to a single day, in which phage is selectively bound, the host is infected and grown in liquid culture for an hour in LB containing ampicillin, methicillin, and glucose, and then superinfected with helper phage.

3.7. Interpretation of Genetic Data

Determining the number of phage after every selective binding phage production step is useful for two reasons. First, it allows one to measure accurately how many particles are used at every step. Second, it is also used to follow the progress of the selection. Depending on the selection criteria, a certain percentage from the initial library is expected to be retained on the beads. As the cycles of selection progress, this number should increase. After a number of rounds the number of retained phage may reach a plateau, suggesting that a population of phage displaying domains with similar binding affinities has been accumulated. At this point, one can choose to stop the selection, or to increase the stringency of the selective step. When a population of phage with similar binding properties is established, biological differences unrelated to the in vitro selective procedure can begin to play a role, and may allow certain phage to take over the population *(1)*. This demonstrates the importance of choosing the

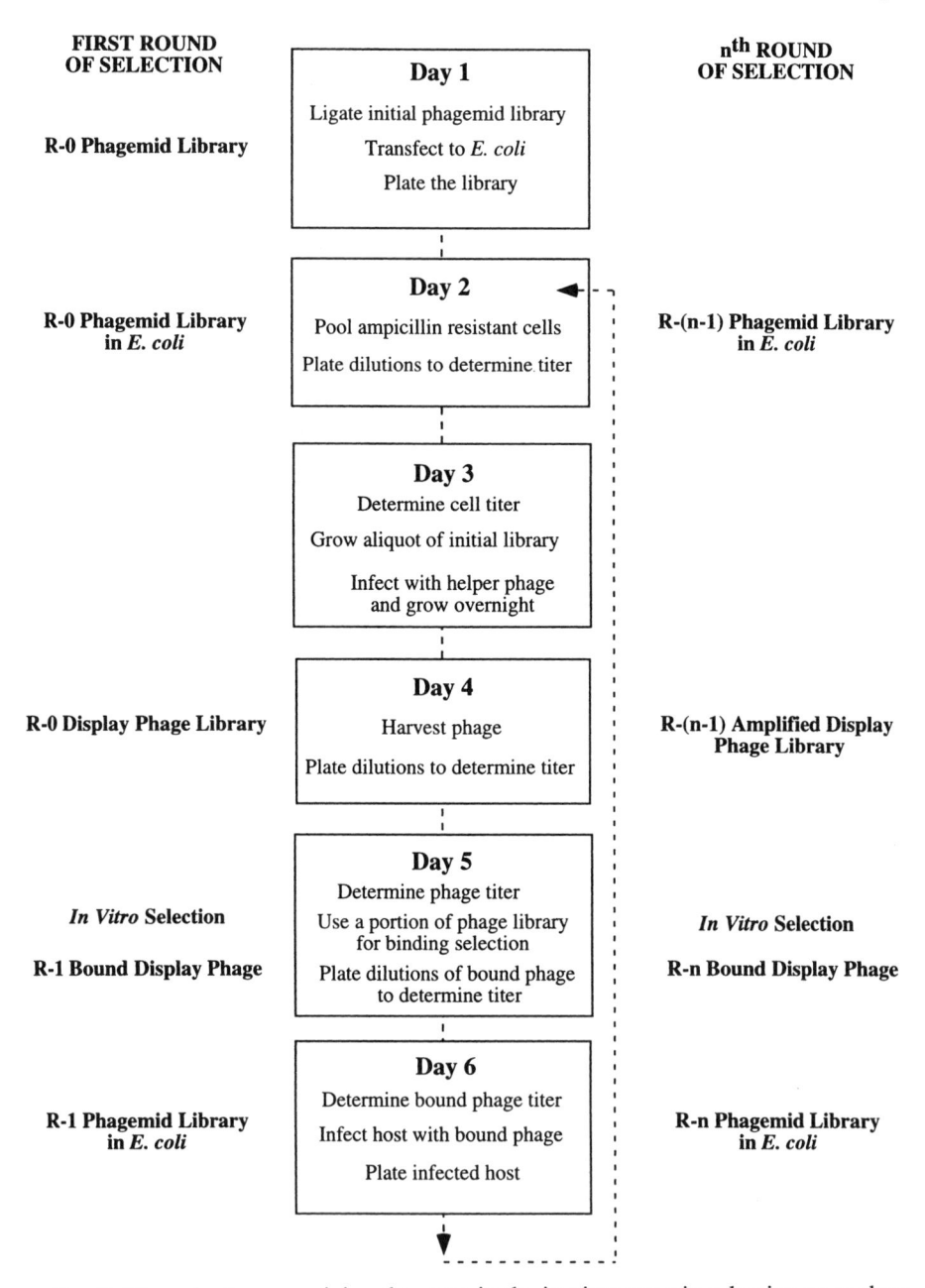

FIRST ROUND OF SELECTION		n^{th} ROUND OF SELECTION
	Day 1 Ligate initial phagemid library Transfect to *E. coli* Plate the library	
R-0 Phagemid Library		
R-0 Phagemid Library in *E. coli*	**Day 2** Pool ampicillin resistant cells Plate dilutions to determine titer	R-(n-1) Phagemid Library in *E. coli*
	Day 3 Determine cell titer Grow aliquot of initial library Infect with helper phage and grow overnight	
R-0 Display Phage Library	**Day 4** Harvest phage Plate dilutions to determine titer	R-(n-1) Amplified Display Phage Library
In Vitro Selection R-1 Bound Display Phage	**Day 5** Determine phage titer Use a portion of phage library for binding selection Plate dilutions of bound phage to determine titer	*In Vitro* Selection R-n Bound Display Phage
R-1 Phagemid Library in *E. coli*	**Day 6** Determine bound phage titer Infect host with bound phage Plate infected host	R-n Phagemid Library in *E. coli*

Fig. 5. Flow chart summarizing the steps in the in vitro genetic selection procedure. The diagram shows the slowest (and most accurate) procedure, where each library is titered before the next step is taken. To save time, titers can be estimated (e.g., during the later rounds of selection), so that the next step of the experiment can be performed as titer plates incubate.

right number of selective cycles. What this number is, depends on the type of experiment, and the nature of the information that is sought. For example, if one is searching for a mutant protein that binds tightly to a certain sequence, a single tightly binding clone obtained after many rounds of selection is a very good result. However, if one were studying the relative importance of certain amino acids in a protein region *(1)*, the most information would be obtained by studying a population of selected phage that had been enriched for particular amino acids at certain positions (that therefore seem to be relevant for binding) but that was still random at other positions (indicating no strong amino acid preference). Ending with a single clone would in this case provide very little information because the relative contribution of each amino acid to binding would not be clear.

The size of the initial library is also important in choosing the number of selection cycles. A small initial library will require many fewer rounds of selection than a very complex library. Because it may be difficult to predict the number of cycles needed, one can simply perform rounds of selection, using the absolute number of phage retained during the binding step as a measure of progress. Once the number of binders levels off, one could stop the selection and analyze the changes in the library, either by pool sequencing or by analyzing individual clones. Since all the steps of the selection are stored as phage preparations and glycerol stocks of host cells, the selection can be continued at any time.

The analysis of clones obtained by genetic selection will suggest certain conclusions, for example, a consensus sequence that allows specific binding to an RNA target. Such conclusions should always be verified by other means, e.g., by biochemical methods (e.g., purification of mutant proteins and determination of binding affinities). It should be noted that during selection, proteins that bind to targets other than the intended one can arise. For example, it is theoretically possible to select a protein variant that binds to the RNA–DNA double stranded region, instead of the actual RNA target sequence. This problem can be minimized by varying the RNA sequence to which the DNA oligo is annealed in the different rounds of selection, or by prebinding to beads carrying a mock target RNA.

4. Notes

1. Next to its function in attachment to the host, the bacteriophage gene III protein also confers resistance to superinfection, even when it has an N-terminal fusion. This means that one bacterial cell can only give rise to particles displaying one type of fusion protein.
2. To be infected, host cells must have pili and must be grown at 37°C. If cells are not grown at 37°C, or if they are left on the bench for too long before being used, they will not be susceptible to infection.

3. 1× TENT is the buffer that we chose to study interactions between the spliceosomal U1A protein and U1A. It was chosen because the RNA–protein complex is relatively stable in that buffer. Any buffer can be used in the in vitro genetic system, and for each protein–RNA pair, one may want to choose the binding buffer based on published or empirically determined optimal binding conditions. However, to minimize RNase activity, it is best to avoid divalent cations if possible. In addition, the inclusion of detergents such as Triton X-100 will help reduce the level of background binding to the beads.
4. Acrylamide is a neurotoxin. Wear gloves and a laboratory coat.
5. Integrated DNA Technologies provides good quality, gel-purified biotinylated oligos. Gel purification is useful to remove unbiotinylated oligos, so that all annealed RNA will attach to the streptavidin beads.
6. We have found that the strains IAL-3 and DH12S behave slightly differently in the phoA assay. Export from IAL-3 is best evaluated using plates containing glucose. Absence of glucose will lead to a very unstable colony phenotype, presumably because the increased synthesis of the fusion protein, which is inefficiently exported, disrupts the bacterial secretion machinery. DH12S is more tolerant of growing in the absence of glucose, and in addition, the endogenous phoA activity seems to increase on glucose-containing plates. Therefore, plates lacking glucose can be used to assess export from DH12S. However, colonies from such plates should not be used for further experimentation. If no secretion is seen with the phoA assay, two things should be checked, before assuming the secretion conditions need to be optimized: the maintenance of the reading frame (sequence of junctions between the RNA-BP gene and the vector) and the absence of stop codons. If folding of the protein is suspected to be the cause of poor export, a reduction in temperature during phage production should be tried. Experiment with temperatures ranging from 30°C to 22°C. In our hands, 30°C is optimal for RRM type proteins.
7. Because helper phage is used to provide the wild-type phage proteins for phage replication (including a wild-type gene III), the preparations of display phage will consist of a mix of four different types of particles. These display either no protein or the RNA-BP on their exterior, and carry either the helper phage genome or the phagemid in their interior. Because the expression of the fusion construct is low, most particles will carry one or less gene III fusion proteins on their outside. Particles containing a helper phage genome in their interior will not give rise to ampicillin resistant colonies. Therefore, helper phage does not interfere with the experiments. The only particles that are relevant are the phagemids that display an RNA-BP on their exterior. These will be bound during the selection, and will then give rise to ampicillin-resistant colonies.
8. For biochemical assays, we make a shorter (radiolabeled) RNA from a plasmid that is linearized just downstream of the target-encoding sequences (at the *Acc*I site). Such RNAs can't anneal to the biotinylated DNA oligonucleotide, and could also be used as competitors in the in vitro genetic experiments.
9. If the gel for purifying the RNA target is not prewarmed, or the RNA sample is not loaded immediately after denaturation, the RNA may basepair and form structures that migrate through the gel with different mobilities.

10. To make a sterile G-50 Sephadex column, plug a 1-mL syringe with siliconized glass wool and autoclave it. Then load it with sterile G50 in TE, and attach a needle at the end. Insert the needle into the lid of a 1.5-mL microtube. Spin in a tabletop centrifuge to pack column to final volume of 0.9–1.0 mL. Replace the microtube with a clean one to catch the eluate.
11. Radiolabeled RNA can be used to measure the binding of the DNA–RNA duplex to the beads by determining the amount of radioactivity transferred. Depending on the quality of the biotinylated oligo, 50–100% of the label should be bound to the beads (*see also* **Note 5**).
12. The 2–5% binders is the percentage we find when strain IAL-3 is used. Percentages of binders obtained from host strain DH12S can be up to fivefold lower. Because the phage displaying the RNA-BP become highly purified in each round of binding, and because very large numbers of phage are easily prepared (10^{12}–10^{13} cfu/mL), the percentage of displaying phage need not be very high. Indeed, too high a percentage may mean that bivalent phage could be present, which could be a problem during selection, as a particles displaying two weak binders may bind as tightly as particles displaying one stronger binder. However, when very large libraries are displayed it may be desirable to optimize the level of display to achieve 10% (e.g., by experimenting with the growth temperature during phage production). Whether this can be done depends on the strain and the displayed protein. Do not attempt to increase display by inducing the expression of the fusion product with IPTG, as this will make the cells sick and result in less display.
13. Extending the washing times after allowing the display phage to bind to beads can be used to select for proteins that release the RNA very slowly.

References

1. Laird-Offringa, I. A. and Belasco, J. G. (1995) Analysis of RNA-binding proteins by in vitro genetic selection: identification of an amino acid residue important for locking U1A onto its RNA target. *Proc. Natl. Acad. Sci. USA* **92**, 11,859–11,863.
2. Jain, C. and Belasco, J. G. (1996) A structural model for the HIV-1 Rev-RRE complex deduced from altered specificity Rev variants isolated by a rapid genetic strategy. *Cell* **87**, 115–125.
3. Harada, K., Martin, S. S., and Frankel, A. D. (1996) Selection of RNA-binding peptides in vivo. *Nature* **380**, 175–179.
4. SenGupta, D. J., Zhang, B., Kraemer, B., Pochart, P., Fields, S., and Wickens, M. (1996) A three-hybrid system to detect RNA–protein interactions in vivo. *Proc. Natl. Acad. Sci. USA* **93**, 8496–8501.
5. Tuerk, C. and Gold, L. (1990) Systematic evolution of ligands by exponential enrichment: RNA ligands to bacteriophage T4 DNA polymerase. *Science* **249**, 505–510.
6. Tuerk, C. (1997) Using the SELEX combinatorial chemistry process to find high affinity nucleic acid ligands to target molecules. *Methods Mol. Biol.* **67**, 219–230.
7. Smith, G. P. (1985) Filamentous fusion phage: novel expression vectors that display cloned antigens on the virion surface. *Science* **228**, 1315–1317.
8. Smith, G. P. (1993) Surface display and peptide libraries. *Gene* **128**, 1–2.

9. Smith, G. P. and Scott, J. K. (1993) Libraries of peptides and proteins displayed on filamentous phage. *Methods Enzymol.* **217**, 228–257.
10. Mikawa, Y. G., Maruyama, I. N., and Brenner, S. (1996) Surface display of proteins on bacteriophage lambda heads. *J. Mol. Biol.* **262**, 21–30.
11. Francisco, J. A. and Georgiou, G. (1994) The expression of recombinant proteins on the external surface of Escherichia coli. Biotechnological applications. *Ann. NY Acad. Sci.* **745**, 372–382.
12. Gennity, J., Goldstein, J., and Inouye, M. (1990) Signal peptide mutants of *Escherichia coli*. *J. Bioenerg. Biomembr.* **22**, 233–269.
13. Boyd, D. and Beckwith, J. (1990) The role of charged amino acids in the localization of secreted and membrane proteins. *Cell* **62**, 1031–1033.
14. Model, P. and Russel, M. (1990) Prokaryotic secretion. *Cell* **61**, 739–741.
15. Laird-Offringa, I. A. and Belasco, J. G. (1996) In vitro genetic analysis of RNA-binding proteins using phage display libraries. *Methods Enzymol.* **267**, 149–168.
16. San Millan, J. L., Boyd, D., Dalbey, R., Wickner, W., and Beckwith, J. (1989) Use of phoA fusions to study the topology of the *Escherichia coli* inner membrane protein leader peptidase. *J. Bacteriol.* **171**, 5536–5541.
17. Nagai, K., Oubridge, C., Ito, N., Avis, J., and Evans, P. (1995) The RNP domain: a sequence-specific RNA-binding domain involved in processing and transport of RNA. *Trends Biochem. Sci.* **20**, 235–240.
18. Burd, C. G. and Dreyfuss, G. (1994) Conserved structures and diversity of functions of RNA-binding proteins. *Science* **265**, 615–621.
19. Hoogenboom, H. R., Griffiths, A. D., Johnson, K. S., Chiswell, D. J., Hudson, P., and Winter, G. (1991) Multi-subunit proteins on the surface of filamentous phage: methodologies for displaying antibody (Fab) heavy and light chains. *Nucleic Acids Res.* **19**, 4133–4137.
20. Studier, F. W., Rosenberg, A. H., Dunn, J. J., and Dubendorff, J. W. (1990) Use of T7 RNA polymerase to direct expression of cloned genes. *Methods Enzymol.* **185**, 60–89.
21. Evan, G. I., Lewis, G. K., Ramsay, G., and Bishop, J. M. (1985) Isolation of monoclonal antibodies specific for human c-myc proto-oncogene product. *Mol. Cell. Biol.* **5**, 3610–3616.
22. Hanahan, D., Jessee, J., and Bloom, F. R. (1991) Plasmid transformation of *Escherichia coli* and other bacteria. *Methods Enzymol.* **204**, 63–113.
23. Vieira, J. and Messing, J. (1987) Production of single-stranded plasmid DNA. *Methods Enzymol.* **153**, 3–11.
24. Maniatis, T., Fritsch, E. F., and Sambrook, J. (1982) *Molecular Cloning, A Laboratory Manual*, Cold Spring Harbor Laboratory Press, Cold Spring Harbor, New York.
25. Manoil, C. (1991) Analysis of membrane protein topology using alkaline phosphatase and beta-galactosidase gene fusions. *Methods Cell. Biol.* **34**, 61–75.

16

In Vitro Selection of Aptamers from RNA Libraries

Daniel J. Kenan and Jack D. Keene

1. Introduction

Methods of iterative nucleic acid selection and amplification were enabled by the invention of the polymerase chain reaction (PCR). Thus, the ability to amplify as few as a single DNA or RNA molecule made it possible to diversify sequences and to partition the desirable from the undesirable subset (**Fig. 1A**) *(1,2)*. Initially, this approach was practiced in vivo using biological amplification and selection, by diversification of plasmid sequences and iterative growth against a selective marker *(3)*. The power of in vitro RNA selection from a randomized combinatorial library was demonstrated by Tuerk and Gold *(4)* using T4 DNA polymerase and the R17 phage coat protein. These investigators called their iterative RNA selection procedure "SELEX." Ellington and Szostak *(5)* also derived RNA ligands against organic dyes using iterative in vitro selection. Moreover, the demonstration that RNAs could be selected that bind to proteins and compounds with no known role in RNA-binding in vivo led to the concept of aptamers. An aptamer is a folded RNA that forms a shape that coincidentally fits against another surface, to which it is "apt" to bind. Generally, RNA ligands that bind to naturally occurring RNA-binding domains of proteins are not considered to be aptamers. Unexpectedly, an RNA ligand selected against an antibody generated by immunization with a peptide demonstrated that RNA and protein could crossreact at the level of antibody recognition *(6)*. This led to the suggestion that such interactions may play a role under natural conditions in vivo in which RNAs may displace proteins as part of cellular interaction networks *(7)*. Artificial DNA and RNA ligands selected in vitro often have high affinity for the target against which they were selected, opening the potential for using them as functional inhibitors. Although such aptamers have a variety of possible uses, their value as therapeutic pharmaceu-

From: *Methods in Molecular Biology, Vol. 118: RNA-Protein Interaction Protocols*
Edited by: S. Haynes © Humana Press Inc., Totowa, NJ

A

```
                              Libraries:
                              Stem-loops N10, N13 or
         Forward primer       Linear N25, N40, N68, etc.
         ────────→         ⌒
                           ─────────────

Step 1:                    │  Primer Extension
                           ▼    (Klenow)
         T7 Pro
         ═══════════════════════════════→
         ←───────────────

Step 2:                    │  T7 RNA polymerase
                           ▼     transcription
                                                        Repeat until
              RNA Pool  ⌒                               selected RNA
                       ─────────→            Protein binding and      population binds
Step 3:                    │  immunoprecipitation       with desired
                           ▼                            characteristics
              Bound RNA ⌒
                       ─────────→
                                 Reverse primer
                                 ←────────────
Step 4:                    │  Reverse transcription
                           ▼     and PCR
         T7 Pro               with both primers
         ═══════════════════════════════
         ←───────────────

                                    │
                                    ▼
                                  Clone
                                    │
                                    ▼
                                **Sequence**
```

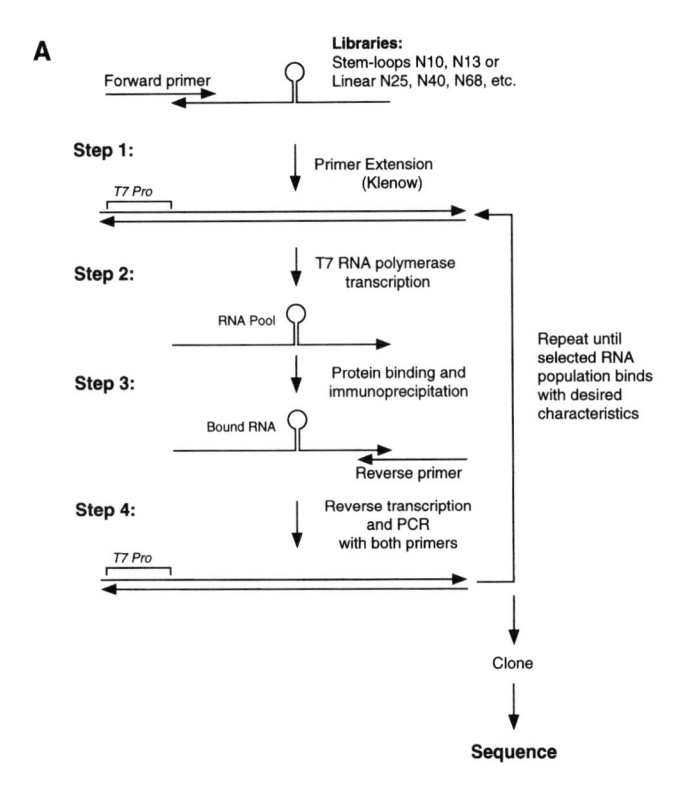

Fig. 1. **(A)** A schematic representation of the randomized RNA selection protocol. The sequences of the primers and template are given in **(B)**. The T7 promoter region is labeled T7 pro. DNA and RNA are represented by double lines and single lines, respectively. The number of iterative cycles used varies with the nature of the target molecule and can allow evolution of the target set using error-prone PCR. **(B)** The primer set illustrated was developed for a particular application and contains a length of random nucleotides (indicated by N) and flanking sequences that may not be optimal for other applications. It is shown for illustrative purposes only. The sequences flanking the random segment are somewhat arbitrary but need to be chosen carefully to avoid amplification artifacts. For preparation of the initial library, equimolar amounts of the N25 library and Forward oligonucleotides are annealed and extended using the Klenow enzyme. A double-stranded DNA is formed that serves as a transcription template in the presence of T7 RNA polymerase. The 5' terminus of the N25 library oligonucleotide is designed to give the same RNA 3' terminus as that generated by transcription from *Bam*HI-digested templates in successive cycles of selection. RNA molecules recovered from the selection procedure are converted to cDNA by reverse transcription using the Reverse Primer, followed by PCR using both the Forward and Reverse Primers. The amplified cDNA is converted to a transcription template by *Bam*HI digestion, which restores the correct RNA 3' terminus upon subsequent transcription by T7 RNA polymerase.

B *Primer Set*

Forward 5'-CGCGGATCCTAATACGACTCACTATAGGGCCACCAACGACATT
N25 Library 5'-GGATCCATGGCACTATTTATCAACNNNNNNNNNNNNNNNNNNNNNNNNNAATGTCGTTGGTGGCCC
Reverse 5'-CCCGACACCCGGATCCATGGCACTATTTATATCAA

Preparation of Double Stranded DNA Transcription Template and RNA Pool

5'-CGCGGATCCTAATACGACTCACTATAGGGCCACCAACGACATT
                                       CCCGGTGGTTGCTGTCTAANNNNNNNNNNNNNNNNNNNNNNNNNCAACTATATTTATCACGGGTACCTAG-5'

→ **Primer Extension (Klenow)**

5'-CGCGGATCCTAATACGACTCACTATAGGGCCACCAACGACATTNNNNNNNNNNNNNNNNNNNNNNNNNGTTGATATAAATAGTGCCATGGATC
   GCGCCTAGGATTATGCTGAGTGATATCCCCGGTGGTTGCTGTAANNNNNNNNNNNNNNNNNNNNNNNNNCAACTATATTTATCACGGGTACCTAG-5'

→ **Transcription (T7 RNA Polymerase)**

5'-GGGGCCACCAACGACAUNNNNNNNNNNNNNNNNNNNNNNNNNGUUGAUAAAUAGUGCCAUGGAUC

Reverse Transcription and PCR Amplification of Selected RNAs and Resynthesis of RNA pool

5'-GGGGCCACCAACGACAUNNNNNNNNNNNNNNNNNNNNNNNNNGUUGAUAAAUAGUGCCAUGGAUC
                                           AACTATATTTATCACGGGTACCTAGGCGCCCACAGCCC-5'

→ **Reverse Transcription & PCR**

5'-CGCGGATCCTAATACGACTCACTATAGGGCCACCAACGACATTNNNNNNNNNNNNNNNNNNNNNNNNNGTTGATATAAATAGTGCCATGGATCGATCCGCGGGTGTCGGG
   GCGCCTAGGATTATGCTGAGTGATATCCCCGGTGGTTGCTGTAANNNNNNNNNNNNNNNNNNNNNNNNNCAACTATATTTATCACGGGTACCTAGGCGCCCACAGCCC-5'

→ **BamHI Digestion**

5'-GATCCTAATACGACTCACTATAGGGCCACCAACGACATTNNNNNNNNNNNNNNNNNNNNNNNNNGTTGATATAAATAGTGCCATG
       GATTATGCTGAGTGATATCCCCGGTGGTTGCTGTAANNNNNNNNNNNNNNNNNNNNNNNNNCAACTATATTTATCACGGGTACCTAG-5'

→ **Transcription (T7 RNA Polymerase)**

5'-GGGGCCACCAACGACAUNNNNNNNNNNNNNNNNNNNNNNNNNGUUGAUAUAAAUAGUGCCCAUGGAUC

Fig. 1B

ticals remains to be demonstrated. On the other hand, for target validation and for use in deriving structure–activity relationships, aptamers have potential as discovery tools. Thus, in both academic and commercial settings, these methods are being applied to a variety of cellular and pathogenic target molecules.

The original experiments by Oliphant et al. *(1)* were intended to derive clues to the DNA-binding specificity of transcription factors. RNA-binding specificity has been examined by in vitro selection using various proteins with known sequence binding preferences *(4,8,9)*. However, there are very few examples in which the RNA-binding specificity of a protein has been determined *de novo*. Levine et al. *(10)* determined that the mammalian ELAV proteins prefer to bind to AU-rich elements in the 3' untranslated regions of mRNAs of protooncogenes and cytokines using in vitro RNA selection from a randomized 25 nucleotide combinatorial RNA library. This RNA selection result provided a vital clue that allowed these investigators to test the authentic mRNAs and to map the binding sites directly to the AU-rich elements involved in mRNA stability. Buckanovich and Darnell *(11)* derived a similar clue that the Nova-I K-type RNA-binding protein preferred sequences similar to those in the 3' untranslated regions of the glycine receptor mRNA using a combinatorial RNA library. Direct binding of Nova-1 to the authentic mRNA was then demonstrated. Iterative methods to select RNAs and DNAs from naturally occurring sequences have also been developed *(12,13)*.

Some of the obvious applications of in vitro selected aptamers include their use as probes for detecting the presence of a target molecule in an unknown setting, or for purification of the target by affinity selection. In addition, aptamers have the potential to inhibit enzymes or to block receptors. Theoretically, both agonists and antagonists can be derived for such functional studies. The advantage of DNA and RNA aptamers over small organic molecules is the potential to express the nucleic acid ligands in cells or animals. Thus, as gene therapy tools, aptamers may be used like antisense RNAs as direct inhibitors of cellular or pathogenic functions. In the postgenomic era, it is predicted that aptamers will have significant value in functional studies to provide information on structure–function properties *(14–16)*. Therefore, high-throughput screening of large libraries of potential binding ligands can give rise to multiple aptamers from which to derive functional information.

In this chapter, methods for selection of RNA ligands from a randomized RNA library are described, together with the necessary reagents. The protocol proceeds in three phases as illustrated in **Fig. 1A**: (1) conversion of library oligonucleotides into double-stranded DNA transcription templates; (2) transcription of RNA and multiple cycles of binding, partitioning, amplification, and retranscription; and (3) cloning selected RNAs as double-stranded cDNA into an appropriate vector for downstream studies such as sequencing and target binding. These protocols are fit for using any protein or other organic molecule, as well as whole cells, in the selection procedure. A variety of immobilized surfaces such as plastic, nitrocellulose, or antibodies bound to Staph A beads are amenable to these methods. In Chapter 17, iterative selection methods based upon naturally occurring DNAs or RNAs are described.

2. Materials
2.1. Purification of T7 RNA Polymerase

1. Plasmid pAR1219 in bacterial strain BL21 (strain number 39563, ATCC, Manassas, VA).
2. LB medium and plates.
3. Ampicillin.
4. Isopropyl β-D-thiogalactopyranoside (IPTG): Prepare a 1 M solution in dH_2O.
5. Lysis buffer: 20 mM Tris-HCl, pH 8.0, 100 mM NaCl, 5 mg/mL lysozyme, 125 µM phenylmethylsulfonyl fluoride (PMSF). Prepare fresh as needed, as PMSF cannot be stored in aqueous solutions. PMSF is toxic; take precautions to avoid inhalation and skin contact. It is inactivated by incubation at room temperature for several hours in aqueous alkaline solutions (>pH 8.6).
6. Polyethyleneimine (Sigma, St. Louis MO, USA).
7. Ammonium sulfate.
8. Buffer B: 20 mM Tris-HCl, pH 8.0, 10% glycerol, 125 µM PMSF. Prepare fresh as needed.
9. 1 M NaCl in buffer B. Prepare fresh as needed.
10. 10 mM Tris-HCl, pH 8.1, 10% glycerol.
11. Fast-Flow S Sepharose column and Fast-Flow DEAE Sepharose column (Pharmacia, Piscataway NJ, USA; 50-mL columns are more than adequate).
12. Dialysis buffer: 20 mM Tris-HCl, pH 8.0, 0.1 mM dithiothreitol (DTT), 125 µM PMSF, 50% glycerol. Prepare fresh as needed.
13. Bio-Rad Protein Assay kit (Bio-Rad, Hercules, CA, USA).
14. Sonicator.

2.2. Conversion of Library Oligonucleotides into Double-Stranded DNA Transcription Templates

1. Custom degenerate oligonucleotides are ordered from commercial sources and diluted in dH_2O to the proper concentration (*see* **Note 1**). Three oligonucleotides are required (**Fig. 1B**): (1) a "Forward" oligonucleotide, providing a T7 promoter sequence and a 5'-terminal restriction endonuclease cloning site; (2) a "Reverse" oligonucleotide, providing a 3'-terminal cloning site and serving as a reverse transcription primer; and (3) a library oligonucleotide providing a randomized region and primer annealing sequences for both the Forward and Reverse primers. The Forward and Reverse oligonucleotides are used in all subsequent nucleic acid amplification steps. Prepare a 100 µM stock of the library oligonucleotide, 5 µM and 100 µM stocks of the Forward oligonucleotide, and a 5 µM stock of the Reverse oligonucleotide.

2. 5× Klenow buffer: 0.25 M Tris-HCl, pH 7.2, 50 mM MgSO$_4$, 0.5 mM DTT.
3. Klenow enzyme, 3'->5' exo$^-$, at 5 U/μL (New England Biolabs, Beverly, MA, USA).
4. dNTP mix: 10 mM each of dATP, dCTP, dGTP, and dTTP in dH$_2$O, prepared from 100 mM stock solutions (Pharmacia, Piscataway, NJ, USA).
5. 3 M Sodium acetate, pH 5.2.
6. 100% and 70% Ethanol.
7. TE8: 10 mM Tris-HCl, pH 8.0, 1 mM EDTA.
8. 10× TBE: 890 mM Tris base, 890 mM boric acid, 20 mM EDTA.
9. 40% Acrylamide stock solution: 38 % acrylamide, 2 % bisacrylamide.
10. 10% Ammonium persulfate (APS) in dH$_2$O; N,N',N',N'-tetramethylethylene-diamine (TEMED).
11. DNA loading dye: 20% Ficoll 400, 0.1 M EDTA, pH 8.0, 1.0% sodium dodecyl sulfate, 0.25% bromphenol blue.
12. Polyacrylamide gel electrophoresis apparatus with 1.5-mm spacers and comb.
13. 0.5 M Sodium acetate, pH 5.2.
14. Dry ice/ethanol bath.
15. Siliconized glass wool.
16. RNase-free glass rods (*see* **Note 1**).

2.3. Transcription, Selection and Amplification

1. 10× T7 RNA polymerase buffer: 400 mM Tris-HCl, pH 8.1, 60 mM MgCl$_2$, 50 mM DTT, 10 mM Spermidine, 0.1% v/v Triton X-100 (*see* **ref. 17**).
2. 100 mM stock solutions of ATP, CTP, GTP, and UTP (Pharmacia, Piscataway, NJ, USA).
3. RQ1 DNase at 1 U/μL (Promega, Madison, WI, USA).
4. Urea loading buffer: 8 M urea, 1 × TBE, 0.025% bromphenol blue, 0.025% xylene cyanol.
5. Protein A beads (Sigma, St. Louis, MO, USA).
6. NT2 buffer: 50 mM Tris-HCl, pH 7.4, 150 mM NaCl, 1 mM MgCl$_2$, 0.05% Nonidet P-40 (NP40).
7. Purified or crude RNA binding protein.
8. RNA binding buffer: 50 mM Tris-HCl, pH 7.4, 100 mM NaCl, 4 mM EDTA, 2 mM MgCl$_2$, 0.05% NP-40, 0.4% Vanadyl ribonucleoside complex, 200 U/mL RNasin (Promega, Madison, WI, USA), 100 μg/mL poly A, 100 μg/mL tRNA, 50 μg/mL acetylated bovine serum albumin (BSA) (*see* **Note 2**).
9. 2× Proteinase K buffer: 20 mM Tris-HCl, pH 7.8, 10 mM EDTA, 1% sodium dodecyl sulfate (SDS).
10. 10 mg/mL Proteinase K.
11. PCI: Phenol:chloroform:isoamyl alcohol (50:48:2). Preequilibrate the phenol with 100 mM Tris-HCl, pH 8.0.
12. 10 mg/mL Yeast tRNA in dH$_2$O.
13. 0.25-mL Thin-walled PCR tubes.
14. AMV or Superscript II (Life Technologies, Gaithersburg, MD, USA) Reverse Transcriptase (200 U/μL) and the appropriate 10× reaction buffer.

In Vitro Selection of RNA Aptamers

15. dNTP mix: 10 mM each of dATP, dCTP, dGTP, and dTTP in dH$_2$O.
16. *Taq* polymerase (5 U/μL) and commercial 10× reaction buffer.
17. *Bam*HI (10 U/μL).
18. rNTP mix: 10 mM each ATP, CTP, GTP, and UTP in dH$_2$O.
19. [γ-^{32}P] ATP or [α-^{32}P] UTP.

2.4. Subcloning Selected RNA Species

1. Kit for blunt-cloning PCR products, such as the Prime PCR Cloner kit (5 Prime -> 3 Prime, Boulder, CO, USA).
2. pSP64 or other vector containing a *Bam*HI cloning site and lacking a T7 promoter.

3. Methods
3.1. Purification of T7 RNA Polymerase

This protocol requires large quantities of T7 RNA polymerase, which would be prohibitively expensive if purchased commercially. However, the purification of the enzyme is relatively straightforward. This protocol uses a bacterial strain containing a cloned T7 RNA polymerase gene under the control of an inducible promoter, and is taken from the procedures developed by Milligan and Uhlenbeck *(17)*.

1. If necessary, transform pAR1219 into BL21 cells *(18)*. Inoculate an overnight broth culture containing 100 μg/mL of ampicillin from a single fresh colony of BL21(pAR1219).
2. The next morning, inoculate the overnight culture 1:100 into 500 mL of LB containing 100 μg/mL of ampicillin. Grow to an OD$_{600}$ of 0.6–0.8. Induce expression of T7 RNA polymerase by adding IPTG to a final concentration of 100 μM.
3. Harvest the cells after an additional 3 h of growth by centrifugation at 5000g for 10 min.
4. Resuspend the cell paste in 10 mL of lysis buffer. Lyse the cells by flash-freezing in liquid nitrogen or dry ice/ethanol, and thawing rapidly in a 37°C water bath.
5. Sonicate for 30 s, twice, to reduce sample viscosity.
6. Add polyethyleneimine to 1% final concentration and rock for 10 min at 4°C.
7. Centrifuge at 15,000g for 20 min at 4°C and harvest the supernatant, carefully avoiding debris in the pellet.
8. Add ammonium sulfate to 50% saturation and rock gently for 30 min at 4°C. Collect the precipitate by centrifugation at 15,000g for 20 min at 4°C. Discard the supernatant.
9. Dissolve the pellet in approx 5 mL of cold Buffer B. Have the columns and buffers prechilled. Load onto a Fast-Flow S Sepharose column and wash with 5–10 column volumes of Buffer B, until the OD$_{280}$ of the column effluent returns to baseline.
10. Elute the bound proteins with a linear gradient of 0–1 M NaCl in Buffer B, collecting 5-mL fractions.
11. Identify the fractions containing T7 RNA polymerase by SDS–polyacrylamide gel electrophoresis (SDS–PAGE) and Coomassie staining. Prepare a 10% poly-

acrylamide gel (*see* Chapters 23 and 31) and load 10 μL of each fraction on the gel. T7 RNA polymerase has a molecular mass of 98,850 Daltons and should easily be visible by Coomassie staining.

12. Pool the peak fractions and dilute into a final volume of 50 mL with 10 m*M* Tris-HCl, pH 8.1, 10% glycerol. Load the sample onto a Fast-Flow DEAE Sepharose column (Pharmacia) and elute with a 0–1 *M* NaCl gradient as before.
13. Assay the eluate by SDS–PAGE and pool peak fractions as done in **step 11**.
14. Dialyze the sample against a total of 2 L of dialysis buffer over at least four changes at 4°C.
15. Determine protein concentration by Bio-Rad Protein Assay and dilute the enzyme to 1 mg/mL; the purified enzyme has a specific activity between 300,000 and 500,000 U/mg *(19)*. Store in aliquots at –20°C. The enzyme is stable for years if fresh DTT is added every 6 mo *(17)*. Prior to use, assess the transcriptional activity and nuclease contamination by standard assays *(18)*. If nuclease contamination is a problem, additional purification steps can be undertaken *(17)* but are rarely necessary.

3.2. Conversion of Library Oligonucleotides into Double-Stranded DNA Transcription Templates

The library oligonucleotides must be converted into double stranded DNA prior to serving as templates for transcription by T7 RNA polymerase. In practice, it is technically difficult to manipulate more than 1×10^{15} molecules in an RNA selection, equivalent to 1.7 nmol of template. It is thus convenient to dilute the library oligonucleotide stocks to 100 pmol/μL (100 μ*M*). Because preparation of the initial libraries is laborious, scale up the preparation and store aliquots of both DNA and RNA forms of the library for future use. The protocol below starts with a fivefold scaleup (8.5 nmol) of the double-stranded DNA preparation.

1. Anneal the Forward and Library oligonucleotides (as 100 μ*M* stock solutions) by combining the following components in a 1.5-mL microfuge tube: 200 μL of 5× Klenow buffer, 85 μL of Forward primer, 85 μL of Library oligonucleotide N25, and 590 μL of dH$_2$O (final volume is 960 μL).
2. Heat to 75°C for 15 min. Transfer to a beaker containing 500 mL of water at 75°C and allow to cool slowly to room temperature. Centrifuge briefly in a microcentrifuge to collect the condensate.
3. Extend the annealed primers by the addition of DNA polymerase. To the annealing reaction, add 20 μL of dNTP mix and 20 μL of Klenow enzyme. Incubate at 37°C for 1 h.
4. Precipitate the reaction with ethanol. Add 1/10 vol (100 μL) of 3 *M* sodium acetate, pH 5.2. Divide the sample into four aliquots of 275 μL each and add 600 μL of 100% ethanol. Incubate for 10 min in a dry ice–ethanol bath and then centrifuge at ≥ 12,000*g* for 20 min. Decant the supernatant, wash the DNA pellet carefully with 70% ethanol, and centrifuge at ≥ 12,000*g* for 5 min. Decant the supernatant and air dry the pellet. Avoid overdrying the DNA pellet.

5. Purify the full-length, double-stranded library products by gel electrophoresis to remove unextended primers and aberrant primer extension products. Prepare a 15% nondenaturing TBE polyacrylamide gel. A typical size is 15 cm × 15 cm × 1.5 mm with a single large well (*see* **Note 3**). For 50 mL of gel mix, combine 18.75 mL of 40% acrylamide stock and 5 mL of 10× TBE, and bring to 50 mL with dH$_2$O. Add 500 µL of 10% APS and 27 µL of TEMED immediately before pouring the gel. Resuspend the DNA pellet in 200 µL of TE8. Add 1/10 vol of DNA loading dye and apply the sample to the gel. Use 1 × TBE for the running buffer and electrophorese the sample until the xylene cyanol dye is close to the bottom of the gel.
6. Visualize the nucleic acids by ultraviolet shadowing (*see* Chapter 1), and excise the dominant band representing intact double stranded DNA. Place the gel fragments into a 15 mL polypropylene screw-capped tube and pulverize with an RNase-free glass rod. Add 1–2 mL of 0.5 *M* sodium acetate and elute the DNA by shaking overnight at 37°C.
7. Centrifuge the sample at approximately 1000*g* for 5 min in a low-speed centrifuge and collect the supernatant. Filter through siliconized glass wool to remove residual acrylamide fragments. If necessary, repeat the elution once to increase recovery. Add 2 vol of ethanol, precipitate, and recover the DNA as described in **step 4**.
8. Dissolve the gel-purified library DNA in approximately 100 µL of TE8. Dilute 2 µL into 100 µL of TE8 and determine the OD$_{260}$. Convert to molar concentration using the value of 50 µg/mL/OD$_{260}$, together with the molecular weight of the double-stranded DNA (62,440 g/mol for the 95 nt product in **Fig. 1B**). Dilute the DNA to 1.7 nmol/100 µL (1.7 nmol is approx 106 µg for the DNA product shown in **Fig. 1B**). Divide the preparation into 100-µL aliquots and store at –80°C for future experiments.

3.3. Transcription, Selection and Amplification
3.3.1. Transcription Reaction

1. Transcribe the double-stranded DNA template using T7 RNA polymerase. Combine 1 aliquot (100 µL; 1.7 nmol) of DNA library, 1 mL of 10 × T7 RNA polymerase buffer, 100 µL of ATP, 100 µL of CTP, 100 µL of GTP, 100 µL of TTP (all ribonucleotide stocks at 100 m*M*), 1 mL of T7 RNA polymerase, and 7.5 mL of dH$_2$O (final volume is 10 mL). Mix gently by inversion and incubate at 37°C for 2 h (*see* **Note 4**).
2. Add 100 µL of RQ1 DNase to the transcription reaction and continue incubation for an additional 15 min at 37°C (*see* **Note 5**).
3. Purify full-length RNA transcripts on a denaturing polyacrylamide gel (*see* **Note 6**). Prepare a denaturing (8 *M* urea) 15% polyacrylamide gel as described in Chapter 1. Resuspend the RNA pellet in 200 µL of urea loading buffer and load it onto the gel. Excise and elute the full-length RNA transcripts as described above (**Subheading 3.2., steps 6** and **7**).

4. Dissolve the gel purified RNA in approx 100 μL of TE8. Dilute 2 μL into 100 μL total of TE8 and determine the OD_{260}. Convert to molar concentration using the value of 40 μg/mL/OD_{260}, together with the molecular weight of the single-stranded RNA (23,630 g/mol for the 69 nt product in **Fig. 1B**). Dilute to 17 nmol/100 μL (17 nmol is approx 402 μg for the RNA product shown in **Fig. 1B**). Divide the preparation into 100 μL aliquots and store at –80°C for future experiments. One such aliquot thus contains, on average, 10 RNA copies of each of the 1×10^{15} DNA templates that were sampled in **Subheading 3.2., step 1**.

3.3.2. Iterative RNA Selection

The initial round of RNA selection is unique in that the binding reaction requires a large volume to accommodate the RNA library. Assuming at least a 1000-fold enrichment of specific species at each round of selection, the complexity of the selected RNA pool is expected to decrease sufficiently to perform all subsequent reactions in modest volumes. Also, the nature of the binding reaction depends on the target. In all cases, conditions that are optimal for the target being screened should be used.

The mode of partitioning the library is flexible. Approaches that have been employed include filter binding, immunoprecipitation, mobility shift in native gels, and column binding (*see* Chapters 9, 10, and 15). Filter binding has the advantage that quantitative data can readily be obtained during the course of the experiments. Immunoprecipitation has the advantage that relatively complex or unpurified target proteins can be assessed. The remainder of this method will focus on immunoprecipitation. The binding conditions given are somewhat arbitrary and will need to be altered to accommodate different proteins.

1. Prepare Protein A Sepharose beads by swelling in an excess of NT2 buffer for 10 min. Wash 3× in 1 mL of NT2 buffer, pelleting the beads by centrifugation at 12,000*g* for 10 s. Prepare sufficient beads to provide 2 mg per each target protein being screened, as well as an equal number of beads to be used in preclearing the RNA library.
2. Add an appropriate volume of antibody to the beads. The amount of antibody to use depends on the target protein and the antibody (*see* **Note 7**). In all cases, it is necessary to provide at least a slight excess of antibody relative to the target being screened. Incubate at 4°C for 1 h to bind the antibody to beads. Wash 3× in 1 mL each of NT2 buffer.
3. Preclear the library by eliminating all RNA species that can bind directly to Protein A Sepharose beads plus antibody. Aliquot half of the beads plus antibody and add an equal volume of RNA binding buffer. For the first round, incubate beads plus antibody with a 17 nmol aliquot of the RNA library for 30 min at 4°C. In subsequent rounds, one quarter of a 20-μL transcription reaction is used. Pellet the beads by centrifugation and harvest the supernatant containing the precleared RNA. Discard the beads.

In Vitro Selection of RNA Aptamers

4. Aliquot the remaining half of the beads plus antibody into a clean tube and add a preparation of target protein. Typically, between 1 and 10 μg of protein will be used per screen. Note that 10 μg of protein is equivalent to 0.2 nmol for a 50-kDa protein. This is acceptable even for the first round, as the goal is for all binding reactions to be performed in RNA excess. The target protein may be purified, or it may be a minor component of a cell extract. After incubating for 10 min at 4°C, wash the beads 3× in 1 mL each of NT2 buffer.
5. After the last wash, add the precleared RNA prepared in **step 3**. Incubate for 10 min at 4°C. Wash the beads 5× using 1 mL of NT2 buffer for each wash.
6. Leave 100 μL of NT2 covering the beads after the last wash. Add 100 μL of dH$_2$O, 200 μL of 2× proteinase K buffer, and 5 μL of proteinase K and incubate for 15 min at 37°C.
7. Add 400 μL of PCI, vortex mix for 30 s, and centrifuge in a microfuge at 12,000g for 1 min.
8. Transfer the aqueous phase to a clean tube, add 2 μL of yeast tRNA, 40 μL of 3 M sodium acetate, pH 5.2, and 1100 μL of 100% EtOH. Mix thoroughly. Incubate for 10 min in a dry ice–ethanol bath and then centrifuge at ≥ 12,000g for 20 min. Decant the supernatant, wash with 70% ethanol, and air dry the RNA pellet. Resuspend the pellet in 13 μL of dH$_2$O.

3.3.3. Reverse Transcription and Amplification of Selected RNA

1. Mix the following components in a 0.25–mL thin-walled tube: 3 μL of Reverse oligonucleotide, 2 μL of 10X reverse transcriptase buffer, 2 μL of dNTP mix, 13 μL of selected RNA, and 0.2 μL of reverse transcriptase. Incubate for 5 min at 55°C, then for 1 h at 42°C.
2. Regenerate the double-stranded DNA transcription template by PCR using the reverse transcription product as a template. Mix the following components in a 0.25-mL thin-walled tube: 75.5 μL of dH$_2$O, 10 μL of 10X *Taq* buffer, 2 μL of dNTP mix, 3 μL of 5 μM Forward primer, 3 μL of reverse primer, 6 μL of template (the product of the reverse transcriptase reaction), and 0.5 μL of *Taq* polymerase (final volume 100 μL). Amplify using the following cycle parameters: 25 cycles of 0.5 min at 94°C, 0.5 min at 50°C, and 0.5 min at 72°C; then 1 cycle of 7 min at 72°C. Note that the cycle parameters will depend on the primer set being used and different primers may need optimization.
3. Add 1 μL of *Bam*HI to the PCR reaction and incubate for 1 h at 37°C to regenerate the 3' end of the original library. Note that *Bam*HI works adequately in PCR buffer.
4. Extract the reaction with an equal volume of PCI. Add 1/10 vol of 3 M sodium acetate, pH 5.2, and 2 vol of 100% ethanol. Incubate for 10 min in a dry ice-ethanol bath and then centrifuge at ≥ 12,000g for 20 min. Decant the supernatant and wash the DNA pellet in 70% ethanol, as described in **Subheading 3.2., step 4**. Air dry the DNA pellet and resuspend the DNA in 15 μL of dH$_2$O.
5. Transcribe the amplified DNA as follows: combine 2 μL of 10× T7 transcription buffer, 8 μL of rNTP mix, 3 μL of PCR amplified DNA template, 1 μL of RNasin, 4 μL of dH$_2$O, and 2 μL of T7 RNA polymerase (final volume 20 μL). Incubate at 37°C for 1 h. Add 1.5 μL of RQ1 DNase and incubate for an additional 10 min at 37°C.

6. Add 78.5 µL of dH$_2$O and 10 µL of 3 M sodium acetate, pH 5.2. Extract the reaction with an equal volume of PCI, add 2.5 vol of ethanol to the aqueous material, precipitate, and wash the RNA pellet in 70% ethanol. Air dry the pellet, then resuspend the RNA in 20 µL of dH$_2$O. After the first round, approx 5 µL (25%) of the RNA transcripts are used in each subsequent round of binding.
7. **Subheading 3.3., steps 2** and **3** are repeated, including the preclearing of the selected RNA, until sufficient enrichment of the specific binding pool has occurred. Enrichment can be monitored by measuring the binding efficiency (percent bound) of the selected RNA pool at each round (*see* **Notes 8–10**). We generally gel purify the RNA transcripts at each round of selection, although that may not be necessary for all applications (*see* **Note 6**).

3.4. Subcloning Selected RNA Species

1. We have employed many different approaches to subcloning the selected species. Probably the most efficient method currently available is to blunt clone the PCR products using a commercial kit such as the Prime PCR Cloner kit (5 Prime –> 3 Prime, Boulder, CO, USA), following the manufacturer's protocol. Inserts in the resulting recombinant plasmids can then be sequenced according to various methods. However, these clones cannot be used subsequently for in vitro transcription to produce single RNA species for binding studies. Because the clones contain two T7 promoters (one in the pNoTA/T7 vector and one carried within the library sequences), in vitro transcription would yield heterogeneous transcripts. Therefore, clones of interest must be subcloned into pSP64 (using *Bam*HI), or into any other vector that lacks a T7 promoter.
2. Sequence the inserts from 50–100 clones. The simplest approach to elucidating consensus sequences is to import all of the randomized inserts into a word processing program such as Microsoft Word. All sequences should be viewed in a monospaced font such as Courier so that sequences can be accurately aligned. Take care to flip all sequences to the 5'→ 3' orientation with respect to the RNA, as approximately half of the blunt-cloned fragments will be in the reverse orientation. Note that the reverse complements of any reverse-oriented inserts are the sequences of interest. The drag and drop feature of Word makes it relatively straightforward to organize the sequences so that patterns can be discerned by eye. We have found no computer programs that do a better job at pattern recognition than the human brain. Keep in mind that the nonrandom flanking sequences also exert an influence and should be included in the final analysis. We find it convenient to indicate constant sequences with lowercase letters and randomized sequences with uppercase letters. Also, consider the possibility that the consensus sequence will be noncontiguous and may contain basepaired elements that are conserved in structure but not sequence.

4. Notes

1. Perform all steps using the following precautions to prevent contamination by RNase and environmental sources of nucleic acids: baked glassware, diethylpyrocarbonate (DEPC)-treated solutions, and aerosol-resistant micropipet tips. The dH_2O used for all solutions and reactions must be pretreated with DEPC. Note that Tris solutions cannot be treated with DEPC and should instead be made up with DEPC-treated water and then autoclaved.
2. This buffer is a starting point, and may need to be modified and optimized depending on the particular target protein being studied.
3. Specialty combs can be purchased or custom manufactured. Alternatively, a standard comb can be modified by careful taping.
4. Generation of a milky white precipitate (probably magnesium–pyrophosphate complex; *see* **ref. *17***) is normal and is compatible with an efficient transcription reaction. The precipitate is readily removed by centrifugation and/or addition of EDTA.
5. This step degrades the DNA template prior to gel purification of the RNA, resulting in unambiguous visualization and excision of the desired RNA products.
6. Full-length RNA products are gel purified initially and at each round of selection to remove aberrant high and low molecular weight products. Failure to gel purify often results in the selection of aberrant and irrelevant RNA species that have accumulated various mutations conferring an amplification advantage.
7. Pilot titration experiments should be performed to determine the saturation of antibody binding to target. A fixed amount of target protein can be immunoprecipitated with increasing amounts of antibody to determine the saturation point. Selections are then performed in slight antibody excess. It is best not to use a great excess of antibody to protein, as antibodies can also select RNA aptamers from the library.
8. To label the RNA pool, either perform successive phosphatase and kinase reactions using [γ-^{32}P] ATP *(18)* or perform a transcription reaction (**Subheading 3.3., step 1**) supplemented with [α-^{32}P] UTP. The amount of label is not critical; all that is required is sufficient specific activity so that the input and bound counts can be monitored by scintillation counting, autoradiography, or phosphorimage analysis. For this analysis, set up parallel binding reactions using labeled RNA from each round of selection. Measure the input counts into each binding reaction, and measure the counts retained by the target protein. Plot the fraction of the input counts that were bound at each cycle of selection. Typically, the binding curve plateaus somewhere between 50% and 100% bound, although some targets may plateau with much less fractional binding.
9. The stringency of selection may be controlled in at least three ways. (1) The target protein concentration can be decreased in subsequent rounds, conferring a theoretical advantage to the species with the greatest binding efficiencies. (2) The ionic concentration can be altered, typically by increasing the concentration

of NaCl in later rounds. (3) Chaotropic reagents such as urea can be included in later rounds. The latter two approaches have been observed to decrease nonspecific binding (16). For all three approaches, it is helpful to apply increasing stringency only in the later rounds, typically in cases where background binding has proven problematic.

10. With regard to the number of iterations required to achieve a useful consensus, the binding characteristics of the protein probably represent the primary determining factor. It should be noted that selection with many cycles can generate a few or single "winners" from the original pool of RNAs. Information of biological value may be lost by taking the selection process to an extreme point. Although many proteins select RNAs containing a single recognizable consensus sequence, sometimes the sequences fall into two or more distinct classes.

References

1. Oliphant, A. R., Nussbaum, A. I., and Struhl, K. (1986) Cloning of random-sequence oligodeoxynucleotides for determining consensus sequences. *Gene* **44,** 177–183.
2. Oliphant, A. R. and Struhl, K. (1987) The use of random-sequence oligonucleotides for determining consensus sequences. *Methods Enzymol.* **155,** 568–581.
3. Horwitz, M. S. Z. and Loeb, L. A. (1986) Promoters selected from random DNA sequences. *Proc. Natl. Acad. Sci. USA* **83,** 7405–7409.
4. Tuerk, C. and Gold, L. (1990) Systematic evolution of ligands by exponential enrichment: RNA ligands to bacteriophage T4 DNA polymerase. *Science* **249,** 505–510.
5. Ellington, A. D. and Szostak, J. W. (1990) In vitro selection of RNA molecules that bind specific ligands. *Nature* **346,** 818–822.
6. Tsai, D. E., Kenan, D. J., and Keene, J. D. (1992) In vitro selection of an RNA epitope immunologically cross-reactive with a peptide. *Proc. Natl. Acad. Sci. USA* **89,** 8864–8868.
7. Keene, J. D. (1996) RNA surfaces as functional mimetics of proteins. *Chem. Biol.* **3,** 505–514.
8. Bartel, D. P., Zapp, M. L., Green, M. R., and Szostak, J. W. (1991) HIV-1 Rev regulation involves recognition of non-Watson–Crick base pairs in viral RNA. *Cell* **67,** 529–536.
9. Tsai, D. E., Harper, D. S., and Keene, J. D. (1991) U1-snRNP-A protein selects a ten nucleotide consensus sequence from a degenerate RNA pool presented in various structural contexts. *Nucleic Acids Res.* **19,** 4931–4936.
10. Levine, T. D., Gao, F., Andrews, L., King, P. H., and Keene, J. D. (1993) Hel-N1: an autoimmune RNA-binding protein with specificity for 3' uridylate-rich untranslated regions of growth factor mRNAs. *Mol. Cell. Biol.* **13,** 3494–3504.
11. Buckanovich, R. J. and Darnell, R. B. (1997) The neuronal RNA binding protein Nova-1 recognizes specific RNA targets in vitro and in vivo. *Mol. Cell. Biol.* **17,** 3194–3201.

12. Gao, F. B., Carson, C. C., Levine, T., and Keene J. D. (1994) Selection of a subset of mRNAs from combinatorial 3' untranslated region libraries using neuronal RNA-binding protein Hel-N1. *Proc. Natl. Acad. Sci. USA* **91,** 11,207–11,211.
13. Gold, L., Brown, D., He, Y., Shtatland, T., Singer, B. S., and Wu, Y. (1997) From oligonucleotide shapes to genomic SELEX: novel biological regulatory loops. *Proc. Natl. Acad. Sci. USA* **94,** 59–64.
14. Kenan, D. J., Tsai, D. E., and Keene, J. D. (1994) Exploring molecular diversity with combinatorial shape libraries. *Trends Biochem. Sci.* **19,** 57–64.
15. Lander, E. S. (1996) The new genomics: global views of biology. *Science* **274,** 536–539.
16. Czarnik, A. W. and Keene, J. D. (1998) Combinatorial chemistry: a primer. *Curr. Biol.* **8,** R705–R707.
17. Milligan, J. F. and Uhlenbeck, O. C. (1989) Synthesis of small RNAs using T7 RNA polymerase. *Methods Enzymol.* **180,** 51–62.
18. Sambrook, J., Fritsch, E. F., and Maniatis, T. (1989) *Molecular Cloning: A Laboratory Manual.* Cold Spring Harbor Laboratory Press, Cold Spring Harbor, NY.
19. Chamberlin, M. and Ring, J. (1973) Characterization of T7-specific ribonucleic acid polymerase. 1. General properties of the enzymatic reaction and the template specificity of the enzyme. *J. Biol. Chem.* **248,** 2235–2244.

17

Identification of Specific Protein–RNA Target Sites Using Libraries of Natural Sequences

Lucy G. Andrews and Jack D. Keene

1. Introduction

RNA-binding proteins are involved in a variety of regulatory and developmental processes such as RNA processing, transport, and translation and are integral components of ribosomes, spliceosomes, nucleoli, and other ribonucleoprotein particles. Proteins have been shown to interact directly with mRNA at all stages from the nascent transcript through capping, polyadenylation and splicing, nuclear export, translation initiation, and translocation of ribosomes along the message as well as during degradation of the mRNA (for reviews *see* **refs. *1–5***). It is becoming increasingly clear that many RNA-binding proteins regulate gene expression through their interactions with mRNAs at various stages of development. The classic example is that of the iron-response element mRNA-binding protein, aconitase, which was shown by Klausner and co-workers to regulate translation of ferritin mRNA by binding to a stem-loop structure in its 5' untranslated region (UTR) (reviewed in **ref. *6***). Additional examples include the regulated stability of transferrin, histone, and various cytokine and proto-oncogene mRNAs that are controlled by the binding of specific proteins to consensus sequences in their 3' untranslated regions (3' UTR) *(7–9)*. In addition, studies of tobacco mosaic virus *(10)*, the *Drosophila* developmental gene *hunchback (11)*, 15-lipogenase *(12)*, and various cytokine mRNAs *(13)* have implicated RNA-binding proteins in modulation of translational efficiency by interaction with sequences within the 3' UTR. For most putative RNA-binding proteins, however, the cellular targets, and therefore the functions, are unknown.

RNA-binding proteins fall into various classes according to the structure of their RNA-binding domains and the amino acid composition of these domains.

From: *Methods in Molecular Biology, Vol. 118: RNA-Protein Interaction Protocols*
Edited by: S. Haynes © Humana Press Inc., Totowa, NJ

The largest known family of RNA-binding proteins is defined by one or more RNA recognition motifs (RRMs) of 80–90 amino acids containing conserved hexamer and octamer (RNP consensus) sequences that are assumed to interact with RNA via solvent-exposed aromatic and basic residues (for review *see* **ref. 14**). Hundreds of members of the RRM family of RNA-binding proteins have been reported, most of which are widely expressed and phylogenetically conserved. Well-characterized examples include the La, U1-70K snRNP, and U1-A snRNP proteins. An important tissue-specific subgroup of the RRM family is involved in differentiation of neurons and includes Drosophila ELAV (embryonic lethal abnormal vision) *(15)*, human homologues Hel-N1/Hel-N2 (human ELAV-like neuronal proteins), HuD *(16)*, HuC/PLE21 *(17)*, and HuR *(18)*, all of which contain three RRMs. As with all RRM family members, in order to understand the functions of the ELAV proteins in cellular metabolism, it is essential to identify the target RNAs with which they interact. The identification of RNA target sequences in mRNA has proven difficult owing to the inabundance of individual mRNAs and the relative instability of RNA itself. A combinatorial strategy to address the problem of RNA targeting has been developed in our laboratory using the U1-A snRNP protein *(19)* and in the studies described in this chapter we have used Hel-N1 to illustrate this approach with mRNA. To date the ELAV-family proteins are the only examples of RNA-binding proteins of unknown binding specificity, for which the cellular target RNAs have been revealed using in vitro combinatorial RNA selection. However in vitro selection procedures are generally applicable to identification of specific RNA target sequences of other RNA-binding proteins. For this chapter, we provide a detailed protocol for selection by proteins of sequences from 3' UTR libraries derived from natural mRNAs.

Selection of ligands from random RNA libraries serves as a first step in identifying natural ligands of RNA-binding proteins and antibodies. The detailed procedure may be found in Chapter 16 of this volume as well as elsewhere (e.g., *see* **refs. 20** and **22**). Randomized RNA selection is useful in assigning a characteristically preferred sequence that can be bound by a protein of interest; however, it is limited in defining RNA ligand sequences because it usually selects RNAs that do not exist biologically. Database searches using the consensus sequence obtained from the selection protocol outlined in Chapter 16 may help in identification of natural RNAs that are potential ligands of the protein of interest depending on the parameters of the algorithms used. For example, the sequences selected by the ELAV protein, Hel-N1, from an RNA library with 25 randomized positions yielded short stretches of uridylates separated by one or two nucleotides (usually purines) *(21)*. Such sequences are found in the 3' untranslated regions (UTRs) of short-lived mRNAs such as those of cytokines and proto-oncogenes and are associated with the instability of these mRNAs. Therefore we tested transcripts from plasmid templates containing the 3' UTRs of c-*myc*, c-*fos*, and GM-CSF for binding to Hel-N1. Upon finding that the full-length 3' UTRs bound tightly to Hel-N1, we further delineated the minimal binding sequences within these 3' UTRs and found that they are similar to those derived by our randomized RNA selection experiments.

2. Materials

1. Poly(A)+ RNA from Clontech or isolated from total cellular RNA in the laboratory.
2. Enzymes: Superscript II reverse transcriptase (RT) and buffer (Life Technologies, Rockville, MD, USA), *E. coli* RNase H (Life Technologies, Rockville, MD, USA), terminal deoxynucleotide transferase (TdTase) and buffer (Life Technologies, Rockville, MD, USA), RNasin RNase inhibitor (Promega, Madison, WI, USA), T7 RNA polymerase and buffer, *Bam*HI, RQ 1 DNase (Promega, Madison, WI, USA), *Taq* polymerase.
3. Synthetic oligonucleotides:
   (One example uses the following sequences:
   Forward primer [oligo-dG]: 5'-ACCAGGATCCTAATACGACTCACTATA[G]$_{11}$-3'
   Reverse primer [oligo-T]: 5'-CATGGAATTCGGATCC[T]$_{15}$-3').
4. dNTP mix: 10 mM each, dATP, dCTP, dGTP, and TTP.
5. RTP mix: 10 mM each, ATP, CTP, GTP, and UTP.
6. 0.1 M dithiothreitol (DTT).
7. Nuclease-free 0.5- and 1.5-mL microcentifuge tubes.
8. Nuclease-free water for transcription and RNA-binding reactions, i.e., treated with diethylpyrocarbonate (DEPC) (Sigma, St. Louis, MO, USA).
9. Nuclease-free sterile micropipet tips.
10. G-50 Sephadex columns (5 Prime-3 Prime, Boulder, CO, USA).
11. Protein A–Sepharose beads (P3391, Sigma, St. Louis, MO, USA).
12. 3 M Potassium acetate for ethanol precipitation of RT reaction products.
13. Ethanol, 100% and 70% in water.
14. Agarose for electrophoresis.
15. 7.5 M Ammonium acetate.
16. 0.1 M Na$_2$EDTA (ethylenediaminetetraacetic acid), pH 8.
17. PCI = Phenol:chloroform:isoamyl alcohol (25:24:1 [v/v/v]) saturated with 10 mM Tris-HCl, pH 8.0, 1 mM EDTA, pH 8.0 (TE buffer, pH 8.0).
18. NT2 buffer: 50 mM Tris-HCl, pH 7.4, 150 mM NaCl, 0.05% NP40, 1 mM MgCl$_2$.
19. KNET buffer: 20 mM KCl, 80 mM NaCl, 2 mM EGTA (ethylene glycol-*bis-N,N,N',N'*-tetraacetic acid), 50 mM Tris-HCl, pH 7.4, 0.05% NP-40, 5 mg/mL poly(A) RNA (Sigma), 1 mM MgCl$_2$, 2.5% polyvinyl alcohol, 0.2% VRC (vanadyl ribonucleoside complex), 0.1 mg/mL bovine serum albumin, 0.5 mg/mL yeast tRNA, 10 mM DTT, and 80 U/mL Rnasin. Make fresh as required.
20. 3 M Sodium acetate, pH 5.2.

3. Methods

3.1. Selection of Ligands from Libraries of 3' Untranslated Region Sequences Derived from Poly(A)+ RNA

We have devised an in vitro RNA selection protocol that has proven useful in identifying cellular targets of proteins known to bind mRNAs *(23)*. Originally, this approach was developed to address the inaccessibility of Hel-N1 to antibodies, which limits the ability to coimmunoprecipitate them from cell extracts with the RNAs to which they bind in vivo. Therefore, this approach has high potential for many RNA-binding proteins that bind multiple target species. In brief, this method (**Fig. 1**) consists of reverse transcription of poly(A)+ RNA from one of many sources, e.g., human brain, medulloblastoma cell lines, mixed-stage *C. elegans*, etc., using an oligo-T primer, followed by C-tailing of the cDNA transcripts with terminal deoxynucleotide transferase (TdTase). The 3' oligo-dC enables polymerase chain reaction (PCR) amplification of the cDNA library using a forward primer that includes the T7 RNA polymerase promoter sequence and oligo-dG at the 3' end and a reverse primer ending in oligo-T (*see* **Notes 1 and 2**).

To identify cellular mRNAs to which a protein of interest can bind, the following steps are undertaken:

1. Generation of cDNA from poly(A)+ RNA.
2. C-tailing of cDNA.
3. PCR amplification of cDNA.
4. Transcription of amplified DNAs into RNA.
5. Iterative binding and selection of RNAs by protein (recovery of bound subset, reverse transcription, amplification by PCR, transcription, binding, etc.).

3.2. Generation of cDNA from poly(A)+ RNA

1. Add 1–5 µg of poly(A)+ RNA in 13 µL of DEPC-treated water to a sterile RNase-free 0.5-mL microcentrifuge tube.
2. Add 1 µL (0.5 µg) of oligo-T primer to the tube and mix gently.
3. Heat the mixture to 70°C for 10 min and incubate on ice for 1 min. Collect the contents of the tube by brief centrifugation and add 2 µL of 10× reverse transcriptase (RT) buffer, 1 µL of 10 m*M* dNTP mix, 2 µL of 0.1 *M* DTT, and 1 µL of Superscript II RT at 200 U/µL in a total volume of 20 µL.
4. Mix gently and collect the reaction by brief centrifugation. Incubate at room temperature for 10 min.
5. Transfer the tube to a 42°C water bath or heat block and incubate for 50 min.
6. Terminate the reaction by incubating the tube at 70°C for 15 min. Place on ice.
7. Collect the reaction by brief centrifugation. Add 1 µL of *E. coli* RNase H (2 U/µL) to the tube and incubate for 20 min at 37°C.
8. Purify reverse transcription reaction products by passing twice through G-50 Sephadex columns to remove completely the unincorporated nucleotides.

Ligand Selection from Natural Libraries

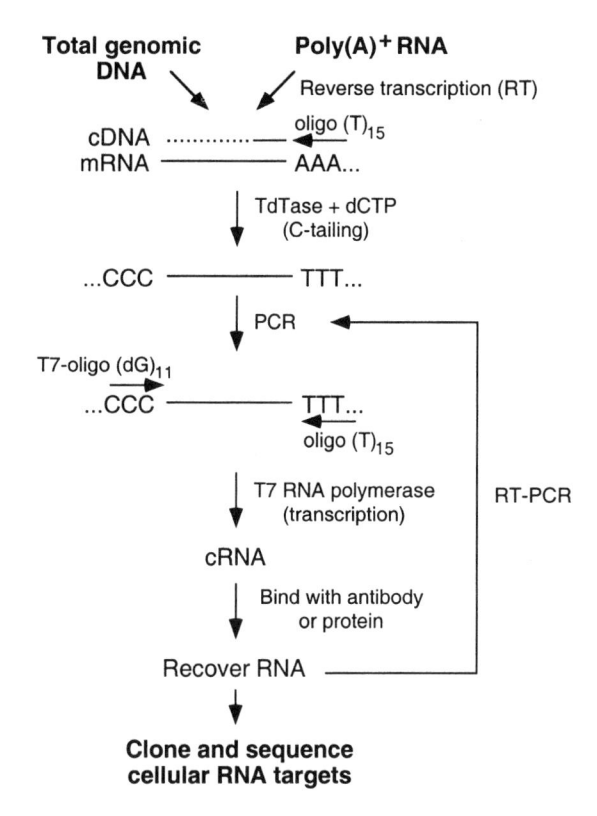

```
Total genomic          Poly(A)+ RNA
     DNA           ↙  Reverse transcription (RT)
              ↘
     cDNA  ············  oligo (T)₁₅
     mRNA  ──────────  AAA...
              ↓ TdTase + dCTP
                (C-tailing)
   ...CCC ────────── TTT...
              ↓ PCR  ◄──────────┐
T7-oligo (dG)₁₁                  │
    ────►                        │
   ...CCC ────────── TTT...      │
              ←─ oligo (T)₁₅     │
              ↓ T7 RNA polymerase   RT-PCR
                (transcription)  │
              cRNA               │
              ↓ Bind with antibody
                or protein       │
          Recover RNA ───────────┘
              ↓
        Clone and sequence
        cellular RNA targets
```

Fig. 1. General method for in vitro selection of natural RNA sequences derived from cellular genomes (23). Either cellular DNA or cDNA can be incorporated into libraries for synthesis of RNA followed by partitioning and iterative selection. Binding targets can be any protein or target molecule whether or not it is a known RNA-binding molecule. It should be noted that mutation of the naturally derived sequences is an undesirable outcome of this protocol.

9. Ethanol-precipitate the single-stranded cDNAs using 0.1 vol of 3 M potassium acetate rather than sodium or ammonium because, according to the manufacturer, these cations inhibit TdTase.

3.3. C-Tailing of cDNA Library

1. Resuspend the cDNA in water. It is advisable to use about one fourth of the cDNA library for each reaction. The manufacturer recommends a final concentration of 40 pmol/mL of 3' termini.
2. Assemble the terminal deoxynucleotide transferase (TdT) reaction in a 1.5 mL microcentrifuge tube by adding 10 µL of 5× TdT buffer, 25 µL of 100 µM dCTP, the cDNA, water to 48.5 µL, and lastly 1.5 µL (15 U) of TdTase.
3. Incubate at 37°C for 30 min.

4. Stop the reaction by placing the tube on ice and adding 10 µL of 0.1 M Na$_2$EDTA, pH 8.0.
5. Add 60 µL of buffer-saturated phenol:chloroform:isoamyl alcohol (25:24:1 [v/v/v]). Vortex thoroughly. Centrifuge for 5 min at 15,000g at room temperature to separate the phases.
6. Transfer the upper aqueous phase to a G-50 Sephadex centrifuge column to remove unincorporated dCTP.
7. Precipitate the DNA by adding 0.5 vol of 7.5 M ammonium acetate followed by 2.5 vol of absolute ethanol. Centrifuge at 14,000g at room temperature for 30 min. Carefully remove the supernatant.
8. Resuspend the pellet in 50 µL of water for use as template in PCR reaction.
9. Store the DNA at –20°C.

3.4. PCR Amplification of cDNA Library

1. Assemble the PCR by making a master mix of the first five components and aliquoting into the appropriate number of reaction tubes. Add template, then polymerase using aerosol barrier tips. Reactions should include a control lacking template. Reaction components are 84.5 µL of sterile H$_2$O (DEPC has been reported to inhibit PCR), 10 µL of 10× commercial PCR buffer, 2 µL of dNTP mix, 1 µL of oligo-dG primer at 0.1 µg/µL, 1 µL of oligo-T primer at 0.1 µg/µL, 1 µL of C-tailed cDNA, and 0.5 µL of *Taq* polymerase at 5 U/mL.
2. Run 25 cycles of: 1 min at 94°C, 1 min at 50°C, 2 min at 72°C followed by one cycle of 7 min at 72°C.
3. Remove the oil (if used) using 24 parts chloroform:1 part isoamyl alcohol.
4. Add 1 µL of *Bam*HI at 20 U/µL and incubate for 1 h at 37°C to reduce any concatemers that may have formed during PCR (*see* **Note 3**).
5. Extract with PCI.
6. Precipitate with 0.1 vol of 3 M sodium acetate, pH 5.2, and 2.5 vol of 100% ethanol in a dry ice–ethanol bath for 20 min.
7. Centrifuge at 12,000g in a microfuge at 4°C for 10 min.
8. Wash the pellet with 70% ethanol and centrifuge briefly.
9. Dry the pellet and resuspend in 15 µL of H$_2$O.
10. Separate the PCR products on a 1% agarose gel and cut out of the gel the DNA fragments ranging from 200 to 800 nucleotides (*see* **Note 4**).
11. Purify the size-selected DNA for transcription by electroelution. (If an electroelution apparatus is not available, refer to **ref. 24**).
12. Extract the eluted DNA with PCI and precipitate by addition of 0.1 vol of 3 M sodium acetate, pH 5.2, and 2 vol of 100% ethanol. Leave on ice for 10 min. Recover DNA by centrifugation at 12,000g for 15 min at 4°C. Wash with 70% ethanol, dry the pellet, and resuspend it in 20 µL of DEPC-H$_2$O.

3.5. Transcription of cDNA into 3' UTR Library

1. Assemble the transcription reaction (*see* **Note 5**) by mixing (in a 0.5-mL tube) 0.1 µg of PCR-amplified cDNA (5 µL of the total of 20 µL), 4 µL of 5× commer-

cial T7 or SP6 RNA polymerase buffer, 8 µL of RTP mix, 1 µL of RNasin at 20–40 U/µL, and 1 µL of T7 RNA polymerase.
2. Bring to a total volume of 20 µL with DEPC-H$_2$O.
3. Incubate at 37°C for 1 h.
4. Add 1.5 µL of RQ 1 DNase at 1 U/µL. Incubate at 37°C for 10 min.
5. Precipitate and wash and dry the pellet as in **Subheading 3.4.** above.
6. Resuspend the RNA in 20 µL of DEPC-treated H$_2$O.
7. Run 2 µL on a 3% agarose gel to quantify.

The RNA product of this transcription reaction ideally represents all the 3' UTRs of the original poly(A)+ RNA. This 3' UTR library is used as a substrate for in vitro selection essentially as described by Gao et al. *(23)* and detailed below.

3.6. Iterative Binding and Selection of 3' UTRs by Protein
3.6.1. Binding and Recovery of Subset of Library
1. Use 1 mg of Protein A beads for each reaction. Wash 3× with 1 mL of NT2 buffer in a 1.5-mL microcentrifuge tube.
2. Leave 100 µL of buffer on the pellet after the third wash. Add approx 2 µL of antiserum to each tube.
3. Store on ice at least 10 min with occasional vortexing of the tubes.
4. Wash 3× with 1 mL of NT2. (If binding with an antibody only rather than using the antibody to attach a protein of interest to the beads, skip **steps 5** and **6**.)
5. Leave 100 µL of buffer on the pellet after the third wash. Add about 5 µL of purified protein at 0.5 µg/mL (or approx 30–50 µL/reaction of lysate of cells expressing the protein) to each tube. Leave on ice at least 10 min with occasional mixing (*see* **Note 6**).
6. Wash beads + antibody + protein 3× in NT2.
7. Resuspend in 100 µL of freshly made RNA-binding buffer (KNET) (*see* **Note 7**).
8. Add 4–5 µL of RNA (*see* **Note 8**).
9. Incubate at room temperature for 5 min with mixing.
10. Wash 5× in 1 mL of NT2 (*see* **Note 9**).
11. Leave 100 µL of the final wash. Add 100 µL of H$_2$O.
12. Add 200 µL of PCI, vortex for 30 s, and centrifuge at 12,000*g* for 1 min. Recover the aqueous supernatant avoiding the interface.
13. Add 2 µL of 1 *M* MgCl$_2$ (or 2 µL of 10 m*M* tRNA), 20 µL of 3 *M* sodium acetate, and 700 µL of 100% ethanol to the aqueous supernatant.
14. Precipitate in a dry ice/ethanol bath for 20 min. Centrifuge at 12,000*g* at 4°C for 10 min, wash with 70% ethanol, and dry the pellet.
15. Resuspend in 14 µL of DEPC-H$_2$O.

3.6.2. Reverse Transcription of Selected 3' UTRs
1. Add to the 14 µL RNA + DEPC-H$_2$O (from **step 15** above) 1 µL of Reverse primer (oligo-T) at 0.1 µg/µL, 2 µL of 10× RT buffer, 1 µL of dNTP mix, 2 µL of 0.1 *M* DTT, and 1 µL of Superscript II RT.

2. Incubate for 5 min at room temperature, then for 50 min at 42°C.
3. Terminate the reaction by incubating the tube at 70°C for 15 min. Place on ice.
4. Collect the reaction by brief centrifugation. Add 1 µL of *E. coli* RNase H (2 U/µL) to the tube and incubate for 20 min at 37°C.
5. Purify the reverse transcription reaction products by passing twice through G-50 Sephadex columns to completely remove the unincorporated nucleotides.
6. Use 6 µL of cDNA product as the template for PCR and repeat **Subheading 3.3.**, **steps 3–5** two or more times as needed.

Alternatively, a single binding reaction may be performed under high stringency conditions such as increased salt concentration (we have used 350 mM) or the addition of urea to the wash buffer (optimal concentration of urea varies with the protein of interest from 0.5 M up to 4 M). The advantage of performing a single high-stringency binding of RNA (other than the obvious savings of time and effort) is that PCR artifacts are minimized.

Comparison of binding of the selected subset of 3' UTRs with that of the original 3' UTR library may be accomplished by synthesizing labeled transcripts of each and comparing the radioactivity precipitable with the protein of interest from equivalent amounts of each transcription. When the end-point has been reached and significantly more radioactivity is immunoprecipitated from the radiolabeled subset transcripts than from transcripts of the original library, the subset is cloned into a plasmid such as pGEM or pSP64 using the restriction enzyme cleavage sites built into the PCR primers. Individual clones are purified, sequenced, and analyzed for matches to natural sequences in the databases using the BLAST program *(26)* which is available on the world wide web at http://www.ncbi.nlm.nih.gov/BLAST/. Verification of individual 3' UTR ligands and measurement of relative affinities of binding can be accomplished by synthesizing labeled transcripts and performing binding experiments as described above. Iterative binding and selection with Hel-N1 followed by cloning and sequencing of the selection products yielded a specific subset of cDNAs present in the original libraries (*see* **Note 10**).

To search vast regions of mRNA sequence space, we have developed the procedure described here for identifying natural target RNA(s) of RNA-binding proteins, including antibodies. This method has been used successfully to find the ligands of RNA-binding proteins that bind within the 3' UTR of mRNAs *(23)*. Selection from libraries representing full-length mRNAs is also possible. One must proceed with caution and always test the biologically relevant sequences for direct binding once they have been identified. Depending upon how much is known about the biological function of the protein or antibody of interest, it may be unnecessary to perform combinatorial RNA selection; e.g., if the protein is the homologue of an RNA-binding protein whose target has been identified, one could proceed directly to analysis of binding to

Ligand Selection from Natural Libraries

labeled transcripts of the corresponding target RNA. On the other hand, finding a subset of preferred sequences from a pool of random RNAs may prove useful in identifying a larger set of different biological target RNAs for a given protein *(21,23)*. For example, we have proposed that many cellular proteins not considered conventional RNA-binding proteins interact with RNA in vivo as part of homeostatic regulatory networks such as signaling pathways *(27)*. In every case, however, the suspected target RNAs must be tested individually.

Although we have presented exclusively in vitro methods here, in vivo analysis of RNA-binding proteins and their ligands is a logical next step, but technically more difficult. Approaches we have taken to isolate RNP complexes containing proteins of interest from cell lysates include gradient centrifugation and co-immunoprecipitation (with and without UV crosslinking) followed by Northern blotting, RT-PCR, and RNase protection assays using probes for RNAs shown to bind in vitro. Ultimately, cDNA subset libraries representing a structurally or functionally related set of mRNA targets is the goal. Without question new methodologies are urgently needed to identify and examine the functions of multitargeted messenger RNA-binding proteins.

4. Notes

1. It is important that the entire procedure be performed using RNase-free tubes and solutions prepared with DEPC-treated water. A positive control sample involving a previously characterized protein or antibody and/or a blank sample as a negative control should be carried through all procedures.
2. Whereas these cDNA libraries may be considered mRNA libraries, in practice the transcripts are mostly incomplete and biased to 3' UTR sequences (partly due to the size selection step in **Subheading 3.2.3.**); as such, they may be less useful for proteins whose binding sites are in the 5' regions of mRNA. On the other hand, recent experiments in which *C. elegans* RNA has been reverse-transcribed into a cDNA library using Superscript II RT indicate that clones expressing full-length mRNA can be generated (C. C. Carson and J. D. Keene, unpublished). This was shown by PCR amplification of the RT product using a forward primer complementary to 5' leader sequences found in many *C. elegans* genes. In addition, preliminary studies (B. Lipes and J. D. Keene, unpublished) indicate that autoimmune sera can be used to select single mRNAs from a library of full-length messages generated using the CapFinder amplification technology (CLONTECH Laboratories, Inc., Palo Alto, CA, USA).
3. *Bam*HI works well in PCR buffer. Use whatever restriction enzyme cleaves at the site you have built into your PCR primers.
4. Size selection within this range limits sequences to within about 800 nucleotides of the 3' end of mRNAs.
5. The remainder of the PCR product may be stored frozen and used as template to produce control RNA transcripts for comparison with selected RNA transcripts, for selection by another protein or for further amplification as needed.

6. A sufficient amount of protein should be added to saturate antibody so that no antibody-specific RNAs will be selected. RNA libraries may also be precleared, at the first or at each round, by exposure to beads + antibody.
7. This multicomponent buffer contains RNase inhibitors and nonspecific RNA competitors to ensure specificity of binding as well as pH and ionic conditions that are optimal for the protein of interest. Components may be optimized for your particular protein; e.g., tRNAs have been reported to fold and bind better in the presence of K^+ and NH_2^+ than in Na^+ *(25)*. Alternatively, binding reactions may be extremely simplified by using NT2 with 100 U/mL RNasin and 10 mM DTT.
8. RNA should be in molar excess to ensure that only high-affinity sequences bind. An estimation based on relative sizes of the protein and RNA molecules, expected yield of the transcription reaction, and binding capacity of the Protein A beads should be sufficient.
9. One can increase the stringency by increasing the salt concentration or washing with 0.5 M urea in NT2 during the third (or last) round of selection. RRM proteins are particularly stable in urea (up to 4 M in some cases).
10. The cloned sequences from Gao et al. *(23)* were characterized as 76% AU-rich and were located 50–100 nucleotides upstream of the poly(A) tail of identifiable messages. Of 100 of these cDNAs that were sequenced, approx 90% were not found in the GCG databases. The remaining 10%, which did match 3' UTRs in the databases, could be characterized as associated with cell growth and differentiation *(23)*. These include kinases (*src*-like kinase, protein kinase C, p21^{cdc42Hs}-binding tyrosine kinase, etc.), growth factors (bFGF, TGF, various interleukins, etc.), cell cycle control factors (CDC2) and translation factors (eIF-2-β), CREB2, and steroid sulfatase. Interestingly, two ribosomal protein messages were selected from the medulloblastoma mRNA library by Hel-N1. These targets were presumably retained by virtue of a stretch of adenylates downstream of the AU-rich binding site to which the oligo-T primer was able to anneal prior to PCR amplification.

Acknowledgments

The methods outlined in this chapter have involved the contributions of several students and postdoctoral individuals in the Keene laboratory over a number of years. We would like to give special thanks for the contributions of Donald Tsai, FenBiao Gao, and Barbara Lipes to this protocol and to Trygve Tollefsbol and Barbara Lipes for careful reading of the manuscript.

References

1. Bernstein, P. L. and Ross, J. (1989) Poly(A), poly(A) binding protein and the regulation of mRNA stability. *Trends Biochem. Sci.* **14,** 373–377.
2. Bingham, P. M., Chou, T. B., Mims, I., and Zachar, Z. (1988) On/off regulation of gene expression at the level of splicing. *Trends Genet.* **4,** 134–138.
3. Clawson, G. A., Feldherr, C. M., and Smuckler, E. A. (1985) Nucleocytoplasmic RNA transport. *Mol. Cell. Biochem.* **67,** 87–99.

4. McCarthy, J. E. and Kollmus, H. (1995) Cytoplasmic mRNA–protein interactions. *Trends Biochem. Sci.* **20,** 191–197.
5. Frankel, A. D., Mattaj, I. W., and Rio, D. C. (1991) RNA–protein interactions. *Cell* **67,** 1041–1046.
6. Melefors, O. and Hentze, M. W., (1993) Translational regulation by mRNA/protein interactions in eukaryotic cells: ferritin and beyond. *Bioessays* **15,** 85–90.
7. Casey, J. L., Hentze, M. W., Koeller, D. M., Caughman, S. W., Rouault, T. A., Klausner, R. D., and Harford, J. B. (1988) Iron-responsive elements: regulatory RNA sequences that control mRNA levels and translation. *Science* **240,** 924–928.
8. Pandey, N. B. and Marzluff, W. F. (1987) The stem-loop structure at the 3' end of the histone mRNA is necessary and sufficient for regulation of histone mRNA stability. *Mol. Cell. Biol.* **7,** 4557–4559.
9. Shaw, G. and Kamen, R. (1986) A conserved AU sequence from the 3' untranslated region of GM-CSF mRNA mediates selective mRNA degradation. *Cell* **46,** 659–667.
10. Gallie, D. R. and Walbot, V. (1990) RNA pseudoknot domain of tobacco mosaic virus functionally substitutes for a poly(A) tail in plant and animal cells. *Genes Dev.* **4,** 1149–1157.
11. Murata, Y. and Wharton, R. P. (1995) Binding of *pumilio* to maternal *hunchback* mRNA is required for posterior patterning in *Drosophila* embryos. *Cell* **80,** 747–756.
12. Ostareck-Lederer, A., Ostareck, D. H., Standart, N., and Thiele, B. J. (1994) Transcription of 15-lipoxygenase mRNA is inhibited by a protein that binds to a repeated sequence in the 3' untranslated region. *EMBO J.* **13,** 1476–1481.
13. Kruys, V. M., Wathelet, M. G., and Huez, G. A. (1988) Identification of a translation inhibitory element (TIE) in the 3' untranslated region of the human interferon-β mRNA. *Gene* **72,** 191–200.
14. Kenan, D. J., Query, C. C., and Keene, J. D. (1991) RNA recognition: towards identifying determinants of specificity. *Trends Biochem. Sci.* **16,** 214–220.
15. Robinow, S., Campos, A. R., Yao, K.-M., and White, K. (1988) The *elav* gene product of *Drosophila*, required in neurons, has the RNP consensus motifs. *Science* **242,** 1570–1572.
16. Szabo, A., Dalmau, J., Manley, G., Rosenfeld, M., Wong, E., Henson, J., Posner, J. B., and Furneaux, H. M. (1991) HuD, a paraneoplastic encephalomyelitis antigen, contains RNA-binding domains and is homologous to sex-lethal. *Cell* **67,** 325–333.
17. Sakai, K., Gofuku, M., Kitagawa, Y., Ogasawara, T., Hirose, G., Yamazaki, M., Koh, C-H., Yanagisawa, N., and Steinman, L. (1994) A hippocampal protein associated with paraneoplastic neurologic syndrome and small lung carcinoma. *Biochem. Biophys. Res. Commun.* **199,** 1200–1208.
18. Ma, W.-J., Cheng, S., Campbell, C., Wright, A., and Furneaux, H. (1996). Cloning and characterization of HuR, a ubiquitously expressed Elav-like protein. *J. Biol. Chem.* **271,** 1–8.
19. Tsai, D. E., Harper, D. S., and Keene, J. D. (1991) U1-snRNP-A protein selects a ten nucleotide consensus sequence from a degenerate RNA pool presented in various structural contexts. *Nucleic Acids Res.* **19,** 4931–4936.

20. Andrews, L. G. and Keene, J. D. (1997) Interactions of proteins with specific sequences in RNA, in *mRNA Formation and Function* (Richter, J. D., ed.). Academic Press, San Diego, pp. 237–261.
21. Levine, T. D., Gao, F-B., King, P. H., Andrews, L. G., and Keene, J. D. (1993) Hel-N1: an autoimmune RNA-binding protein with specificity for 3' uridylate-rich untranslated regions of growth factor mRNAs. *Mol. Cell. Biol.* **13,** 3494–3504.
22. Keene, J. D. (1996) Randomization and selection of RNA to identify targets for RRM RNA-binding proteins and antibodies. *Methods Enzymol.* **267,** 367–383.
23. Gao, F.-B., Carson, C. C., Levine, T. D., and Keene, J. D. (1994) Selection of a subset of mRNAs from combinatorial 3' untranslated region libraries using neuronal RNA-binding protein, Hel-N1. *Proc. Natl. Acad. Sci. USA* **91,** 11,207–11,211.
24. Sambrook, J., Fritsch, E. F., and Maniatis, T. (1989) *Molecular Cloning: A Laboratory Manual*, 2nd ed., Cold Spring Harbor Laboratory Press, Cold Spring Harbor, New York.
25. Wang, Y.-X., Lu, M., and Draper, D. E. (1993) Specific ammonium ion requirement for functional ribosomal RNA tertiary structure. *Biochemistry* **32,** 12,279–12,282.
26. Altschul, S. F., Gish, W., Miller, W., Myers, E. W., and Lipman, D. J. (1990) Basic local alignment search tool. *J. Mol. Biol.* **215,** 403–410.
27. Keene, J. D. (1996) RNA surfaces as functional mimetics of proteins. *Chem. Biol.* **3,** 505–513.

18

Northwestern Screening of Expression Libraries

Paramjeet S. Bagga and Jeffrey Wilusz

1. Introduction

In the cell, all RNAs are involved in interactions with proteins that influence many aspects of their processing, localization, and expression (reviewed in **refs. *1*** and ***2***). In order to gain mechanistic insights into the role of proteins in these processes, one must *(1)* define the sequence elements involved RNA processing and regulation of transcript function; *(2)* identify proteins that interact with these elements; and *(3)* create biochemical, molecular, and/or antibody reagents. Detailed insights into RNA–protein interactions are greatly facilitated by obtaining cDNA clones of the proteins of interest. In systems not amenable to genetic approaches, the isolation of cDNA clones is often preceded by the labor-intensive development of peptide sequence or antibody reagents. In this chapter, we describe a rapid Northwestern screening approach to identify cDNA clones for vertebrate RNA binding proteins that requires only prior knowledge of the RNA element.

Northwestern technology, the identification of immobilized proteins using RNA probes, is a potentially powerful tool to obtain cDNA clones of proteins that specifically interact with defined RNA sequences. The rationale for this screening approach is based on the ligand-binding technique originally used for cloning DNA-binding transcription factors *(3)*. Furthermore, Northwestern approaches have also been successfully used to identify proteins by probing gels that have been blotted onto nitrocellulose membranes or similar solid supports (e.g., **refs. *4*** and ***5***). This method is rapid, inexpensive, and does not require the biochemical purification of the protein, preparation of antibodies, knowledge of partial amino acid/nucleic acid sequences, or location of the gene. Basic information about the specific RNA element of interest that may interact with protein is sufficient to initiate a Northwestern screen. The screen utilizes

From: *Methods in Molecular Biology, Vol. 118: RNA-Protein Interaction Protocols*
Edited by: S. Haynes © Humana Press Inc., Totowa, NJ

245

short RNA probes to isolate cDNA library clones expressing proteins that engage in sequence-specific interactions with the probe. Generally speaking, expression cDNA libraries produce Isopropyl-β-D-thiogalactopyranoside (IPTG)-inducible fusion proteins whose N-terminal portion is encoded by the vector sequence and C-terminal portion is encoded by an open reading frame in the cloned cDNA. The fusion proteins from the plated libraries are immobilized on nitrocellulose membranes and screened for specific interactions with the labeled RNA probe to identify the protein of interest and ultimately the cDNA clone that produces that protein.

1.1. Parameters to Consider in Northwestern Screening

Designing a successful Northwestern screen requires consideration of the following important parameters:

1.1.1. Identification of a High-Affinity Binding Site on the RNA

The first step in the development of a successful Northwestern screen is the identification of a minimal RNA element. We usually use supporting assays such as ultraviolet (UV)-crosslinking *(6)*, gel shift *(7)* or filter binding to identify proteins that interact with an RNA element of interest prior to initiating a Northwestern screen (*see* Chapters 3, 6, 9, and 10). This allows us to determine the minimal RNA sequences required for binding as well as optimal buffer and salt conditions to be used in the screen. One can, of course, initiate a Northwestern screen without having any prior knowledge of a protein that interacts with an RNA of interest. It should be noted, however, that to detect an RNA protein interaction by Northwestern assay, it must have a binding constant of >10^9 M^{-1}. It is advisable, therefore, to ensure that one can detect a protein–RNA interaction using one or more of the conventional solution-based assays outlined above before embarking on a Northwestern screen. Once a protein–RNA interaction is determined using one of these assays, the minimal requirements for high-affinity binding can be determined by deletion and site-directed mutagenesis.

1.1.2. Conformation, Stability and Binding Properties of the Immoblized Protein

The RNA binding protein, immobilized on transfer membranes, must be able to bind stably and efficiently to the RNA probe. Binding to the transfer membrane may cause conformational changes in the protein that prohibit RNA interactions. A denaturation/renaturation protocol has been successfully applied to several immobilized RNA binding proteins to increase the signal obtained in Northwestern analysis (e.g., **ref.** *8*). Furthermore, proteins must effectively bind to RNA as monomers (or possibly homomultimers) for the screen to be successful. Finally, it should be noted that several RNA–protein interactions require cooperative interactions to bind RNA with high affinity (e.g., **ref. 9**). It would be very difficult, if not impossible, to isolate selective RNA binding proteins of such complexes by Northwestern screening.

1.1.3. Construction and Handling of RNA Probes

RNA probes of high specific activity should be made using standard phage polymerase technology (*10*; *see* Chapter 1). We have not investigated the use of nonradioactive RNA probes in Northwestern assays because of the need for high sensitivity of detection of protein–RNA interactions. In general, probes should be small RNAs containing the minimal elements required for high-affinity protein binding to minimize unwanted interactions. Probes containing appropriately spaced multimers of the binding site can also be prepared to increase the likelihood of interaction with the protein of interest. Probes must remain generally intact throughout the assay, so high-quality, RNase-free reagents and careful handling are required from start to finish.

1.1.4. Increasing Selectivity of the Interaction of Immobilized Protein with the RNA Probe

One major difficulty in studying RNA binding proteins is the large number of single-strand nucleic acid binding activities found in cell lysates. To ensure selectivity in Northwestern screening assays, it is recommended that discriminating conditions for the probe binding and membrane washing steps be determined prior to starting the screen. Filter binding/dot blot assays using nitrocellulose membranes spotted with either total cell extracts containing the protein of interest (or preferably partially purified protein) are recommended for such studies. RNA probes can then be incubated with these dot blots under a variety of conditions (salt, pH, heparin, competitor RNAs) to determine optimal conditions for binding and washing to maximize sensitivity. In our experience, the monovalent salt concentration present in the binding and washing buffers is often a very useful determinant of the selectivity of a Northwestern screen. In general, higher salt concentrations at which specific, stable complexes are retained on the nitrocellulose after washings are recommended to avoid nonspecific RNA–protein interactions which otherwise lead to detection of false-positives.

1.1.5. Signal-to-Noise Ratio

Any successful screen requires a low level background hybridization of the RNA probe in addition to a high level of specific binding. Background hybridizations can be minimized by using short RNA probes, gel purifying probes to remove rNTPs, preventing the activity of RNases during the assay, selective washes, and ensuring that the probe does not interact with *E. coli* proteins released during phage lysis and immobilized on nitrocellulose filters.

1.2. Time Considerations

Careful consideration of the parameters discussed previously will help determine the feasibility of using a Northwestern approach to obtain cDNA clones of an RNA binding protein of interest, as well as maximize chances for a successful screen. The literature contains numerous examples of successful Northwestern assays to detect RNA–protein interactions *(4,5,11,12)*. Before deciding on whether or not to try a Northwestern screening approach to obtain cDNA clones, one final point should be emphasized. The approach outlined below can generally be fully evaluated for a specific application in a reasonable amount of time (approx 2 wk). For the Northwestern screening procedure described below, we recommend that everything from phage plating through hybridization and washing be completed in one day.

2. Materials
2.1. Expression Libraries

Bacteriophage λ based vectors have been extensively used to construct cDNA expression libraries. Most of these constructions have used λgt11, or its close relatives λZAP or λORF8. These vectors usually carry the *lacZ* gene with a multiple cloning site in its coding region that is suitable for cDNA insertion. Inserts in the appropriate orientation and reading frame are expressed as fusion proteins upon induction of the *lacZ* promoter by IPTG. Improved versions of these vectors (e.g., λTriplEx [Clontech, Palo Alto, CA, USA]) allow translation of all three possible reading frames to ensure expression of the correct protein from the cDNA insert. In addition to providing a means of identifying recombinants by blue-white screening, fusion of the recombinant protein to the β-galactosidase peptide has also been found to increase the stability of the fusion product in some cases (e.g., **ref. *13***). Presence of the 5' untranslated region of the *E. coli ompA* gene in the λTriplEx vector also helps stabilize recombinant transcripts, resulting in better levels of expression. Premade expression libraries are commercially available from several sources, including Stratagene (La Jolla, CA, USA) and Clontech (Palo Alto, CA, USA).

2.2. Host Strains for λ Expression Libraries

The appropriate *E. coli* host strains are usually supplied with the library and vary with the vector. It is important to use only the recommended host strain for a given vector. For instance, *E. coli* strain XL1-Blue is used as a host for plating λZAP and λTriplEx based libraries for several reasons. XL1-Blue contains the episome F' which has the Δ*M15 lacZ* and *lacIq* genes. While the Δ*M15 lacZ* gene is required for *lacZ* α-complementation to select recombinant clones by blue-white screening, *lacIq* codes for *lac* repressor which blocks transcription from the *lacZ* promoter in the absence of the inducer, IPTG. The latter is important for expression control of fusion proteins that could be toxic to the *E. coli* host. The F' episome also contains genes for F' pili required for filamentous phage infection. Conversion of recombinant λ phage clones to phagemids requires superinfection with a filamentous helper phage.

2.3. Library Plating

1. NZY broth: Dissolve 5 g of NaCl, 2 g of MgSO$_4$·7H$_2$O, 5 g of yeast extract, and 10 g of NZ amine (casein hydrolysate) in a final volume of 1 L. Autoclave.
2. LB broth: Dissolve 10 g of Bacto tryptone, 5 g of Bacto-yeast extract, and 10 g of NaCl in a final volume of 1 L. Adjust to pH 7.5. Autoclave.
3. NZY top agar: Add 0.7% agarose to NZY broth. Autoclave.
4. NZY plates: Add 15 g of agar to 1 L of NZY broth. Autoclave and pour ~70 mL per 150-mm Petri dish or ~30 mL per 100-mm Petri dish.
5. Antibiotics: Add appropriate antibiotics to media to the following concentrations when required: ampicillin: 50 µg/mL; tetracycline: 12.5 µg/mL.
6. SM buffer: Dissolve 5.8 g of NaCl, 2 g of MgSO$_4$ · 7H$_2$O, 50 mL of 1 *M* Tris-HCl, pH 7.5, and 5 mL of 2% gelatin in a final volume of 1 L. Autoclave.
7. 0.5 *M* IPTG: Dissolve 2.3 g of IPTG in 20 mL of distilled water. Filter sterilize.
8. X-Gal: Dissolve 2.5 g in 10 mL of dimethyl formamide.
9. 20% Maltose: (Sigma, St. Louis, MO, USA) Dissolve in distilled water. Filter sterilize.
10. Library plating equipment: Use 15-mL (17 × 100 mm) plastic tubes (Falcon 2057) and standard 150-mm plastic Petri dishes.

2.5. Nitrocellulose Filters

Available commercially (e.g., Schleicher & Schuell, Keene, NH, USA) in a variety of sizes to fit any application. Use 137-mm filters for 150-mm Petri dishes. Supported nitrocellulose (BA-S85) is more durable than standard membranes and is recommended. An 18-gage needle (Becton Dickinson, Franklin Lakes, NJ, USA) and a ballpoint pen will be needed for filter identification/orientation.

2.6. Preparation of ^{32}P Labeled RNA Probes

In general, standard procedures can be used to synthesize RNA probes from DNA constructs containing a phage promoter inserted upstream of a minimal RNA binding element(s) (*10*; *see* Chapter 1). RNA probes should be purified on acrylamide gels containing 7 *M* urea using standard vertical gel electrophoresis equipment (*18*) prior to use. RNAs from gel slices are eluted over night into HSCB buffer (25 m*M* Tris-HCl, pH 7.5, 400 m*M* NaCl, 0.1% SDS).

2.7. Northwestern Screening

1. Blocking buffer: 5 mg/mL *Torula* yeast core RNA (Sigma, St. Louis, MO, USA), 10 m*M* Tris-HCl, pH 7.8, and 150 m*M* NaCl.
2. Rinsing buffer: 10 m*M* Tris-HCl, pH 7.8.
3. Binding buffer: 10 m*M* Tris-HCl, pH 7.8, 1 m*M* EDTA, 0.02% Ficoll, 0.02% polyvinylpyrrolidone (PVP), 0.02% bovine serum albumin (BSA) (ultrapure, Roche Molecular Biochemicals, Indianapolis, IN, USA), and a predetermined concentration of NaCl (usually approx 50 m*M*).
4. Washing buffer: 10 m*M* Tris-HCl, pH 7.8, 1 m*M* EDTA, 0.02% Ficoll, 0.02% PVP, 0.02% BSA, and a predetermined concentration of NaCl (or ~100 m*M*).
5. 2% Ficoll (10× stock): 2 g of Ficoll (type 400) dissolved in 100 mL of distilled water. Autoclave at 110°C for 15 min. Store at 4°C for 1 mo.
6. 2% PVP (10× stock): 2 g of polyvinylpyrrolidone (M_r 40,000) dissolved in 100 mL of distilled water. Autoclave at 110°C for 15 min. Store at 4°C for 1 mo.
7. 2% BSA (10× stock): 2 g of ultrapure grade bovine serum albumin (Roche Molecular Biochemicals, Indianapolis, IN, USA) dissolved in 100 mL of distilled water. Filter sterilize. Store at –20°C.
8. 50 mg/mL Yeast core RNA (10× Stock): Dissolve 5 g of *Torula* yeast core RNA (Sigma, St. Louis, MO, USA) in 100 mL of distilled water. Treat with proteinase K (0.25 mg/mL at 37°C for 1 h), extract with phenol:chloroform followed by ethanol precipitation to remove contaminating nucleases.

2.8. Handling of Nitrocellulose Filters During Screening

RNA probe incubation steps are performed in 170 mm × 90 mm crystallizing dishes (Fisher Scientific, Pittsburgh, PA, USA). Pyrex baking dishes can be used for subsequent washing steps. Agitation during both the binding and washing steps is provided by a gyrotory shaker. Care should be taken to minimize personal exposure to radioactivity as well as contamination of laboratory surfaces. Whatman 3MM chromatography paper is convenient for drying and exposing nitrocellulose filters.

2.9. Autoradiography Equipment

Standard autoradiography equipment (e.g., X-ray film, cassettes, dark room, and developer) will be required.

3. Methods

3.1 Handling and Storage of λ Phage Libraries

Most premade libraries are shipped in 7% dimethyl sulfoxide (DMSO). DMSO does not affect the performance of the phage and need not be removed. For long-term storage, the library should be divided into 50-µL aliquots and stored at –80°C. Repeated freeze–thaw cycles should be avoided. Dilutions of the library should be made in S*M* buffer and working aliquots can be stored at 4°C up to 2 mo without a noticeable loss of titer.

It is recommended to verify the titer of premade libraries before proceeding with the screening. A titer of $\geq 10^8$ plaque-forming units (pfu)/mL and a total of $\geq 10^6$ independent clones (depending upon the complexity of the expressed genome represented) are indications of a good representative cDNA library. It is generally a good idea to confirm the recombination frequency of the library as well by the blue-white screening protocol provided by the manufacturer. A good representative library should have at least 75% recombinant clones.

3.2. Preparation of the Host Plating Culture

Inoculate a single colony of the appropriate host *E. coli* strain into 50 mL of LB supplemented with 10 m*M* MgSO$_4$ and 0.2% maltose in a sterile 125-mL flask. MgSO$_4$ and maltose treatment allows for optimal adsorption of the λ phage to host bacteria. Incubate at 30°C overnight with shaking at 120 rpm until the OD$_{600}$ of the culture reaches 2.0. Lower temperature and limited aeration are required to prevent overgrowth of the overnight culture. Adherence of virus to dead cells in the culture in lytic phase will lower the phage titer and thus the chance of a successful screen.

Centrifuge the cells in a sterile conical tube at 2000 rpm for 5 min. Decant the supernatant and gently resuspend the cell pellet in 10 m*M* MgSO$_4$ solution to obtain an OD$_{600}$ of 0.5–1.0. It is important not to vortex, but still ensure that the cells are fully resuspended. This cell suspension can be stored at 4°C for up to 1 wk without loss of viability.

3.3. Plating the Phage Expression Library

1. To infect the host cells, mix 1 mL of the above MgSO$_4$-treated bacterial suspension with an appropriate volume (~5 × 10^4 pfu) of the phage library aliquoted in each of the 20 sterile test tubes (Falcon 2057 or equivalent). Incubate at 37°C for 15–20 min to allow phage adsorption to the host bacterial cells.
2. Add 9 mL of molten NZY top agar to one tube at a time and spread immediately and evenly onto 150-mm NZY agar Petri dishes prewarmed to 37°C. Swirling of the plates quickly after pouring the top agar will help in the even distribution of the bacterial lawn. The melted top NZY agar should be cooled to 45°C before use. Higher temperatures will kill the host bacteria. Because each tube contains 5 × 10^4 pfu, plating of phage–host mixtures onto twenty 150-mm plates will result in a total number of 1 × 10^6 plaques. This amount of phage will generally allow a representative screening of the entire library of cDNA clones. Allow the top agar to harden at room temperature for 10 min and then incubate the plates inverted at 37°C for 3–5 h until pinpoint plaques appear (*see* **Note 1**). It is important not to stack the plates in the incubator. This will ensure an even temperature on each plate and hence simultaneous appearance of plaques in all plates.

3.4. Induction of Proteins with IPTG and Transfer to Nitrocellulose Filters

1. Number nitrocellulose membrane filter circles (137 mm diameter) with a lead pencil or ballpoint pen on the side of the filter that will not touch the surface of the plates. While the plates are incubating, soak the circles in 10 mM IPTG for 30 min and air dry for 5 min on 3MM Whatman chromatography paper. It is important that the membranes are slightly damp (not dry or dripping wet) when applied to the surface of the plates. Use of clean gloves and sterile forceps for handling filters is recommended.
2. Place the IPTG-treated nitrocellulose filter circles onto the plates containing pinpoint plaques with the help of sterile forceps. Incubate the plates with the filters in place for 4 h at 37°C to allow induction of recombinant proteins.
3. Remove the plates from the incubator and mark at least three asymmetric locations close to the edges by puncturing the membrane filters and agarose under it with an 18-gage needle. Remove the membrane filter circles from the plate very carefully. Avoid picking up any top agarose along with the nitrocellulose circles. If it is not possible to remove the filters cleanly, refrigerate the plates with the membrane in place for 15 min before trying again. After removal, air-dry the nitrocellulose membranes for 5 min on 3MM Whatman chromatography paper and then soak in blocking buffer for 1 h at 37°C with agitation as described below. Wrap the Petri plates in plastic wrap and store at 4°C for later use (*see* **Note 2**).

3.5. Preparation of RNA Probes

1. Prepare RNA probes by in vitro transcription reactions using T7/T3/SP6 RNA polymerase according to the enzyme manufacturer's recommendation (*see* Chapter 1). We routinely label with [α-^{32}P]UTP for radiolabeling so that approx 1 out of every 10 uridylate residues in the final RNA is derived from the radioisotope source (*see* **Notes 4** and **7**).
2. Following transcription, RNAs are phenol extracted and gel purified on a 5% acrylamide gel containing 7 M urea using standard vertical electrophoresis equipment. The position of the RNA on the gel is determined by autoradiography and the RNA is excised from the gel using a single-edged razor blade. Intact gel slices are incubated overnight in HSCB buffer to elute the RNA. Eluted RNAs are extracted with phenol:chloroform and quantitated using a scintillation counter.

3.6. Incubation of RNA Probes with Membranes

1. Nitrocellulose membranes containing the transferred phage proteins are prehybridized in blocking buffer to block the nonspecific binding sites. The blocking and hybridization buffers are very viscous. Therefore, special precautions must be taken to avoid bubbles and to ensure uniform contact of all the membranes with these solutions. We have found that about 100 mL of solution in a 170 × 90 mm crystallizing dish is sufficient for ten 137-mm diameter nitrocellulose circles. The dish should be baked at 200°C overnight before use to destroy any contaminating ribonucleases. It is important to submerge one membrane

circle at a time in the solution to ensure even contact of all the filters with the solution. Incubate the membranes in blocking solution at 37°C with gentle agitation on a platform shaker (40–50 rpm) for 1 h.

2. After the incubation in blocking buffer, briefly rinse the membranes twice with 500 mL of rinsing buffer and transfer immediately to 100 mL of binding buffer supplemented with labeled RNA probe. Incubate at room temperature with gentle agitation on a platform shaker (40–50 rpm) for 1 h (*see* **Note 3**).

3.7. Washing of Membranes

Wash the membranes with 250 mL (per 10 filters) of washing buffer in a 170 × 90 mm crystallizing dish. Repeat washings until the background levels of radioactivity are minimal (assessed using a handheld Geiger counter). It is sometimes helpful to compare the background noise on the filters with a negative control membrane that has been lifted from a plate containing an uninfected bacterial lawn.

3.8. Autoradiography and Isolation of Positive Clones

Thoroughly dry the membranes on 3MM Whatman chromatography paper. Tape the filters individually to the paper, cover the membranes and the filter paper backing with plastic wrap, and expose to Kodak XAR film at –70°C using an intensifying screen. The exposure times will vary with the concentration of probe used and the strength of the specific signal. Generally when the concentration of the RNA probe used is in the range of 10^7 cpm/mL, 8–16 h of exposure time is sufficient to observe positive "spots." If desired, duplicate plaque lifts can be taken at **step 3.4** above to assist in the determination of potential positive clones. To obtain duplicate plaque lifts, place fresh IPTG-soaked filters onto the phage lawns and incubate at 37°C for an additional 3–4 h. In our experience, however, we have not found such duplicate filters to be very helpful.

Develop the film and align it with the membrane circles to locate positive spots representing the position of potentially positive plaques on the stored plates (*see* **Notes 4–8**). Use the broad end of a sterilized Pasteur pipet to pick the selected plaques from the plates. To elute phage from these plugs, place them separately into tubes containing 1 mL of SM buffer and two drops of chloroform. Eluted phage should be stored in the refrigerator for future use.

3.9. Identification of True Positives

To confirm the identity of a selected phage plaque as a true positive, a secondary screening is an absolute necessity. After overnight elution of the phage particles from the agar plug, replate the new phage stock to obtain up to 200 plaques on a 100-mm NZY agar plate. Rescreen these plaques using the protocols described above. True positive clones should contain an amplified number of positives on these secondary filters. Pick a well-isolated positive plaque from these secondary screens. If necessary, repeat the screening a third time with approx 50 phage/plate to obtain a well-isolated phage plaque. Follow the manufacturer's protocol to excise the phagemid containing the cloned insert from the λ vector and proceed with characterization of your cDNA clone using standard protocols.

4. Notes

1. The time of incubation required for appearance of plaques, as well as the overall diameter of the plaques themselves, will depend on the density of host cell suspension. Higher cell density ($OD_{600} = 1.0$) will result in small diameter of plaques in a given period of time while larger plaques will be obtained when suspensions with lower cell density ($OD_{600} = 0.5$) are used as the host.
2. The dishes containing the plated expression library can be safely stored at 4°C for 2–3 wk without any fungal contamination. This also precludes the need to autoclave the nitrocellulose membranes. However, placing a small piece of 3MM Whatman paper saturated with chloroform inside the plates will prevent any contamination of the bacterial lawn until the results from autoradiography are obtained and can be acted upon.
3. It is vital to monitor the integrity of the RNA probe during the binding step. This can be done by analysis of a 400-μL aliquot of the binding mixture (taken after it is incubated with the filters for the prescribed time) on a denaturing polyacrylamide gel. Extensive degradation of the probe will reduce your signal strength and result in failure of the screen. Use only high-quality reagents and RNase-free water for all steps. Common sources of RNase contamination are commercially obtained yeast *Torula* core RNA and sources of BSA other than Roche Molecular Biochemical's ultrapure grade. It is absolutely necessary to treat the yeast RNA with proteinase K followed by phenol:chloroform extractions and ethanol precipitation as described above before using it in the blocking buffer.
4. Identification of false-positive clones is one of the common problems of any library screening process. Although the majority of the undesired clones picked due to artifacts can be eliminated by secondary and tertiary screenings, dealing with a large number of false-positives could be labor intensive and disheartening. Random sticking of the probe to the membrane, as well as nonspecific RNA–protein interactions, are usually the causes of such problems. One way to reduce nonspecific interactions is to keep the size of the RNA probe as small as possible. Ideal RNA probes consist of the specific RNA binding site of the protein with only minimal flanking sequences, thus allowing only specific interactions with the protein of interest. Stringent binding and washing conditions can also be very helpful to eliminate selection of undesired clones due to nonspecific RNA–protein interactions. High salt concentrations in the appropriate binding and washing solutions will create such conditions if they are tolerated by the desired protein–RNA interaction. It is necessary to determine empirically the highest salt concentrations that will allow specific interactions of the protein with the RNA probe.

5. High levels of background noise can also significantly affect the success of a library screen. Several factors will determine the background levels of radioactivity on the membranes. Insufficient blocking can lead to such problems. Make sure that there is enough volume of blocking buffer and that each and every membrane is properly submerged. Another possible cause of high background could be nonspecific interactions between the RNA probe and the proteins released by the host bacteria following phage lysis. It is therefore necessary to experimentally test such a possibility by a dot blot assay using an immobilized total host cell extract and the labeled RNA probe. Use of a small probe and high salt concentrations (described above) during hybridization and in washing buffers can eliminate such low-level RNA–protein interactions. Repeated stringent washings are also very helpful. It may be helpful to include a negative control test strip, which has been lifted from a plate of uninfected bacterial lawn, along with the plaque lift membranes to determine the appropriate amount of washings.
6. A high signal-to-noise ratio is essential for a successful library screening. However, if the signal itself is very weak, controlling the background noise will not be sufficient to allow identification of the positive clones. Several parameters determine the strength of the signal. A high plaque density can lead to decrease in the signal intensity. Thus, it is necessary to plate no more than 5×10^4 pfu per 150-mm plate. Sometimes it might be necessary to plate even fewer phage per plate to obtain a reasonable signal. The amount of protein in each plaque (and hence the strength of the signal) will depend also on both the plaque diameter as well as the number of host cells lysed in unit area (which is a function of the host cell density). Signal intensity is a function of the strength of the RNA–protein interaction being screened. In case the protein of interest binds weakly to its specific binding site, it may be advisable to generate multiple tandem binding sites in the probe. This could increase the number of protein molecules bound to the probe, thus resulting in a stronger signal. However, creation of such artificial RNA molecules could lead to the formation of undesirable RNA secondary structures that might inhibit the protein binding or alter the binding specificity, resulting in increased background noise.
7. In general, an RNA probe with higher specific activity will help the signal to be detected more quickly.
8. If available, it is helpful to include a nitrocellulose test strip on which the partially purified RNA-binding protein of interest is immobilized to monitor the screening procedure. This "positive" control allows screening results to be interpreted with confidence.

References

1. Dreyfuss, G., Hentze, M., and Lamond, A. I. (1996) From transcript to protein. *Cell* **85,** 963–972.
2. Nagai, K. (1996) RNA–protein interactions. *Curr. Biol.* **6**, 53–61.
3. Singh, H., Clerc, R. G., and LeBowitz, J. H. (1989) Molecular cloning of sequence-specific DNA binding proteins using recognition site probes. *Biotechniques* **7**, 252.

4. Blackwell, J. L. and Brinton, M. A. (1995) BHK cell proteins that bind to the 3' stem-loop structure of the west nile virus genome RNA. *J. Virol.* **69**, 5650–5658.
5. Park, Y. W. and Katze, M. G. (1995) Translational control by influenza virus. *J. Biol. Chem.* **270**, 28433–28439.
6. Wilusz, J. and Shenk, T. (1988) A 64 kd nuclear protein binds to RNA segments that include the AAUAAA polyadenylation motif. *Cell* **52**, 221–228.
7. Bagga, P. S., Ford, L. F., Chen. F., and Wilusz, J. (1995) The G-rich auxiliary downstream element has distinct sequence and position requirements and mediates efficient 3' end pre-mRNA processing through a trans-acting factor. *Nucleic Acids Res.* **23**, 1625–1631.
8. Vinson, C. R., LaMarco, K. L., Johnson, P. F., Landschulz, W. H., and McKnight, S. L. (1988) In situ detection of sequence-specific DNA binding activity specified by a recombinant bacteriophage. *Genes Dev.* **2**, 801–806.
9. MacDonald, C. C., Wilusz, J., and Shenk, T. (1994) The 64 kilodalton subunit of CstF polyadenylation factor binds to pre-mRNAs downstream of the cleavage site and influences cleavage site location. *Mol. Cell. Biol.* **14**, 6647–6654.
10. Melton, D. A., Krieg, P. A., Rebagliati, M. R., Maniatis, T., Zinn, K., and Green, M. R. (1984) Efficient in vitro synthesis of biologically active RNA and RNA hybridization probes from plasmids containing a bacteriophage SP6 promoter. *Nucleic Acids Res.* **12**, 7035–7056.
11. Qian, Z. and Wilusz, J. (1993) Cloning of a cDNA encoding an RNA binding protein by screening expression libraries using a Northwestern strategy. *Anal. Biochem.* **212**, 547–554.
12. Qian, Z. and Wilusz, J. (1994) GRSF-1: a poly(A) + mRNA binding protein which interacts with a conserved G-rich element. *Nucleic Acids Res.* **22**, 2334–2343.
13. Stanley, K. K. (1983) Solubilization and immune-detection of β-galactosidase hybrid proteins carrying foreign antigenic determinants. *Nucleic Acids Res.* **11**, 4077–4092.

19

Screening Expression Libraries with Solution-Based Assay

Philippa J. Webster and Paul M. Macdonald

1. Introduction

To date, the most common approach to screening an expression library for an RNA-binding protein is one based on a procedure that was originally developed for cloning DNA-binding proteins *(1,2)*: cDNA-encoded proteins from λ plaques are immobilized on nitrocellulose or nylon filters, which are then probed with a labeled RNA containing multimerized copies of the binding site of interest *(3; see also* Chapter 18). This technique has been used to clone a number of RNA-binding proteins, and has proven particularly useful for isolating proteins containing double-stranded RNA-binding motifs (e.g., **refs. 4–6**). However, some RNA-binding proteins no longer bind RNA efficiently once they are attached to filters; these obviously cannot be identified using a filter-based approach. As the isolation of an RNA-binding protein through the direct screening of an expression library can represent an enormous savings in time over its biochemical purification, it is profitable to consider alternate ways of screening libraries that don't rely on the binding of proteins to filters.

This chapter presents a solution-based approach to expression screening. The screen is performed by analyzing protein extracts of pools of bacterially expressed cDNAs from a plasmid library using gel-shift or ultraviolet (UV)-crosslinking assays. The native extracts are never denatured or immobilized on a filter but rather are tested directly, which may allow the protein of interest to retain its RNA-binding properties. Once a pool containing the specific binding activity is found, it is subdivided into less complex pools, and the procedure is reiterated until a single cDNA encoding the activity has been identified. We have used this approach successfully to isolate an RNA-binding protein which cannot be identified on filter lifts *(7)*. Here we provide an overview of the

From: *Methods in Molecular Biology, Vol. 118: RNA-Protein Interaction Protocols*
Edited by: S. Haynes © Humana Press Inc., Totowa, NJ

solution-based procedure, including a protocol for creating protein extracts from pools of expressed cDNAs. Detailed descriptions of gel-shift and UV-crosslinking assays can be found in Chapters 3, 6, and 10 of this volume.

1.1. Is this the best procedure for isolating your protein?

Creating protein extracts from pools of expressed cDNAs is simple; however, screening them can be quite labor intensive, as each pool will typically contain only a few hundred clones. For example, using pools of 250 cDNAs, you will have to perform 400 assays to screen a total of 100,000 clones. This type of expression cloning will consequently work best for relatively abundant proteins (allowing you to screen fewer pools) and/or proteins that show a strong signal in your assay (allowing your pools to be more complex).

In comparison, a filter-based approach will easily allow you to examine a million clones or more in a short period of time. Therefore, before embarking on a solution-based screen, it is usually worth attempting to isolate the protein of interest by screening filters. Alternatively, you can use Northwestern assays of tissue or cell extracts (*see* Chapter 18) to test a variety of binding conditions and assess if your protein retains its ability to bind RNA once it is immobilized on a filter. Keep in mind, however, that if the protein is not abundant, it may be difficult to visualize on a Northwestern blot but still be possible to detect when expressed at high levels in a λ plaque.

As with other expression cloning strategies, the approach described here can be successful only if the activity of the desired protein does not depend on posttranslational modifications or interactions with other proteins not provided by the host bacterial cells.

1.2. Choosing a Library

An ideal cDNA library for expression screening is one cloned into a λ phagemid vector which can be used to generate both λ and plasmid expression libraries *(8,9)*; in this way the same library can be used for a variety of different screens. Commercially available λ phagemid vectors include Lambda ZAP (Stratagene, La Jolla, CA, USA) and Lambda TriplEx (Clontech, Palo Alto, CA, USA). The cDNA inserts are cloned downstream of a *lacZ* translational start site, allowing them to be expressed as inducible *lacZ*-fusion proteins. It is best if the inserts are a mixture of random-primed and oligo-dT-primed cDNAs, cloned in an oriented fashion to maximize the number of in-frame fusions with the *lacZ* open reading frame. In Lambda ZAP, only one third of the clones will be in frame; in Lambda TriplEx, inserts are expressed in all three reading frames, but lower levels of each protein are made.

If the library is to be made in a plasmid vector, similar considerations apply to the cDNA synthesis and cloning. A wide variety of expression plasmids with various features are available; although the protocol described here uses IPTG induction of *lacZ*-fusion proteins, the technique can easily be adapted to other expression systems.

1.3. Choosing and Modifying an Assay

Gel-shift and UV-crosslinking assays are currently the most common solution-based procedures for analyzing RNA-binding proteins. In establishing an appropriate assay for your screen, it is helpful to perform a small pilot experiment employing the same assay and conditions that were initially used to identify your protein of interest, but substituting bacterial extracts for the tissue or cell extract in which the RNA-binding activity was found. The bacterial extracts may generate nonspecific shifts or bands; such spurious signals can often be reduced by adjusting the type and amount of cold competitor in the assay. However, many RNA probes show persistent nonspecific binding to bacterial proteins. In such cases, UV-crosslinking may be the most useful assay, as it can reveal specific binding in the presence of numerous background bands (**Fig. 1**). Once the new conditions are established, use them to test your original tissue or cell extract to verify that the protein of interest still binds.

Design the assay to use the maximum quantity of extract possible in a given volume (typically this will be 5 µL of bacterial extract in a 10 µL reaction). Note that in the protocol that follows, the suggested components of the bacterial extract buffer can be modified to complement the final assay conditions.

When working out the assay, use both the experimental probe and a negative control probe to assess nonspecific background binding. Ideally the negative control will be a sequence that is highly similar to the experimental probe, but contains point mutations that inactivate the binding sites. For the full-scale screen, however, it is not necessary to test each bacterial extract with both an experimental and a negative control probe; extracts that do not contain a cDNA-encoded binding activity will serve as a negative comparison for those that do. (Omitting the negative control will halve the number of primary assays you need to perform.) Once you have identified a pool that contains a unique binding activity, retest this pool with both the experimental and the negative control probe to verify that the binding is specific.

If you are screening an expression library that contains randomly primed inserts, the average molecular weight of the resulting proteins will be relatively low; the percentage of the acrylamide in the gel used for your assay should be adjusted to maximize separation in the appropriate size range.

Finally, the assay you choose will need to be streamlined to allow for the processing of many samples. Determine the minimal procedure that clearly identifies the protein of interest in your tissue or cell extract. If possible, adapt the protocol so that the reactions can be performed in microtiter trays instead

protein extract	A	A	B	B
probe	+	−	+	−

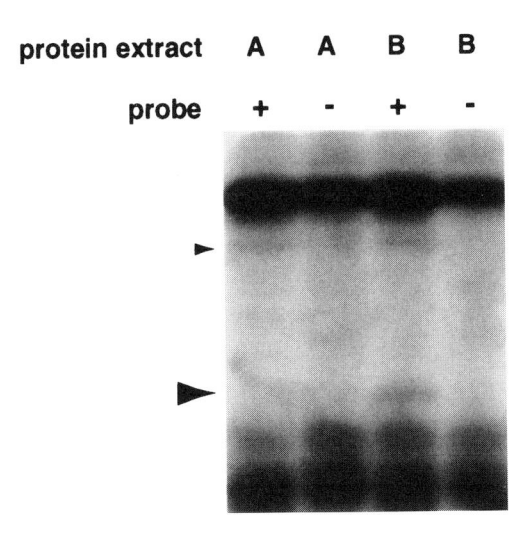

Fig. 1. Autoradiograph showing UV-crosslinking assays of protein extracts from pools of bacteria expressing *Drosophila* ovarian cDNAs. Data from only two pools are shown. Probes are in vitro transcribed, radiolabeled RNAs of multimerized copies of either a wild-type binding site (+) or a mutated binding site (−) for bruno, a *Drosophila* ovarian RNA-binding protein that plays a role in the translational repression of *oskar* mRNA *(10)*. Each pool was inoculated with 250 bacteria carrying different plasmid clones; only one in three of the clones encodes an in-frame lacZ-fusion protein, and therefore each extract is predicted to contain approx 80 cDNA-encoded *Drosophila* proteins. Extracts from all pools display substantial nonspecific binding to both of the probes; all extracts also contain a protein that binds more strongly to the wild-type than the mutant probe (small arrow). The bands that are common to every extract tested are presumed to be *E. coli* proteins. Pool B, however, additionally contains a protein that binds only the wild-type and not the mutant probe (large arrow) and is not seen in any other pool (compare with pool A). Following further rounds of purification and molecular characterization, this protein was shown to be a fragment of bruno *(7)*. Approximately one million cpm of ^{32}P-labeled RNA probe were used in each UV-crosslinking assay. The samples were separated by sodium dodecyl sulfate-polyacrylamide gel (SDS-PAGE) on a 15% gel; the fragment of bruno (large arrow) is 20 kDa. The X-ray film was exposed for 14 h at −80°C with two intensifying screens.

of individual tubes, reduce the sample sizes so that you can run a large number of narrow lanes on each gel, and prepare master mixes of all possible components to decrease the number of pipetting steps.

2. Materials

1. 2× TY: 16 g of tryptone, 10 g of yeast extract, and 5 g of NaCl per 1 L of water. Sterilize by autoclaving. (Other bacterial growth media can be used, but *see* **Note 4**.)

2. Antibiotic that will select for the plasmid to be maintained.
3. 1 M Isopropyl-β-D-thiogalactopyranoside (IPTG). Make up a stock solution in distilled water and store in 1-mL aliquots at –20°C.
4. 1 M Dithiothreitol (DTT): Make up a stock solution in distilled water and store 100-µL aliquots at –20°C. DTT is very temperature sensitive, and should be kept on ice during use.
5. 100 mM Phenylmethylsulfonyl fluoride (PMSF): ***PMSF is a highly toxic cholinesterase inhibitor and should be handled with gloves at all times; a mask should be worn when weighing out the powdered form.*** Make up a stock solution in anhydrous isopropanol using a stir plate or rotator (it can take up to 30 min to go into solution at room temperature). PMSF is stable for 9 mo at room temperature when completely anhydrous, but breaks down rapidly in the presence of water, and thus it is often recommended that a solution be prepared fresh daily. The half-life of PMSF at 25°C in an aqueous solution at pH 7.0 is 110 min; the half-life at pH 8.0 is 35 min *(11)*. To detoxify PMSF before disposal, add water and a drop of NaOH and let sit overnight.
6. Lysozyme. Store powdered at –20°C (*see* **Note 5**).
7. Bacterial lysis buffer: 150 mM NaCl, 50 mM Tris-HCl, pH 8.0, 1 mM EDTA, and 1% Nonidet P-40 or Igepal CA-630 (a nonionic detergent equivalent to Nonidet P-40 available from Sigma, St. Louis, MO, USA). Store at 4°C. Before use, chill a 10-mL aliquot on ice and add 10 µL of 1 M DTT, 100 µL of 100 mM PMSF and approx 20 mg of dry lysozyme. Mix well, store on ice, and use immediately. (The salt and buffer components of this solution can be adjusted to complement particular assay conditions; *see* **Subheading 1.3**. The final extract will be in 0.5× bacterial lysis buffer, 20% glycerol.)
8. 40% Glycerol in water. Sterilize by autoclaving. Store at 4°C, and chill on ice before use.

3. Methods
3.1. Screening Protein Extracts from Pools of Expressed cDNAs

1. Transform a plasmid cDNA expression library into an appropriate strain of *E. coli*. Grow the bacteria under antibiotic selection in liquid medium by shaking at 37°C until the OD$_{600}$ equals 0.5–1.0. Store the culture at 4°C. Immediately before use, titer the library by plating out a dilution series on selective media and counting colonies.
2. Label culture tubes containing 2 mL of 2× TY plus antibiotic, and inoculate each tube with 250 bacteria (*see* **Note 1**). Shake tubes at 37°C until the OD$_{600}$ equals 0.5.
3. Induce the expression of the *lacZ*-fusion proteins by adding IPTG to a final concentration of 10 mM, and continue shaking at 37°C for 3 h to overnight (*see* **Note 2**.)
4. Label a microfuge tube to correspond with each culture. Centrifuge 1.5 mL of each culture in a microfuge at 14,000g for 1 min at room temperature. Store the remaining 0.5 mL of culture at 4°C (*see* **Note 3**.)
5. Discard the supernatant and place the pellet on ice.
6. Resuspend cell pellets in 30 µL of bacterial lysis buffer by vortexing (*see* **Note 4**.)

7. Incubate on ice 15 min, by which time the lysed suspension should be extremely viscous (*see* **Note 5**.)
8. Centrifuge in a microfuge at 14,000g at 4°C for 15 min (*see* **Note 6**.)
9. While centrifuging, set up labeled tubes on ice containing 20 µL of 40% glycerol for each sample.
10. Transfer the sample tubes from the centrifuge onto ice. Pipet 20 µL of supernatant from each sample into a tube containing the glycerol, and mix by pipetting up and down. Keep everything as cold as possible. Discard the pellet and remainder of sample.
11. Immediately after mixing with glycerol, freeze each extract in a dry ice/ethanol bath or liquid nitrogen. (Be sure the labels on the tubes survive this step!) Store at –70°C. To use extracts, thaw on ice. Freeze in a dry ice/ethanol bath again before returning to –70°C (*see* **Note 7**).
12. Assay extracts with the binding site probe (*see* **Subheading 1.3.**).
13. Once a pool containing a binding activity of interest has been identified, verify the specificity of the binding by retesting the extract with both the experimental probe and a negative control probe.
14. Titer the remaining 0.5 mL of culture from which the extract was made, and inoculate 40 cultures with approx 25 bacteria each (*see* **Note 1**). Shake the tubes at 37°C until the OD_{600} equals 0.5.
15. Repeat **steps 3–12**. Once a pool containing the binding activity has been identified, plate the corresponding culture out so that individual colonies can be picked, and grow up 20 new pools inoculated with five colonies each. Repeat **steps 3–12**. For the final round, inoculate 20 cultures with individual colonies.

4. Notes

1. The pool size you choose is important. The protein from the cDNA of interest needs to be abundant enough in the bacterial extract to be detectable in your assay; however, the less complex your pools are, the more you will have to screen. The number and complexity of pools in subsequent rounds of screening (*see* **steps 14** and **15** above) will depend on the complexity of the primary pool. Factors to consider in designing your pool size include how many of the clones in the library are expected to express in-frame fusion proteins (this can range from all of the clones to only one in six, depending on which vector was used and whether the cDNAs were inserted in an oriented fashion or not); the level of expression of the cDNA-encoded proteins (vectors that express all three reading frames produce a lower level of each protein than ones that express only a single reading frame); and the sensitivity of your assay. As a guideline, we recommend starting with pools that have between 50 and 100 in-frame cDNAs being expressed. The pool size of 250 given as an example in this protocol was used to clone an RNA-binding protein from a library in which one third of the clones were in-frame, producing approx 80 proteins in each pool (*see* **Fig. 1**).

You may want to grow only as many cultures as you can conveniently assay at one time; by performing the assays in batches in an ongoing fashion, you will screen only as many primary pools as it takes to find a single positive.
2. It is difficult to predict the stability of the cDNA-encoded lacZ-fusion proteins, although they are often quite stable. It is probably ideal to grow the cultures to saturation following induction, and then process them right away or store them at 4°C overnight.
3. The cultures should survive quite well for a month at 4°C. If you don't expect to complete your assays in this time, or if you want to be able to test your pools for other binding proteins in the future, you can make a long-term stock of each culture by adding sterile glycerol to a final concentration of 15% and storing them at –70°C.
4. The volume suggested for resuspending the cell pellet is the approximate minimal volume necessary for lysing and clearing the pellet from a 1.5-mL culture grown in 2× TY, a rich medium that allows the bacteria to grow to a relatively high density. The pellet should be resuspended in a smaller volume if a less rich medium (such as LB) is used, and the volume of 40% glycerol added (*see* **steps 9** and **10**) should also be adjusted accordingly. The objective is to create a protein extract that is highly concentrated.
5. If the cells do not lyse efficiently, a common cause is the integrity of the lysozyme. Be sure it is stored dry at –20°C, and not as a stock solution, and that it is added to the lysis buffer just before use. If you continue to have problems, order a fresh bottle of lysozyme.
6. If it is available, a microfuge with a refrigeration unit that will consistently maintain the temperature of the samples at 4°C is preferable to an ordinary microfuge placed in a 4°C room, which can get quite warm during a long centrifugation.
7. It may be useful to know whether the RNA-binding activity in your tissue or cell extract will survive being stored at –70°C, and whether it will remain active through repeated cycles of freeze–thawing; the storage conditions of the extracts can be adjusted accordingly. It is not clear how relevant this information is, however, as the stability of *lacZ*-fusion proteins will vary (as mentioned above, they are frequently quite stable); in addition, a clone that shows the binding activity of interest may only be expressing a fragment of the full-length biological protein, and therefore can manifest quite different physical properties. As creating the extracts is relatively rapid, it is also possible to assay them immediately after they are made without relying on storage at all.

References

1. Vinson, C. R., LaMarco, K. L., Johnson, P. F., Landschulz, W. H., and McKnight, S. L. (1988) In situ detection of sequence-specific DNA binding activity specified by a recombinant bacteriophage. *Genes Dev.* **2,** 801–806.
2. Singh, H., Clerc, R. G., and LeBowitz, J. H. (1989) Molecular cloning of sequence-specific DNA binding proteins using recognition site probes. *Biotechniques* **7,** 252–261.

3. Qian, Z. and Wilusz J. (1993) Cloning of a cDNA encoding an RNA binding protein by screening expression libraries using a northwestern strategy. *Anal. Biochem.* **212,** 547–554.
4. Bass, B. L., Hurst, S. R., and Singer, J. D. (1994) Binding properties of newly identified Xenopus proteins containing dsRNA-binding motifs. *Curr. Biol.* **4,** 301–314.
5. Lee, K., Fajardo, M. A., and Braun, R. E. (1996) A testis cytoplasmic RNA-binding protein that has the properties of a translational repressor. *Mol. Cell Biol.* **16,** 3023–3034.
6. Castiglia, D., Scaturro, M., Nastasi, T., Cestelli, A., and Di Liegro, I. (1996) PIPPin, a putative RNA-binding protein specifically expressed in the rat brain. *Biochem. Biophys. Res. Commun.* **218,** 390–394.
7. Webster, P. J., Liang, L., Berg, C. A., Lako, P., and Macdonald, P. M. (1997) Translational repressor bruno plays multiple roles in development and is widely conserved. *Genes Dev.* **11,** 2510–2521.
8. Short, J. M., Fernandez, J. M., Sorge, J. A., and Huse, W. D. (1988) Lambda ZAP: a bacteriophage lambda expression vector with in vivo excision properties. *Nucleic Acids Res.* **16,** 7583–7600.
9. Elledge, S. J., Mulligan, J. T., Ramer, S. W., Spottswood, M., and Davis, R. W. (1991) Lambda YES: a multifunctional cDNA expression vector for the isolation of genes by complementation of yeast and Escherichia coli mutations. *Proc. Natl. Acad. Sci. USA* **88,** 1731–1735.
10. Kim-Ha, J., Kerr, K., and Macdonald P. M. (1995) Translational regulation of *oskar* mRNA by bruno, an ovarian RNA-binding protein, is essential. *Cell* **81,** 403–412.
11. James, G. T. (1978) Inactivation of the protease inhibitor phenylmethylsulfonyl fluoride in buffers. *Anal. Biochem.* **86,** 574–579.

20

An Immunoprecipitation-RNA:rPCR Method for the In Vivo Isolation of Ribonucleoprotein Complexes

Edward Chu, John C. Schmitz, Jingfang Ju, and Sitki M. Copur

1. Introduction

Within the last few years, it has become well established that a given nucleic acid binding protein has the potential to interact specifically with more than one target nucleic acid sequence. Various immunoprecipitation techniques have been developed to isolate specific DNA–protein complexes from intact cells *(1–4)*. These studies have allowed for the precise identification of in vivo target genes for a given DNA binding protein. Moreover, they have provided important insights as to the critical nucleotide elements required for DNA binding.

With regard to RNA–protein complexes, immunoprecipitation procedures have been developed to identify and characterize the components of small ribonucleoproteins (RNPs) in mammalian cells *(5–8)*. It has been well documented that the small RNPs range in abundance from 10^3 to 10^8 molecules/cell. This relatively high expression within cells has allowed for their ready detection by immunoprecipitation. However, this technique is inadequate in terms of its sensitivity to identify the mRNA components of RNP complexes that are expressed at much lower levels in cells.

In this chapter, we outline an immunoprecipitation-RNA:random polymerase chain reaction (rPCR) method (**Fig. 1**) that was developed to identify and characterize the cellular RNA components of RNP complexes. This technique is based on the immunoprecipitation procedure of Lerner and Steitz *(6,7)* and modified by incorporating the RNA:rPCR technique of Froussard *(9,10)*. The rPCR method was originally developed for the random amplification of whole cDNA sequences derived from microscale amounts of RNA transcripts. The strategy of coupling the rPCR amplification method to an immunoprecipitation procedure has several important advantages over the immunoprecipita-

Immunoprecipitation-RNA:rPCR Method

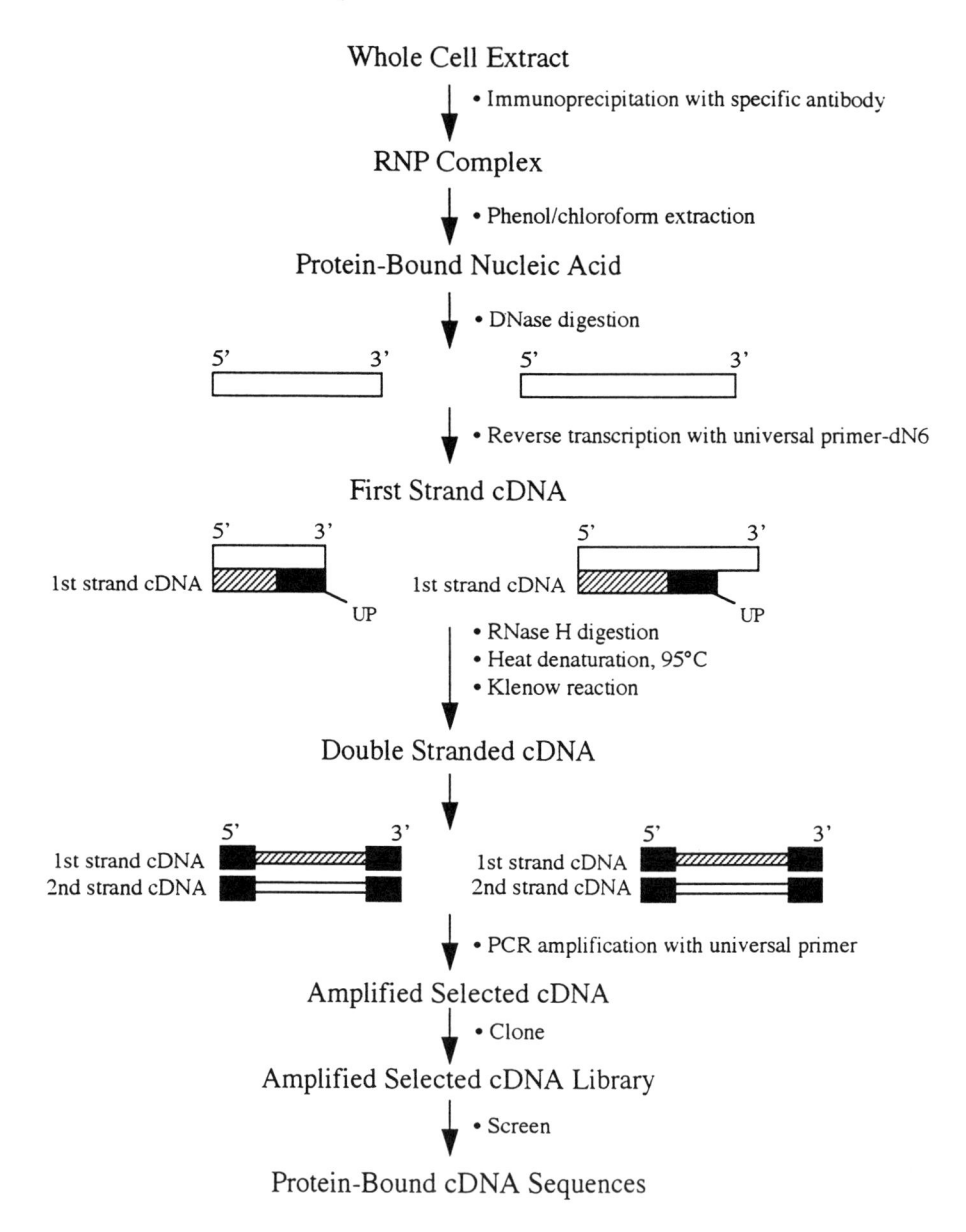

Fig. 1. Immunoprecipitation-RNA:rPCR method. Strategy to isolate cellular RNP complexes. Hatched bar represents cDNA sequence, and solid bar represents universal primer sequence.

tion method originally described by Lerner and Steitz *(6,7)* and the immunoprecipitation-RT:PCR method previously developed in our laboratory *(11)*. First, using the rPCR step, picomole amounts of immunoprecipitated RNA can be amplified and employed as the starting material. Second, concern over the efficiency of recovery at each step can be greatly minimized with the use of the amplification reaction. Consequently, the selection process can be designed to allow for maximal stringency. Third, the use of universal random primers in both the reverse transcription and PCR amplification reactions allows for the isolation of protein-bound cellular RNAs whose sequences are previously unknown. Finally, the amplified selected cDNAs can be directly cloned into either a plasmid or phage library, and their respective sequences can then be identified.

2. Materials
2.1. Preparation of Cell Extract
1. Sterile 1× phosphate-buffered saline (PBS) (Biofluids; Rockville, MD, USA).
2. NET-2 buffer: 50 mM Tris-HCl, pH 7.4, 150 mM NaCl, and 0.05% (v/v) Nonidet P-40.
3. Prime RNase Inhibitor (5-Prime, 3-Prime; Boulder, CO, USA).

2.2. Immunoprecipitation
1. Protein A agarose (Gibco-BRL; Gaithersburg, MD, USA).
2. Carrier yeast tRNA at 1 mg/mL (US Biochemical; Cleveland, OH, USA).
3. Prime RNase Inhibitor (5-Prime, 3-Prime; Boulder, CO, USA).
4. NET-2 buffer (as described previously).
5. RQ1 DNase (Promega; Madison, WI, USA).
6. Phenol/chloroform (1:1). Store protected from light.
7. Chloroform/isoamyl alcohol (24:1).
8. 3 M Sodium acetate, pH 5.2 (5-Prime, 3-Prime; Boulder, CO, USA).
9. Glycogen (Boehringer Mannheim; Indianapolis, IN, USA).
10. Ethanol and 80% ethanol.
11. Diethylpyrocarbonate (DEPC) water (5-Prime, 3-Prime; Boulder, CO, USA).

2.3. RNA:rPCR Method
1. Reverse transcription reaction: 50 mM Tris-HCl, pH 8.3, 10 mM DTT, 3 mM MgCl$_2$, 0.5 mM of each dNTP, 1 U of RNase Block I, and 50 U of MMLV reverse transcriptase. All reagents were obtained from Stratagene (La Jolla, CA, USA).
2. Universal primer-dN6: 5'-GCCGCTCGAGTGCAGAATTCNNNNNN-3'.
3. RNase H (Gibco-BRL; Gaithersburg, MD, USA).
4. Second-strand cDNA synthesis: 10× Klenow buffer and Klenow fragment of DNA polymerase I (Gibco-BRL); dNTP mix containing 25 mM of each dNTP.

5. Phenol/chloroform (1:1). Store protected from light.
6. Chloroform/isoamyl alcohol (24:1).
7. G-50, D-RF RNase-free centrifuge columns (5-Prime, 3-Prime).
8. PCR amplification reaction. All reagents obtained from Perkin-Elmer (Norwalk, CT, USA): 10× PCR buffer (100 mM Tris-HCl, pH 8.3, 500 mM KCl, 15 mM MgCl$_2$, 0.1% gelatin); dNTPs (10 mM each); *Taq* DNA polymerase; mineral oil.
9. Universal primer: 5'-GCCGCTCGAGTGCAGAATTC-3'.
10. DEPC water (5-Prime, 3-Prime).
11. Electrophoresis running buffer (1× TBE): dilute 10× TBE (Biofluids) 10-fold with deionized water before use.
12. 10× Gel loading buffer for the nondenaturing agarose gel: 0.25% bromophenol blue, 0.25% xylene cyanol, and 15% (w/v) Ficoll type 400, dissolved in sterile water.

2.4. Construction of cDNA Plasmid Library and Transformation

1. Promega PCR Prep kit (Promega).
2. *Eco*RI restriction enzyme and React 3 buffer (Gibco-BRL).
3. pGEM-7Z (Promega).
4. Calf intestinal alkaline phosphatase (CIAP) 10× buffer and CIAP (Promega).
5. 0.5 M EDTA.
6. Phenol/chloroform (1:1). Store protected from light.
7. Chloroform/isoamyl alcohol (24:1).
8. 3 M Sodium acetate, pH 5.2 (5-Prime, 3-Prime).
9. Ethanol and 80% ethanol. Store at –20°C.
10. 10× T4 DNA ligase buffer and T4 DNA ligase (Promega).
11. JM109 competent cells (Promega).
12. Falcon 2059 polypropylene tubes.
13. Super optimum catabolite (SOC) medium (Gibco-BRL).
14. LB/ampicillin (100 µg/mL) plates (Gibco-BRL).

2.5. Colony Screening Method

1. TE buffer: 10 mM Tris-HCl, pH 7.4, 0.1 mM EDTA.
2. PCR amplification reaction: as listed in **Subheading 2.3**.
3. SP6 promoter primer (Promega) and T7 promoter primer (Promega).
4. Phenol/chloroform (1:1). Store the solution protected from light.
5. Chloroform/isoamyl alcohol (24:1).
6. 3 M Sodium acetate, pH 5.2 (5-Prime, 3-Prime).
7. Preparation of nondenaturing agarose gel: as described in **Subheading 2.3., step 11**.
8. Electrophoresis running buffer (1× TBE): dilute 10× TBE (Biofluids) 10-fold with deionized water prior to use.
9. 10× Gel loading buffer for the nondenaturing agarose gel: as described in **Subheading 2.3., step 13**.
10. *fmol* DNA Cycle Sequencing System (Promega).

2.6. Other Items

1. Minigel horizontal electrophoresis apparatus (Horizon 58, Gibco-BRL).
2. UV transilluminator (Fotodyne; Hartland, WI, USA).
3. DNA thermal cycler 480 (Perkin-Elmer).
4. High-speed refrigerated centrifuge (RC-5 Superspeed; Sorvall, Newtown, CT, USA).
5. Refrigerated microcentrifuge (Micromax RF; International Equipment, Needham Heights, MA, USA).
6. Benchtop refrigerated centrifuge (Sorvall RT-7).
7. Speed-Vac concentrator (Speed-Vac Plus; Savant, Holbrook, NY, USA).

3. Methods

3.1. Preparation of Cell Extract

1. Set up three or four 150 cm^2 tissue-culture flasks of cells of interest (*see* **Notes 1** and **2**).
2. Wash the flasks 3× with 10–15 mL of ice-cold PBS. Harvest the cells with a rubber policeman, resuspend in 10 mL of ice-cold PBS, and place in a sterile 50-mL conical tube.
3. Centrifuge for 10 min, 1000*g* at 4°C.
4. Resuspend the cell pellet in 0.5 mL of NET-2 buffer and 100 U of Prime RNase Inhibitor. It is important to prepare this buffer immediately before each use.
5. Sonicate cells on ice with three 5–10-s bursts using a Branson Sonifier (setting 2). Place in an autoclaved 15-mL Corning glass tube.
6. Centrifuge the homogenate at 10,000*g* for 10 min at 4°C. The supernatant is the source of antigen, and it will be used in the immunoprecipitation reaction. Although it can be stored at –80°C until future use, it is preferable to use it fresh.

3.2. Immunoprecipitation

1. To preclear the whole cell extract, incubate 1–1.5 mg of total cellular protein with an equal volume of Protein A agarose for 30 min on a rotator in a 4°C cold room. It is important to gently tumble this incubation.
2. Centrifuge at 10,000*g* for 2 min at 4°C to remove the Protein A agarose and save the supernatant portion for the next step.
3. Incubate the precleared extract at 4°C with the appropriate antibody (*see* **Notes 3** and **4**), 40 µL of 1 mg/mL carrier yeast tRNA, and 30 µL of Prime RNase Inhibitor for 1 h (*see* **Note 5**). Again, gently tumble the sample during the incubation.
4. Add Protein A agarose at an extract–Protein A agarose ratio of 2:1, and incubate, with gentle tumbling, for an additional 30 min. at 4°C.
5. Centrifuge the Protein A agarose–immune complex precipitate at 10,000*g* for 2–3 min at 4°C, and wash at least four or five times with NET-2 buffer (350 µL each wash) (*see* **Note 3**).
6. After the final wash, add 350 µL of NET-2 buffer and 350 µL of phenol/chloroform (1:1) to the pellet. Vortex vigorously for 1 min.
7. Centrifuge at 10,000*g* for 5 min at 4°C.

8. Incubate the aqueous layer with 10 U of RQ1 DNase (Promega; Madison, WI, USA) for 15 min at 37°C. This step will remove any contaminating DNA.
9. Extract the aqueous phase with an equal volume of phenol/chloroform (1:1).
10. Centrifuge at 10,000g for 5 min at 4°C.
11. Extract the aqueous phase with an equal volume of chloroform/isoamyl alcohol (24:1).
12. Centrifuge at 10,000g for 5 min at 4°C.
13. Precipitate the RNA with 0.1 vol of 3 M sodium acetate, 2.5 vol of ethanol, and 20 µg of glycogen. The glycogen is included to facilitate the precipitation of very low amounts of nucleic acid (*see* **Note 6**). Place the sample in a dry ice/ethanol bath for at least 30 min.
14. Centrifuge at 10,000g for 30 min at 4°C.
15. Wash the pellet with 1 mL of 80% ethanol.
16. Centrifuge at 10,000g for 5 min at 4°C.
17. Dry the pellet in a Speed-Vac concentrator and then reconstitute it in 15–20 µL DEPC water. The RNA can now be stored at –80°C until further use (*see* **Note 7**).

3.3. RNA:rPCR Method

1. Heat half of the immunoprecipitated RNA sample at 65°C for 5 min and chill rapidly on ice.
2. Subject the sample to reverse transcription (RT) in a reaction (final volume, 50 µL) containing 50 mM Tris-HCl, pH 8.3, 10 mM DTT, 3 mM MgCl$_2$, 500 µM of each dNTP, 1 U of RNase Block I, 50 U of MMLV reverse transcriptase (*see* **Note 8**), and 150 ng of the universal primer-dN6, 5'-GCCGCTCGAGTGCAGAATTCNN-NNNN-3'.
3. Incubate the RT reaction at 42°C for 1 h. At the end of the reaction, heat the sample at 95°C for 3 min, and then cool rapidly on ice.
4. Add 1 µL of RNase H and incubate at 37°C for 15 min. This step will cleave the RNA strand.
5. Heat sample at 95°C for 3 min and cool on ice.
6. For synthesis of second-strand cDNA, incubate 41.5 µL of the first strand cDNA reaction with 100 ng of the universal primer-dN6 (5'-GCCGCTCGAGTGCAGAAT-TCNNNNNN-3') at 95°C for 3 min and quick chill on ice. To this, add 5 µL of 10× Klenow buffer, 1 µL of 25 mM dNTPs, 0.5 µL of 100 mM DTT, and 1 µL (6 U) of Klenow fragment of DNA polymerase I (final volume, 50 µL) and incubate at 37°C for 30 min.
7. Extract sample with phenol/chloroform (1:1) and chloroform/isoamyl alcohol (24:1) as described in **Subheading 3.2.**, and then purify sample on a G-50, D-RF centrifuge column to eliminate excess universal primer-dN6.
8. Use one-tenth of the double stranded cDNA sample for the PCR amplification reaction, which is performed as described by Saiki et al. (*12*). To the cDNA, add 5 µL of 10× PCR buffer, 2.5 µL of dNTP (10 mM each), 100–200 ng of the universal primer (5'-GCCGCTCGAGTGCAGAATTC-3'), and DEPC water to a final volume of 50 µL. Add 1.5 µL of *Taq* DNA polymerase, and overlay the

reaction with 35 µL of mineral oil. Amplify for 40 cycles at 94°C for 1 min, 55°C for 1 min, and 72°C for 3 min.
9. At the completion of the PCR reaction, a PCR chase reaction *(13)* is performed to ensure the synthesis of intact double-stranded cDNA. To 30 µL of the PCR reaction, add 30 µL of 10× PCR buffer, 6 µL of dNTPs, 100 ng of universal primer, 3 µL of *Taq* DNA polymerase, and DEPC water to a final volume of 300 µL (*see* **Note 9**).
10. Mix gently and perform three cycles of amplification at 94°C for 30 s, 50°C for 1 min, and 72°C for 2 min. Following this step, do an additional four cycles of amplification at 50°C for 1 min and 72°C for 5 min.
11. Extract the PCR-chased material with an equal volume of phenol/chloroform (1:1) and precipitate with 30 µL of 3 *M* sodium acetate and 750 µL of ethanol. Resuspend in 20–30 µL of DEPC water.
12. Analyze one-tenth of the amplified PCR products on a 1% nondenaturing agarose gel. To prepare the gel, mix 1 g of electrophoresis-grade agarose (Gibco-BRL) in 100 mL of 1× TBE buffer and microwave for 1–1.5 min. Add 2 µL of 25 mg/mL ethidium bromide to solution and let cool to 65°C. Pour the agarose mix into a Horizon 58 minigel apparatus (*see* **Note 10**).

3.4. Construction of cDNA Plasmid Library and Transformation

1. Remove the contaminating universal primers from the remaining amplified cDNA products with the Promega PCR Prep kit (Promega).
2. Digest the purified amplified cDNA products with 15 µL of 10× React 3 buffer and 10 µL of restriction enzyme *Eco*RI (100 U) for 2 h at 37°C in a final volume of 150 µL.
3. At the same time, digest 500 ng of plasmid pGEM-7Z with *Eco*RI (10 U) for 2 h at 37°C in a final volume of 10 µL. Following digestion of pGEM-7Z with *Eco*RI, incubate the digested DNA with 10 µL of CIAP 10 × reaction buffer, 2 µL of CIAP, and DEPC water to a final volume of 100 µL for 30 min at 37°C. To stop the dephosphorylation reaction, add 2 µL of 0.5 *M* EDTA and heat at 65°C for 20 min.
4. Extract both *Eco*RI-digested DNA samples with an equal volume of phenol/chloroform (1:1).
5. Centrifuge at 10,000*g* for 5 min at 4°C.
6. Extract the aqueous phase with an equal volume of chloroform/isoamyl alcohol (24:1).
7. Centrifuge at 10,000*g* for 5 min at 4°C.
8. Precipitate the DNAs with 0.1 vol of 3 *M* sodium acetate and 2.5 vol of ethanol. Place the samples in a dry ice/ethanol bath for 30 min.
9. Centrifuge at 10,000*g* for 30 min at 4°C
10. Wash the pellets with 1 mL of 80% ethanol.
11. Centrifuge at 10,000*g* for 5 min at 4°C.
12. Dry the pellets in a Speed-Vac and reconstitute the samples in 10–12 µL of DEPC water.
13. For construction of the cDNA plasmid library, ligate 1–5 ng of the *Eco*RI-digested amplified cDNA with 100 ng of the *Eco*RI-digested, dephosphorylated

pGEM-7Z, 1.5 µL of 10× T4 DNA ligase buffer, and 1.5 µL of T4 DNA ligase in a final volume of 15 µL.
14. Incubate the ligation reaction at room temperature for 3–3.5 h. This sample can be stored at 4°C until used in the transformation reaction.
15. For the transformation reaction, carefully aliquot 100 µL of JM109 competent cells into a Falcon 2059 tube that already contains 5 µL of the ligation reaction and place the tube on ice for 30 min.
16. Heat shock the cells for 2 min at 42°C in a water bath with gentle movement.
17. Immediately place the tube in ice for 2 min.
18. Add 1 mL of SOC medium that had been prewarmed to 37°C.
19. Incubate for 1 h at 37°C with shaking (approx 150–200 rpm).
20. Plate 100 µL of each transformation culture onto LB plates containing 100 µg/mL ampicillin.
21. Incubate the plates overnight at 37°C.

3.5. Colony Screening

1. To identify insert-containing colonies, use the colony PCR method. Transfer a portion of each colony with a toothpick into 100 µL of TE.
2. Heat at 100°C for 10 min. Store the template at –20°C for future use.
3. For the PCR amplification reaction, incubate 10 µL of the template, 10 µL of 10× PCR buffer, 8 µL of dNTPs, 100 ng of SP6 promoter primer, 100 ng of T7 promoter primer, 1 µL of *Taq* DNA polymerase, and sterile water to a final volume of 100 µL. Overlay the reaction with 35 µL of mineral oil.
4. Amplify for 35 cycles at 94°C for 40 s, 50°C for 1 min, and 72°C for 2 min.
5. Extract the PCR sample with an equal volume of phenol/chloroform (1:1) and precipitate the DNA with 10 µL of 3 *M* sodium acetate and 750 µL of ethanol.
6. Analyze the PCR products on a 1% nondenaturing agarose gel.
7. For insert-containing colonies, perform a nucleotide sequence analysis using the fmol PCR sequencing method.

4. Notes

1. The goal of this procedure is to isolate and identify cellular RNAs in RNPs. Thus, to minimize the risk of RNA degradation during this procedure, it is absolutely critical that all of the reagents employed be of the highest molecular biology quality and RNase-free. Careful attention to detail must be paid to all of the technical aspects in this method. We have also found that sterile plasticware is preferable to autoclaved glassware in terms of minimizing the risk of RNase contamination.
2. Cultured cells are the preferred source of extracts from which to immunoprecipitate RNP complexes. Although tissues can be used, the levels of ribonuclease are usually high and may result in the rapid degradation of RNA during the extraction and immunoprecipitation step. With regard to the minimal amount of cells required for this method, it is best to start off with approx $1–5 \times 10^8$ cells. Once the whole cell extract has been isolated, we recommend proceeding immediately

on to the immunoprecipitation procedure. Although the extract can be stored at −80°C until future use, there is an increased risk for degradation of both RNA and protein to occur during storage.
3. In the immunoprecipitation step, two important control experiments are required. The first is to determine the optimal concentration of antibody that is required to completely immunoprecipitate the RNPs. The second is to determine the concentration of NaCl in the NET-2 washing step that will minimize the incidence of nonspecific immunoprecipitation and still allow for specific isolation of RNPs. It has been shown that stable RNP complexes are generally stable in the presence of high salt concentrations. We recommend performing a titration of NaCl from 100 m*M* up to 600 mM.
4. In addition to the control experiments outlined in **Note 3**, it is important to demonstrate the specificity of binding by a given protein to cellular RNAs. For these studies, an immunoprecipitation reaction using preimmune serum and/or an unrelated antiserum is required.
5. Carrier yeast tRNA and a potent RNase inhibitor such as Prime RNase Inhibitor from 5-Prime, 3-Prime are included to minimize the risk of RNA degradation during immunoprecipitation. We have found that their presence does not inhibit the efficiency of immunoprecipitation.
6. The amount of RNA present in the immunoprecipitated RNP may be in the submicromolar level. For this reason, glycogen is included in the ethanol precipitation step to enhance the efficiency of RNA precipitation.
7. Once the RNA has been isolated by immunoprecipitation, it can be stored at −80°C. However, we have found it best to proceed immediately with the reverse transcription step. Once in either the single-strand or double-strand cDNA form, the sample is stable, and it can be stored in −20°C until its future use in the PCR amplification reaction.
8. We have observed that the use of an RNase H(−) reverse transcriptase is critical for efficient and optimal reverse transcription of the isolated RNA sample. Our initial experiments employed the MMLV reverse transcriptase from Stratagene but we have subsequently found that the SuperScript RNase H⁻ II from Gibco-BRL gives equally good results. In addition, it is important to incubate the reverse transcription reaction at 42°C instead of 37°C, the usually recommended temperature, as no single-strand cDNA products have been obtained at 37°C. The reverse transcriptase enzymes from both Stratagene and Gibco-BRL are stable at 42°C.
9. One of the potential problems with the PCR amplification reaction is that some of the PCR DNA products may be partially single-stranded. To ensure the presence of homogeneous double-stranded DNA, a PCR chase reaction is performed to allow for proper denaturation of hairpin structures and heteroduplexes that may result from interactions of terminal amplification sequences. This step is necessary especially since the PCR products, once they are purified, are to be cloned into the pGEM-7Z plasmid.
10. Once the PCR chase reaction is performed, the DNA products should be analyzed on an agarose gel to confirm the success of the immunoprecipitation-

RNA:PCR method to isolate cellular RNP complexes. As revealed by ethidium bromide staining, a smear of amplification products should be observed usually ranging in size from 0.1 to 1.5 kilobases. For comparison, rPCR on total eukaryotic RNA yields DNA fragments approx 0.6 kilobases.

References

1. Gilmour, D. S. and Lis, J. T. (1984) Detecting protein-DNA interactions in vivo: distribution of RNA polymerase on specific bacterial genes. *Proc. Natl. Acad. Sci. USA* **81,** 4275–4279.
2. Gould, A. P., Brookman, J. J., Strutt, D. I., and White, R. A. H. (1990) Targets of homeotic gene control in Drosophila. *Nature* **348,** 308–311.
3. Dedon, P. C., Soults, J. A. Allis, C. D., and Gorovsky, M. A. (1991) Formaldehyde cross-linking and immunoprecipitation demonstrate developmental changes in H1 association with transcriptionally active genes. *Mol. Cell. Biol.* **11,** 1729–1733.
4. Bigler, J. and Eisenman, R. N. (1994) Isolation of a thyroid hormone-responsive gene by immunoprecipitation of thyroid hormone receptor-DNA complexes. *Mol. Cell. Biol.* **14,** 7621–7632.
5. Kessler, S. W. (1975) Rapid isolation of antigens from cells with a staphylococcal protein A-antibody adsorbent: parameters of the interaction of antibody-antigen complexes with protein A. *J. Immunol.* **115,** 619–1624.
6. Lerner, M. R. and Steitz, J. A. (1979) Antibodies to small nuclear RNAs complexed with proteins are produced by patients with systemic lupus erythematosus. *Proc. Natl. Acad. Sci. USA* **76,** 5495–5499.
7. Steitz, J. A. (1989) Immunoprecipitation of ribonucleoproteins using autoantibodies. *Methods Enzymol.* **180,** 468–481.
8. Matter, L., Schopfer, K., Wilhelm, J. A., Nyffenegger, T., Parisot, R. F., and DeRobertis, E. M. (1982) Molecular characterization of ribonucleoprotein antigens bound by antinuclear antibodies. A diagnostic evaluation. *Arthritis Rheum.* **25,** 1278–1283.
9. Froussard, P. (1993) rPCR: a powerful tool for random amplification of whole RNA sequences. *PCR Methods Appl.* **2,** 185–190.
10. Froussard, P. (1992) A random-PCR method (rPCR) to construct whole cDNA library from low amounts of RNA. *Nucleic Acids Res.* **20,** 2900.
11. Chu, E., Voeller, D. M., Jones, K. L., Takechi, T., Maley, G. F., Maley, F., Segal, S., and Allegra, C. J. (1994) Identification of a thymidylate synthase ribonucleoprotein complex in human colon cancer cells. *Mol. Cell. Biol.* **14,** 207–213.
12. Saiki, R., Gelfand, D., Stoffel, S., Saike, R., Gelfand, D., Stoffel, S., Scharf, S. J., Higuchi, R., Horn, G. T., Mullis, K. B., and Erlich, H. A. (1988) Primer-directed enzymatic amplification of DNA with a thermostable DNA polymerase. *Science* **239,** 487–491.
13. Davis, L., Kuehl, M., and Battey, J. (1994) Methods in Molecular Biology, 2nd ed. Appleton & Lange, Norwalk, CT, pp. 302–304.

21

Purification and Depletion of RNP Particles by Antisense Affinity Chromatography

Benjamin J. Blencowe and Angus I. Lamond

1. Introduction

Antisense oligonucleotides made of 2'-*O*-alkyl RNA are useful reagents for the study of the structure and function of ribonucleoprotein (RNP) particles. The nuclease resistant properties of these oligonucleotides, coupled with their ability to form specific and stable hybrids with targeted RNA sequences in crude cellular extracts, makes them well suited for applications involving the biochemical characterization of RNP particles. 2'-*O*-alkyl RNA oligonucleotides were initially used in "antisense masking" experiments to investigate the function of individual snRNA domains in the pre-mRNA splicing process *(1–3)*. The coupling of biotin residues to these oligonucleotides subsequently allowed their use as affinity "hooks" for the purification and removal of specific RNP particles from cellular extracts *(2,3)*. The purification of specific RNPs by this methodology has resulted in the identification of new RNP proteins *(4–10)*. The ability to deplete targeted RNPs has allowed the function of individual snRNP and non-snRNP splicing factors to be investigated *(11–17)*. The antisense affinity technology is thus a powerful alternative to conventional chromatographic methods and, in principle, is applicable to any RNP particle that contains a specific sequence accessible to oligonucleotide binding.

Subheading 3.1. of this chapter describes a method for the single-step antisense affinity selection (AAS) of snRNP proteins. The procedure is rapid and allows proteins associated with relatively abundant snRNPs to be detected starting from milligram quantities of unfractionated and unradiolabeled nuclear extracts. **Subheading 3.2.** describes an updated version of a method for the antisense-affinity depletion (AAD) of RNP particles that was previously published in this series *(18)*.

2. Materials
2.1 Purification of RNP Particles by AAS

Buffers should be prepared fresh and prechilled at 4°C. Pefabloc (Boehringer, Indianapolis, IN, USA) and dithiothreitol (DTT) should be added to the buffers immediately before use.

1. Oligonucleotides made of biotinylated 2'-*O*-methyl RNA can be prepared on a standard DNA oligonucleotide synthesizer *(19,20)*. 2'-*O*-methyl RNA phosphoramidites, which were primarily used to prepare oligonucleotides for the AAS and AAD procedures, can be purchased from Glen Research (Sterling, VA, USA). Biotinylated phosphoramidites for coupling on solid phase can also be purchased from Glen Research or Cambridge Research Biochemicals (Wilmington, DE, USA). An alternative biotinylation method is to incorporate an amino linker in the oligonucleotide and then couple biotin postsynthetically *(21)*. Reagents for postbiotinylation are obtained from Pierce (Rockford, IL, USA). An alternative to 2'-*O*-methyl RNA is to use 2'-*O*-allyl RNA oligonucleotides (Boehringer, Indianapolis, IN, USA), which have reduced non-specific protein-binding properties *(22)*.
2. Streptavidin agarose from Sigma (St. Louis, MO, USA) is recommended as it gives relatively low levels of non-specific binding.
3. Nuclear extracts should, preferably, be prepared from fresh grown HeLa cells (*23*; **Subheading 3.2.**). However, nuclear extracts can be prepared from purchased frozen cells. Two recommended suppliers are Cellex Biosciences (Minneapolis, MN, USA) and the Computer Cell Culture Centre (Mons, Belgium).
4. Dounce homogenizer (all glass-type) and pestles Type A and B from Wheaton (Millville, NJ, USA).
5. 1.0-mL Microdialysis cups (12,000 mol wt cutoff) are available from Sartorius (Edgewood, NY, USA).
6. 100 mM NaCl wash buffer (WB100): 0.1 M NaCl, 20 mM HEPES, pH 7.9, 0.05% Nonidet P-40 (NP40), 5 mM Pefablock (Boehringer, Indianapolis, IN, USA), and 0.5 mM DTT.
7. 250 mM NaCl wash buffer (WB250): same as WB100 except 250 mM NaCl.
8. 1 M MgCl$_2$.
9. 0.1 M ATP, pH 7.0 (Pharmacia, Piscataway NJ, USA).
10. 0.5 M Creatine phosphate, pH 7.0 (Sigma, St. Louis, MO, USA).
11. 3 L Chilled, sterile distilled H$_2$O.
12. 8 M Urea.
13. 1 M NaCl wash buffer (WB1000): same as WB250 except 1 M NaCl.

2.2. Depletion of RNP Particles by AAD

Refer to **Subheading 2.1.** for a description of the synthesis of biotinylated 2'-*O*-alkyl RNA oligonucleotides. Buffers for the preparation of snRNP-

Purification and Depletion of RNP Particles

depleted HeLa cell nuclear extracts should be made fresh and prechilled at 4°C. Buffers D, MD 0.1, and MD 0.6 should be prepared in 3-L amounts; the amount of buffers A and S will depend on the number of cells used for the nuclear extract preparation (*see* **Subheading 3.2.**). Phenylmethylsulfonyl fluoride (PMSF) and DTT should be added to the buffers immediately before use.

1. Dialysis tubing (purchased from Spectrum Medical Industries, Los Angeles, CA, USA) having 12,000–14,000 mol wt cutoff is needed in both 10 mm and 25 mm widths.
2. HeLa cells and dounce homogenizers (as in **Subheading 2.1.**)
3. Buffer A: 10 mM HEPES, pH 7.9, 1.5 mM MgCl$_2$, 10 mM KCl, and 0.5 mM DTT.
4. Buffer S: 20 mM HEPES, pH 7.9, 10% glycerol, 0.42 M KCl, 1.5 mM MgCl$_2$, 0.2 mM EDTA, 0.5 mM DTT, and 0.5 mM PMSF.
5. Buffer D: 20 mM HEPES, pH 7.9, 20% glycerol, 0.1 M KCl, 0.2 mM EDTA, 0.5 mM DTT, and 0.5 mM PMSF.
6. MD 0.1: buffer D containing 0.1 M KCl and 10% glycerol.
7. MD 0.6: buffer D containing 0.6 M KCl and 10% glycerol.
8. 250 mM KCl wash buffer (WB250): 20 mM HEPES, pH 7.9, 0.01% NP40, 0.5 mM DTT, and 250 mM KCl.
9. 0.1 M ATP, pH. 7.0 (Pharmacia, Piscataway NJ, USA)
10. 0.5 M creatine phosphate, pH. 7.0 (Sigma, St. Louis, MO, USA)
11. Streptavidin agarose "blocking buffer": Prepare by addition of 100 mg/mL of glycogen, 1 mg/mL bovine serum albumin (BSA) and 100 mg/mL of tRNA (final concentrations) to WB250 and streptavidin agarose.
12. 5% NP40: prepare in H$_2$O.

3. Methods
3.1. Purification of snRNPs

The steps described below are divided into two stages, which can be carried out on separate days: **Subheading 3.1.1.**, preparation of a precleared nuclear extract, and **Subheading 3.1.2.**, RNP selection.

3.1.1. Preparation of Precleared Nuclear Extract (PCNE).

Streptavidin agarose is notorious for its nonspecific binding properties. To avoid nonspecific binding of proteins during the AAS procedure, it is necessary to perform extensive preclearing steps using streptavidin agarose. Typically, preclearing is carried out on at least 5 mL of NE in 15-mL Falcon tubes. All steps should be carried out at 4°C, unless stated otherwise. Streptavidin agarose should be centrifuged at 1300g; avoid higher speeds as this may damage the agarose beads.

1. Prewash the streptavidin agarose by mixing with an equal volume of WB250. Centrifuge the beads and remove the buffer.

2. Add an equal volume of fresh WB250. Divide the beads into five aliquots, each aliquot corresponding to at least 0.6 volumes (=packed bead volume) of the total amount of (undiluted) NE to be precleared. For example, if preclearing 5 mL of NE, prepare five aliquots of streptavidin agarose, each containing at least 3 mL of packed beads. Centrifuge and remove all the buffer from the beads.
3. Add nuclear extract diluted 1:1 with WB100 to one of the aliquots.
4. Rotate tube slowly on a rotating wheel for 1 h.
5. Centrifuge for 2 min at 1300g.
6. Remove the supernatant and add it to a fresh aliquot of streptavidin agarose.
7. Repeat **steps 4–6** three more times (= five rounds of preclearing).
8. Centrifuge the PCNE twice to ensure complete removal of the streptavidin agarose.
9. If not proceeding to the next stage, the PCNE should be aliquoted (e.g., 1-mL aliquots), snap-frozen in liquid nitrogen, and stored at –80°C. Note: it is possible to regenerate the streptavidin agarose for future preclearing steps (*see* **step 3.3** below).

3.1.2. AAS (see Notes 1–4)

1. Quick-thaw the requisite number of PCNE aliquots by warming briefly at 30°C. AAS from 1 mL of PCNE is described.
2. In a 1.5-mL microfuge tube, prepare a 70-μL premix containing the following components: biotinylated antisense 2'-*O*-methyl RNA oligonucleotide (1–5 nmol; *see* **Note 9**), 15 μL of 0.1 M ATP, 10 μL of 0.5 M creatine phosphate, 2 μL of 1 M MgCl$_2$, H$_2$O to 70 μL.
3. Add the 70-μL premix to 0.9 mL of PCNE, and mix gently.
4. Add 30 μL of 5 M NaCl to bring the PCNE mixture to a final concentration of 250 mM NaCl.
5. Incubate at 30°C for up to 1 h (*see* **Note 8**).
6. Centrifuge the extract in a microfuge at 13,000g for 30 s to remove non-specific precipitate.
7. Transfer the supernatant to a fresh tube containing 0.15 mL of streptavidin agarose (packed bead volume) that has been prewashed twice in WB250.
8. Mix the suspension on a rotating wheel for 90 min at 4°C.
9. Centrifuge the streptavidin agarose in a microfuge at 1300g.
10. Remove the supernatant and resuspend the streptavidin agarose in 0.9 mL of WB250. Rotate for 5 min.
11. Repeat **steps 9** and **10** three times. Remove one tenth (~115 mL) of the extract-bead slurry and save for analysis of co-selected RNA. This will be necessary to determine the efficiency and specificity of the selection. Procedures for the analysis of affinity selected RNAs are described in **ref. 2**.
12. Centrifuge the remaining beads and remove the supernatant.
13. Resuspend the beads in 0.8 mL of 8 M urea. Rotate at room temperature for 30 min.
14. Centrifuge for 30 s at 9000g (10,000 rpm) in a microcentrifuge and remove the supernatant into a fresh tube.
15. Repeat **step 14** to ensure complete removal of the streptavidin agarose.

Purification and Depletion of RNP Particles

16. Remove the urea from the sample by dialyzing against 3 L of distilled H_2O in 1-mL Sartorius dialysis cups.
17. Concentrate the sample in a rotary Speed-Vac.
18. Resuspend the protein pellet in Laemmli loading buffer. Analyze by sodium dodecyl sulfate-polyacrylamide gel electrophoresis (SDS-PAGE) (*see* Chapter 31). Proteins can readily be detected by silver staining (*see* Chapter 3).

3.1.3. Regeneration of Streptavidin Agarose.

Streptavidin agarose can be regenerated for future use in pre-clearing.

1. Resuspend the streptavidin agarose used for preclearing in at least 5 vol of 8 *M* urea.
2. Rotate for 30 min.
3. Centrifuge and remove the urea supernatant.
4. Add several volumes of WB100.
5. Rotate for 10 min, then centrifuge and remove the supernatant.
6. Repeat **steps 4** and **5** three more times.
7. Store the beads in WB100 containing 0.01% NaN_3 at 4°C.
8. Supplement the regenerated beads with 1/10 vol of fresh streptavidin agarose before reuse.

3.2. Depletion of snRNPs

Set out below is a detailed step-by-step protocol for obtaining snRNP depletion from HeLa cell nuclear extracts. The protocol can be conveniently divided into two stages, carried out on separate days. Alternatively, it is possible to follow the entire protocol in one day, if cells for preparing nuclear extracts are already prepared. Specific conditions in the protocol, which directly affect the efficiency of snRNP depletion, are outlined in the **Notes** section. These general conditions should be adhered to if the protocol is to be applied to depleting RNPs from other types of in vitro assay system.

3.2.1. Preparation of "High Salt" Nuclear Extract (HSNE).

Prepare a nuclear lysate. The method we use for nuclear lysate preparation is essentially the same as that described by Dignam et al. *(23)*, but incorporates several changes designed to maximize the efficiency of snRNP depletion. All of the steps described below should be carried out at 4°C, unless stated otherwise.

1. Harvest HeLa cells in early-mid log phase. Alternatively, cells can be purchased as a frozen pellet (*see* **Materials**). Frozen cell pellets should be thawed quickly at 30°C, immediately before use.
2. Determine the packed cell volume (PCV). Wash by adding 5× PCV of buffer A and then collect the cells by centrifugation at 4°C for 10 min at 750*g* using a Sorvall SS34 rotor. Repeat this wash step once.

3. Resuspend the washed cell pellet in 2× PCV (i.e., 2× the original PCV) of buffer A. Transfer to a 40 mL Dounce homogenizer and lyse the cells by 10 strokes using an A-type pestle. Monitor cell lysis by viewing the lysate under an inverted microscope. Additional dounce strokes may be required to fully release nuclei. In our experience, as many as 20–30 dounce strokes may be required.
4. Transfer the lysed cell suspension to 50 mL Oakridge-type centrifuge tubes. Pellet the nuclei by centrifugation (at 4°C) for 5 min at 750g using a Sorvall SS34 rotor.
5. Discard the resulting supernatant. Take care not to disturb the pelleted nuclei. Centrifuge the nuclear pellet again, in the same rotor, for an additional 20 min at 25,100g.
6. After the second centrifugation, discard the supernatant and resuspend the nuclear pellet in buffer S. Use 4.5 mL of buffer S per 10^9 cells. This is one change we have made to the original Dignam et al. protocol *(23)*, where 3 mL of buffer C per 10^9 cells was used (*see* **Note 10**).
7. Transfer the resuspended nuclei to a 40-mL Dounce homogenizer. Lyse the nuclei by 10 strokes with a B-type pestle. Monitor the extent of nuclear lysis under an inverted microscope and apply additional Dounce strokes until intact nuclei are no longer observed; 20–30 strokes may be required for complete lysis of nuclei. Transfer the resulting lysate to 15-mL Falcon tubes and then rotate slowly for 30 min at 4°C.
8. After rotation, centrifuge the lysate in 50-mL Oakridge tubes (4°C) for 30 min at 25,100g in a Sorvall SS34 rotor. Transfer the resulting supernatant to 25-mm dialysis tubing.
9. Dialyze the nuclear lysate against 3 L (3 × 1 L changes) of MD 0.1 buffer for 3.5 h. This step should result in considerable precipitation from the extract.
10. Remove the precipitate by centrifugation in 15-mL Falcon tubes at 600g for 10 min in a benchtop centrifuge. Discard the precipitate.
11. Dialyze the "cleared" supernatant against 3 L (3 × 1 L changes) of MD 0.6 buffer for 1.5 h. This is the HSNE.
12. The HSNE can be quick-frozen in liquid nitrogen and stored at –80°C at this stage. It is convenient to store the HSNE in 1.0-mL aliquots.

3.2.2. AAD (see **Notes 5–15**).

1. If using frozen HSNE, thaw by warming briefly in a 30°C water bath.
2. Add to the HSNE: predetermined quantity of biotinylated antisense oligonucleotide (*see* **Note 9**), 1.5 mM ATP, 5 mM creatine phosphate, and 0.05% NP40 (final concentrations). At least 1.5 mL of HSNE should be used for depletion if regular dialysis tubing is to be employed, otherwise excessive dilution of the extract leading to a loss of activity may occur. Depletion of smaller volumes can be performed by using 1.0-mL microdialysis cups. We typically perform depletion of snRNPs starting from 2–5 mL of HSNE.

Oligonucleotide, ATP, creatine phosphate, and NP40 should be added in less than 10% of the total extract volume, also to avoid dilution. Thus, per mL of HSNE add: 16 μL of 0.1 M ATP, 11 μL of 0.5 M creatine phosphate, 10 μL of 5% NP40,

and oligonucleotide (in less than 40–50 µL). The amount of oligonucleotide will depend on the particular snRNP to be depleted (*see* **Note 9**).
3. Incubate extracts at 30°C for 30 min (U1 and U2 snRNPs) or 1 h (U4/U6 and U5 snRNPs).
4. Before or during incubation of the HSNE with oligonucleotide, streptavidin agarose should be prepared for the depletion step. Streptavidin agarose is first preblocked in "blocking buffer" (*see* **Subheading 2.2., step 11**) to saturate nonspecific binding sites. The volume of beads should correspond to about 50% of the extract volume: thus for 2 mL of extract to be depleted, use ~0.5 mL of streptavidin agarose (packed bead volume) per round of depletion. Prepare enough streptavidin agarose for two rounds of depletion (=1 mL of beads per milliliter of HSNE; *see* below). Preblock the beads by incubating with blocking buffer. The volume of blocking buffer should be ~30% the packed bead volume.
5. Remove the preblocking mixture by centrifugation (1300g) for 1 min in a benchtop microfuge, then wash the beads by rotating in 3 vol of fresh WB250. Repeat two more times. Aliquot beads into separate tubes for each depletion.
6. Remove all of the WB250 from the streptavidin agarose by centrifuging twice (1300g) for 1 min. This is important to avoid unnecessary dilution of the extract in **step 8**.
7. Add the nuclear extract after preincubation with biotinylated oligonucleotide to the pelleted streptavidin agarose. If depletions are being carried out on a 2-mL scale, it is convenient to divide the extract into two halves and incubate each 1 mL with 0.5 mL of pelleted streptavidin agarose in 2-mL Eppendorf tubes. The extract can later be repooled before the final dialysis step.
8. Incubate the nuclear extracts with streptavidin agarose for 45 min while rotating at 0°C. Note it is ***very*** important to carry out this incubation step while the Eppendorf tubes containing extract and streptavidin agarose are surrounded by a jacket of crushed ice. This can be achieved by placing Eppendorf tubes (sealed tightly with parafilm) in 50-mL Falcon tubes packed with crushed ice and rotating in a 4°C room. It should be noted that the combination of 0.6 M KCl and streptavidin agarose is detrimental to the extract. Quick handling at this step of the protocol is recommended.
9. After the 45-min incubation, remove the streptavidin agarose by centrifugation (1300g) for 1 min and then repeat **steps 8** and **9** with a fresh round of streptavidin agarose.
10. Remove the streptavidin agarose from the extracts by centrifuging twice (1300g, 1 min). Pool the depleted extracts if divided previously.
11. Dialyze the extract against D buffer (3 × 1 L changes) for 1.5–2 h; 10 mm width dialysis tubing is best suited for this purpose.
12. Aliquot the depleted extracts (e.g., 50-µL aliquots), snap-freeze in liquid nitrogen, and store at –80°C.

Assays for the characterization of snRNP depleted nuclear extracts have been described in detail previously (e.g., *see* **ref. 11**) and are not detailed here.

4. Notes

1. The AAS strategy was optimized for obtaining specific snRNP selection, rather than maximal selection efficiency. For this purpose, oligonucleotides made of 2'-*O*-methyl RNA containing inosine (instead of guanosine) to base pair with cytosine are recommended. It has been observed previously that incorporation of inosine into 2'-*O*-alkyl RNA oligonucleotides results in more specific selection than can be obtained with the corresponding guanosine-containing oligonucleotides. For further details on oligonucleotide design, *see also* **Note 6**.

2. The AAS method provides an alternative to procedures employing column and/or immunoaffinity chromatography for RNP purification. This is particularly useful in situations where RNP particles are difficult to fractionate or specific antibodies with high affinity are not available. However, it is important to consider that different procedures will yield different results and, if possible, should be compared in parallel. For example, specific RNP proteins may be disrupted by antibody and/or oligonucleotide binding, resulting in the appearance of a different protein profile. Furthermore, oligonucleotide binding may result in additional proteins being selected if, e.g., the antisense oligonucleotide-target RNA hybrid or the oligonucleotide itself binds to one or more proteins in the extract (e.g., *see* **Fig. 1**). An important control in this regard is to select from PCNE that has been pretreated extensively with RNase to remove the target RNA, prior to performing affinity selection. In general, the protein composition of RNP particles should be analyzed by as many different procedures as possible. It may also be informative to compare protein profiles obtained by the AAS procedure by antisense oligonucleotides complementary to different sequences within the targeted RNP.

3. The success of the AAS procedure will depend on several parameters, particularly the abundance of the targeted RNA and the specificity and efficiency of the selection. We have successfully applied the procedure to relatively abundant snRNPs involved in pre-mRNA splicing, including U1, U2, and U4/U6 snRNPs (**Fig. 1**; **ref. 4**; B. Blencowe, unpublished data). The procedure should also be applicable to less abundant snRNPs, keeping in mind that non-specific protein binding to streptavidin agarose may result in a reduced signal relative to background. Prior enrichment of the targeted RNA by density centrifugation or column fractionation may be necessary if high backgrounds are encountered. Streptavidin-coated magnetic beads or "neutravidin" beads (Pierce, Rockford, IL, USA) have reduced non-specific binding properties and may be suitable for performing selection from RNP-enriched fractions *(10)*. However, streptavidin-coated magnetic beads do not have as high a binding capacity as streptavidin agarose and are not recommended for the selection of abundant RNPs from unfractionated extracts.

4. The AAS procedure described in this chapter is designed to allow the rapid purification and analysis of proteins associated with targeted RNPs. If recovery of the native RNP particle in an active form is desired, it is feasible to elute the particle from streptavidin beads using a displacement oligonucleotide that effectively competes for hybridization to the 2'-*O*-methyl RNA oligonucleotide used

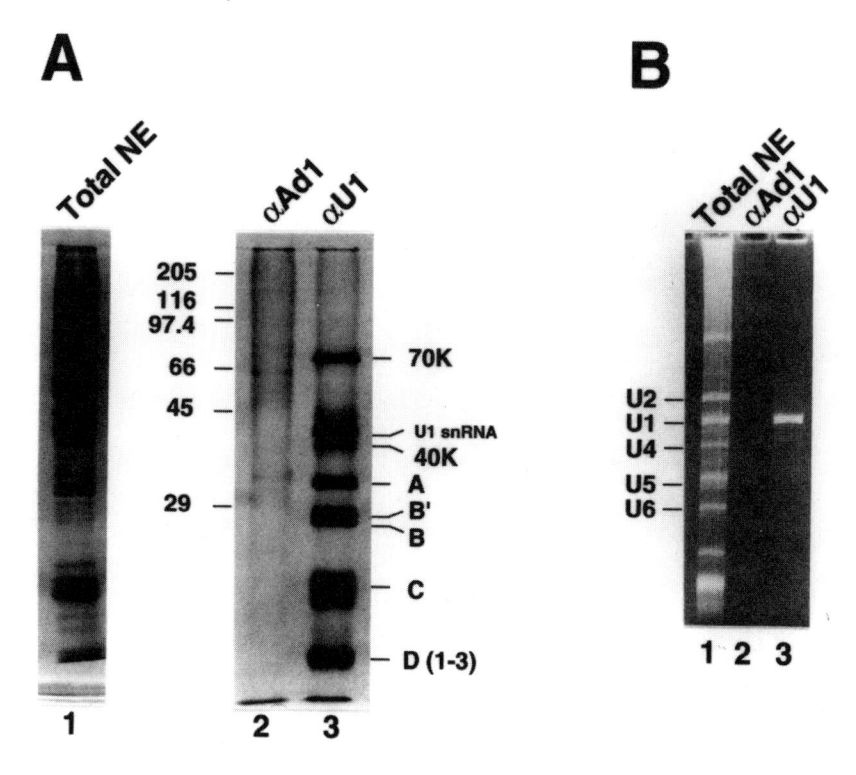

Fig. 1. Single-step purification of U1 snRNP by the AAS method. U1 snRNP was affinity selected from precleared nuclear extract using a biotinylated 2'-O-methyl RNA oligonucleotide complementary to U1 snRNA (residues 1–11). The antisense oligonucleotide contained inosine residues in place of guanosine (*refer to* **Note 1**) and two 3'-biotin residues. Protein recovered from streptavidin agarose beads following AAS was analyzed in a 12% SDS polyacrylamide gel stained with silver (**A**), and RNA recovered in parallel was analyzed in a 10% denaturing polyacrylamide-urea gel stained with ethidium bromide (**B**). Precleared nuclear extract prior to AAS is shown in *lanes 1*. Selection with a 14-mer, control, biotinylated 2'-O-methyl oligonucleotide (complementary to the adenovirus major-late intron) is shown in *lanes 2* and selection with the biotinylated anti-U1 snRNA oligonucleotide is shown in *lanes 3*. Size markers in panel A are indicated in kilodaltons. Note that the selection resulted in the recovery of U1 snRNA, as confirmed by RNase treatment (data not shown). Proteins of the apparent molecular masses expected for previously identified U1 snRNP-specific proteins (A, C and 70K) and common snRNP "Sm" proteins (B, B' and D-G) were detected. The identity of 70K and Sm proteins was confirmed by immunoblotting using specific antibodies (data not shown). An additional protein of 40 kDa was also detected. This latter protein binds specifically and directly to the anti-U1 snRNA oligonucleotide used for AAS, indicating that it is not an integral U1 snRNP component (B. Blencowe, unpublished observations).

for AAS. This approach has been successfully applied to the purification of active telomerase from Euplotes extracts *(10)*.
5. The AAD strategy described in this chapter was optimized for the depletion of targeted snRNPs from HeLa cell nuclear splicing extracts. To apply this procedure to a different type of in vitro assay system, RNP complex or cellular extract, it may be necessary to change certain conditions to preserve functional activity and optimize efficiency. **Notes 6–15** describe some key parameters in the protocol that influence depletion efficiency. These should be taken into consideration when modifying the protocol for a different system.
6. The efficiency of RNP depletion will depend on the ability of a stable hybrid to form between targeted RNA sequence and biotinylated 2'-*O*-alkyl RNA oligonucleotide. For this purpose, oligonucleotides containing guanosine rather than inosine should be used. As previously mentioned, oligonucleotides containing inosine give lower levels of non-specific RNP selection than corresponding guanosine-containing oligonucleotides, but are not as effective for obtaining high levels of depletion using the AAD protocol (*see* **Note 1**). For RNPs containing only limited regions of RNA accessible to oligonucleotide binding, it may be necessary to stabilize hybrid formation further by incorporating the modified nucleoside 2-aminoadenine. This modified base forms three hydrogen bonds with uracil and therefore is especially useful for targeting short A:U-rich sequences. Oligonucleotides containing 2-aminoadenine have allowed highly efficient depletion of U5 snRNP, which otherwise cannot be efficiently depleted using normal adenosine-containing 2'-*O*-alkyl RNA oligonucleotides *(12)*.
7. Following oligonucleotide binding, the efficiency of snRNP removal will depend on the accessibility of oligonucleotide-biotin residues to streptavidin agarose. This later step can be influenced to some extent by the location of biotin residues on the oligonucleotide *(3)*. Therefore, incorporation of biotin at both termini of the probe may be helpful.
8. Generally, oligonucleotide incubation times of up to 1 h are sufficient for depletion of spliceosomal snRNPs from HeLa cell nuclear extracts. Incubation times may be shorter for some RNPs and should ideally be determined empirically for each specific case.
9. Depletion of more abundant snRNPs (such as U1 and U2) requires 3–4 nmol oligonucleotide/mL extract *(11)*. whereas the less abundant snRNPs U4/U6 and U5 require 1–2 nmol of oligonucleotide *(11,12,14)*. The exact concentration of oligonucleotide is best determined by carrying out a small scale titration experiment reproducing the conditions described previously. Depletions are monitored by separating RNA isolated from nuclear extracts on denaturing polyacrylamide/urea gels (for examples and methods on how to characterize depletion levels, refer to Barabino et al., 1990 *(11*; *see also* Chapter 1; **Fig. 2**).
10. Increasing the volume of the nuclear suspension in the modified buffer C (= buffer S) serves to reduce the protein concentration of the nuclear lysate, facilitating more efficient snRNP removal. Increasing the buffer S volume of the nuclear lysate to 1.5× does not appear to significantly affect the specific activity of the

	Ctrl∆	U4/U6∆
U2 —	●	●
U1 —	●	●
U4 —	●	
U5 —	●	●
U6 —	●	‒
	1	2

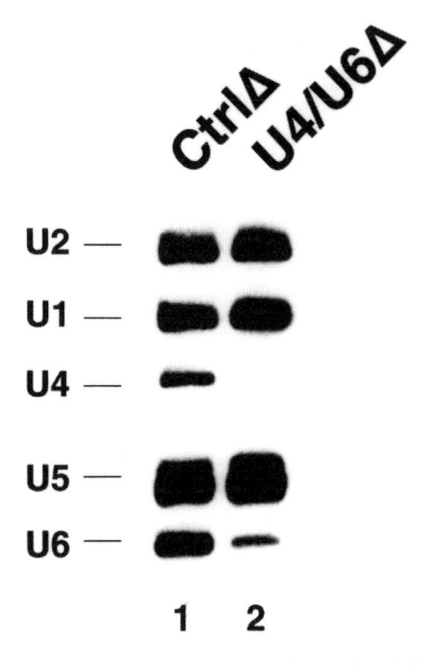

Fig. 2. Depletion of U4/U6 snRNP by the AAD method. RNA was isolated from mock-depleted and U4/U6 snRNP-depleted extracts, separated in a 10% polyacrylamide-urea gel, transferred to a nylon membrane, and detected by Northern hybridization using snRNA-specific riboprobes. *Lane 1* shows RNA recovered from a mock-depleted extract (no oligonucleotide used during AAD) and *lane 2* shows RNA recovered from an extract depleted of U4/U6 snRNP using a biotinylated 2'-O-methyl RNA oligonucleotide complementary to U6 snRNA (residues 82–101; *see* **ref. 11**). Note that this oligonucleotide quantitatively depletes the U4/U6 snRNP, but not a fraction of U6 that is unbound to U4. For additional information on the use of the AAD method for the depletion of spliceosomal snRNPs, refer to **refs. 11, 12,** and **14**.

nuclear extract for RNA processing activities such as pre-mRNA splicing.
11. Buffers containing KCl appear to work better for obtaining efficient snRNP depletion compared to buffers containing NaCl. Oligonucleotide binding in HeLa cell nuclear extracts is more efficient at relatively high KCl concentrations (0.5–0.6 M KCl).
12. A glycerol concentration higher than 10% markedly reduces the efficiency of depletion.
13. The time of incubation with streptavidin agarose will depend on the particular RNP to be depleted and must be determined empirically. Times longer than 45 min per round of streptavidin agarose incubation (i.e., 1.5 h total) should be avoided owing to the non-specific inhibitory effect of streptavidin agarose when incubated in the HSNE under high salt conditions.

14. The loss of functional activity due to non-specific inhibition can be attributed to several causes. It is most important to begin with fresh cells that give high levels of functional activity. This can be controlled for by preparing a normal "Dignam" nuclear extract *(18)*. in parallel with the depleted extracts. In general, manipulation times during the AAD protocol should be kept to a minimum, and dilution of the extract should be avoided. It is also important that samples to be frozen are "snap frozen" in liquid nitrogen.
15. When assaying for splicing activity, it is often worthwhile titrating the optimum concentration of $MgCl_2$, as this can vary between extract preparations. Optimized splicing assay conditions for snRNP depleted HeLa cell nuclear extracts have been described previously *(17)*.

Acknowledgments

We thank Dr. Brian Sproat for his many contributions to the development of the methods described in this chapter. We also thank Silvia Barabino for her contributions to the development of the AAD method and Ursula Ryder for proofreading the manuscript.

References

1. Lamond, A. I., Sproat, B., Ryder, U., and Hamm, J. (1989) Probing the structure and function of U2 snRNP with antisense oligonucleotides made of 2'-OMe RNA. *Cell* **58,** 383–390.
2. Blencowe, B. J., Sproat, B. S., Ryder, U., Barabino, S., and Lamond, A. I. (1989) Antisense probing of the human U4/U6 snRNP with biotinylated 2'-OMe RNA oligonucleotides. *Cell* **59,** 531–539.
3. Barabino, S. M., Sproat, B. S., Ryder, U., Blencowe, B. J., and Lamond, A. I. (1989) Mapping U2 snRNP-pre-mRNA interactions using biotinylated antisense oligonucleotides. *EMBO J.* **8,** 4171–4178.
4. Blencowe, B. J. (1991) The application of antisense technology to the study of mammalian pre-mRNA splicing factors. Ph.D. Thesis, University of London.
5. Wassarman, D. A. and Steitz, J. A. (1991) Structural analyses of the 7SK ribonucleoprotein (RNP), the most abundant human small RNP of unknown function. *Mol. Cell. Biol.* **11,** 3432–3445.
6. Groning, K., Palfi, Z., Gupta, S., Cross, M., Wolff, T., and Bindereif, A. (1991) A new U6 small nuclear ribonucleoprotein-specific protein conserved between *cis*- and *trans*-splicing systems. *Mol. Cell. Biol.* **11,** 2026–2034.
7. Palfi, Z., Gunzl, A., Cross, M., and Bindereif, A. (1991) Affinity purification of *Trypanosoma brucei* small nuclear ribonucleoproteins reveals common and specific protein components. *Proc. Natl. Acad. Sci. USA* **88,** 9097–9101.
8. Smith, H. O., Tabiti, K., Schaffner, G., Soldati, D., Albrecht, U., and Birnstiel, M. L. (1991) Two-step affinity purification of U7 small nuclear ribonucleoprotein. *Proc. Natl. Acad. Sci. USA* **88,** 9784–9788.
9. Wassarman, K. M. and Steitz, J. A. (1992) The low-abundance U11 and U12 small nuclear ribonucleoproteins (snRNPs). *Mol. Cell. Biol.* **12,** 1276–85.

10. Lingner, J., and Cech, T. R. (1996) Purification of telomerase from Euplotes aediculatus: requirement of a primer 3' overhang. *Proc. Natl. Acad. Sci. USA* **93**, 10,712–10,717.
11. Barabino, S. M., Blencowe, B. J., Ryder, U., Sproat, B. S., and Lamond, A. I. (1990) Targeted snRNP depletion reveals an additional role for mammalian U1 snRNP. *Cell* **63**, 293–302.
12. Lamm, G. M., Blencowe, B. J., Sproat, B. S., Iribarren, A. M., Ryder, U., and Lamond, A. I. (1991) Antisense probes containing 2-aminoadenosine allow efficient depletion of U5 snRNP from HeLa splicing extracts. *Nucleic Acids Res.* **19**, 3193–3198.
13. Wolff, T. and Bindereif, A. (1992) Reconstituted mammalian U4/U6 snRNP complements splicing: a mutational analysis. *EMBO J.* **11**, 345–359.
14. Blencowe, B. J., Carmo-Fonseca, M., Behrens, S. E., Luhrmann, R., and Lamond, A. I. (1993) Interaction of the human autoantigen p150 with splicing snRNPs. *J. Cell Sci.* **105**, 685–697.
15. Crispino, J. D., Blencowe, B. J., and Sharp, P. A. (1994) Complementation by SR proteins of pre-mRNA splicing reactions depleted of U1 snRNP. *Science* **265**, 1866–1869.
16. Segault, V., Will, C. L., Sproat, B. S., and Luhrmann, R. (1995) In vitro reconstitution of mammalian U2 and U5 snRNPs active in splicing: Sm proteins are functionally interchangeable and are essential for the formation of functional U2 and U5 snRNPs. *EMBO J.* **14**, 4010–4021.
17. Blencowe, B. J., Issner, R., Nickerson, J. A., and Sharp, P. A. (1998) A coactivator of pre-mRNA splicing. *Genes Dev.* **12**, 996–1009.
18. Blencowe, B. J. and Barabino, S. M. (1995) Antisense affinity depletion of RNP particles. *Methods Mol. Biol.* **37**, 67–76.
19. Sproat, B. S., Lamond, A. I., Beijer, B., Neuner, P., and Ryder, U. (1989) Highly efficient chemical synthesis of 2'-*O*-methyloligoribonucleotides. *Nucleic Acids Res.* **17**, 3373–3386.
20. Sproat, B. S. (1993) Synthesis of 2'-*O*-alkyloligoribonucleotides. *Methods Mol. Biol.* **20**, 115–141.
21. Pieles, U., Sproat, B. S., and Lamm, G. M. (1990) A protected biotin containing deoxycytidine building block for solid phase synthesis of biotinylated oligonucleotides. *Nucleic Acids Res.* **18**, 4355–4360.
22. Iribarren, A. M., Sproat, B. S., Neuner, P., Sulston, I., Ryder, U., and Lamond, A. I. (1990) 2'-*O*-alkyl oligoribonucleotides as antisense probes. *Proc. Natl. Acad. Sci. USA* **87**, 7747–7751.
23. Dignam, J. D., Lebowitz, R. M., and Roeder, R. G. (1983) Accurate transcription initiation by RNA polymerase II in a soluble extract from isolated mammalian nuclei. *Nucleic Acids Res.* **11**, 1475–1489.

22

Purification of U Small Nuclear Ribonucleoprotein Particles

Berthold Kastner and Reinhard Lührmann

1. Introduction

The spliceosome is the catalytic entity that removes the introns from the primary transcripts in eukaryotes. The spliceosome consists of four small nuclear ribonucleoproteins (snRNPs), named U1,U2, U4/U6, and U5 snRNP, and numerous non-snRNP proteins. Each snRNP consists of one (U1, U2, and U5 snRNP) or two (U4/U6 snRNP) RNA molecules and a large number of different proteins. Within the U4/U6 snRNP, U4 and U6 snRNA are basepaired together. Under conditions of in vitro splicing reactions in HeLa cell nuclear extracts (i.e., at about 150 mM salt concentration), the spliceosomal snRNPs are organized in three RNP forms: 12S U1, 17S U2, and 25S [U4/U6.U5] tri-snRNP. These snRNP complexes may be considered as functional subunits of the spliceosome.

The snRNPs from HeLa cells contain about 50 distinct proteins (**Fig. 1**). These can be classified into two groups: the common proteins, which are associated with each of the snRNPs U1, U2, U4, and U5, and the specific proteins that are present at one snRNP species only (for review *see* **refs.** *1–4*).

The procedures given here for the preparation of snRNPs have been in use in our laboratory for some years. Essential aspects of the methods to be described include the following:

1. The salt concentration must always be exactly right, because the binding of the specific proteins is frequently salt labile, as is the 17S U2 snRNP and the [U4/U6.U5] tri-snRNP complex.
2. All snRNPs possess a characteristic structural feature, an m$_3$G group ("cap") at the 5' end of the RNA. This allowed the development of chromatographic proce-

From: *Methods in Molecular Biology, Vol. 118: RNA-Protein Interaction Protocols*
Edited by: S. Haynes © Humana Press Inc., Totowa, NJ

NAME	app.M_R kDa	Presence in snRNP particles: 12S U1	12S U2	17S U2	20S U5	25S U4/U6.U5	12S U4/U6
G	9	○	○	○	○	○	○
F	11	○	○	○	○	○	○
E	12	○	○	○	○	○	○
D1	16	○	○	○	○	○	○
D2	16.5	○	○	○	○	○	○
D3	18	○	○	○	○	○	○
B	28	○	○	○	○	○	○
B′	29	○	○	○	○	○	○
69k/tls	69	◍					
C	22	●					
A	34	●					
70K	70	●					
B″	28.5		●	●			
A′	31		●	●			
	14			●			
	16			●			
	33			●			
	35			●			
	92			●			
SF3a	60			●			
SF3a	66			●			
SF3a	110			●			
SF3b	53			●			
SF3b	120			●			
SF3b	150			●			
SF3b	160			●			
	15				●	●	
	40				●	●	
	52				●		
	65				●	●	
	100				●	●	
	102				●	●	
	110				●	●	
	116				●	●	
	200				●	●	
	220				●	●	
	15.5					●	●
	20					●	●
	60					●	●
	90					●	●
	27					●	
	61					●	
	63					●	

Fig. 1. Protein composition of 12S U1 snRNP, 12S and 17S U2 snRNP, 20S U5 snRNP, the 25S [U4/U6.U5] tri-snRNP complex, and the 12S U4/U6 snRNP. The apparent molecular weights of the common proteins (*light dots*) and the specific proteins (*dark dots*) are indicated.

dures employing immobilized antibodies against the m_3G cap; these selectively bind the snRNPs, which are subsequently eluted by displacement with free m_3G or cross-reacting m^7G.
3. Despite their large size and very similar structures, the snRNPs can be separated from each other by high-performance liquid chromatography (HPLC) using the anion exchanger Mono Q.

2. Materials

2.1. Preparation of HeLa Nuclear Extract

1. HeLa cells grown in suspension culture.
2. Phosphate-buffered saline (PBS) Earle: 130 mM NaCl, 20 mM K_2HPO_4/KH_2PO_4, pH 7.4.
3. Buffer A: 10 mM HEPES-KOH, pH 8.0, 10 mM KCl, 1.5 mM $MgCl_2$, 0.5 mM 1,4-dithioerythritol (DTE).
4. Buffer C: 20 mM HEPES-KOH, pH 8.0, 420 mM NaCl, 1.5 mM $MgCl_2$, 0.5 mM DTE, 0.5 mM phenylmethylsulfonyl fluoride (PMSF), 0.2 mM EDTA, pH 8.0, 25% (v/v) glycerol.
5. Buffer G: 20 mM HEPES-KOH, pH 8.0, 150 mM KCl, 1.5 mM $MgCl_2$, 0.5 mM DTE, 0.5 mM PMSF, 5% (v/v) glycerol.
6. A 40-mL Dounce homogenizer with a tight-fitting (type S) pestle (Kontes glass).

2.2. Glycerol Gradient Centrifugation

1. Buffer C*: 20 mM HEPES-KOH, pH 8.0, 420 mM NaCl, 1.5 mM $MgCl_2$, 0.5 mM DTE, 0.5 mM PMSF, 0.2 mM EDTA, pH 8.0.
2. Buffer G: *see* **Subheading 2.1**.

2.3. Purification of U Small Nuclear Ribonucleoproteins by Anti-m₃G Immunoaffinity Chromatography

1. Anti-m_3G immunoaffinity column: about 7 mg of anti-m_3G IgG (Euro-Diagnostica, Arnhem, The Netherlands) is coupled to 1 mL of swollen CNBr-activated Sepharose by the standard procedure provided by Pharmacia (Piscataway, NJ, USA) (*see* **refs. 5–7**).
2. Buffer C: *see* **Subheading 2.1**.
3. Buffer C low: same composition as buffer C, but only 5% glycerol.
4. Buffer F: same composition as buffer C low, except that the concentration of NaCl is 250 mM.

2.4. Purification of U1 Small Nuclear Ribonucleoproteins by Mono Q Chromatography

1. Buffer Q-0: 20 mM Tris-HCl, pH 7.0, 1.5 mM $MgCl_2$, 0.5 mM DTE, and 0.5 mM PMSF.
2. Buffer Q-50: buffer Q-0 plus 50 mM KCl.

3. Buffer Q-1000: buffer Q-0 plus 1000 m*M* KCl.
4. A 1-mL Mono Q ion-exchange column (Pharmacia) mounted to a programmable chromatography system such as the Pharmacia FPLC-System.

2.5. Purification of the [U4/U6.U5] tri-snRNP Complex

1. An immunoaffinity column with the monoclonal H386 IgG-antibody (Euro-Diagnostica, Arnhem, The Netherlands).
2. A 32-mer peptide with the sequence DRDRERRRSHRSERERRRDRDRDRDR-DREHKR comprising the epitope of the H386 antibody. It can be chemically synthesized by standard procedures.
3. Buffer G: *see* **Subheading 2.1**.
4. Phosphate buffer: 10 m*M* sodium phosphate, pH 7.2.

2.6. Purification of 17S U2 Small Nuclear Ribonucleoproteins

1. Anti-m3G-immunoaffinity column: *see* **Subheading 2.3**.
2. Buffer G: *see* **Subheading 2.1**.

3. Methods

3.1. Preparation of HeLa Nuclear Extract

1. Allow HeLa S3 cells to grow in suspension culture in minimum essential medium for suspension cultures (S-MEM) supplemented with 5% (v/v) newborn calf serum, 50 µg/mL of penicillin, and 100 µg/mL of streptomycin at 37°C, keeping the cells at a density between 2.5 and 5×10^5/mL medium at logarithmic growth rate. To harvest sufficient amounts of U snRNPs, the number of cells has to be increased to 5×10^9 (corresponding to about 8–10 L of medium). Alternatively, Hela cells can be purchased commercially (e.g., 4C, Mons, Belgium) (*see* **Note 1**).
2. Harvest the cells by centrifugation at 4°C with a large scale swing-out rotor for 10 min at 1000*g*.
3. Resuspend the cells in PBS-Earle using 20 mL/10^9 cells and pellet again in a Sorvall HB4 rotor for 10 min at 1000*g*.
4. Determine the volume of the cell pellet and resuspend in 5 vol of buffer A.
5. Allow the cells to swell for 10 min, pellet again (*see* **step 3**), and resuspend in 2 vol of buffer A.
6. Break the cells by 10 strokes in the Dounce homogenizer.
7. Remove the cytoplasm by two successive centrifugations in a Sorvall SS 34 rotor for 10 min at 1000*g* and then 20 min at 25,000*g*.
8. Resuspend the pellet (cell nuclei) in buffer C; use 3 mL of buffer C per 10^9 cells.
9. Break the nuclei by 10 strokes in the Dounce homogenizer.
10. Transfer the resulting suspension into a beaker and stir carefully with a magnetic stirrer on ice for 30 min.
11. Remove the nuclear membrane by centrifugation in an SS 34 rotor at 25,000*g* for 30 min (*see* **Note 2**).

12. The resultant supernatant is the nuclear extract obtained under high-salt conditions (420 m*M*). The nuclear extract can be stored at –80°C after quick freezing in liquid nitrogen.

To obtain an extract active in splicing, dialyze to a lower salt concentration (150 m*M*) with 100 vol of buffer G for 4–5 h. Nuclear extracts prepared in this way are active when they make up approx 40–60% of the final volume in the usual splicing assay *(8)*; 5×10^9 cells yield about 18 mL of low-salt nuclear extract.

3.2. Glycerol Gradient Centrifugation

Owing to the different sizes of the snRNPs, glycerol gradient centrifugation can be used for separation of the various snRNP species (**Fig. 2**). Prefractionation of the nuclear extract by gradient centrifugation strongly increases the purity of the final snRNP product. For fractionation of the labile 17S U2 snRNP and the 25S [U4/U6.U5] tri-snRNP low-salt concentrations should be used in the gradients.

1. Pour 10–30% linear glycerol gradients at room temperature either with buffer C* ("high-salt gradient") or buffer G ("low-salt gradient") using sterilized SW 28 tubes. By using the BioComp Gradient Master (Fredericton, NB, Canada) linear gradients can easily be obtained in a very fast and reproducible manner.
2. Store the gradients to even out any irregularities for at least 1 h (maximum overnight) at 4°C.
3. Load the nuclear extract onto the gradient carefully (use an Eppendorf pipet). It is possible to load up to 8 mL of extract onto a 30-mL SW 28 gradient.
4. Start the centrifugation using a low acceleration rate and run it for 17 h at 27,000 rpm ($10^5 g$). Stop the centrifugation without the brake.
5. Harvest the gradients with a automatic fractionation device, or manually in 1.5-mL fractions from top to bottom, using an Eppendorf pipet.
6. To analyze the U snRNP content, take one tenth of each fraction, extract with 1:1 phenol/chloroform, and precipitate the RNA from the aqueous phase by adding 0.1 vol of sodium acetate (3 *M*, pH 4.8) and 3 vol of ethanol. Check the RNA by gel electrophoresis (*see* Chapter 1) followed by silver staining using standard procedures *(5)*.

3.3. Purification of U Small Nuclear Ribonucleoproteins by Anti-m₃G Immunoaffinity Chromatography

1. Equilibrate an anti-m₃G immunoaffinity column by washing with about 5 column volumes of buffer C. For 15 mL of nuclear extract, a column with a bed volume of at least 5 mL must be used to achieve sufficient retardation of U snRNPs (*see* **Notes 3** and **4**).
2. Apply nuclear extract prepared under high-salt conditions to the affinity column at about 1.5 mL/h.

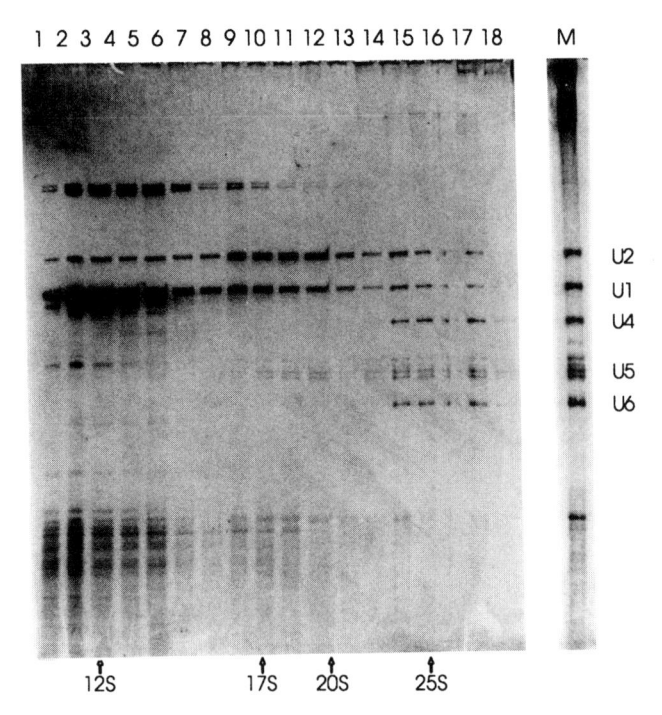

Fig. 2. Sedimentation of U snRNP particles. HeLa cell nuclear splicing extract was sedimented in a linear, 10–30% glycerol gradient at low salt concentrations from left to right. Small nuclear RNAs in each fraction were isolated as described in the text and analyzed on a 10% polyacrylamide-urea gel. The RNA is visualized by silver staining *(5)*. The positions of some sedimentation coefficients are indicated. M is a U snRNA marker; about 5 µg of snRNA is loaded. HeLa snRNAs are 189 (U2), 165 (U1), 146 (U4), 103–122 (U5), and 106 (U6) nucleotides long.

3. Elute nonspecifically bound components of the extract with about 6 column volumes of buffer C low.
4. Elute the specifically bound U snRNPs using 2 column volumes of 15 mM m^{7}G nucleoside (Pharmacia) dissolved either in high-salt buffer (buffer C low) or at more moderate salt concentrations (buffer F). If buffer F is used, the U4, U5, and U6 snRNAs can be eluted as complete [U4/U6.U5] tri-snRNP complexes. From 15 mL of nuclear extract, about 4 mg of U snRNPs (determined by the method of Bearden *[9]*) can be isolated. The snRNPs can be stored at –80°C after quick freezing in liquid nitrogen.
5. Remove the antibody-bound m^{7}G by passing 1 column volume of a solution of 6 M urea in buffer C low over the column.

6. Regenerate the affinity column by washing with 20 column volumes of buffer C low. For long-term storage, add NaN$_3$ to give a final concentration of 0.02% (v/v).

3.4. Purification of U1 Small Nuclear Ribonucleoproteins by Mono Q Chromatography

1. Wash the chromatographic system, which includes a 50-mL "superloop" and a Mono Q HR 5/5 column (1-mL bed volume), with a volume of buffer Q-1000 equal to 20× the total volume of the system.
2. Wash and equilibrate the column with the same volume as in **step 1** with buffer Q-50. Determine the absorbance at 280 nm. The value obtained is the zero point for subsequent absorbance measurements.
3. Dilute the U snRNPs obtained by anti-m$_3$G chromatography with buffer Q-0 so as to bring the concentration of univalent ions below 200 mM.
4. Load U snRNPs (1–40 mg) onto the Mono Q column, using the superloop, with a flow rate of 2 mL/min. The pressure should not exceed 3.0 MPa.
5. Wash with buffer Q-50 until the absorbance reaches zero (usually 2 column volumes).
6. Elute the snRNPs at a flow rate of 1 mL/min. Program the FPLC-controller for the following KCl gradient. Start with 50 mM KCl (buffer Q-50). Increase the amount of buffer Q-l000 by 5.4% per minute for 4 min, 1% per minute for 30 min, and then 4.2% per minute for 10 min. Finally, elute the column with pure buffer Q-1000 for 4 min. During the elution, collect 1-mL fractions. The U1 snRNPs elute in the first main peak at 350–370 mM KCl.
7. Determine the concentration of U1 snRNPs in the fractions by measurement of their absorbance at 280 nm (1 A_{280} unit is about 0.35 mg/mL) and analyze the RNA and protein content by standard PAGE procedures (*see* Chapters 1, 23, and 31) *(5)*.
8. Contaminating 20S U5 snRNPs can be separated from the 12S U1 snRNPs by glycerol density gradient centrifugation immediately after the Mono Q chromatography (*see also* **ref. 10**).

3.5 Purification of the [U4/U6.U5] tri-snRNP Complex

We use two methods for isolation of the [U4/U6.U5] tri-snRNP complex. One procedure involves the fractionation by glycerol gradient centrifugation of total snRNPs as obtained by immunoaffinity chromatography with the cap-specific antibody. This relatively fast method results in good yields but less pure tri-snRNP particles. The second procedure employs a second immunoaffinity chromatography column with the H386 antibody. This antibody was originally raised against the U1 snRNP-specific 70K protein *(11)*, but it also exhibits strong cross-reactivity with a 100-kDa protein that is a component of the [U4/U6.U5] tri-snRNP complex *(12)*. This method produces relatively pure tri-snRNP complexes but with more effort and lower yields than the first method. In the following protocol the immunaffinity chromatography with the immobilized H386 antibody is described (*see* **Note 5**).

1. Pool the 25S fractions from a nuclear extract fractionated on a glycerol gradient and pour slowly (1 mL/min) onto an H386 immunoaffinity column. Up to 7 mL, corresponding to four fractions, containing about 100–150 μg of U snRNPs, can be loaded onto a column of 2-mL bed volume.
2. Elute nonspecifically bound components of the extract with about 20 column volumes of buffer G.
3. Elute the specifically bound U snRNPs with 5 column volumes of a 0.01 m*M* solution in buffer G of the competing 32-mer peptide which corresponds to the primary epitope of H386 (*see* **ref. 12**). Collect 500-μL fractions and analyze one-tenth of each fraction for RNA and protein content using standard procedures (*see* Chapters 1, 23, and 31).
4. If necessary, separate minor amounts of coretarded and coeluted 12S U1 snRNPs from the 25S [U4/U6.U5] tri-snRNP complexes by glycerol gradient centrifugation as described in **Subheading 3.2.**
5. Elute the antibody-bound peptide with 5 column volumes of phosphate buffer (10 m*M* sodium phosphate, pH 7.2) and then with 5 column volumes of 3.5 *M* MgCl$_2$ in the same buffer.
6. Regenerate the affinity column by washing with 10 column volumes of buffer G. For long-term storage, add NaN$_3$ to give a final concentration of 0.02% (w/v).

3.6. Purification of 17S U2 Small Nuclear Ribonucleoproteins

The 17S U2 snRNP is difficult to isolate with its complete set of proteins. This is because most of the U2-specific proteins dissociate from the U2 snRNP particle at salt concentrations above 200 m*M*, and, in addition, the m$_3$G cap of the U2 RNA is much less accessible for antibody binding in the 17S form of the U2 snRNP. Therefore, all experimental steps must be performed at low ionic strength and, prior to immunoaffinity isolation of 17S U2 snRNP with the cap-specific antibody, most of the other snRNPs with well-exposed cap structures should be removed *(13)*. For isolation from nuclear extract a prefractionation of 17S U2 snRNP is done by glycerol gradient centrifugation.

1. Pool the 17S fractions from a glycerol gradient-fractionated nuclear extract. Remove small quantities of contaminating 12S U1 snRNPs, U5 snRNPs, and [U4/U6.U5] tri-snRNP complexes by passage over an H386 immunoaffinity column.
2. Apply the H386 flow-through, which contains the 17S U2 snRNPs, to an anti-m$_3$G immunoaffinity column.
3. Elute nonspecifically bound components with 20 column volumes of buffer G.
4. Elute the 17S U2 snRNPs with m^{7}G and analyze as described in **Subheading 3.3**.
5. Regenerate the affinity columns as described in **Subheading 3.3**.

4. Notes

1. For the preparation of nuclear extracts and the purification of U snRNPs, the following precautions are essential, to keep all solutions and glassware free of

RNase contamination and activity. All glassware must be washed thoroughly, rinsed extensively with distilled water, dried thoroughly, and then heated for at least 1 h at 250°C. All buffers and solutions should be autoclaved and then stored, if appropriate, at 4°C. Sterile gloves should be worn during all steps. All steps should be performed at 4°C unless otherwise specified.
2. Nuclear extract prepared according to **protocol 3.1** contains lipids that tend to clog chromatography columns. Most of the lipid vesicles can be removed by ultracentrifugation at 165,000g for 20 min in a Beckman Ti 70 rotor and subsequent filtration through a 5-µm and a 1.2-µm membrane filter. The lipid layer floating on top after ultracentrifugation is very instable and mixes readily with the cleared nuclear extract directly below it. This can be prevented by pipetting hexane on top of the extract before centrifugation and then collecting the nuclear extract after centrifugation with a syringe equipped with a long needle.
3. Immunoaffinity purification protocols are very sensitive to any kind of variations in the pH and salt concentration. For this reason, the pH of all buffers should be monitored very carefully, and buffer solutions should be checked regularly for absence of any precipitate.
4. The following precautions should be taken using immunoaffinity columns. For long-term storage of the antibody columns, NaN_3 must be added to the storage buffer. Antibodies must never be incubated with denaturing solutions such as urea or 3.5 M $MgCl_2$ for longer than 4–5 h; shorter periods are preferable. Affinity columns should always be kept at 4°C.
5. For immobilization of the H386 antibody, which is a monoclonal antibody of the IgM subclass, incubate approx 2 mg of an IgG-antibody specific for mouse IgM (Sigma) with 1 mL packed volume of Protein A–Sepharose. After removing unbound antibodies by washing with 0.1 M Tris-HCl, pH 8.0, incubate with 4 mg of H386 IgM antibody per 1 mL of packed Sepharose beads. Wash with 0.2 M sodium borate, pH 9.0, and covalently attach the antibodies to the Protein A–Sepharose by adding 10 vol of 0.2 mM dimethylpimel imidate (DMP). After stopping the coupling reaction with 0.2 M ethanolamine-HCl, pH 8.0, and washing with PBS, the affinity matrix can be transferred to a column *(5)*.

References

1. Moore, M. J., Query, C. C., and Sharp, P. A. (1993) Splicing of precursors to messenger RNAs by the spliceosome, in *RNA World* (Gesteland, R. F. and Atkins, J. F., eds.), Cold Spring Harbor Press, Cold Spring Harbor, NY, pp. 303–358.
2. Krämer, A. (1996) The structure and function of proteins involved in mammalian pre-mRNA splicing. *Annu. Rev. Biochem.* **65,** 367–409.
3. Lührmann, R., Kastner, B., and Bach, M. (1990) Structure of spliceosomal snRNPs and their role in pre-mRNA splicing. *Biochim. Biophys. Acta Gene Struct. Expression* **1087,** 265–292.
4. Will, C. L. and Lührmann, R. (1997) snRNP structure and function, in *Eukaryotic mRNA Processing* (Krainer, A. R., ed.), IRC Press, Oxford, pp. 130–173.

5. Will, C. L., Kastner, B., and Lührmann, B. (1993) Analysis of ribonucleoprotein interactions, in *RNA Processing — A Practical Approach*, vol. 2 (Higgens, S. J. and Hames, B. D., eds.), IRL Press, Oxford, pp. 141–177.
6. Bochnig, P., Reuter, R., Bringmann, P., and Lührmann, R. (1987) A monoclonal antibody against 2,2,7-trimethylguanosine that reacts with intact U snRNPs as well as with 7-methylguanosine capped RNAs. *Eur. J. Biochem.* **168,** 461–467.
7. Bringmann, P., Rinke, J., Appel, B., Reuter, R., and Lührmann R. (1983) Purification of snRNPs U1, U2, U4, U5 and U6 with 2,2,7-trimethylguanosine-specific antibody and definition of their constituent proteins reacting with anti-Sm and anti-(U1)RNP antisera. *EMBO J.* **2,** 1129–1135.
8. Krainer, A. R., Maniatis, T., Ruskin, B., and Green, M. (1984) Normal and mutant human β-globin pre-mRNAs are faithfully and efficiently spliced in vitro. *Cell* **36,** 993–1005.
9. Bearden, J. C. (1978) Quantitation of submicrogram quantities of proteins by an improved protein-dye binding assay. *Biochim. Biophys. Acta* **553,** 525–529.
10. Bach, M., Winkelmann, G., and Lührmann, R. (1989) 20 S small nuclear ribonucleoprotein U5 shows a surprisingly complex protein composition. *Proc. Natl. Acad. Sci. USA* **86,** 6038–6042.
11. Reuter, R. and Lührmann, R. (1986) Immunisation of mice with purified U1 small nuclear ribonucleoprotein (RNP) induces a pattern of antibody specificities characteristic of the anti-Sm and anti-RNP autoimmune response of patients with lupus erythematosus, as measured by monoclonal antibodies. *Proc. Natl. Acad. Sci. USA* **83,** 8689–8693.
12. Behrens, S. E. and Lührmann, R. (1991) Immunoaffinity purification of a [U4/U6. U5] tri-snRNP complex from human cells. *Genes Dev.* **5,** 1429–1452.
13. Behrens, S. E., Tyc, K., Kastner, B., Reichelt, J., and Lührmann, R. (1992) Small nuclear ribonucleoprotein (RNP) U2 contains numerous additional proteins and has a bipartite RNP structure under splicing conditions. *Mol. Cell. Biol.* **13,** 307–309.

23

Preparation of Heterogeneous Nuclear Ribonucleoprotein Complexes

Maurice S. Swanson and Gideon Dreyfuss

1. Introduction

Proper formation of different types of ribonucleoprotein (RNP)–RNA complexes is critical for the correct expression of genetic information at the posttranscriptional level. In the nucleus, RNA polymerase II transcripts bind to an abundant class of nuclear pre-mRNA/mRNA-binding proteins, the heterogeneous nuclear ribonucleoproteins (hnRNPs), to form hnRNP complexes *(1,2)*. The molecular architecture of these complexes must be exceedingly dynamic because each step in the pre-mRNA processing pathway alters the structure of the underlying RNA substrate. Therefore, hnRNP complexes within the cell, although probably unique for each transcript, are not uniform structures, such as a ribosomal subunit or a particular small nuclear ribonucleoprotein particle (snRNP), but are composed of a heterogeneous mixture of protein–RNA structures.

1.1. Principle of the Immunopurification Procedure to Isolate hnRNA Complexes

HnRNP complexes have been isolated using several different experimental techniques, including sucrose gradient sedimentation *(1)*. However, here we will discuss the use of anti-hnRNP monoclonal antibodies to immunopurify hnRNP complexes directly from ^{35}S-labeled cell lysates *(3)*. This immunopurification procedure is based on the observation that hnRNP complexes are held together primarily by protein–RNA interactions, and not protein–protein interactions, as nuclease treatment results in complex dissociation *(1)*. Therefore, the immunopurification procedure described in **Subheading 3.2.** is ideally suited to the isolation of hnRNP complexes associated with large pre-mRNAs/

mRNAs because it is specific, rapid, minimizes RNA degradation and proteolysis, and is readily adapted to almost any cell type. Although we will only describe the isolation and characterization of human hnRNP complexes from HeLa cells, other human cell lines as well as nonmammalian cell cultures have been used successfully to isolate hnRNP complexes *(5)*.

1.2. Examples of hnRNP Complexes from Human Cells

Following immunopurification of hnRNP complexes using an anti-hnRNP A1 monoclonal antibody (4B10), the protein composition of these complexes can be examined by sodium dodecyl sulfate-polyacrylamide gel electrophoresis (SDS-PAGE). This analysis reveals that hnRNP complexes are composed of numerous proteins (**Fig. 1**, hnRNP complex). These complexes are purified in the presence of a nonionic detergent that minimizes disruption of strong protein–protein and protein–RNA interactions. In contrast, if strong ionic detergents are used for immunopurification, then the predominant protein detected is the hnRNP A1 protein (**Fig. 1**, hnRNP A1). Analysis of the hnRNP complex by two-dimensional gel electrophoresis illustrates that this large macromolecular structure is composed of a complex array of many major (≥ 20) and minor proteins (**Fig. 2**). Individual groups of hnRNPs (hnRNP A1-U) were originally categorized based on their nonequilibrium pH gel electrophoresis (NEPHGE) migration properties *(3)*. Subsequent isolation of individual hnRNPs and cDNA/gene sequencing has confirmed that most of the proteins within each group are transcribed from a single gene, and the multiple protein isoforms within each group are generated by pre-mRNA splicing and/or post-translational modification *(1,2)*.

2. Materials

2.1. Cells and Metabolic Labeling of Proteins

1. HeLa S3 cell growth: Dulbecco's modified Eagle medium (DMEM; Gibco-BRL, Gaithersburg, MD, USA) supplemented with 10% calf serum and 1% penicillin/streptomycin. Use 10 mL per 10-cm plate.
2. HeLa S3 cell labeling: DMEM minus methionine and cystine (Gibco-BRL), 10% DMEM, 5% fetal calf serum (FCS), 1% penicillin/streptomycin, and 20 µCi/mL of ^{35}TRANS (ICN, Costa Mesa, CA, USA).

2.2. Antibodies and Protein A-Sepharose

1. Antibodies: Any monoclonal antibody (mAb) that is monospecific, and has a high affinity, for a characterized hnRNP protein may be employed for immunopurification. Monoclonal antibodies that have been used successfully to immunopurify human hnRNP complexes include 4B10 (anti-hnRNP A1), 4F4 (anti-hnRNP C1/C2), and 4D11 (anti-hnRNP L) *(3,4)*. These antibodies are not available commercially, but may be obtained by contacting the authors.
2. Protein A–Sepharose (Pharmacia Biotech, Piscataway, NJ, USA) (*see* **Note 1**).

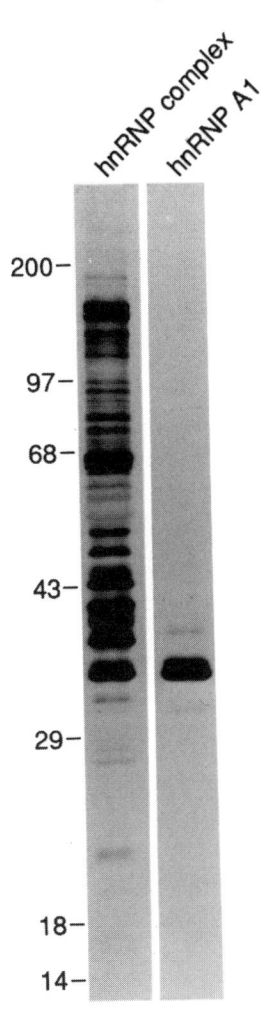

Fig. 1. One-dimensional SDS-PAGE analysis of immunopurified hnRNP complexes and the hnRNP A1 protein.

2.3. Immunopurification Buffers

1. Immunopurification buffer 1 (IPB1): RSB–100 (10 mM Tris-HCl, pH 7.4, 100 mM NaCl, 2.5 mM MgCl$_2$), 0.5% Triton X-100, 1 µg/mL leupeptin, 1 µg/mL pepstatin, 0.5% (v/v) aprotinin (see **Note 2**).
2. Immunopurification buffer 2 (IPB2): same as IPB1 but omit Triton X-100.
3. Immunopurification buffer 3 (IPB3): phosphate-buffered saline (PBS), 1 mM EDTA, 1% Empigen BB (Calbiochem, La Jolla, CA, USA), 0.1 mM dithiothreitol (DTT) (see **Note 3**).

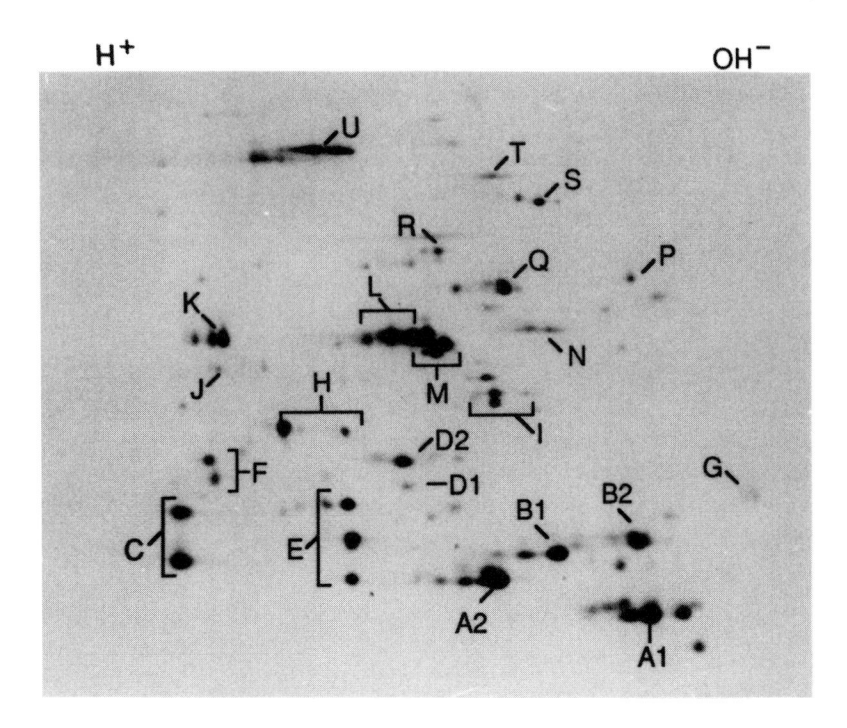

Fig. 2. Two-dimensional NEPHGE gel of isolated hnRNP complexes. Individual hnRNPs (A1–U) are indicated.

4. Immunopurification buffer 4 (IPB4): PBS, 1 mM EDTA, 1% Triton X-100, 0.5% deoxycholic acid, 0.1% SDS, 0.5% aprotinin. Make IPB4 as a 5× stock and store at 4°C for <1 mo.

2.4. One-Dimensional Gel Electrophoresis

1. Reagents: 10% (w/v) SDS (autoclave and store at room temperature); 3% (w/v) ammonium persulfate (APS, store in 1-mL aliquots at –20°C); *N,N,N'N'*-tetramethylenediamine (TEMED).
2. Acrylamide stock for the running gel: dissolve 167.5 g (33.5%) of acrylamide and 1.5 g (0.3%) of *bis*-acrylamide (w/v) in 450 mL of water, stir until dissolved, bring up to 500 mL with water, and filter through Whatman no. 1 paper. Store in an amber bottle at 4°C. Stable for at least two months.
3. Running buffer: 1 M Tris-HCl, pH 9.1. Store at 4°C. Stable for at least 3 mo.
4. Acrylamide stock for the stacking gel: dissolve 60 g (30%) of acrylamide and 0.88 g (0.44%) of *bis*-acrylamide (w/v) in 175 mL of water, stir until dissolved, bring up to 200 mL with water, and filter through Whatman no. 1 paper. Store in an amber bottle at 4°C. Stable for at least 2 mo.
5. Stacking buffer: 0.5 M Tris-HCl, pH 6.8. Store at 4°C. Stable for at least 3 mo.

6. 2× Loading buffer: 5 mL of 0.5 M Tris-HCl, pH 6.8, 8 mL of 10% SDS, 4 mL of redistilled glycerol, 2 mL of β-mercaptoethanol, 2 mg of bromophenol blue; bring to 20 mL with water. Store at –20°C in a non-frost-free freezer. Stable for at least 4 mo.
7. 4× Electrophoresis tank buffer: 96 g of Tris base, 460.8 g of glycine, 32 g of SDS dissolved in water to a final volume of 8 L. Store carboy at room temperature.
8. Apparatus: Any vertical electrophoresis slab gel unit (such as the Hoefer SE400, Pharmacia Biotech) will work.

2.5. Two-Dimensional Nonequilibrium pH Gel Electrophoresis

1. Reagents: ultrapure urea (ICN, Aurora, OH, USA); 10% Nonidet P-40 (NP40) (Sigma); 40% Bio-Lyte 3/10 ampholytes (Bio-Rad Laboratories, Hercules, CA, USA); 10% APS; TEMED.
2. Isofocusing acrylamide stock: Dissolve 28.38 g of acrylamide and 1.62 g of *bis*-acrylamide in 90 mL of water by stirring and then bring volume up to 100 mL. Filter through Whatman no. 1 paper and store in an amber bottle for up to 3 mo.
3. Top chamber solution — 0.01 M phosphoric acid: Add 1 mL of 85% phosphoric acid to 1470 mL of water.
4. Bottom chamber solution — 0.02 M NaOH: Add 12 mL of 1 M NaOH to 600 mL of water and degas for 5 min using a house or pump vacuum.
5. Sample buffer: Combine 1.43 g of ultrapure urea, 0.5 mL of 10% Nonidet P-40, and 0.125 mL of 40% ampholines, and add water to bring the volume to 2.93 mL. Store 100-µL aliquots at –70°C up to 1 yr.
6. 95% Sample extrusion buffer: Mix 12.7 mL of 0.5 M Tris-HCl, pH 6.8, 8 mL of glycerol, 10 mL of 0.1% bromophenol blue, and 20 mL of 10% SDS. Bring the volume to 95 mL with water and store at room temperature. Immediately before use, add β-mercaptoethanol to 5% final concentration.
7. Other equipment: tube gel electrophoresis apparatus (Hoefer GT1 tube gel unit, Pharmacia Biotech), slab gel electrophoresis apparatus, test tube rack (use a 13 × 100 mm rack with three rungs if available), 200-µL Dade Accupettes (place a mark at 11.5 cm from the bottom), tuberculin syringe, modeling clay.

3. Methods
3.1. Preparation of Nucleoplasm

1. Grow HeLa cells to subconfluency. For labeling with ^{35}S-methionine, remove the growth medium and wash the plates once with sterile PBS. Add 5 mL of cell labeling medium per 10-cm plate, and allow cells to grow overnight (8–12 h). All subsequent procedures are performed at 4°C using 1.5-mL microcentrifuge tubes. Each immunopurification requires approximately one half plate of cells.
2. Wash the plates twice with 5 mL of PBS per 10-cm plate, and following the addition of 1.5 mL of IPB1, scrape thoroughly with a rubber policeman. Pass the cell suspension four times through a 25-gage needle attached to a tuberculin syringe. Avoid foaming the suspension (*see* **Note 4**).

3. Centrifuge the lysate at ~3000g (6000 rpm in an Eppendorf 5415C microfuge) for 3 min.
4. Resuspend the nuclear pellet in 0.5 mL of IPB2 and sonicate 3× for 5 s each time (allow the tube to sit on the ice bath for 15 s between each sonication step) using a microtip at a setting of ~3 (the optimal setting must be determined empirically for each type of sonicator).
5. Overlay the sonicate on a 0.5-mL 30% sucrose cushion (prepared in IPB2) and centrifuge at ~4800g (7000 rpm in an Eppendorf 5415C microfuge) for 15 min.
6. Remove the nucleoplasm (~0.5 mL), which is above the sucrose cushion, and add Triton X-100 to a final concentration of 0.5% (*see* **Note 5**).

3.2. Immunopurification of hnRNP Complexes and Gel Analysis

This method involves purification of the hnRNP complex using a monoclonal antibody against an hnRNP. Immunopurification is performed in the presence of a nonionic detergent to prevent disruption of strong protein–protein and protein–RNA interactions. This procedure should be performed rapidly to minimize nuclease activity; therefore the monoclonal antibody is prebound to Protein A–Sepharose beads and the incubation of these beads with nucleoplasm is limited to 15 min.

1. All steps are performed at 4°C. Aliquot 1 mL of IPB1 into a microfuge tube and then add 1–5 µL of ascites fluid (*see* **Note 6**), 3 µL of rabbit antimouse IgG (*see* **Note 1**), and 50 µL of Protein A–Sepharose beads (50% suspension in PBS). Mix and place on a nutator (*see* **Note 7**) for 1 h at 4°C.
2. Wash the beads 3× with 1 mL of IPB1 by vortexing followed by centrifugation in a microfuge for 3 s. Discard the supernatants.
3. To each microfuge tube containing ~25 µL packed antibody-bound beads, add 250 µL of labeled nucleoplasm (**Subheading 3.1., step 6**) and mix on a nutator for 15 min at 4°C.
4. Wash the beads (as described in **step 2**) 5× with 1 mL of IPB1 and once with 1 mL of RSB-100.
5. Draw off any excess liquid with a tuberculin syringe equipped with a 27-gage needle (*see* **Note 8**).
6. For analysis by one-dimensional gel electrophoresis (SDS-PAGE, *see* **Subheading 3.4.**), add 40 µL of 2× electrophoresis loading buffer, heat at 100°C for 3 min, centrifuge for 2 min in a microfuge at maximum speed, and load 20 µL per lane (1.5-cm gel, 15-well comb).
7. For analysis by two-dimensional gel electrophoresis (*see* **Subheading 3.5.**), add 20 µL of sample buffer and 2 µL of 20 mM DTT to the beads. Mix and incubate for 5 min at room temperature. Centrifuge in a microfuge for 30 s and recentrifuge the supernatant for 30 s. Load 18–20 µL per tube gel and perform electrophoresis as described in **Subheading 3.5., steps 1–15** (*see* **Note 9**).

3.3. Immunopurification of Individual hnRNP Proteins and Gel Analysis

It is often informative to immunopurify only the protein recognized by the monoclonal antibody especially if the protein has not yet been characterized as a component of the hnRNP complex. This can be done two ways. The hnRNP protein can be directly immunopurified from labeled cell extracts or from immunopurified hnRNP complexes. Both procedures are described in **Subheading 3.3.**, **steps 1** and **2**.

1. For immunopurification from labeled cell extracts: Follow the procedure described in **Subheading 3.1.**, **steps 1** and **2**, but use either IPB3 or IPB4 instead of IPB1, and then directly sonicate the cell lysate as described in **Subheading 3.1.**, **step 4**. Centrifuge the lysate in a microfuge at maximum speed for 10 min to clarify. Immunopurifications are performed as described in **Subheading 3.2.** except that IPB3 or IPB4 is substituted for IPB1.
2. For immunopurification from immunopurified hnRNP complexes: Immunopurify hnRNP complexes as described in **Subheading 3.2.**, **step 1–5**. Dissociate the complexes by adding 50–100 µL of 1% SDS in RSB-100 and heating at 100°C for 3 min, followed by cooling to room temperature and 10-fold dilution in IPB4 (to reduce the SDS concentration to ~ 0.1%). Immunopurify at 4°C as described in **Subheading 3.2.** using IPB4 instead of IPB1.

3.4. One-Dimensional Gel Electrophoresis

The discontinuous buffer gel system that is described in detail in **Subheading 3.4.**, **steps 1–5** has several important features including better resolution of both low and high molecular weight hnRNPs and resistance to cracking during drying on vacuum gel dryers (*6*).

1. Preparation of a 12.5% running gel: Combine 14.8 mL of acrylamide/*bis* stock for running gel, 15.2 mL of running buffer, 8.3 mL of water, and 0.4 mL of 10% SDS, mix and then add 1 mL of 3% APS and 20 µL of TEMED. Pour the running gel solution immediately into a slab gel apparatus, overlay with 0.1% SDS, and allow to polymerize.
2. For the stacking gel, combine 1.3 mL of acrylamide/*bis* stock for stacking gel, 2.5 mL of 0.5 *M* Tris-HCl, pH 6.8, 6.1 mL of water, and 0.1 mL of 10% SDS, mix and add 0.1 mL of APS and 25 µL of TEMED. Pour the stacking gel immediately into the apparatus with a comb previously inserted about 1 cm above the polymerized running gel.
3. For the Hoefer SE400 slab gel electrophoresis apparatus, 20 µL of proteins (**Subheading 3.2.**, **step 6**) are stacked at 80 V and the gel is then run at 150 V for ~5 h or until the bromophenol blue dye has run off the bottom of the running gel.
4. Fixation, staining, and fluorography: Gels are fixed and stained overnight in 10% acetic acid, 25% isopropanol, and 0.025% (w/v) Coomassie blue (*see* **Note 10**). Gels are destained in 10% acetic acid for 2 h.

5. For fluorography, soak the gel in dimethyl sulfoxide (DMSO) for 15 min, pour off the DMSO and add fresh DMSO for another 15 min, pour off the DMSO and add a solution containing 22% 2,5-diphenyloxazole (PPO) in DMSO for 1 h, pour off the PPO/DMSO, and then wash with running tap water for 1 h (*see* **Note 11**). Dry the gel for 1.5 h using a heated vacuum gel dryer.

3.5. Two-Dimensional Gel Electrophoresis

Because the hnRNP complex consists of numerous proteins, a more accurate representation of the complexity of these complexes is obtained by two-dimensional gel electrophoresis. NEPHGE is used for the first dimension because the human hnRNP complex contains numerous basic proteins. The following section describes a convenient technique that uses an inexpensive apparatus for running NEPHGE gels.

1. Preparation of the tube gel for the first dimension: Mix in a 10-mL beaker containing a flea stir bar 1.38 g of ultrapure urea, 0.5 mL of 10% NP40, 125 µL of 40% ampholytes, 0.49 mL of water, and 0.33 mL of isofocusing acrylamide stock. This amount is sufficient for 9 gels. Cover the beaker with parafilm and allow the acrylamide solution to mix for ~30 min until all of the urea has dissolved. Do not heat the acrylamide solution (to insulate the solution, place a piece of styrofoam between the beaker and the stirrer platform).
2. While the acrylamide solution is stirring, flatten the modeling clay out on a clean benchtop and cover with parafilm. Place two rubber bands around the outside of a test tube rack and press the rack into modeling clay. Firmly attach a 200-µL pipet (marker at 11.5 cm from the bottom) to the tuberculin syringe.
3. When the gel solution is ready (the urea has dissolved), add 4 µL of 10% APS and 5 µL of TEMED, mix, and draw the acrylamide solution to just above the 11.5 cm mark. Keeping the syringe attached, thread the capillary pipet through the rubber bands attached to the test tube rack and stab firmly into the clay base. Adjust the liquid level to exactly 11.5 cm and check for any air bubbles.
4. After all the gels have been cast, layer 15 µL of 8 *M* urea on top of each gel and cover the pipets with a parafilm tent. Allow 4 h to polymerize.
5. Gel loading: following gel polymerization, shake off the urea overlay. Use gels that are of equal length.
6. Add the degassed NaOH solution to the lower chamber of the tube gel apparatus.
7. Use a 5-mL syringe equipped with a bent 21-gage needle to remove bubbles in the pipet bottom.
8. Insert pipets into the tube holder equipped with rubber adapters, place the tube holder into the lower chamber, and fill the upper chamber with the 0.01 *M* phosphoric acid solution.
9. Prior to loading the samples, wash out tubes with the upper chamber solution using an elongated (sequencing gel type) pipetman tip. Load the gels, then reverse the electrodes and run the gels at 400 V for 4 h.

10. Gel extrusion and storage: Following electrophoresis, extrude each gel using a 20-mL syringe and slow, even pressure into 4.75 mL of sample buffer in a 15-mL conical tissue-culture tube (*see* **Note 12**).
11. In the hood, add 250 µL of β-mercaptoethanol to each tube, mix gently, allow to equilibrate for 5 min at room temperature, quick freeze in a dry ice–ethanol bath, and store at –80°C.
12. Loading of the gel for the second dimension: pour a 12.5% gel using 1.5 mm spacers as described in **Subheading 3.4., steps 1–3**, except insert a single lane preparative comb ~0.5 cm into the stacking gel.
13. Quickly thaw the first dimensional gel in a 37°C water bath with gentle mixing (the gel is very fragile).
14. Carefully pour the gel onto the preparative comb positioned above a beaker, center, and straighten the gel using blunt forceps. Position the gel between the two glass plates of the second dimension slab gel and gently push the first dimension gel until it is positioned directly on top of the stacking gel.
15. Run the second dimension until the bromophenol blue dye has run off the end of the gel. Following electrophoresis, the gel is fixed, stained, and processed for fluorography (*see* **Subheading 3.4., steps 4–5**).

4. Notes

1. The Protein A–Sepharose should be purchased as a dry powder, hydrated in 50 mL of PBS in a conical centrifuge tube, and allowed to settle (repeat this wash step once), and finally stored in PBS + 0.05% azide at 4°C for no longer than 3 mo. It is important to perform a mock immunodepletion control (Protein A–Sepharose with a control nonimmune antibody) for each experiment because high background signals are generally traceable to a bad lot of Protein A–Sepharose. If the monoclonal antibody is of the IgG1 subclass, then a rabbit antimouse IgG (Organon Teknika, Durham, NC, USA) can be used. Protein G–Sepharose can also be employed, but for several cell types we have found that this immunoadsorbent often results in the immunopurification of nonspecific proteins in mock immunodepletion controls.
2. Aprotinin is purchased as a sterile liquid (Sigma, St. Louis, MO, USA).
3. IPB3 contains 1% Empigen BB which has been found to give the best immunopurification results for individual proteins immunopurified from cell lysates.
4. Care must be employed to avoid spraying the ^{35}S-labeled cell extract when passing the cell lysate through the 25-gage needle. It is very easy to contaminate the surrounding area. We use a dedicated area for processing these extracts.
5. Once the nucleoplasm has been prepared, it should be used for immunopurification as soon as possible to avoid excessive nuclease degradation of the hnRNP complex.
6. Most ascites fluids contain 2–6 mg/mL of monoclonal antibody. The optimal amount required for efficient immunopurification should be determined for each ascites lot.

7. Any device that efficiently mixes small volumes in a microfuge tube will work although the nutator (Clay Adams, Parsippany, NY, USA) is especially useful for this procedure.
8. The tuberculin syringe with attached 27-gage needle is used to remove most of the fluid prior to addition of loading or sample buffer. However, do not leave the beads dry for longer than a few minutes.
9. Loading the two-dimensional gel requires care and some practice as the sample buffer does not contain a dye, but the density of the sample is greater than the top chamber buffer so that you can visually follow the loading of the sample.
10. Dissolve Coomassie blue in water and isopropanol prior to adding the acetic acid.
11. The PPO/DMSO solution should be prepared in a hood. This solution can be used for many gels or until the PPO starts to precipitate out of the DMSO. The PPO can be recrystallized or the PPO/DMSO solution can be disposed of as hazardous chemical waste.
12. Extrusion of the gel from the 200-µL capillary pipet is tricky. Attach the syringe and insert the capillary pipet into a 15-mL conical tube that contains the sample buffer. Apply relatively strong pressure until the gel begins to move, submerge the end of the pipet into the sample buffer, and then use very light pressure until the gel has popped out of the tube. Excess pressure will result in gel breakage.

References

1. Dreyfuss, G., Matunis, M. J., Piñol-Roma, S., and Burd, C. G. (1993) HnRNP proteins and the biogenesis of mRNA. *Annu. Rev. Biochem.* **62**, 289–321.
2. Swanson, M. S. (1995) Functions of nuclear pre-mRNA/mRNA-binding proteins. In *Pre-mRNA Processing* (Lamond, A. I., ed.), Springer-Verlag, New York, pp. 17–33.
3. Piñol-Roma, S., Choi, Y. D., Matunis, M. J., and Dreyfuss, G. (1988) Immunopurification of heterogeneous nuclear ribonucleoprotein particles reveals an assortment of RNA-binding proteins. *Genes Dev.* **2**, 215–227.
4. Piñol-Roma, S., Swanson, M. S., Gall, J. G., and Dreyfuss, G. (1989) A novel heterogeneous nuclear RNP protein with a unique distribution on nascent transcripts. *J Cell Biol.* **109**, 2575–2587.
5. Matunis, M. J., Matunis, E. L., and Dreyfuss, G. (1992) Isolation of hnRNP complexes from *Drosophila melanogaster*. *J Cell Biol.* **116**, 245–255.
6. Dreyfuss, G., Adam, S. A., and Choi, Y. D. (1984) Physical change in cytoplasmic messenger ribonucleoproteins in cells treated with inhibitors of mRNA transcription. *Mol. Cell. Biol.* **4**, 415–423.

24

Preparation of Hela Cell Nuclear and Cytosolic S100 Extracts for In Vitro Splicing

Akila Mayeda and Adrian R. Krainer

1. Introduction

Following the initial discovery of split genes in 1977, it took several years before in vitro systems were successfully developed to study the biochemistry of pre-mRNA splicing. The first systems relied on coupling of transcription and splicing in whole-cell extracts and were fairly inefficient, because of the different optima for these two reactions *(1,2)*. It was later shown that these reactions could be uncoupled *(3,4)*, but obtaining discrete pre-mRNAs in useful amounts remained an obstacle until in vitro transcription with bacteriophage RNA polymerases *(3)* was adopted for this purpose. Another useful development was a nuclear extract preparation procedure that was initially developed for in vitro transcription studies *(5)*. No splicing was detected in this study. However, the same extract preparation procedure, in conjunction with pre-mRNAs transcribed from cloned genes by SP6 RNA polymerase, was used to define optimal conditions for in vitro splicing *(6)*. This system results in relatively efficient and accurate splicing, and is now in wide use, with slight variations from laboratory to laboratory. Variations in extract preparation include primarily the use of slightly different buffers and salts for nuclear extraction or dialysis.

HeLa cells are used most frequently, but the same procedure has been used to prepare splicing-competent extracts from other cells that grow in spinner culture. Many cell lines fail to yield active extracts, and this is sometimes due to the presence of ribonucleases, proteases, or other nonspecific inhibitors. This problem appears to be exacerbated when animal tissues are used as the source. Studies of in vitro transcription have been much more successful in this regard, and the reason may be that the splicing apparatus includes highly sensitive, multisubunit snRNP components. In addition, owing to the very large number of components required to assemble an active spliceosome, it appears that limiting components are lost when attempts are made to obtain more purified nuclei, e.g., from animal tissues. A very similar extract preparation procedure using HeLa cells has been used to carry out accurate polyadenylation (*8; see* Chapter 32), and coupling of splicing and polyadenylation can be studied in this system *(9)*. The extract preparation procedure described here is also useful to study splicing of the recently discovered nonconventional (AT–AC) introns, in addition to the major splicing pathway *(10,11)*. A similar procedure has been used to prepare splicing extracts from *Drosophila* cell lines *(12)*. An efficient system for yeast pre-mRNA splicing has been widely used, and is based on whole-cell extracts from *S. cerevisiae* (*13; see* Chapter 26). In contrast, pre-mRNA splicing in plant extracts has not yet been reported.

Here we provide a detailed protocol for the preparation of pre-mRNA splicing-competent extracts from HeLa cells. It is based on the method for preparation of transcription-competent extracts *(5,6)* with slight variations to improve splicing efficiency *(7)*, and has been used in our laboratory for many years with reproducible results. The same preparation yields two kinds of extract: a nuclear salt wash and a cytosolic S100. The nuclear extract is fully competent for splicing of numerous pre-mRNAs. The S100 extract contains many splicing factors, but is not competent for splicing because it has limiting amounts of SR proteins, which are required for splicing. However, this extract can be complemented by one or more SR proteins to give efficient splicing *(14)*. These two systems (nuclear extract and S100 plus SR proteins) differ in interesting and useful ways, in part because of the somewhat different ratios of various splicing factors *(15–17)*. Both systems have been widely used to study constitutive splicing and/or alternative splicing of different pre-mRNAs.

The protocol described here is for extract preparation on a medium scale. The procedure can be easily scaled up using appropriate glassware and rotors, although cell lysis is more difficult with larger Dounce homogenizers. A procedure for nuclear extract preparation on a very small scale has been described, which should be useful for surveying many different cell lines or cells grown under many different conditions *(18)*.

2. Materials

All reagents should be prepared with high quality autoclaved water, e.g., Milli Q (Millipore, Bedford, MA, USA) or double-distilled water. Sterilization of all components used for cell growth is carried out by autoclaving or, in the case of thermolabile materials, by filtration through 0.22-μm filters (Millipore).

2.1. HeLa Cell Suspension Culture

1. Medium: Joklik's modification of Eagle's minimal essential medium for suspension culture, or equivalent (ICN Pharmaceuticals, Costa Mesa, CA, USA; cat. no.

10-323-24). This powdered medium already contains all required additions except serum. A 5× stock is prepared by dissolving the powder in Milli Q water and is sterilized by filtration. The working growth medium is prepared by dilution with filter sterilized water and addition of sterile calf serum (Gibco-BRL, Grand Island, NY, USA) to 50 mL/L. The pH should be approx 7.0.

2. HeLa cells: Use a strain adapted to growth in suspension culture, e.g., the S-3 strain (available from ATCC, Manassas, VA, USA).

2.2. Preparation of Extracts

Reagents 1 and 2 are stock solutions that should be stored at –20°C and added to reagents 3–6 immediately prior to use.

1. 1 M Dithiothreitol (DTT).
2. 20 mg/mL (115 mM) Phenylmethanesulfonyl fluoride (PMSF) dissolved in ethanol.
3. Phosphate-buffered saline (PBS) solution: 137 mM NaCl, 2.7 mM KCl, 8 mM Na$_2$HPO$_4$, 1.5 mM KH$_2$PO$_4$, 0.5 mM MgCl$_2$.
4. Buffer A: 10 mM HEPES-KOH, pH 8.0, 10 mM KCl, 1.5 mM MgCl$_2$, 1 mM DTT.
5. Buffer B: 0.3 M HEPES-KOH, pH 8.0, 1.4 M KCl, 30 mM MgCl$_2$.
6. Buffer C: 20 mM HEPES-KOH, pH 8.0, 0.6 M KCl, 1.5 mM MgCl$_2$, 0.2 mM EDTA, 25% (v/v) glycerol, 0.5 mM PMSF, 1 mM DTT.
7. Buffer D: 20 mM HEPES-KOH, pH 8.0, 100 mM KCl, 0.2 mM EDTA, 20% (v/v) glycerol, 0.5 mM PMSF, 1 mM DTT.
8. 40-mL Dounce glass homogenizer with loose-fitting pestle (Kontes, Vineland, NJ, USA; 0.003–0.006 in. clearance).
9. Dialysis tubing with 12–14,000 nominal mol wt cutoff (Spectrum, Houston, TX, USA).

3. Methods

3.1. Suspension Culture of HeLa Cells

The cells are grown in spinner flasks at 37°C. The cell density should be maintained daily between 2×10^5 and 5×10^5. The doubling time is approx 24 h (*see* **Note 1**). It is important to use magnetic stirrers that do not generate heat at the surface, or to avoid direct contact between the stirring plate and the bottom of the flask. The cell density can be checked with a hemocytometer or Coulter counter.

3.2. Preparation of HeLa Cell Nuclear Extract

1. Harvest the cells at the logarithmic growth stage (4–6 × 10^5 cells/mL, do not exceed 1 × 10^6 cells/mL) by low-speed centrifugation at ~1800g (e.g., Sorvall H-6000A rotor, 2500 rpm, 10 min).
2. Wash the cells by gently resuspending the pellet in ice-cold PBS, centrifuge again, and measure the packed cell volume (PCV). All subsequent steps should be carried out on ice or in the cold room, and all centrifugation steps are carried out at 4°C.

3. Gently resuspend the cell pellet in 5× PCVs of buffer A and keep on ice for 10 min to swell the cells in the hypotonic buffer.
4. Pellet the cells by centrifugation at ~1,800g (e.g., Sorvall H6000-A, 2500 rpm, 10 min) and resuspend in 2× PCVs of buffer A. The cell pellet approximately doubles in volume due to swelling; use the original PCV in the calculation.
5. Homogenize with approx 10 strokes in a Dounce glass homogenizer. The number of strokes may vary depending on the homogenizer but is fairly consistent if the same set is used (*see* **Note 2**). The extent of cell lysis can be measured on a hemocytometer using trypan blue staining. Aim for ~90% lysis, but do not use more strokes than necessary. Note that the HeLa nuclei are very large and not much smaller than the intact cells.
6. Transfer the lysate into high-speed centrifuge tubes and centrifuge first at low speed, using appropriate adapters (~1200g, e.g., Sorvall H6000-A, 2000 rpm, 10 min); remove the supernatant carefully into a graduated cylinder with a pipet. (The supernatant is further processed to prepare the S100 extract, *see* **Subheading 3.3.**)
7. Centrifuge the pellet at higher speed in the same tubes (~33,000g, e.g., Sorvall SS-34, 16,500 rpm, 20 min); remove the supernatant carefully with a pipet and discard it.
8. The packed nuclei volume (PNV) is usually 9–15 mL for a 12-L culture of HeLa cells. Add 2 mL of buffer C and dislodge the pellet by swirling, but do not disperse the pellet. Transfer the pellet and liquid into a graduated 50-mL disposable tube (Corning Labware, Corning, NY, USA). Estimate the PNV by subtracting 2 mL from the measured volume, and add more buffer C (which contains 0.6 M KCl) to obtain a final KCl concentration of 0.24 M. The acceptable range is 0.20–0.25 M KCl (*see* **Note 3**).
9. Transfer the pellet and buffer (do not resuspend yet, or the solution will become sticky and difficult to transfer without losses) into a clean 40-mL glass Dounce. Suspend by several strokes with a loose-fitting pestle. A total of 10 strokes should be sufficient to make a homogeneous suspension.
10. Transfer the homogenate into one or more screw-capped high-speed centrifuge tubes and mix gently by rocking for 30–45 min.
11. Pellet the salt-washed nuclei by high-speed centrifugation (~33,000g, e.g., Sorvall SS-34, 16,500 rpm, 30 min) and carefully transfer the supernatant into a disposable tube, avoiding contamination with material from the pellet.
12. Dialyze the supernatant twice against buffer D (usually 1–2 L). One of the dialysis steps is done overnight and the other for at least 4 h.
13. Remove the cloudy precipitate formed during dialysis by high-speed centrifugation (~33,000g, e.g., Sorvall SS-34, 16,500 rpm, 20 min) and carefully withdraw the supernatant, which is the nuclear extract, into a disposable tube.
14. Aliquot the extract into microcentrifuge tubes, in 0.1–1-mL portions, as desired. Quick freeze by dropping the tubes into liquid nitrogen, and store at –70°C or lower. The usual yield of nuclear extract, starting from a 12-L spinner culture, is 8–10 mL with a total protein concentration of 10–15 mg/mL. The extract remains active for several years.

3.3. Preparation of HeLa Cell Cytosolic S100 Extract

While waiting for one of the centrifugation steps in **Subheading 3.2.**, the supernatant obtained in **Subheading 3.2.6.** can be processed further to obtain an S100 extract, which contains cytosolic components as well as many, but not all, nuclear components, which are extracted in the hypotonic buffer A.

1. Record the volume of supernatant, add 0.11 vol of buffer B, and mix gently.
2. Spin in an ultracentrifuge at ~100,000g (38,000 rpm, Beckman 60Ti rotor or 36,500 rpm, Beckman 45Ti rotor) for 1 h.
3. Transfer the supernatant into a disposable tube.
4. Dialyze twice against 2 L of buffer D, once overnight and once for at least 4 h. The volume of extract is reduced considerably during dialysis because of the glycerol in buffer D, thus resulting in a concentrated extract.
5. Remove insoluble material by high-speed centrifugation (~33,000g, e.g., Sorvall SS-34 rotor, 16,500 rpm) for 20 min.
6. Aliquot the extract into microcentrifuge tubes, in 0.1–1-mL portions, as desired. Quick freeze and store as described in **Subheading 3.2.**, **step 14**. The usual yield of cytosolic S100 extract, starting from a 12-L spinner culture, is 40–55 mL with a total protein concentration of 10–15 mg/mL. The extract remains active for several years (*see* **Note 4**).

4. Notes

1. The use of healthy cells under optimal growth conditions is essential. Slowly dividing cells (longer than 24 h doubling time) or overgrown cells should not be used. Cultures that show a lot of dead cells and debris also fail to yield good extracts, and reflect problems with the medium, serum batch, and/or mechanical injury due to the spinner flask setup.
2. The cell lysis process is also critical. Mild homogenization keeps the isolated nuclei intact and prevents excessive leakage of splicing factors. Avoid using a very tight-fitting pestle and limit the number of strokes to prevent excessive damage to nuclei.
3. The salt extraction of nuclei in buffer C is also a critical step. The final KCl concentration should be 0.20–0.25 M, and the buffer should be mixed rapidly to prevent transient localized exposure to higher salt concentration. Lower salt concentrations fail to extract splicing factors efficiently, whereas higher salt concentrations disrupt chromatin, resulting in a substantial increase in viscosity, and in extraction of inhibitory components. The final salt concentration can be checked by measuring the conductivity, if desired.
4. The activity of S100 extracts is not as reproducible as that of nuclear extracts, and some batches may fail to be complemented efficiently by SR proteins. It may be necessary to prepare and test more than one batch of S100 extract for particular purposes.

References

1. Kole, R. and Weissman, S. M. (1982) Accurate in vitro splicing of human β-globin RNA. *Nucleic Acids Res.* **10**, 5429–5445.
2. Padgett, R. A., Hardy, S. F., and Sharp, P. A. (1983) Splicing of adenovirus RNA in a cell-free transcription system. *Proc. Natl. Acad. Sci. USA* **80**, 5230–5234.
3. Green, M. R., Maniatis, T., and Melton, D. A. (1983) Human β-globin pre-mRNA synthesized in vitro is accurately spliced in Xenopus oocyte nuclei. *Cell* **32**, 681–694.
4. Hernandez, N. and Keller, W. (1983) Splicing of in vitro synthesized messenger RNA precursors in HeLa cell extracts. *Cell* **35**, 89–99.
5. Dignam, J. D., Lebovitz, R. M., and Roeder, R. G. (1983) Accurate transcription initiation by RNA polymerase II in a soluble extract from isolated mammalian nuclei. *Nucleic Acids Res.* **11**, 1475–1489.
6. Krainer, A. R., Maniatis, T., Ruskin, B., and Green, M. R. (1984) Normal and mutant human β-globin pre-mRNAs are faithfully and efficiently spliced in vitro. *Cell* **36**, 993–1005.
7. Solnick, D. (1985) Alternative splicing caused by RNA secondary structure. *Cell* **43**, 667–676.
8. Moore, C. L. and Sharp, P. A. (1985) Accurate cleavage and polyadenylation of exogenous RNA substrate. *Cell* **41**, 845–855.
9. Niwa, M., Rose, S. D., and Berget, S. M. (1990) In vitro polyadenylation is stimulated by the presence of an upstream intron. *Genes Dev.* **4**, 1552–1559.
10. Tarn, W. Y. and Steitz, J. A. (1996) A novel spliceosome containing U11, U12, and U5 snRNPs excises a minor class (AT-AC) intron in vitro. *Cell* **84**, 801–811.
11. Wu, Q. and Krainer, A. R. (1996) U1-mediated exon definition interactions between AT-AC and GT-AG introns. *Science* **274**, 1005–1008.
12. Rio, D. C. (1988) Accurate and efficient pre-mRNA splicing in *Drosophila* cell-free extracts. *Proc. Natl. Acad. Sci. USA* **85**, 2904–2908.
13. Lin, R.-J., Newman, A. J., Cheng, S.-C., and Abelson, J. (1985) Yeast mRNA splicing in vitro. *J. Biol. Chem.* **260**, 14,780–14,792.
14. Krainer, A. R., Conway, G. C., and Kozak, D. (1990) Purification and characterization of pre-mRNA spicing factor SF2 from HeLa cells. *Genes Dev.* **4**, 1158–1171.
15. Krainer, A. R., Conway, G. C. and Kozak, D. (1990) The essential pre-mRNA splicing factor SF2 influences 5' splice site selection by activating proximal sites. *Cell* **62**, 35–42.
16. Mayeda, A. and Krainer, A. R. (1992) Regulation of alternative pre-mRNA splicing by hnRNP A1 and splicing factor SF2. *Cell* **68**, 365–375.
17. Mayeda, A., Helfman, D. M., and Krainer, A. R. (1993) Modulation of exon skipping and inclusion by heterogeneous nuclear ribonucleoprotein A1 and pre-mRNA splicing factor SF2/ASF. *Mol. Cell. Biol.* **13**, 2993–3001.
18. Lee, K. A., Bindereif, A., and Green, M. R. (1988) A small-scale procedure for preparation of nuclear extracts that support efficient transcription and pre-mRNA splicing. *Gene Anal. Tech.* **5**, 22–31.

25

Mammalian In Vitro Splicing Assays

Akila Mayeda and Adrian R. Krainer

1. Introduction

Splicing reactions are typically carried out using nuclear extracts, S100 extracts complemented with SR proteins, or partially purified fractions derived from the crude extracts. The extract preparation procedures are described in Chapter 24. Extracts derived from HeLa cells are used most commonly (*see* **Note 1**). The pre-mRNA substrates are usually prepared by in vitro runoff transcription with a bacteriophage polymerase (*see* Chapter 1). The intermediates and products of splicing are most conveniently visualized by urea/polyacrylamide gel electrophoresis (urea-PAGE) and autoradiography, which requires the use of labeled pre-mRNA substrate. The protocols provided here are based on **ref. 1**, with modifications introduced in **refs. 2,3**, and have been routinely used in our laboratory for many years.

Many different pre-mRNAs, both general and alternative splicing substrates, can be spliced in HeLa cell extracts, although slightly different reaction conditions may be required for optimal processing of particular substrates (*see* **Note 2**). In general, the in vitro splicing reaction is remarkably accurate. For example, with β-globin pre-mRNA only the correct splice sites are chosen, and cryptic splice sites are ignored. Point mutations that inactivate or weaken the splice sites and activate cryptic splice sites have similar effects in vivo and in vitro *(1)*. With other substrates, inappropriate exon skipping or unexpected use of cryptic splice sites have been observed in vitro, but similar products can also be detected by transient expression in transfected cells. Regulatory sequences, such as exonic splicing enhancers, have similar effects in vivo and in vitro. More complex regulatory splicing events may be more difficult to reproduce in vitro, but this is largely attributable to the current difficulty in obtaining active extracts from relevant tissue sources.

2. Materials

All reagents should be prepared with high-quality autoclaved water, e.g., Milli Q (Millipore, Bedford, MA, USA) or double-distilled water. All commercially available chemicals should be ultrapure grade or special grade for molecular biology. As the in vitro splicing assays involve femtomole quantities of RNA, special care should be exercised to avoid ribonuclease contamination of solutions and surfaces that will be in contact with RNA.

2.1. Pre-mRNA Substrates

The substrate is prepared by runoff transcription of a linearized plasmid in which the gene of interest, or a portion thereof, is subcloned downstream of a bacteriophage promoter (*see* **Note 3** and Chapter 1). Commonly used bacteriophage RNA polymerases are derived from SP6, T7, and T3 bacteriophages. Purified or cloned polymerases, as well as vectors containing appropriate promoters and polylinkers, are commercially available. For example, minigenes subcloned in the plasmid pSP64 or 65 are transcribed with SP6 RNA polymerase (Promega, Madison, WI, USA). The plasmid should be linearized with a restriction enzyme; 5' overhangs may be preferable, as some 3' overhangs may result in antisense end-to-end transcription. Capped pre-mRNAs are more stable and are spliced more efficiently than uncapped pre-mRNAs *(1,4)*. The most convenient and efficient method for capping the pre-mRNA substrate is to prime transcription with a dinucleotide primer, ^{7}mGpppG or GpppG *(5)*. Usually one, but in specialized instances two or more ^{32}P-labeled ribonucleotides are included in the transcription reaction to label the pre-mRNA uniformly to the desired specific activity. Kits for efficient in vitro transcription of capped, labeled pre-mRNAs are available commercially (Ambion, Austin, TX, USA), and standard procedures have been described (e.g., *6,7*). The transcribed pre-mRNA is purified by phenol extraction and ethanol precipitation, and the yield and concentration of RNA are determined on the basis of the incorporated label, as measured, e.g., by trichloracetic acid (TCA)-precipitable counts *(8)*. Since bacteriophage RNA polymerases are highly specific for their cognate promoters, gel purification of the transcript is usually not necessary. However, if a particular template yields heterogeneous transcripts, full-length pre-mRNA should be purified by preparative denaturing polyacrylamide gel electrophoresis *(8)*.

2.2. Reagents

1. 25× ATP/CP mixture: 12.5 mM ATP, 0.5 M creatine phosphate (CP). Prepare a working solution with 100 mM ATP stock, pH 7.5 (Pharmacia, Piscataway, NJ, USA; cat. no. 27-2056-01) and creatine phosphate (Calbiochem, San Diego, CA, USA; cat. no. 2380). Aliquot in ~0.2 mL portions in microcentrufuge tubes and store at –20°C.

2. 80 mM MgCl$_2$. Prepare a 1 M stock solution using the highest available grade of MgCl$_2$ powder. Dilute to prepare the working solution, aliquot in ~0.2-mL portions in microcentrifuge tubes, and store at 4°C or –20°C.
3. 0.4 M HEPES-KOH, pH 7.3. Do not autoclave. Sterilize by filtration through a 0.2-µm filter and store at 4°C.
4. 13% (w/v) polyvinyl alcohol (PVA). Use low molecular weight PVA (Sigma, St Louis, MO, USA; cat. no. P-8136). To dissolve easily, suspend with Milli-Q water in a screw-capped glass bottle and autoclave for 10 min. Aliquot in ~1-mL portions in microcentrifuge tubes and store at –20°C.
5. Splicing stop solution: 0.3 M sodium acetate, pH 5.2, 0.1% (w/v) sodium dodecyl sulfate (SDS), 62.5 µg/mL tRNA (e.g. Sigma, cat. no. R-9001). Store at room temperature or at 4°C.
6. Phenol saturated with Tris-HCl, pH 8.0: available commercially (e.g., Boehringer Mannheim, Indianapolis, IN, USA; cat. no. 100997), or prepare from high-grade phenol (Ultra Pure or Molecular Biology grade) as follows *(8)*. Add an equal volume of 0.5 M Tris base, and 8-quinolinol (8-hydroxyquinoline, Sigma) to 0.2% (w/v), thaw out at 40–50°C, and mix well. Let the phases separate, remove the upper phase, add an equal volume of 0.1 M Tris-HCl, pH 8.0, and mix well. Again remove the upper phase and store at 4°C in an amber bottle. Do not use phenol mixed with chloroform, as the chloroform and the PVA form a very large interphase. Hydroxyquinoline is added as an antioxidant and its yellow color facilitates visualization of the phases.
7. RNA dye mixture: 90% (v/v) formamide, 50 mM Tris-HCl, pH 7.5, 1 mM EDTA, 0.1% (w/v) bromophenol blue, 0.1% (w/v) xylene cyanol FF. Aliquot in ~1-mL portions in microcentrifuge tubes and store at –20°C.
9. Acrylamide/urea gel stock solution: 19% (w/v) acrylamide, 1% (w/v) *bis*-acrylamide, 7 M urea, 89 mM Tris base, 89 mM boric acid, and 2 mM EDTA. Dilute this stock solution to the running percentage of acrylamide (usually 4–10%) with the same solution lacking acrylamide. Warm up to 37°C and degas under vacuum prior to polymerization with ammonium persulfate (133 µL of 10% [w/v] per 20 mL) and *N,N,N',N'*-tetramethylethylenediamine (TEMED) (10 µL per 20 mL) as described *(8)*. The percentage of acrylamide to be used is chosen according to the expected size of the spliced products *(8)*. Higher acrylamide concentrations can be used to make the lariat molecules migrate above the longer, linear pre-mRNA *(9,10)*. This procedure facilitates the identification of the intermediate and product lariat molecules, and results in increased sensitivity because the lariats also migrate above the smear of degraded RNA that is usually seen on long autoradiography exposures.

3. Methods
3.1. In Vitro Splicing Reaction
1. Prepare a fresh batch of splicing buffer mixture, calculating the total volume required for the desired number of reactions (plus an allowance for

measuring errors). Keep on ice in a microcentrifuge tube. The contents per individual reaction are as follows: 1.0 µL of 25× ATP/CP mixture, 1.0 µL of 80 mM MgCl$_2$, 1.25 µL of 0.4 M HEPES-KOH, pH 7.3, 5.0 µL of 13% PVA (add last), 20 fmol (usually 0.1–0.4 µL) of ^{32}P-labeled pre-mRNA, and Milli-Q water to 10 µL (total). 13% PVA is very viscous and should be added last and mixed gently by pipetting up and down with a P-20 pipet. Avoid foaming and do not vortex. Centrifuge very briefly (about 3 s) after mixing (see **Note 2**).

2. Thaw out the required number of aliquots of frozen extract. The extract is thawed out at room temperature and immediately placed on ice. Unused extract can be refrozen at –70°C. Repeated freezing and thawing does not usually result in loss of splicing activity.
3. Set up the splicing reactions (25 µL in microcentrifuge tubes). While holding each tube on ice, carefully pipet a total of 15 µL of HeLa cell nuclear extract (or S100 extract plus SR proteins in buffer D) plus buffer D (see Chapter 24). Depending on the quality and concentration of the extracts, efficient splicing usually requires 5–10 µL of the nuclear extract or 5–10 µL of S100 extract plus 15–20 pmol of SR proteins (*11,12*; see Chapter 31). Add 10 µL of splicing buffer mixture (see **Subheading 3.1., step 1**) using a fresh tip and mix all the reagents gently by pipetting up and down with a P-20 pipet (do not vortex).
4. Centrifuge very briefly (about 3 s) and incubate at 30°C for 1–4 h. (The kinetics and efficiency of splicing depend on the particular pre-mRNA; see **Note 2**.)
5. Add 0.2 mL of splicing stop solution.
6. Add 0.2 mL of Tris-saturated phenol and vortex for 1–2 min immediately.
7. Microfuge for 5 min, and transfer supernatant (aqueous phase) into a fresh tube. Avoid carry over of organic and interphase material.
8. Add 0.5 mL of ethanol and vortex. Keep on ice or freeze for at least 10 min, or store overnight if desired.

3.2. Analysis of Splicing Products by Denaturing PAGE

Standard materials, reagents, and apparatus for denaturing PAGE are as described (*8*). Small gel sizes (15–20 cm × 15–20 cm) are adequate for routine splicing assays. Longer gels can be used if necessary to separate multiple RNA species that are close in size (see **Note 4**). Thin gels (less than 0.5 mm) result in sharper bands by autoradiography.

1. Microfuge the ethanol precipitates of the extracted RNAs (see **Subheading 3.1., step 8**) for 15 min and carefully remove the ethanol without disturbing the pellet. If the ethanol is thoroughly removed, e.g., using a P-200 pipet or fine-tip Pasteur pipet, an aqueous ethanol rinse and drying under vacuum are not necessary.
2. Add 3–4 µL of RNA dye mixture and dissolve each pellet by pipeting up and down or by vortexing.
3. Heat at 80–85°C for 5–10 min.
4. The denaturing polyacrylamide gel should be prerun for at least 30 min prior to applying the samples. Rinse out the sample wells of the gel with a syringe to

Mammalian In Vitro Splicing Assays

remove the urea that has diffused out into the wells, just prior to applying the sample. Load the heated samples into individual wells without cooling.
5. Run the gel until the marker dyes have migrated the desired distance. A useful table for the correlation of dye electrophoretic mobilities with those of single stranded nucleic acids on denaturing polyacrylamide gels of different percentages is available *(8)*. The gels are usually run at constant voltage at 35–45 V/cm of gel length.
6. Separate the glass plates using a thin spatula, and carefully pull the upper plate off.
7. Place a sheet of used X-ray film or 3MM paper (Whatman, Hillsboro, OR, USA) on top of the gel and apply gentle pressure so that the gel becomes firmly attached. Flip the plate over and slowly lift up the glass plate from one side, making sure that the gel sticks to the film or paper. Cover gel with clear film wrap. Gel fixation and drying are not necessary, although they may result in slightly higher sensitivity and resolution.
8. Autoradiography is usually done with an intensifying screen at –70°C. Exposure times are usually from 2 h to overnight, depending on the specific activity of the RNA.

4. Notes

1. The intrinsic activity of splicing extracts can vary from batch to batch. Therefore, it is a good idea to use a well characterized control pre-mRNA that splices efficiently, such as β-globin pre-mRNA *(1,9)*. It may be necessary to prepare and test multiple batches of extracts to obtain satisfactory activity.
2. The in vitro splicing conditions described in this protocol were optimized with β-globin pre-mRNA *(1–3)*, and are not necessarily optimal for other pre-mRNA substrates. To establish optimal conditions for other pre-mRNAs, it is necessary to determine the effect of varying one parameter at a time. In our experience, the most critical parameters are: the divalent and monovalent cation concentrations, the extract volume (adjust with buffer D), and the incubation time. For instance, lower $MgCl_2$ concentrations (1.0 mM rather than 3.2 mM $MgCl_2$) are optimal for β-tropomyosin pre-mRNA splicing *(13,14)*. Generally, splicing of shorter introns requires higher salt concentrations, in the range of 40–100 mM KCl *(7,15)*. δ-Crystallin pre-mRNA splices with faster kinetics than β-globin pre-mRNA, with an optimal incubation time of 1 h *(16)*, compared to 4 h *(1)*. In the case of SV40 pre-mRNA, the efficiency of splicing is improved by addition of a different monovalent cation in the splicing reactions, i.e., 12.6 mM $(NH_4)_2SO_4$ plus 20 mM KCl instead of 60 mM KCl *(17)*. PVA and similar hydrophilic polymers generally improve splicing efficiency by an excluded volume effect *(1)*.
3. The choice of pre-mRNA substrate and its primary structure are critical for successful in vitro splicing assays. In most cases, minigene template constructs are chosen that have a minimal number of introns (usually one or two) and are less than about 2 kbp in length. More complex substrates result in numerous intermediates and products that may be difficult to analyze directly by autoradiography of labeled RNA. Furthermore, with longer RNAs there may be problems in

obtaining discrete full-length transcripts, and degradation in the crude extract is more likely. For complex genes, it is usually feasible to construct simplified minigenes with the relevant exons and introns, or portions thereof. In this case, it is recommended that the splicing patterns first be checked in vivo, if possible, e.g., by transient transfection *(8)*. Long introns are usually truncated around the middle, as the signals involved in splicing catalysis are located at both intron ends. Exons are generally short, but can often tolerate removal of sequences from the outer ends. However, important positive or negative regulatory signals that can have pronounced effects on the splicing efficiency of particular introns are common, although difficult to predict solely by sequence inspection, and they may be present anywhere along the introns or exons. It is possible to assay splicing of specific exons within a complex, long multiexon pre-mRNA, by indirect analysis of the in vitro spliced products, e.g., by reverse transcriptase-polymerase chain reaction (RT-PCR), RNase protection, S1 mapping, or primer extension *(8)*. In fact, such procedures are usually required when new substrates are analyzed for the first time, to establish the accuracy of the splicing reaction.

4. When identifying intermediates and products for substrates that have not been characterized previously, sequencing of the RT-PCR or primer extension products corresponding to spliced mRNA is recommended. Identification of lariat molecules is facilitated by the fact that they migrate more slowly than linear molecules in high-percentage polyacrylamide gels *(9,10)*. The branch site can be mapped because it blocks extension of a downstream primer by reverse transcriptase *(1,9)*; more precise mapping can be carried out using RNA fingerprinting methods *(9,10)*. Once the identity of the intermediates and products of splicing has been established, these molecules can usually be recognized by direct denaturing PAGE and autoradiography analysis on the basis of size.

References

1. Krainer, A. R., Maniatis, T., Ruskin, B., and Green, M. R. (1984) Normal and mutant human β-globin pre-mRNAs are faithfully and efficiently spliced in vitro. *Cell* **36,** 993–1005.
2. Mayeda, A. and Ohshima, Y. (1990) β-globin transcripts carrying a single intron with three adjacent nucleotides of 5' exon are efficiently spliced in vitro irrespective of intron position or surrounding exon sequences. *Nucleic Acids Res.* **18,** 4671–4676.
3. Mayeda, A., Hayase, Y., Inoue, H., Ohtsuka, E., and Ohshima, Y. (1990) Surveying *cis*-acting sequences of pre-mRNA by adding antisense 2'-*O*-methyl oligoribonucleotides to a splicing reaction. *J. Biochem.* **108,** 399–405.
4. Konarska, M. M., Padgett, R. A., and Sharp, P. A. (1984) Recognition of cap structure in splicing in vitro of mRNA precursors. *Cell* **38,** 731–736.
5. Contreras, R., Cheroutre, H., Degrave, W., and Fiers, W. (1982) Simple, efficient in vitro synthesis of capped RNA useful for direct expression of cloned eukaryotic genes. *Nucleic Acids Res.* **10,** 6353–6362.
6. Krainer, A. R. and Maniatis, T. (1985) Multiple factors including the small nuclear ribonucleoproteins U1 and U2 are necessary for pre-mRNA splicing in vitro. *Cell* **42,** 725–736.

7. Mayeda, A. and Ohshima, Y. (1988) Short donor site sequences inserted within the intron of β-globin pre-mRNA serve for splicing in vitro. *Mol. Cell. Biol.* **8,** 4484–4491.
8. Sambrook, J., Fritsch, E. F., and Maniatis, T. (1989) *Molecular Cloning*, 2nd ed. Cold Spring Harbor Press, Cold Spring Harbor, New York.
9. Ruskin, B., Krainer, A. R., Maniatis, T., and Green, M. R. (1984) Excision of an intact intron as a novel lariat structure during pre-mRNA splicing in vitro. *Cell* **38,** 317–331.
10. Grabowski, P. J., Padgett, R. A., and Sharp, P. A. (1984) Messenger RNA splicing in vitro: an excised intervening sequence and a potential intermediate. *Cell* **37,** 415–427.
11. Krainer, A. R., Conway, G. C., and Kozak, D. (1990) The essential pre-mRNA splicing factor SF2 influences 5' splice site selection by activating proximal sites. *Cell* **62,** 35–42.
12. Screaton, G. R., Cáceres, J. F., Mayeda, A., Bell, M. V., Plebanski, M., Jackson, D. G. Bell, J. I., and Krainer, A. R. (1995) Identification and characterization of three members of the human SR family of pre-mRNA splicing factors. *EMBO J.* **14,** 4336–4349.
13. Helfman, D. M., Ricci, W. M., and Finn, L. A. (1988) Alternative splicing of tropomyosin pre-mRNAs in vitro and in vivo. *Genes Dev.* **2,** 1627–1638.
14. Mayeda, A., Helfman, D. M., and Krainer, A. R. (1993) Modulation of exon skipping and inclusion by heterogeneous nuclear ribonucleoprotein A1 and pre-mRNA splicing factor SF2/ASF. *Mol. Cell. Biol.* **13,** 2993–3001.
15. Schmitt, P., Gattoni, R., Keohavong, P., and Stévenin, J. (1987) Alternative splicing of E1A transcripts of adenovirus requires appropriate ionic conditions in vitro. *Cell* **50,** 31–39.
16. Ohno, M., Sakamoto, H., and Shimura, Y. (1987) Preferential excision of the 5' proximal intron from RNA precursors with two introns as mediated by the cap structure. *Proc. Natl. Acad. Sci. USA* **84,** 5187–5191.
17. Noble, J. C. S., Pan, Z.-Q., Prives, C., and Manley, J. L. (1987) Splicing of SV40 early pre-mRNA to large T and small t mRNAs utilizes different patterns of lariat branch sites. *Cell* **50,** 227–236.

26

Yeast Pre-mRNA Splicing Extracts

Stephanie W. Ruby

1. Introduction

Splicing of eukaryotic precursor messenger RNAs (pre-mRNAs) excises the intron from the precursor and ligates the two exons together to produce the mature mRNA. It occurs via a two-step mechanism (**Fig. 1**) (reviewed in **ref. *1***). In the first step the 2' hydroxyl group of an intronic adenylyl residue initiates a transesterfication reaction at the 5' splice site. The result is the cleavage of the 5' splice site phosphodiester bond and the formation of a new, 2'-5' phosphodiester bond between the adenylyl residue and the 5' end of the intron. Because the intronic adenylyl residue has both 2'-5' and 3'-5' phosphodiester bonds, it is commonly referred to as the branch point nucleotide. The first splicing step yields two intermediates, "free" exon 1 and the lariat intermediate, which are not covalently linked together. In the second step, the 3' hydroxyl of the "free" exon 1 initiates a second transesterification reaction at the 3' splice site, resulting in the ligation of the two exons together to form the mRNA and the release of the lariat intron.

Splicing occurs in the nucleus in a large ribonucleoprotein, the spliceosome, which is composed of five small nuclear ribonucleoproteins (snRNPs) and numerous non-snRNP proteins (reviewed in **refs. *1–4***). Each snRNP is composed of one small nuclear RNA (snRNA) and several proteins. It is estimated that a total of at least 100 proteins are required for splicing in vivo, with most of these proteins probably having functions auxiliary to the actual catalysis of splicing. The catalysis of the two transesterification reactions is thought to be the function of at least one of the snRNAs, U6.

1.1. Splicing In Vitro

The elucidation of the mechanism of splicing as well as the identification and analysis of the factors required for splicing have progressed rapidly owing

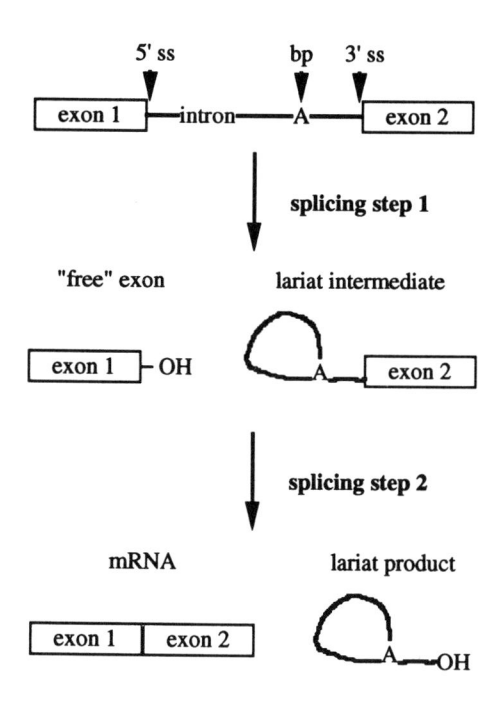

Fig. 1. The two-step mechanism of nuclear pre-mRNA splicing. The 5' and 3' splice sites (5'ss and 3'ss), and the branchpoint nucleotide (bp) of the pre-mRNA are indicated by *arrowheads*.

to the development and use of the in vitro splicing assay. The assay has two components that require some effort for the investigator to prepare: a single species of radiolabeled pre-mRNA, and a cellular extract that catalyzes splicing of the radiolabeled pre-mRNA.

1.1.1. The Radiolabeled Pre-mRNA

The radiolabeled pre-mRNA is synthesized in vitro using a cloned eukaryotic gene with an intron as the DNA template. Usually an 18–20-bp sequence encoding the promoter for the RNA polymerase from bacteriophage T7 or Sp6 is inserted by recombinant DNA techniques at the 5' end of the cloned gene *(5)*. Transcription termination is effected by "runoff"; a double-strand break is made at the 3' end of the gene by a restriction endonuclease and the RNA polymerase runs off the template at the double-strand break. Thus, in the presence of a radiolabeled ribonucleotide, the RNA polymerase produces a radiolabeled pre-mRNA of a defined size. After synthesis, the radiolabeled pre-mRNA is purified by gel electrophoresis to separate it from the DNA template as well as from smaller transcripts that may result from premature termination.

1.1.2. Whole Cell Extracts from Yeast

Cellular splicing extracts are made much like extracts for eukaryotic in vitro transcription assays: actively growing cells are lysed; the lysate is treated with 0.2–0.4 M KCl to release nuclear, nonhistone proteins; cellular debris and organelles are removed by centrifugal sedimentation, and finally the lysate is dialyzed to lower the salt concentration and to introduce 20% glycerol for stabilizing the lysate to freezing. For extracts from the yeast *Saccharomyces cerevisiae*, it is also necessary to remove or disrupt the tough cell wall. The difficulty in preparing an active splicing extract from yeast cells is to use conditions harsh enough to break or remove the cell wall yet gentle enough to preserve the integrity of the numerous splicing factors. In the first method used to produce yeast splicing extracts *(6)*, the cell wall is digested with the enzyme zymolyase. The resulting yeast spheroplasts are then lysed by homogenization in a hypotonic buffer. This method of preparing whole cell extract is still used and is described in this chapter. Two more recent methods use more drastic conditions to break the cell wall and lyse the cells and yet yield active extracts: grinding the cells in hypotonic buffer with glass beads *(7)* or in buffered 0.2 M KCl with a mortar and pestle while frozen in liquid nitrogen *(8)*. The latter "freeze–fracture" method, the most recent of the three, is also described in this chapter (*see* **Note 1**). Other methods for lysing the yeast cells, such as with a French press, have not yielded active extracts. Nor have attempts to make extracts from isolated yeast nuclei been successful, although extracts from human cells are today produced mainly from isolated HeLa nuclei (*see* Chapter 24).

The complex composition of the spliceosome necessitates careful preparation of cellular extracts to prevent inactivation by proteases or nucleases of any one of the numerous splicing factors. Lysing the large cytoplasmic vacuole of yeast cells releases several proteases *(9)*. Vacuolar breakage undoubtedly occurs to some extent in the original method and probably to an even greater extent in the glass-bead and freeze–fracture methods. One way to reduce protease activity is to use yeast mutant strains that are deficient in some vacuolar proteases *(6)*. In addition, the inclusion of protease inhibitors during extract preparation and/or splicing assays may prevent some proteolysis *(8)*. Similarly the inclusion of the nuclease inhibitor, RNasin, may reduce snRNA degradation during extract preparation, and both snRNA and pre-mRNA degradation during splicing assays *(10)*. The use of genetically altered strains deficient in specific nucleases has not yet been explored.

1.1.3. The Splicing Assay

The splicing assay itself is relatively simple to perform. The radiolabeled pre-mRNA and splicing extract are combined with buffer, magnesium, and

ATP, and incubated at 23–25°C. The reaction is then stopped and the RNAs in the reaction are extracted and fractionated by electrophoresis in a denaturing gel. The radiolabeled RNAs in the gel are visualized by autoradiography.

The type of denaturing gel used is critical to the final detection of the RNA species formed in the splicing reaction. The denaturing gels are similar to DNA sequencing gels in several aspects including their electrophoresis at a current and voltage high enough to generate temperatures within the gel sufficient to prevent intra- and intermolecular basepairing. This allows the RNA species to migrate according to their molecular weights. Nonetheless, the lariat intermediate and lariat product migrate anomalously due to their circular structure. Their mobilities relative to those of the other RNAs vary in different acrylamide percentages and crosslinking ratios. For the actin pre-mRNA described in this chapter, the gel conditions have been optimized to resolve all the RNA species from a splicing reaction (**Fig. 2**). The actin lariat intermediate and lariat product migrate more slowly than the pre-mRNA, mRNA, and "free" exon 1 in a 7.5% acrylamide gel (29:1, acrylamide:*bis*-acrylamide). The anomalous migration of the lariat RNA species can be used as described elsewhere *(11)* to tentatively identify these species when new pre-mRNAs are used for in vitro splicing assays.

1.2. Other Uses of Splicing Extracts

Beyond the basic splicing assay, splicing extracts can be used to study the functions of the numerous splicing factors in several different types of assays (reviewed in **refs.** *1–4*). For yeast, genetics and biochemistry are readily combined in vitro. For example, active splicing extract can be made from a temperature-sensitive (ts) mutant yeast strain grown at the permissive temperature. If the ts mutation is in a gene encoding a splicing factor, then incubation at the nonpermissive temperature may inactivate the mutant, but not wild-type, extract. Mutant extracts that are ts in vitro have been used to study the functions of both the proteins *(10)* and snRNAs *(12)* required for splicing. Alternatively yeast cells can be depleted of a splicing factor in vivo by genetic means and extracts from the depleted strain then analyzed for splicing defects *(7)*. Immunological and other methods can also be used to deplete, inhibit, alter, or substitute splicing factors in the extract *(13–15)*. Finally, extracts are starting materials for biochemical fractionation or purification of splicing factors as described elsewhere *(13,15,16)*.

2. Materials
2.1. Solutions for Splicing Extract Preparation

All solutions are made with distilled water unless otherwise indicated.

Yeast Pre-mRNA Splicing Extracts

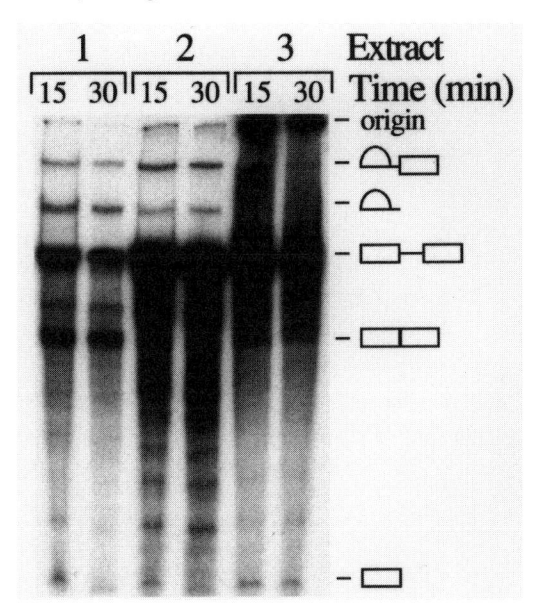

Fig. 2. The good, the bad, and the ugly splicing extracts. Whole cell extracts from three different yeast strains were prepared by the freeze–fracture method. The extracts were then assayed for activity in the presence of buffer, ATP, and a radiolabeled actin pre-mRNA at 23°C. At 15 and 30 min after the start of the reactions, samples were removed and the RNAs in the samples were extracted. The RNAs were fractionated by electrophoresis in a denaturing acrylamide gel and visualized by autoradiography. Shown here is an autoradiogram with Kodak XAR-Xomat film exposed to the gel for 5 h with two intensifier screens at –70°C. The 597 nt pre-mRNA, the intermediates (92 nt "free" exon 1 and 505 nt lariat intermediate), and products (288 nt mRNA and 309 nt lariat product) are designated by the symbols defined in **Fig. 1**. Due to their circular structures, the lariat intermediate and lariat product migrate more slowly than the larger pre-mRNA. Extract 1 (the good) has good splicing activity, extract 2 (the bad) has poor activity as well as nuclease activity, and extract 3 (the ugly) has poor activity as well as contaminants that interfere with either the extraction of the RNAs or their migation in the gel.

1. YPD. For YP, add 20 g of yeast extract and 40 g of bactopeptone to 1 L of deionized or distilled water. Make up a 50% (w/v) solution of dextrose by adding 50 g of anhydrous D-glucose to 70 mL of water and stirring on a hot plate to dissolve the sugar. Bring the volume up to 100 mL with water. Autoclave the YP in a capped, 2800-mL Fernbach flask or 4-L flask, and the 50% dextrose in a screw-capped bottle. Store at room temperature. Add 40 mL of 50% dextrose per liter of YP before using. Addition of the dextrose after autoclaving prevents "caramelizing" that can occur if dextrose is autoclaved in YP for too long.
2. 1 M Tris-HCl at pH 7.6, pH 7.8, and pH 8.0: Add 30.29 g of Tris base (Tris [hydroxymethyl]aminomethane, molecular biological grade) to 125 mL of water

followed by 92 mL of 1 M HCl. Bring the pH to 7.6, 7.8, or 8.0 by titrating with 1 M HCl and the volume to 250 mL by the addition of water. Autoclave then store at room temperature.
3. 4 M NaCl. Add 46.75 g of NaCl to 150 mL, then bring the volume up to 200 mL with water. Autoclave then store at room temperature.
4. 0.5 M ETDA, pH 8.0. Add 18.6 g of ethylenediamine tetracetic acid (EDTA, disodium dihydrate form) to 75 mL of water. Dissolve the EDTA and bring the pH to 8.0 by adding 10 N NaOH. Bring the volume to 100 mL with water. Autoclave, then store at room temperature.
5. 10% (w/v) Sodium dodecyl sulfate (SDS). Add 10 g of SDS (BDH Biochemical available from Hoeffer Scientific Instruments, San Francisco, CA, USA) to 85 mL of water and adjust the volume to 100 mL. Store at room temperature.
6. 50% (v/v) Glycerol. Add glycerin (analytical grade) to 100 mL of water in a graduated cylinder to bring up the volume to 200 mL. Autoclave, then store at room temperature.
7. 1 M MgCl$_2$. Add 20.33 g of magnesium chloride hexahydrate to 90 mL of water, and adjust the volume to 100 mL. Autoclave, then store at room temperature.
8. 2 M KCl. Dissolve 74.55 g in 400 mL of water, then bring the volume up to 500 mL with water. Autoclave, then store at room temperature.
9. 1 M potassium HEPES (HEPES-K$^+$), pH 7.8, at 23°C. Add 119.2 g of N-2-hydroxyethylpiperazine-N'-2-ethanesulfonic acid (Ultrol grade HEPES free acid from Calbiochem, La Jolla, CA, USA) and 20.1 g of KOH to 400 mL of water. Bring the volume to 500 mL with water. Dilute a small sample to 10 mM and check that the pH is 7.5 at 23°C and if necessary adjust with 2 N HCl; when at 4°C, the pH will be 7.8. Autoclave or sterilize by filtration, then store at –20°C. The solution may become a light yellow after autoclaving, but this will not affect the extract.
10. 1 M Dithiothreitol (DTT). Dry DTT is stored at –20°C. Dissolve 3.9 g of DTT in 15 mL of sterile water and bring the volume up to 25 mL with sterile water. Make this up within 24 h of use and keep at 4°C.
11. 0.5 M phenylmethane sulfonate (PMSF). Dissolve 0.87 g of PMSF in 10 mL of dimethyl sulfoxide and store at –20°C (*see* **Note 2**).
12. 1 M Potassium (K) phosphate buffer, pH 7.6. Slowly add 8.85 g of monobasic potassium phosphate anhydrous (KH$_2$PO$_4$), alternating with 75.75 g of dibasic potassium phosphate anhydrous (K$_2$HPO$_4$) to 400 mL water. When the salts are dissolved, check that the pH is 7.6, then adjust the volume to 500 mL and autoclave. Store at room temperature.
13. Zymolyase solution (20 mg/mL in 20 mM K phosphate buffer, pH 7.6, 5% glucose). Store dry zymolyase-100T (100,000 U/g; Seikagaku America, Falmouth, MA, USA) at 4°C. Make up the amount you need just before use as zymolyase is expensive and is not very stable to freezing in solution. For preparing extract from 1 L of cell culture, add 4 µL of sterile 1 M K phosphate buffer pH 7.6, and 50 µL of sterile 50% glucose to 446 µL of sterile water. Add 10 mg of zymolyase. The zymolyase will not completely dissolve but use it as is.

14. 2 M D-Sorbitol. Add 364.4 g to 650 mL of warm water. Stir on a hot plate to dissolve. Bring the volume up to 1 L and autoclave. Store at room temperature.
15. SB buffer: 1 M sorbitol, 50 mM Tris-HCl, pH 7.8, 10 mM MgCl$_2$. Combine 150 mL of sterile 2 M sorbitol, 15 mL of 1 M Tris-HCl, pH 7.8, 3 mL of 1 M MgCl$_2$, and 132 mL of sterile water and store at room temperature. Make SB + 30 mM DTT fresh by adding 1.5 mL of 1 M DTT to 50 mL of SB. Make SB + 3 mM DTT fresh by adding 0.75 mL of 1 M DTT to 250 mL of SB.
16. Isotonic solution: 1 M sorbitol. Dilute 2 M sorbitol to 1 M with water.
17. Lysis solution: 0.1% (w/v) SDS, 50 mM EDTA. Dilute 10% SDS and 0.5 M EDTA, pH 8.0.
18. Buffer A: 10 mM HEPES-K$^+$, pH 7.8, 1.5 mM MgCl$_2$, 20 mM KCl, 0.5 mM DTT, 0.5 mM PMSF. Make up fresh then keep on ice. Add 1 mL of 1 M HEPES-K$^+$, pH 7.8, 150 µL of 1 M MgCl$_2$, 1 mL of 2 M KCl, and 50 µL of 1 M DTT to 97 mL of sterile water. Just before use, add 100 µL of 0.5 M PMSF (optional; *see* **Notes 2** and **3**). Note that PMSF has a half-life of about 30 min in an aqueous solution. When PMSF is added, a small amount will precipitate but this has no effect on the extract.
19. Buffer D: 20 mM HEPES, pH 7.8, 0.2 mM EDTA, 50 mM KCl, 20% glycerol (v/v), 1 mM DTT, 0.5 mM PMSF. For preparing extract from 1 L of cell culture, prepare 2 L of buffer D the day before the extract will be made. Add 40 mL of 1 M HEPES, pH 7.9, 0.8 mL of 0.5 M EDTA, pH 8.0, and 0.50 mL of 2 M KCl to a 2-L graduated cylinder and bring the volume up to 1.8 L with water. Add glycerol (analytical grade glycerin) to bring the volume to 2 L. Autoclave, then place in a cold room to cool overnight. Just before dialysis, add 1 mL of 1 M DTT and 1 mL of 0.5 M PMSF (optional; *see* **Notes 2** and **3**).
20. AGK buffer: 10 mM HEPES-K$^+$, pH 7.8, 1.5 mM MgCl$_2$, 200 mM KCl, 10% (v/v) glycerol, 0.5 mM DTT, and 0.5 mM PMSF. Prepare fresh that day by adding 138 mL of sterile H$_2$O, 40 mL of 50% glycerol, 20 mL of 2 M KCl, 300 µL of 1 M MgCl$_2$, 2 mL of 1 M HEPES-K$^+$, pH 7.9, and 100 µL of 1 M DTT in a sterile bottle and put on ice. Just before using, add 200 µL of 0.5 M PMSF (optional; *see* **Notes 2** and **3**).

2.2. Solutions for In Vitro Transcription and Splicing Assays

1. Concentrated Tris–borate–EDTA (10× TBE): 890 mM Tris-borate, 25 mM EDTA, pH 8.3. Dissolve 216 g of Tris base (molecular biological grade), 110 g of boric acid (crystalline anhydrous), and 18.6 g of disodium dihydrate EDTA in 1600 mL of water. Check that the pH is 8.3. Bring the volume up to 2 L with water. Store at room temperature in a carboy.
2. 1× TBE: 89 mM Tris-borate, 2.5 mM EDTA, pH 8.3. Dilute 10× TBE to 1× TBE using either distilled or deionized water. Store at room temperature in a carboy.
3. 1× TE: 10 mM Tris-HCl, pH 7.6, 1 mM EDTA. Add 5 mL of 1 M Tris-HCl, pH 7.6, and 1 mL of 0.5 M EDTA, pH 8.0, to 494 mL of sterile water. Store at room temperature.
4. 30% (w/v) PEG$_{8000}$. Add 30 g of polyethylene glycol 8000 mol wt (PEG$_{8000}$ from Eastman Kodak, Rochester, NY, USA) to 65–70 mL of warm (50°C) water. If

necessary, continue stirring on a hot plate to dissolve. Bring volume up to 100 mL with water and autoclave. After the solution cools, dispense into 1- and 10-mL aliquots and store at –20°C.
5. 1 M Spermidine. Dissolve 1.27 g of spermidine trihydrochloride in 3.5 mL of sterile water. Adjust the pH to 7.6 by the addition of 10 N NaOH, then bring the volume to 5 mL with water. Store at –20°C.
6. 10× transcription buffer: 0.4 M Tris-HCl, pH 7.8, 60 mM MgCl$_2$, 40 mM spermidine. Add 800 µL of 1 M Tris-HCl, pH 7.8, 120 µL of 1 M MgCl$_2$, and 80 µL of 1 M spermidine to 1 mL of sterile water. Store in aliquots of 500 µL at –20°C. DTT is added separately when the transcription reaction is set up.
7. 0.1 M DTT. Dilute 1 M DTT to 0.1 M with water and store 500–750 µL aliquots at –20°C.
8. Ribonucleotide stocks for in vitro transcription (10× NTP: 5 mM each ATP, GTP, and CTP and 1 mM UTP, pH 7.0), and for splicing assays (100 mM ATP). Stock solutions of about 100 mM of each ribonucleotide are prepared and then used to make the 10× NTP and 100 mM ATP solutions as follows. Add 5 mL of water to 500 mg of each ribonucleotide (the sodium form of ATP, GTP, CTP, and UTP from Amersham Pharmacia Biotech, Piscataway, NJ, USA). Bring the pH of each ribonucleotide carefully to 6.8–7.2 by adding approx 50 µL of 5 N NaOH followed by 100–850 µL of 1 N NaOH in 20-µL aliquots. ATP requires the most NaOH and UTP the least NaOH to bring the solution to pH 7.0. Monitor the pH by placing a drop of the ribonucleotide solution on pH paper (pH range 6.5–10). Measure the absorbance of a 500-fold dilution (3 µL into 1.5 mL of water) of each ribonucleotide solution at λ_{max} for each nucleotide (259, 253, 271, and 262 nm for A, G, C, and U respectively) in a quartz cuvet in a spectrophotometer. Calculate the concentration of each stock solution; the absorbance coefficient E_{max} (M^{-1}cm^{-1}) at pH 7.0 for each ribonucleotide is 1.59×10^4, 1.37×10^4, 0.9×10^4, and 1.0×10^4 for A, G, C, and U respectively, and absorbance equals $E_{max}M$ in a cuvet with a path length of 1 cm. Prepare the 10× NTP and 100 mM ATP stocks by dilution in sterile water. Store all ribonucleotide solutions in 200–1000-µL aliquots at –20°C. Thaw the ribonucleotide solutions quickly in a 37°C water bath and keep on ice while using them at the bench.
9. Trichloroacetic acid (TCA). To 500 g of TCA in the bottle from the supplier, add 227 mL of water to make 100% TCA. Add the 100% TCA to water in a graduated cylinder to dilute it 10- or 20-fold respectively to get 10% and 5% TCA (*see* **Note 1**).
10. 3 M Sodium acetate (NaOAc), pH 5.4. Add 40.82 g of sodium acetate trihydrate to 75 mL of water. Add glacial acetic acid to bring the pH to 5.4, and finally bring the volume to 100 mL with water. Autoclave, then store at room temperature.
11. RNA elution buffer: 20 mM Tris-HCl, pH 7.6, 0.5 M NaCl, 10 mM EDTA, 2% (v/v) phenol. In a 50-mL sterile, screw-capped polypropylene graduated centrifuge tube (such as Falcon Brand no. 2070), add 6.24 mL of 4 M NaCl, 1.0 mL of 0.5 M EDTA, pH 8.0, 1 mL of 1 M Tris-HCl, pH 7.6, and 1 mL of phenol (equilibrated with Tris-EDTA or NaOAc-EDTA) (*see* **Note 2** and **step 23** below) and bring volume to 50 mL with sterile water. Store at 4°C.

Yeast Pre-mRNA Splicing Extracts

12. 7.5 M Ammonium acetate (NH$_4$OAc). Add 57.8 g of ammonium acetate to 60 mL of water, stir until dissolved, and adjust the volume to 100 mL. Sterilize with a filter of 0.2 µm pore size. Store at 4°C.
13. 70% (v/v) Ethanol. Pour 95% ethanol to 350 mL in a graduated cylinder, then bring the volume to 475 mL with sterile water. Alternatively, pour absolute ethanol to 350 mL in a graduated cylinder, then bring the volume to 500 mL with sterile water. Store at room temperature.
14. Dye solution: 10% (w/v) each bromophenol blue and xylene cyanol in water. Add 1 g of bromophenol blue and 1 g of xylene cyanol to 7 mL of sterile water in a sterile, disposable, polypropylene screw-capped tube. Bring the volume to 10 mL with sterile water. Store at room temperature. The dyes may not completely dissolve: use as is and vortex vigorously just before removing an aliquot.
15. Deionized formamide. Add 10–50 mL of formamide (molecular biological grade) to a proportional amount (10–50 g) of mixed bed resin AG501-X8 in a sterile beaker and stir for 30 min at room temperature. Check the pH of a small aliquot with a pH meter to be sure that it is 6.8 or above and discard the aliquot. Remove the resin by filtration through no. 1 Whatman paper in a sterilized Buchner funnel into a sterile filter flask. Store in 1–10-mL aliquots at –20°C. Check the pH of aliquots stored longer than 6 mo. If the pH is lower than 6.5, repeat the deionization process.
16. Transcript loading buffer: 98% formamide, 50 mM EDTA, 0.1% each bromophenol blue and xylene cyanol. Combine 50 µL of 0.5 M EDTA, pH 8.0, 5 µL of dye solution, and 445 µL of deionized formamide. Store at –20°C.
17. Proteinase K solution: 2.5 mg proteinase K/mL in 4% (w/v) sodium dodecyl sulfate, 50 mM Tris-HCl, pH 7.8, 0.25 M EDTA. Add 12.5 mg of proteinase K (Boehringer Mannheim, Indianapolis, IN, USA) to 1 mL of water, 250 µL of Tris-HCl, pH 7.8, 2.5 mL of 0.5 M EDTA, and 1.25 mL of 10% SDS. Store at –20°C in about 1-mL aliquots.
18. Carrier tRNA: 10 mg/mL in sterile water. Dissolve 100 mg of *E. coli* tRNA (Boehringer Mannheim, Indianapolis, IN, USA) in 10 mL of sterile water. Store at –20°C in 0.5–1-mL aliquots.
19. Stop buffer for splicing assay. Mix 200 µL of carrier tRNA and 1 mL of proteinase K solution. Store at –20°C in 0.2–0.5-mL aliquots. Thaw at 37°C and keep at 23–25°C to prevent the SDS from precipitating while using the solution at the bench.
20. Load buffer for splicing assay: 0.1× TBE, 8 M urea, 0.1% each bromophenol blue and xylene cyanol. Add 0.22 g of urea (Ultrapure from ICN Biomedicals Inc., Cleveland, OH, USA) to 200 µL of sterile water. Heat at 65°C until the urea dissolves. Add 200 µL of this urea solution to 94 µL of water, 3 µL of 10× TBE, and 3 µL of dye solution. Make up fresh each day.
21. Sodium acetate (NaOAc)/EDTA solutions: 300 mM NaOAc, pH 5.4, 50 mM EDTA; and 50 mM NaOAc, pH 5.4, 10 mM EDTA. Dilute 3 M NaOAc, pH 5.4, and 0.5 M EDTA, pH 8.0, in sterile water. Store at room temperature.
22. 1 M Tris base. Add 121.1 g of Tris base (Tris[hydroxymethyl] aminomethane, molecular biological grade) to 900 mL of warm water. Stir to dissolve, then bring the volume up to 1 L with water. Autoclave, then store at room temperature.

23. Phenol equilibrated with Tris/EDTA. Whenever working with phenol, wear gloves, safety glasses, shoes, and a laboratory coat (see **Note 2**). Perform the following manipulations in the plastic-coated jar in which the phenol is packaged. Heat 500g of molecular biological grade phenol at 65°C until it becomes liquid. Add 0.5 g of 8-hydroxyquinoline. Add 250 mL of 1 M Tris base, then 250 mL of water. Mix vigorously by adding a magnetic stir bar to the jar and putting the jar on a magnetic stir plate for about 10 min. Stop stirring to allow the phases to separate. The lower, phenolic phase will appear bright yellow due to the 8-hydroxyquinoline and will remain slightly cloudy after the phases separate. Remove most of the upper aqueous phase by suction. Add two volumes of 1× TE, pH 7.6, stir the solution for 2–5 min, and allow the phases to separate again. Check that the upper aqueous solution is pH 6.8–7.2. Remove most of the aqueous phase to leave about 2 cm of it. Store for 1–2 d at 4°C. For longer storage periods, freeze at –20°C. When the organic phase turns to light orange, discard the solution.
24. Phenol equilibrated with NaOAc/EDTA. Heat the phenol and add hydroxyquinoline as described in **step 23**. Add 500 mL of 300 mM NaOAc, pH 5.4, 50 mM EDTA to the heated phenol with hydroxyquinoline. Stir then allow the phases separate as in **step 23**. Suction off the upper aqueous phase. Add about 100 mL of fresh 300 mM NaOAc, pH 5.4, 50 mM EDTA and stir vigorously again for about 5 min. Store as in **step 23**.
25. Phenol/chloroform/*iso*-amyl alcohol equilibrated with NaOAc/EDTA. Add 20 mL of *iso*-amyl alcohol to 480 mL of CHCl$_3$ and then add the solution to 500 mL of phenol equilibrated with NaOAc-EDTA as described in **step 24**. Stir vigorously for about 5 min, allow the phases to separate, and store as in **step 23**.
26. Chloroform/*iso*-amyl alcohol (24:1). Working in a chemical fume hood and wearing gloves (see **Note 2**), pour 240 mL of chloroform (CHCl$_3$) into a glass graduated 250-mL cylinder. Bring the volume to 250 mL with *iso*-amyl alcohol. Store in a capped glass bottle in the hood.
27. 1× TBE, 8 M urea. Add 240.2 g of urea (high quality, not ultrapure) to 250 mL of warm water and 50 mL of 10× TBE. Stir on low heat until the urea dissolves. Pour the solution into a graduated cylinder and after it has cooled, bring the volume up to 500 mL with water. Store at 4°C. If some urea precipitates during storage, warm the solution in a 37°C water bath until the urea dissolves.
28. Acrylamide (29:1 acrylamide to *bis*-acrylamide) in 1× TBE, 8 M urea. For a 5% (w/v) acrylamide solution, add 12.1 g of acrylamide (electrophoresis grade) and 0.4 g of *bis*-acrylamide to 250 mL of 1× TBE, 8 M urea and stir until the acrylamide dissolves (see **Note 2**). For a 7.5% (w/v) acrylamide stock, add 18.12 g of acrylamide (electrophoresis grade), and 0.63 g of *bis*-acrylamide to 250 mL of 1× TBE, 8 M urea and stir until acrylamide dissolves. Both acrylamide stocks are stored in screw-capped bottles at 4°C for up to 1 yr. If a precipitate forms during storage, warm the solution in a 37°C water bath until the precipitate dissolves.
29. 10% (w/v) Ammonium persulfate (APS) and $N,N,N'N'$-tetramethylethylenediamine (TEMED). Put 1 g of APS (molecular biological or electrophoresis grade) into a 15-mL graduated, screw-capped, plastic tube and bring volume up

Yeast Pre-mRNA Splicing Extracts

to 10 mL by adding water. Store 10% APS for up to 1 wk at 4°C. Store TEMED at 4°C.

30. Silanizing solution: 15% (v/v) dichloromethylsilane in chloroform. Wearing gloves and working in a chemical fume hood (*see* **Note 2**), pour 170 mL of $CHCl_3$ into a 200-mL glass graduated cylinder. Pour dichloromethylsilane to bring the volume to 200 mL. Store in the fume hood in a glass jar with a ground-glass stopper.
31. 2% (w/v) $NaHCO_3$. Add 160 g of sodium bicarbonate ($NaHCO_3$) to 3.7 L of water. When it is dissolved, bring the volume up to 4 L with water.
32. 15% (v/v) Methanol, 5% (v/v) acetic acid. Pour 150 mL of methanol (analytical grade) into a 1-L graduated cylinder. Add water to 950 mL. Bring the volume to 1 L by adding glacial acetic acid (*see* **Note 2**). Store at room temperature in a screw-capped bottle.

2.3. Equipment and Supplies for Preparing Extracts

1. An incubator capable of rotating large flasks up to 300 rpm and maintaining the temperature at 30°C for culturing yeast cells.
2. Four types of centrifuges are needed: micro-, high-speed, tabletop (or clinical) and ultracentrifuges with the appropriate rotors.
3. Most items including microcentrifuge tubes used for storing extract, synthesizing transcript, and splicing assays do not need to be autoclaved unless otherwise stated. For most manipulations such as handling dialysis tubing, wear *powder-free* gloves.
4. Screw-capped, polycarbonate, 10-mL (Oak Ridge) tubes (Nalgene Brand no. 3118-0010) and screw-capped, thick-walled, polycarbonate 10.4-mL tubes (no. 355603 from Beckman Instruments, Brea, CA, USA) are washed by hand with soap and water, rinsed with sterile water, and used only for preparing splicing extracts. For preparing extract from more than 2 L of cell culture, use 28-mL Oak Ridge tubes (Nalgene Brand no. 3118-0028) and 26.3-mL thick-walled screw-capped tubes (no. 355618 from Beckman Instruments, Brea, CA, USA). The 10.4 and 26.3 mL thick-walled tubes do not require adapters when used in the Beckman Brand 70.1 and 60 rotors respectively.
5. Both glass and metal Dounce tissue grinders (also called Dounce homogenizers; Wheaton or Kontes Brand) can be used for preparing extract. For 1 L of culture use either a glass or metal 7- or 14–15-mL Dounce. For larger cultures, use a 15- or 40-mL Dounce. It is important to use a Dounce with a tight-fitting pestle. Because Dounces vary considerably, order several at once and test all possible combinations of mortars and pestles for the tightest combination. When the mortar and pestle of a 7.5- or 14-mL metal Dounce (Wheaton Brand) are dry, push the pestle down into the mortar. You should feel resistance. Rapidly pull the pestle out of and away from the mortar to hear a sharp "pop." When the bottom chamber of the mortar is filled with water, you should have to strongly push and pull the pestle into and out of the mortar. For glass Dounces, pulling the pestle out of the mortar should produce a dull, low-pitched "pop."

6. Porcelain 5-in. mortars and pestles (Coors Brand no. 60322 mortar and no. 60323 pestle available from VWR Scientific Products, West Chester, PA, USA). For the initial use, treat the mortar and pestle with chromic acid solution such as Chromerge (VWR Scientific Products, West Chester, PA, USA) for 15–30 min and then rinse thoroughly with water. Add AGK buffer to the top of the mortar, place the pestle in the mortar, and allow to stand for about 30 min at room temperature. Finally rinse the mortar and pestle thoroughly with sterile, distilled water and allow to air-dry. After subsequent uses, scrub and rinse the mortar and pestle in deionized water, then rinse with sterile distilled water and air-dry.
7. Dialysis tubing (1/2 in., Spectropor 3 Brand, >3500 Dalton cutoff) and 4 cm tubing clamps. Treat the dialysis tubing at least 1 d in advance as follows: cut into 30 cm long strips, boil in 4 L of 2% sodium bicarbonate for 15 min, and wash 3× with copious amounts of sterile water. Do not let the tubing dry after it is treated. Store the treated dialysis tubing in 70% ethanol in a tightly closed container at 4°C for several years. About 15–30 min before using a piece of tubing, remove it from the ethanol and wipe off excess fluid by running the tubing between gloved fingers. Submerge the tubing in a few hundred milliliters of sterile water and rinse once with more sterile water. Flush the inside of the tubing twice with 10–12 mL of fresh sterile water and then submerge it in cold buffer D.
8. Two 350-mL Dewar flasks.
9. Insulated (cryo) gloves.
10. 0.5–10 L of liquid nitrogen (N_2).
11. An ultra-low-temperature (–70 to –80°C) freezer or liquid nitrogen storage tank for storing the extract.

2.4. Equipment and Supplies for In Vitro Transcription and Splicing Assays

1. [α-^{32}P]uridine triphosphate (UTP) at 3000 Ci/mmole and 10 mCi/mL.
2. Phage T7 or SP6 RNA polymerase. T7 RNA polymerase (expressed from the cloned gene and available from New England Biolabs, Beverly, MA, USA) has 50 U/µL and Sp6 RNA polymerase (expressed from the cloned gene and available from Promega, Madison, MA, USA) has 20 U/µL.
3. Recombinant placental RNase inhibitor (RNasin) available from Promega (Madison, WI, USA) at 40 U/µL.
4. Steel vacuum filtration membrane holder for 24-mm filters (Hoeffer Scientific Instruments, San Francisco, CA, USA).
5. Whatman Brand glass fiber (GF/C) 24-mm filter papers.
6. A power supply such as Dk2-2 model from Dan-Kar, Reading, MA, USA, that can maintain constant voltage up to 1000 V and reach 100 mA of direct current.
7. Two glass plates (4-mm thick): the back plate is 19-cm wide and 16-cm high; and the notched plate is the same size as the back plate, but it has a notch that is 2.2-cm deep and 16.3-cm wide.
8. Teflon combs and spacers. Sheets of Teflon can be obtained from a local plastics supplier and cut by hand with scissors or by machine. For purifying the tran-

script, the spacers and combs are 1.5 mm thick. The comb is 16 cm wide by 3 cm tall and has seven teeth separated by 2 mm. Each tooth is 1.4 cm long and 1.8 cm wide. The well formed by a single tooth will hold a maximum volume of 80 µL. For the splicing assay gel, the comb and spacers are 0.38 mm thick. The comb is 15.5 cm wide and 2.6 cm deep and has 14 teeth separated by 4 mm. Each tooth is 8 mm high and 7 mm wide. The well formed by a single tooth will hold up to 15 µL.
9. Yellow electrical tape (no. 56, catalog no. HD7116-20, from 3M, Minneapolis, MN, USA).
10. Sterile, disposable, capped, polypropylene culture tubes (12 × 75 mm or 17 × 100 mm) and sterile, disposable, screw-capped, graduated, conical, polypropylene centrifuge tubes (50 mL size such as Falcon Brand no. 2070).
11. Empty Quicksep columns from Isolab, Inc., Akron, OH, USA.
12. X-ray film (8 × 10 in.) such as Kodak XAR5 (available from VWR Scientific Products, West Chester, PA, USA), light-tight film cassettes, intensifier screens, a small flashlight or penlight, and film processing capabilities.
13. Silanized, baked glass rods and Corex tubes. Under the chemical fume hood, submerge 10 inch glass rods in silizanizing solution in a glass pan for 5–10 min. Pour the excess silanizing solution back into the container for reuse and allow the glass rods to dry by evaporation in the pan in the hood. Fill 15-mL, thick-walled glass (Corex) tubes with silanizing solution. Pour the solution back into the container for reuse and dry the tubes by evaporation under the hood. Wash the rods and Corex tubes 3× with deionized water, drain, wrap in foil, and bake in a dry air oven at 252°C for at least 4 h.
14. Platform shaker or rocker.
15. Speed-Vac concentrator or lyophilizer.
16. 3MM Whatman brand chromatography paper (46 × 67 cm).
17. Gel dryer (optional).

3. Methods
3.1. Whole Cell Extracts by Homogenization of Spheroplasts

1. Grow a 1-L culture of yeast cells overnight to mid- to late-log phase (*see* **Note 4**). First grow a 10–50 mL culture overnight in a 125- or 250-mL culture flask. Use this culture to inoculate 1 L of YPD the next day. Grow the culture overnight at 250–300 rpm at 30°C to a density of $3–5 \times 10^7$ cells/mL. For the yeast strain EJ101 (*Mat a, trp1, pro1-126, prb1-112, pep4-3,* and *prc1-126*), the generation time at 30°C is 90 min in YPD, so adding 1.5 mL of the fresh overnight culture (at about 10^8 cells/mL) to 1 L of YPD at 7 p.m. will give the appropriate cell density the following morning at 9 a.m.
2. This method will require from 10 to 12 h beginning with the harvesting of the cells and ending with the freezing of the prepared extract. To get the most active extract possible, it is best to work as quickly as possible until the dialysis step (*see* **Note 5**).
3. Prepare the cells for zymolyase treatment. Harvest the cells by centrifugation in a high-speed centrifuge at 1200–1500*g* for 5 min. Resuspend the cell pellet in about

100 mL of SB + 3 mM DTT and centrifuge again. Resuspend the cell pellet in a total of 30 mL of SB + 30 mM DTT. Let the cells stand at room temperature for 15 min. This incubation in 30 mM DTT is critical for subsequent spheroplast formation. Centrifuge the cells at 1500g for 5 min. If desired, determine the weight wet of the cells for **step 6** below. Resuspend the cells in 30 mL of SB + 3 mM DTT and transfer to a sterile 120-mL Erlenmeyer flask.
4. Incubate the cells with zymolyase and monitor spheroplast formation. Remove two 20-µL aliquots of cells; add one to 1 mL of isotonic solution, the other to 1 mL of lysis solution; and monitor cell lysis by reading the absorbance of the two samples in a spectrophotometer at λ_{600}. Add 250 µL of zymolyase-100T solution to the cells in the flask and shake slowly at 50–60 rpm at 30°C for up to 40 min. Every 10–15 min during this time remove two 20-µL aliquots and put into 1 mL each of isotonic and lysing solutions. Read the absorbance of the samples to monitor spheroplast formation. When the absorbance of the sample in lysis solution is 20% or less that of the sample in isotonic solution (*see* **Note 6**), the enzyme treatment is sufficient for subsequent lysis of the spheroplasts by homogenization.
5. All steps from this point onward are done in the cold room to keep the sample cold. Instruments including pipets and rotors, as well as solutions, are precooled overnight in the cold room. Harvest the spheroplasts by centrifugation at 1500g for 5 min at 4°C.
6. Lyse the spheroplasts by homogenization. Using a sterile, silanized glass rod, gently resuspend the spheroplasts in ice-cold buffer A. Use 8 mL of buffer A for 1 L of original culture at $4–5 \times 10^7$ cells/mL, or calculate the amount of buffer A to use as 1.4 mL/g of wet cells determined in **step 2** (*see* **Note 7**). Transfer the spheroplasts to a cold 7- or 14-mL Dounce homogenizer; the suspension should fill only the lower chamber of the mortar in the absence of the pestle. With the mortar resting in ice in an ice bucket, homogenize the suspension with 5–10 strokes (*see* **Note 8**).
7. Treat the homogenate with 0.2 M potassium to release splicing factors from the nucleus. Pour the homogenate into a sterile, 15 mL, graduated, disposable conical centrifuge tube. If there are a large number of air bubbles in the homogenate, centrifuge the tube briefly (2 min) in a cold tabletop centrifuge to eliminate the bubbles. Note the volume of the homogenate to the nearest 0.5 mL, always rounding up the number.

    Pour the homogenate into a 25-mL sterile, glass beaker with a small stirring bar. Place the beaker surrounded with ice in an ice bucket and put the ice bucket on top of a magnetic stir plate; this arrangement keeps the homogenate from being heated by the stir plate during subsequent steps. Set the stir bar to rotate slowly (approx 60–120 rpm) and let it stir for 10 min. Add 0. 9 vol of cold 2 M KCl dropwise from a Pasteur pipet over a 2-min period to a final concentration of 0.2 M. Let the homogenate stir slowly for 30 min.
8. Sediment cellular debris and organelles by two centrifugations. Pour or pipet the homogenate into a 10- or 30-mL screw-capped Oak Ridge tube and pellet large debris by centrifugation in a high-speed centrifuge at 33,000g for 30 min at 4°C.

Yeast Pre-mRNA Splicing Extracts

During this centrifugation step, cool the ultracentrifuge to 4°C. After the high-speed centrifugation is complete, remove the tube from the rotor in the cold room and pipet the supernatant into a thick-walled, polycarbonate, screw-capped tube. Centrifuge the sample in an ultracentrifuge at 100,000g for 1 h at 4°C.

9. Dialyze the extract. Carry the rotor to the cold room and then remove the sample tube, taking care not to perturb the contents in the tube. Carefully place the tube in a rack and remove its cap. Locate the three layers in the sample: a small, top layer containing lipids and lipoproteins; a large, middle layer that is a clear, light yellow-green, and contains most of the splicing activity and some ribosomes; and a small bottom layer, that is a clear, light brown and contains organelles, ribosomes, and large cellular debris. Remove a piece of dialysis tubing from buffer D, squeeze off excess buffer with gloved fingers, and close one end of the tubing with a clip. With a cold, sterile Pasteur pipet, remove the pale, yellow-green middle layer being careful not to take the upper or bottom layer. Pipet the extract into the dialysis tube; you will have about 6–8 mL. Squeeze out excess air from the top of the tube and clip the top of the tube as close as possible to the extract. Place the tube into a sterile beaker with 1 L of buffer D and a sterile, magnetic stir bar. Surround the beaker with ice in an ice bucket and put the ice bucket on top of a magnetic stir plate in the cold room.

   Dialyze the extract for a total of 3 h, replacing the first liter of buffer D with 1 L of fresh buffer after 1.5 h. Dialyzing the extract for longer than 4 h may cause loss of activity. For larger amounts of extract or multiple extract preparations proportionately increase the amount of buffer D for the dialysis.

10. Centrifuge the extract to remove any precipitate that may have formed during dialysis. The volume of the extract will have been reduced by 1–2 mL at the end of the dialysis. Remove the tube from dialysis buffer, and wipe off excess fluid on the outside of tube with a tissue. Remove the top clip and with a cold, sterile Pasteur pipet, take the extract out of the tube and put into either an ice-cold Oak Ridge tube or several ice-cold microcentrifuge tubes. Centrifuge at 12,000–14,000g for 10 min at 4°C.

11. Store the extract. Aliquot 100, 200, or 500 µL of extract per tube into microcentrifuge or cryogenic tubes kept on ice. Flash freeze by dropping the tubes into liquid N_2 in a Dewar flask. Store in liquid N_2 or in a –70°C freezer.

12. One liter of original cell culture will yield about 6 mL of extract with 20–30 mg of protein/mL.

3.2. Whole Cell Extract by the "Freeze–Fracture" Method

1. Grow the yeast cells as described in **Subheading 3.1., step 1**. Harvest the cells by centrifugation in a high-speed centrifuge at 1500g for 10 min at 4°C.
2. Prepare the cells for flash-freezing. Resuspend the cells in 50–100 mL of cold AGK + PMSF buffer and pellet again by centrifugation for 10 min at 1500g at 4°C. From this step forward, keep the sample ice-cold. Pour off the supernatant and resuspend the cells in 20 mL of fresh, cold AGK + PMSF buffer. Transfer to

one sterile, graduated, 50-mL polypropylene centrifuge tube (such as Falcon Brand no. 2070). Pellet the cells by centrifugation in a tabletop or high-speed centrifuge at 1500g for 10 min at 4°C. Pour off the supernatant and estimate the pellet volume to the nearest 0.5 mL. Add 0.4 vol of AGK + PMSF buffer (*see* **Note 7**) and resuspend the cells with a silanized, baked glass rod. The suspension will be very thick. Keep on ice.
3. Flash-freeze the suspension by dripping it from a Pasteur pipet into the liquid N_2 to create frozen pellets from 0.2 to 0.5 cm in diameter. Wear safety glasses while handling liquid N_2. Before freezing the cell suspension, prepare the mortar and pestle as follows. Place the mortar and pestle in an empty, styrofoam ice bucket. Pour liquid N_2 into and outside of the mortar and wait until it stops bubbling violently. Have the mortar about half-filled with liquid N_2 and submerged in about 2 cm of liquid N_2.

   To easily pipet the thick cell suspension, widen the bore of a 5-in. Pasteur pipet by breaking about 1 in. off the narrow end of the pipet. Keeping the widened pipet tip at least 5 cm above the liquid nitrogen (to prevent the suspension from freezing in the pipet), pipet the suspension dropwise into the liquid N_2 in the mortar (*see* **Note 9**).
4. When all the suspension is frozen as pellets, grind the extract. You will need to wear an insulated glove on at least one hand to hold the mortar in place. Gently pound the round, frozen pellets submerged in liquid N_2 to break them into small pieces, then grind the suspension with a circular motion into a granular powder. After the liquid N_2 has evaporated from inside the mortar, grinding is optimal and the suspension can be ground into a very fine powder. The finer the powder, the more active the extract will be. Total time for grinding is usually 20–30 min. Even though the liquid N_2 is evaporated from the inside of the mortar during the last grinding stage, keep adding liquid N_2 to the ice bucket to keep the mortar submerged in 1–2 cm of liquid N_2.
5. Transfer the cell "powder" into a tube and thaw. Freeze a spatula in liquid N_2, then use it to scoop the frozen powder into a screw-capped Oak Ridge tube on the laboratory bench at room temperature. Allow the ground suspension to thaw on ice while intermittently vortexing the tube (*see* **Note 10**).
6. Continue processing the extract as described in **Subheading 3.1., steps 8–12**.
7. One liter of cell culture will yield about 3–4 mL of extract. About 7 h are required to make the extract once the cells are frozen in liquid N_2.

3.3. Splicing Assay of Extract

3.3.1. Preparing Radiolabeled Pre-mRNA for the Splicing Assay

3.3.1.1. SYNTHESIZING THE TRANSCRIPT (*SEE ALSO* CHAPTER 1).

1. A 40-µL transcription reaction usually provides sufficient transcript for 6–10 splicing assays with 10–15 reactions. Use the transcript for 2 wk after it is made. After 2 wk, it is usually too degraded by radiolysis to splice efficiently and to give well-defined bands in the splicing assay.

2. Prepare the DNA template for transcription. Make a stock of linearized DNA by cutting 50 µL of DNA (about 50–100 µg) in a 200-µL reaction with the appropriate restriction endonuclease according to the conditions of the enzyme supplier. Bring the DNA solution to 0.2 M NaCl by adding 10 µL of 4 M NaCl. Extract the solution once with phenol equilibrated with Tris-EDTA and once with CHCl$_3$/*iso*-amyl alcohol. Precipitate the DNA by adding a threefold volume of absolute ethanol. Wash the DNA pellet once with 70% ethanol, dry it under vacuum, and dissolve it in 50 µL of 1× TE. Store the DNA at –20°C.

3. Set up and incubate the transcription reaction. For a 40-µL reaction, add the following reagents to a microcentrifuge tube at room temperature in the order given: 6 µL of sterile water, 4 µL of 0.1 M DTT, 4 µL of 10× transcription buffer, 4 µL of 10× NTP, 4 µL of 30% PEG$_{8000}$, 2 µL of RNasin, 4 µL of DNA (1–2 µg), 8 µL [α-^{32}P]UTP, and 4 µL of T7 RNA polymerase. Incubate for 2 h at 37°C. Add a second 2-µL aliquot of RNA polymerase and incubate for another hour.

4. Determine the incorporation of radiolabel into TCA-precipitable material. Remove 2 µL and add to 10 µL of *E. coli* tRNA (10 mg/mL). Add 500 µL of 10% TCA and incubate on ice for 10 min. Pour the contents of the tube onto a glass fiber filter paper in a vacuum filtering device, rinse the tube with about 1 mL of 5% TCA, and add the rinse to the filter. Rinse the filter twice again with 5 mL each of 5% TCA followed by 5 mL of 95% ethanol. Dry the filter, put it in a scintillation vial with 15 mL of water, and measure the number of Cerenkov counts per minute in a liquid scintillation counter. Expect about 2×10^6 Cerenkov cpm per 2 µL for a good transcription reaction as shown in **Fig. 3** (*see* **Note 11**). The remainder of the transcription reaction can be frozen at –20°C for up to 2 d if necessary.

5. Prepare the sample for gel electrophoresis. Add an equal volume (40 µL) of transcript loading buffer to the reaction. Heat at 65°C for 10 min to denature the RNA and put on ice.

3.3.1.2. Pouring and Running the Denaturing Polyacrylamide Gel

1. Wash the plates and "silanize" the notched plate. While wearing gloves, wash the glass plates and spacers and combs with a mild detergent and rinse with deionized water. Dry the plates and combs and spacers thoroughly with paper towels. Put the notched plate flat and raised (e.g., on an empty pipet tip box) in a chemical fume hood. Wearing gloves (*see* **Note 2**), pipet about 2 mL of silanizing solution onto the plate. Spread the solution over the entire surface with a tissue. The solution will evaporate rapidly (within 1–2 min), leaving a glossy film on the plate. Be careful not to let the solution drip over onto the opposite side of the plate or it will be difficult to subsequently affix the tape to the plate. If some of the solution has gotten on the wrong side of the plate, wipe the region with a tissue soaked in 95% ethanol to remove some of the silane.

2. Assemble the gel plates and spacers together. Lay the back plate flat and raised above the bench. Place the 1.5-mm thick spacers vertically along the sides on of the plate and with one end of the spacers flush with what will be the bottom of the

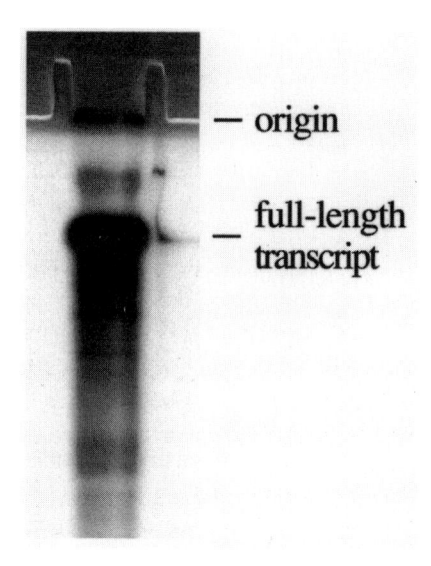

Fig. 3. In vitro synthesized, radiolabeled, actin pre-mRNA. Shown here is an autoradiogram of the denaturing polyacryamide gel used for purifying the 597 nt actin pre-mRNA. In addition to the film being exposed for 2 min at room temperature to the radioactive transcript, the top of the gel is briefly exposed to light to create an outline of the gel well as described in the text.

gel and plate. Lay the notched plate down on top of the spacers with the silanized suface facing inwards. Clamp the sides of the plates midway between top and bottom. At this point remove your gloves to tape the plates together (the tape sticks nearly irreversibly to the gloves). Cut a 20–30-cm piece of yellow electrical tape and tape the bottom of the plates so that the tape wraps evenly around the bottoms of both plates to form a seal. Wrap part of this tape about 2 cm up along the sides of the plates as well. To ensure a good seal rub your fingernail along the tape. Remove one clip from the side of the plate. Cut a ca. 18-cm piece of tape and similarly wrap it along the sides of the plates. Repeat for the other side. Lay the taped plates, notch side up on the benchtop. Put a 25-mL pipet under the back plate to raise the top of the plates about 1 cm above the bench.
3. Pour the gel. While wearing gloves (*see* **Note 2**), pour 70 mL of 5% acrylamide stock into a 125-mL Erlenmeyer flask with a side arm. Add 45 µL of TEMED and mix by swirling. Add 0.45 mL of 10% APS and similarly mix. Pour the solution through the flask's side-arm and between the plates while guiding the solution along one side and to the bottom to avoid the formation of air bubbles. Add the acrylamide solution until it is flush with the top of the notched plate. Position the comb between the plates with the comb teeth submerged in the gel. Put three clamps along the top of the plates to press the comb and plates together. Let the gel poly-

merize while lying at an angle on the bench for at least 1 h. If not used the same day, the gel can be stored for 1–2 d at room temperature with a wet paper towel in a sealed bag.
4. Set up the gel for the electrophoresis. Cut and remove the tape from the bottom of the gel, leaving the tape on the sides of the gel. Pull out the comb and scrape off any excess acrylamide remaining on the back plate. Clamp the gel to the gel box and fill the buffer tanks with 1× TBE. Flush the gel wells with buffer. Begin running the gel at about 60–65 mA with constant voltage being maintained. During the approx 20 min that the gel preruns the gel plates should reach about 40–45°C (and be very warm to the touch). Do not let the gel plates get warmer, or they will crack.
5. Load and run the sample. Turn off the power, quickly flush the well, and then pipet the sample into the gel well. Continue running at 55–60 mA for about 1.5 h until the bromophenol blue has run off the gel. Carefully remove the radioactive buffer from the lower tank and dispose of according to radiation safety procedures. Little radioactivity is in the upper tank, so you can usually place the gel box in a sink and remove the gel plate allowing the upper tank to drain into the sink. Remove the tape and spacers from the plates. Let the gel plates cool to room temperature and separate them. The gel should remain attached to the back plate. Wrap the gel and bottom plate with plastic wrap. If necessary, the wrapped gel-plate can be frozen at –20°C and thawed 24–48 h later to finish processing the transcript.

3.3.1.3. Extracting and Purifying the Transcript

1. Make an autoradiogram of the gel. In a photographic darkroom, place a piece of X-ray film on a clean sheet of paper on a flat surface. Place the gel-plate (gel-side down) on top of the film and begin timing a 2-min exposure. Meanwhile, partially cover the gel-plate with a film cassette leaving the gel wells uncovered. Flash a small flashlight for 1–2 s about 4–5 ft above and away from the gel. This flash will expose the film for an image of the wells. The outline is used to align the gel with the film image to locate the transcript. After the 2-min exposure, develop the film (**Fig. 3**) (*see* **Note 12**).
2. Locate and cut out the gel band containing the transcript. On a light box, align the gel-plate (gel-side up) on the film, and locate the transcript. Using a clean razor blade and wearing gloves, cut out the band through the plastic wrap. Avoid taking excessive amounts of gel; the transcript in **Fig. 3** can be excised in a 2 cm × 1 cm band. Remove the plastic wrap from the band and put the excised band into a sterile disposable polypropylene tube.
3. Use a crush-and-soak method to extract the transcript. Crush the band into tiny pieces with a sterile, silanized glass rod, then add 1.5 mL of transcript extraction buffer. If you add the fluid first, you will not be able to effectively crush the acrylamide. Cap the tube and seal it well. Shake or roll the tube for 4 h at 37°C or overnight at 4°C. Sediment the gel pieces by centrifugation for 10 min at top speed in a tabletop centrifuge at room temperature. Remove the upper aqueous layer containing the radioactive transcript and filter it through the Quicksep column

into a 15-mL Corex tube. Add 1 mL of fresh extraction buffer to the crushed gel, let mix for 15–30 min, and repeat the sedimentation and filtering. About 60–80% of the radioactive transcript will be recovered in the filtered aqueous solution.
4. Extract the transcript with phenol and chloroform. Extract the aqueous solution with a twofold volume of phenol equilibrated with NaOAc-EDTA. Separate the aqueous and organic layers by centifugation at 11,420*g* for 10 min at room temperature in a high-speed centrifuge. Remove the upper, clear aqueous layer and put it into a clean Corex tube. Extract the aqueous solution with a twofold volume of CHCl$_3$/*iso*-amyl alcohol and separate the layers by centrifugation. Remove the upper aqueous layer and put it into a clean Corex tube.
5. Concentrate the transcript by precipitation. Precipitate the transcript by adding a three-fold volume of absolute ethanol and incubating the tube on dry ice for 10 min or at –20°C overnight. Pellet the RNA by centifugation in a high-speed centifuge at 11,420*g* for 10 min at 4°C. The pellet may not be visible but can easily be detected with a geiger counter. Carefully pour off the supernatant and add about 5 mL of 70% ethanol to wash the pellet. If the pellet becomes dislodged, repeat the centrifugation and then pour off the wash. Dry the pellet under vacuum.
6. Reprecipitate the RNA in ammonium acetate and ethanol to remove any remaining salt and for storing the transcript. Dissolve the RNA pellet in 400 µL of sterile water. Put the solution into an Eppendorf tube containing 200 µL of 7.5 *M* NH$_4$OAc and mix. Divide into three tubes with 200 µL each and reprecipitate the RNA by adding a threefold volume (600 µL) of absolute ethanol. Store at –20°C.
7. Use one tube of transcript every 3–5 d. Pellet the transcript in one tube by centrifugation for 10 min at 4°C in a microcentifuge. Wash the pellet with about 750 µL of 70% ethanol, dry it in a Speed-Vac at room temperature, dissolve it in 50 µL of water, and keep it on ice. Remove a 2-µL aliquot of the transcript and put it into 1 mL of water in a 1.7-mL microcentrifuge tube. Put the tube in a scintillation vial without a top, and count the sample in a scintillation counter. The remaining 48 µL of transcript are then diluted with sterile water to give 20,000–40,000 Cerenkov cpm/µL and stored at –20°C. When using the transcript at the bench, keep it on ice.

3.3.2. Basic Splicing Assay

1. Pour the gel before doing the splicing reaction. Set up the gel plates as described in **Subheading 3.3.2.**, **step 2.** except use the 0.38-mm thick spacers and comb. While wearing gloves (*see* **Note 2**), pipet 20–25 mL of 7.5% acrylamide stock in a disposable, polypropylene, 50 mL, screw-capped centrifuge tube. Add 15 µL of TEMED and mix by inverting the capped tube. Add 0.15 mL of 10% APS and similarly mix. Using a 10- or 25-mL pipet, pipet the solution between the plates by putting the pipet onto one side of the plates and guiding the solution along one side and to the bottom to avoid the formation of air bubbles. Add solution until it is flush with the top of the notched plate. Position the comb between the plates with the comb teeth submerged in the gel. Put three clamps along the top of the

plates to hold the comb and plates together. Lay the plate at an angle by resting the bottom plate on the 25-mL pipet on the benchtop. Let the gel polymerize for at least 1 h at room temperature. If not used the same day, the gel can be stored for 1–2 d at room temperature with a wetted paper towel in a sealed bag.
2. Thaw the frozen extract on ice just before using it and keep it on ice (see **Note 13**).
3. While the extract is thawing, begin setting up a splicing reaction. Add the solutions on ice in the following order: 6.8 µL of H$_2$O, 1.5 µL of 1 M potassium phosphate buffer, pH 7.4, 2.5 µL of 30% PEG$_{8000}$, 0.75 µL of 0.1 M MgCl$_2$, 0.5 µL of 100 mM ATP, 0.5 µL of 100 mM DTT, and 2.5 µL of transcript and mix by vortexing.
4. Begin and stop the reaction. Begin the reaction by adding 10 µL of extract to the mix (see **Note 14**). The final concentrations in the 25-µL reaction are 60 mM potassium, 3% (w/v) PEG$_{8000}$, 3 mM MgCl$_2$, 2 mM ATP, 20 mM KCl, 8 mM HEPES, 80 µM EDTA, 2.2 mM DTT, 8% (v/v) glycerol, 0.4 nM pre-mRNA, and 40–80 µg of extract. Mix the reaction quickly by hand and place in a water bath at 23°C. At 15 and 30 min after addition of the extract, remove 10 µL and add it to 6 µL of stop buffer at room temperature. The stopped samples can be kept at room temperature until the last time point of the splicing reaction has been taken, or frozen at –20°C for processing at a later time.
5. Extract the RNA from the sample. Incubate the samples at 37°C for 30 min for the proteinase K to degrade proteins in the extract. Add 200 µL of 50 mM NaOAc, pH 5.4, 10 mM EDTA, then 750 µL of phenol/CHCl$_3$/*iso*-amyl alcohol equilibrated with NaOAc-EDTA (see **Note 2**). Vortex vigorously and centrifuge in a microcentrifuge for 10 min at 12,000–14,000g at room temperature. Remove the aqueous layer with the RNA, avoiding any of the organic (yellow) layer and the white material at the interface of the aqueous and organic layers. Put the aqueous solution into a clean tube with 125 µL of 7.5 M NH$_4$OAc. Precipitate the RNA by adding a threefold volume (about 1000 µL) of absolute ethanol and keeping the tube in dry ice for 10 min or at –20°C for 2 h to overnight.
6. Prepare the RNA for loading onto the gel. Pellet the RNA by centifugation in a microcentrifuge at 12,000–14,000g for 10 min at room temperature. Pour off the supernatant being careful not to dislodge the pellet that should look slightly translucent and be about 1 mm in diameter. Wash the pellet with 70% ethanol and dry it in a Speed-Vac at room temperature. Dissolve the pellet in 12 µL of fresh splicing assay load buffer. Heat the sample at 65°C for 5–7 min and then immediately place on ice. The sample is now ready to load onto the gel.
7. Load and run the gel. While the precipitated samples are drying, set up the gel in the gel box with 1× TBE buffer, flush the wells with buffer, and begin prerunning the gel at 30 mA and about 500 V (constant voltage). After about 15 min, the gel should be very warm to the touch. Just before loading the samples, turn off the power applied to the gel, and flush the gel wells. Load 4–6 µL of sample and run at 25–30 mA and about 600 V. As the current decreases slightly during the prerun, it is necessary to turn up the voltage to maintain 25–30 mA. The electrophoresis time varies according to the migrations of the pre-mRNA, intermediates

and products of the transcript used in the assay, and the concentration of the polyacrylamide gel. For the actin transcript shown in **Fig. 2**, run the gel for about 1 h until the bromophenol blue has run off the gel and the xylene cyanol is 1 in. from the bottom of the gel.

8. Remove the gel and expose it to X-ray film. Remove the gel from the gel box, let the plates cool to room temperature, and then separate the plates. The gel should be left on the back plate. Cut a piece of 3MM Whatman chromatographic paper slightly larger than the gel and place it on the gel. Note that once the paper contacts the gel, it usually remains stuck to the gel, but rub the paper with a gloved hand to firmly affix the gel to the paper. Invert the plate and carefully pull the paper away from the plate with the gel remaining stuck to the paper. Wrap the gel in a single layer of plastic wrap and expose to film with one or two intensifier screens at –70°C in a light-tight film casette for 5–18 h (**Fig. 2**).

9. For multiple film exposures, or exposure to a phosphor screen, dry the gel to prevent diffusion of the RNAs (*see* Chapter 28, **Subheading 3.3.**). Put the gel-paper in a 10-fold volume of 15% methanol, 5% acetic acid and slowly shake it at room temperature for 0.5 h to precipitate the RNAs *in situ* and to remove the urea from the gel. Dry the gel in a gel dryer at 80°C for 0.5–1 h.

3.3.3. Assessing the Activity of an Extract

In general, a yeast extract with excellent activity will splice about 50% of radiolabeled actin pre-mRNA to mRNA in 30 min at 23°C. In some cases quantification of the RNA species is possible (*see*, e.g., **ref. 15**). However, when there is significant nuclease activity in the extract, such quantification is not particularly meaningful.

Figure 2 shows the results of an initial assay to test the activities of three extracts made from different strains by the freeze–fracture method. Extract 1 has good splicing activity as evidenced by the relative levels of pre-mRNA, mRNA, lariate intermediate, and lariat product at both 15 and 30 min. Also note that the lariat intermediate, lariat product, and mRNA are clearly visible as sharp bands. Somewhat less mRNA is recovered at 30 min compared to 15 min, probably due to nuclease activity in the extract. A more extensive kinetic analysis of this extract shows that this extract produces readily detectable levels of intermediates and products within 2 min at 23°C, another indication of good splicing activity (data not shown). Extracts 2 and 3 (**Fig. 2**) are less active than extract 1 and show two other undesirable properties. Extract 2 has significant nuclease activity as seen by the numerous bands migrating faster than the pre-mRNA. Extract 3 has portions of the upper and bottom layers harvested with the middle layer after the ultracentrifugation step. These contaminants may decrease proteolysis during extraction of the RNAs from the splicing reaction and/or interfere with the migration of the RNAs during electrophoresis to cause the "smearing" of the lariat intermediate, lariat product, pre-mRNA, and mRNA in the gel.

4. Notes

1. Which method of extract preparation do you want to use? Several factors, including the types of assays for which the extract will be used, should be considered when deciding upon the method. The original method gives a good yield of extract with relatively low nuclease activity. Furthermore, the extracts generated by this method are the best characterized of those produced by the three methods and they have been used in additional fractionation schemes. The disadvantages of this method are the need to determine the conditions for lysing the spheroplasts (with a Dounce homogenizer) for each particular strain, and the length of time the preparation takes. The freeze–fracture method works well for several different strains and takes less time, but its yields are somewhat lower and the extracts generally have more nuclease activity. However, I have had a higher success rate in preparing active whole cell extracts from several different yeast strains with the freeze fracture method than with the original method. Finally, although not described here, the glass-bead method of cell lysis offers the advantages of allowing extract preparation from relatively small cell cultures as well as from several cultures at the same time.

2. Several chemicals used in the preparation of extract and the splicing assay are harmful and suitable safety precautions include wearing gloves, a laboratory coat, shoes, and safety glasses or goggles. PMSF, acrylamide, and *bis*-acrylamide are toxic, dry, aerosolic powders so when they are weighed out, a face mask should also be worn. Phenol and trichloroacetic acid are caustic and can cause *permanent* damage to the skin and eyes. Phenol, chloroform, and excess silanizing solution should be disposed of according to environmental health safety regulations. The relatively short half-life of PMSF in aqueous solutions makes its disposal easy. Utensils contacting dry or concentrated solutions of PMSF or acrylamide can be rinsed thoroughly with water before disposing of them or placing them in the wash.

3. Additives may affect the activity of an extract. The optimal concentrations of potassium and magnesium for splicing activity are described *(6)* and deviations from these may result in reduced or no splicing activity. Protease inhibitors such as PMSF are optional for either method of preparing extract. For certain yeast strains, such as EJ101, which are genetically designed to be protease-deficient, PMSF can be omitted. Other protease inhibitors such as leupeptin, aprotinin, and benzamidine may be added to the extract during preparation or into the splicing assay *(8)*. The use of RNasin to inhibit degradation of the snRNAs during extract preparation is not normally necessary. Addition of RNasin into the splicing assay, or during other manipulations of the splicing extract, such as in vitro heat-activation of an extract from a temperature-sensitive mutant, may help to maintain snRNA and/or pre-mRNA integrity in some cases *(10)*. Finally, heparin *(17)* or high concentrations of exogenous RNA (*see* Chapter 28) will inhibit splicing.

4. The density of the cell culture is a critical parameter for producing active splicing extracts. Harvest cells in exponential (log) phase growth. Do not use a culture that is in the diauxic shift. For strain EJ101, this occurs at a density of 6×10^7

cells/mL of YPD. During the diauxic shift, the cell doubling rate slows and several physiological changes occur *(18)*. Among these changes, the cell wall begins to become more resistant to zymolyase so that spheroplasting efficiency and subsequent cell lysis are significantly reduced. For either method of extract preparation, the optimal cell density for harvesting a culture of EJ101 is usually $4 \times 10^7 – 5 \times 10^7$ cells/mL of YPD. The optimal cell density of cultures of other strains grown in YPD or other types of medium may be somewhat lower and can be determined by doing a growth curve first to determine the highest density at which the cells can be harvested.

Monitor the growth of the cell culture with a spectrophotometer. Dilute a sample of the culture two- to 10-fold in water and then read at λ_{600}. The dilution is necessary because the absorbance varies linearly with the cell density over a relatively small range (usually from 0.1 to 0.7 U). Determine the relationship of the number of absorbance units to cell concentration for the particular spectrophotometer that you are using because the "absorbance" is actually owing to scattering of the light by the cells. The amount of light scattering detected depends on the geometry of the cuvet relative to the detector, which varies among spectrophotometers. For example, a sample diluted to 1×10^7 cells/mL of water will give an absorbance at λ_{600} of 1 and 2 U in a Beckman DU and Schimadzu UV60 spectrophotometer, respectively.

5. In addition to harvesting the cell culture at optimal density, four factors contribute to getting active extracts by the spheroplast lysis method: efficient spheroplasting with zymolyase; effective Dounce homogenization; keeping the cell solution/lysate cold once the spheroplasts are formed; and working quickly.

6. The first indication that the zymolyase treatment will be effective is the condition of the cells themselves. Just before zymolyase is added to the cells, the cells will have a 10–20% lower absorbance in lysing solution than in isotonic solution. If the cells are going to be refractive to zymolyase treatment, they will not usually show this decrease. During zymolyase treatment at 30°C, the absorbance in lysing solution should drop to 20% of that in isotonic solution. If the cells have not reached this state within 40 min, then add fresh zymolyase, and continue the incubation for another 20 min. If the absorbance still has not dropped sufficiently by this time, it is not worthwhile to continue the preparation. Instead, inoculate another culture and harvest at a lower cell density. The times and temperatures of zymolyase treatment can be modified somewhat however, to suit the yeast strain. To make active extracts from ts mutants grown at the permissive temperature of 23°C, e.g., zymolyase treatment is done at 23°C, usually for 50–60 min for optimal sphereoplasting.

7. The ratio of buffer to cells is also important for getting an active extract. You can increase the amount of buffer by about 10%, but too much buffer used at this step will result in an extract with little or no splicing activity.

8. The Dounce homogenization of the spheroplasts is critical for getting a good extract. Too much lysis at this step will release proteases and nucleases that can inactivate the extract. Too little lysis will also lead to an inactive extract. The

optimal conditions for douncing will vary according to the strain. For strain EJ101, use five to seven very strong and relatively fast strokes of the pestle. For other strains, such as SS330 (*19*), several slowly executed strokes are optimal. Unfortunately, there is not, as yet, a reliable method to monitor cell lysis during the actual douncing. However, if spheroplast formation has been successful most strains will require 10 strokes or less for lysis by Dounce homogenization.

9. If you want to store the frozen suspension and process it into extract at a later time, then do the following. Completely submerge an uncapped 50-mL, sterile, polypropylene, graduated, conical centrifuge tube in liquid N_2 in a 350 mL Dewar flask. Drip the suspension into the liquid N_2 inside the tube. Before removing and capping the tube, punch several small holes into the tube cap to allow liquid N_2 to escape the tube; otherwise the tube can explode. While wearing insulated gloves, screw the cap quickly in place, and store the entire tube in liquid nitrogen.

10. The lysate from some strains can be also thawed quickly at 25–30°C. Thawing on ice works for most strains.

11. If the incorporation of radiolabel into TCA-precipitable material is below 200,000 Cerenkov cpm, you will not get enough transcript. Such low incorporation is most commonly due to one of four factors: (1) the RNA polymerase is not active; (2) not enough DNA template was added; (3) the DNA template has a residual contamination of phenol, salt, or some other substance that affects the polymerase; and (4) the radiolabeled UTP has already decayed more than one half-life. T7 RNA polymerase is quite stable when stored at –20°C. Sp6 polymerase, on the other hand, may become inactive within several months when stored in a frost-free freezer due to the temperature cycling of the frost-free feature. You can try adding some polymerase obtained from a different source to the same reaction and checking the TCA-precipitable counts after an incubation for 1 h at 37°C. If this is the problem, buy new enzyme and store it in an insulated container in a frost-free freezer, or at –70°C. For testing the DNA template, you can add 0.5 µg of a control DNA template to the same reaction and determine whether TCA-precipitable counts can now be obtained. If you now get substantial incorporation, then the concentration of the original template is too low. If you don't get incorporation in the reaction, but you know the control template is active for transcription, then you may have a contaminant in your first DNA template. This may be removed by reextracting and reprecipitating the DNA template. For optimal synthesis, fresh radiolabeled UTP should be used; even if more radioactive UTP is added to compensate for the decay, the decay of even one half-life can lower the incorporation. Another, less commonly occuring problem may be that the ribonucleotides were not at pH 7.0 when you added them to the reaction.

12. If most of the transcript is either larger or smaller than the predicted length in the preparative denaturing gel, then you may have to repeat the transcription reaction. For transcripts larger than predicted, the template DNA is probably not completely cut with the restriction endonuclease. For transcripts smaller than predicted, the cause is most often one of the following: (1) RNasin was not added to the reaction; (2) RNasin was added to the reaction but the DTT concentration

was below 1 mM; (3) premature termination is occurring because of the DNA template — either it is nicked or its sequence is causing the termination; or (4) the DNA template or some other solution is contaminated with excessive nuclease. RNasin acts as an RNase inhibitor by binding reversibly to RNase in a DTT-dependent manner; if the DTT concentration is too low, bound RNase can be released and be active in your reaction.
13. If handled and stored well, a good splicing extract can last for a few years. For optimal activity, a tube of splicing extract is usually subjected to only two freeze–thaw cycles after the extract was made. Mark the tube with a colored dot for each freeze-thaw cycle to keep track. If there is 200 μL or more of extract in the tube, divide the extract from a single tube into several smaller aliquots after the first freeze–thaw. Thaw the extract on ice and keep it on ice for as short a time as possible while reactions are being set up. Flash-freeze the extract in liquid N_2 to return it to storage.
14. A splicing reaction can be started by the addition of either the extract or the transcript, or by shifting the reaction from ice to 23–25°C. The method of starting the reaction may also vary according to the experimental protocol. The temperature at which the reaction is incubated may vary according to the strain or experiment. Extracts from a wild-type strain are usually active over a fairly wide range of temperatures, from 15 to 30°C. Above 26°C, however, nuclease activity may become prohibitive.

Acknowledgments

This work was supported by NSF Grants MCB9104862 and MCB9709915, and a grant from the Dedicated Health Sciences Research Fund of the University of New Mexico Health Sciences Center.

References

1. Moore, M. J., Query, C. C., and Sharp, P. A. (1993) Splicing of precursors to mRNA by the spliceosome, in *The RNA World* (Gesteland, R. F. and J. F. Atkins, eds.), Cold Spring Harbor Laboratory Press, Cold Spring Harbor, New York, pp. 303–358.
2. Guthrie, C. (1991) Messenger RNA splicing in yeast: clues to why the spliceosome is a ribonucleoprotein. *Science* **253**, 157–163.
3. Ruby, S. W. and Abelson, J. (1991) Pre-mRNA splicing in yeast. *Trends Genet.* **7**, 79–85.
4. Kramer, A. (1996) The structure and function of proteins involved in mammalian pre-mRNA splicing. *Annu. Rev. Biochem.* **65**, 367–409.
5. Chamberlin, M., Kingston, R., Gilman, M., Wiggs, J., and deVera, A. (1983) Isolation of bacterial and bacteriophage RNA polymerases and their use in synthesis of RNA in vitro. *Methods Enzymol.* **101**, 540–568.
6. Lin, R. J., Newman, A. J., Cheng, S.-C., and Abelson, J. (1985) Yeast mRNA splicing in vitro. *J. Biol. Chem.* **260**, 14,780–14,792.
7. Seraphin, B. and Rosbash, M. (1989) Identification of functional U1 snRNA-pre-mRNA complexes committed to spliceosome assembly and splicing. *Cell* **59**, 349–358.

8. Umen, J. and Guthrie, C. (1995) A novel role for a U5 snRNP protein in 3' splice site selection. *Genes Dev.* **9**, 855–868.
9. Jones, E. W. (1991) Three proteolytic systems in the yeast *Saccharomyces cerevisiae*. *J. Biol. Chem.* **266**, 7963–7966.
10. Lustig, A. J., Lin, R. J., and Abelson, J. (1986) The yeast RNA gene products are essential for mRNA splicing in vitro. *Cell* **47**, 953–963.
11. Grabowski, P. J., Padgett, R. A., and Sharp, P. A. (1984) Messenger RNA splicing in vitro: an excised intervening sequence and a potential intermediate. *Cell* **37**, 415–427.
12. Zavanelli, M. I. and Ares, M. (1991) Efficient association of U2 snRNPs with pre-mRNA requires an essential U2 RNA structural element. *Genes Dev.* **5**, 2521–2533.
13. Cheng, S.-C. and Abelson, J. (1986) Fractionation and characterization of a yeast mRNA splicing extract. *Proc. Natl. Acad. Sci. USA* **83**, 2387–2391.
14. McPheeters, D. S., Fabrizio, P., and Abelson, J. (1989) *In vitro* reconstitution of functional yeast U2 snRNPs. *Genes Dev.* **3**, 2124–2136.
15. Cheng, S. C., Newman, A. N., Lin, R. J., McFarland, G. D., and Abelson, J. N. (1990) Preparation and fractionation of yeast splicing extract. *Methods Enzymol.* **181**, 89–96.
16. Fabrizio, P., Esser, S., Kastner, B., and Luhrmann, R. (1994) Isolation of *S. cerevisiae* snRNPs: comparison of U1 and U4/U6. U5 to their human counterparts. *Science* **264**, 261–265.
17. Cheng, S.-C. and Abelson, J. (1987) Splicesome assembly in yeast. *Genes Dev.* **1**, 1014–1027.
18. Werner-Washburne, M., Braun, E., Johnston, G. C., and Singer, R. A. (1993) Stationary phase in the yeast *Saccharomyces cerevisiae*. *Microbiol. Rev.* **57**, 383–401.
19. Vijayraghavan, U., Company, M., and Abelson, J. (1989) Isolation and characterization of pre-mRNA splicing mutants of *Saccharyomces cerevisiae*. *Genes Dev.* **3**, 1206–1216.

27

Prespliceosome and Spliceosome Isolation and Analysis

Laura A. Lindsey and Mariano A. Garcia-Blanco

1. Introduction

Spliceosomes are multicomponent enzymes that remove introns from pre-messenger RNAs (pre-mRNAs) *(1–3)* in the reaction known as pre-mRNA splicing. Spliceosomes are ribonucleoprotein (RNP) machines consisting of both RNA and protein components. SnRNPs, composed of small nuclear RNAs (snRNAs) and associated proteins, are critical for the pre-mRNA splicing reaction. In addition to the snRNPs there are several proteins required for splicing, some of which are components of the spliceosome *(4)*. Two different spliceosomes are required to remove two classes of introns. The major class of introns usually have GU and AG dinucleotides at the 5' and 3' splice sites respectively, whereas the minor class of introns, the AT–AC introns, usually have AT and AC dinucleotides at the 5' and 3' splice sites *(5,6)*. Here we focus our attention on the major or conventional spliceosome, which shares the U5 small ribonucleoprotein particle (snRNP) with the AT–AC spliceosome, but otherwise has a unique set of snRNPs: U1, U2, and U4/U6 *(7)*.

1.1. Methods to Identify and Characterize the Spliceosome, a Large Macromolecular Complex

Several different techniques have been used to identify and characterize spliceosomes. All of the methods described below have been used in analytical assays to unravel the components of the spliceosome. Except for nondenaturing polyacrylamide gel electrophoresis, these methods are also amenable for preparative use.

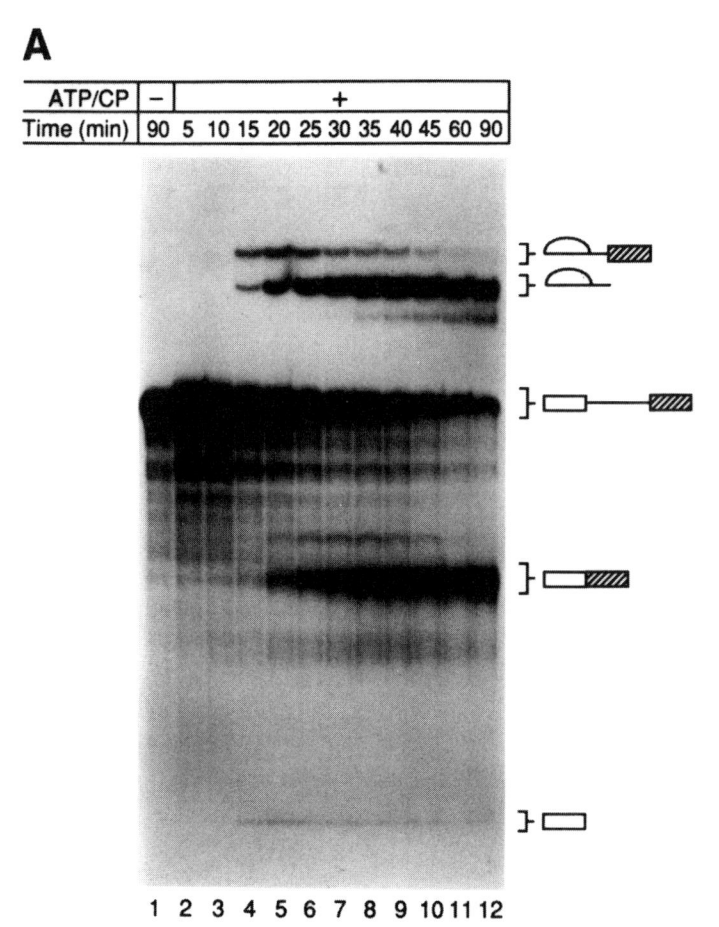

Fig. 1. Analysis of in vitro splicing reactions. (**A**) Autoradiograph of a denaturing gel containing ^{32}P-labeled RNA from in vitro splicing reactions with HeLa nuclear extracts (NE). Splicing reactions were incubated in the absence of ATP and creatine phosphate (CP) (*lane 1*) or in the presence of 1 m*M* ATP and 5 m*M* creatine phosphate (*lanes 2–12*) at 30°C for the indicated amount of time. From *top* to *bottom*, icons indicate the lariat intermediate, lariat product, pre-mRNA, spliced product, and free 5'-exon.

1.1.1. Nondenaturing Polyacrylamide Gel Electrophoresis

It is possible to assay for the production of intermediates and products of in vitro pre-mRNA splicing reactions by denaturing the RNP complexes, extracting the RNA, and subjecting the RNAs to electrophoresis on denaturing polyacrylamide gels (**Fig. 1A**). In this time course, products of the splicing reaction can be seen after 15–20 min (**Fig. 1A**). One can also visualize spliceosomes (complexes B and C) and prespliceosomes (complex A) by using nondenaturing

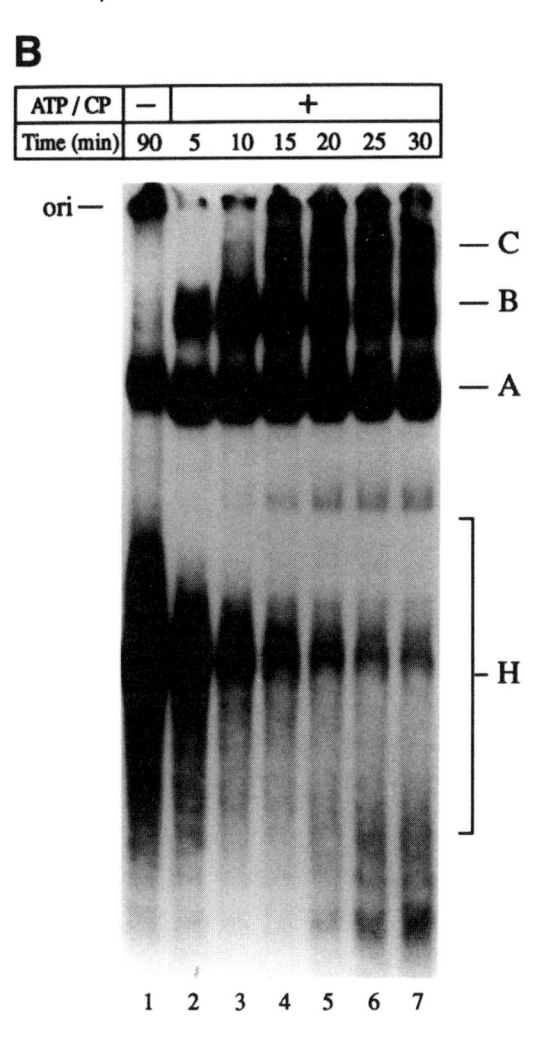

Fig. 1. **(B)** Autoradiograph of a native gel containing complexes formed on ^{32}P-labeled RNA in in vitro splicing reactions with HeLa NE. Splicing complexes were formed in splicing reactions incubated in the absence of ATP and creatine phosphate (*lane 1*), or in the presence of 1 m*M* ATP and 5 m*M* creatine phosphate (*lanes 2–7*) for the indicated amount of time. The splicing complexes prespliceosome **(A)**, spliceosomes **(B** and **C)**, and nonspecific heterogeneous complexes **(H)** are indicated.

polyacrylamide gel electrophoresis *(8–11)* (**Fig. 1B**). These gels can resolve three complexes — A, B, and C — that form specifically with functional pre-mRNAs, and also a heterogeneous smear of complexes (H) that forms with any RNA added to the nuclear extract. The RNP complexes resolved in these gels can be transferred to nylon membranes and probed for protein components such as hnRNP proteins *(9)* or snRNAs *(9,10)*. The nondenaturing gels provide a convenient method to follow the kinetics of spliceosome assembly or the effect of *cis*-acting elements or *trans*-acting factors on this assembly.

1.1.2. Gel Filtration

The laboratory of Tom Maniatis (Harvard University) pioneered the use of gel filtration to purify spliceosomes *(12,13)*. Robin Reed and her colleagues (Harvard Medical School) have made extensive use of this technique to identify an early complex (or family of complexes), complex E, that contains the commitment activity *(14)*. Moreover, this group has used a combination of gel filtration, affinity chromatography, and two-dimensional gel electrophoresis to identify spliceosome associated proteins (SAPs) *(14–17)*. This method separates complex E from spliceosomes; however, it does not resolve prespliceosomes (complex A) from spliceosomes (complexes B and C) *(17)*.

1.1.3. Affinity Chromatography

Grabowski and Sharp *(18)* used affinity chromatography to isolate spliceosomes. The method we describe below is a modification of the methods described in that manuscript. The pre-mRNA is synthesized in the presence of biotin-UTP so that biotin-U residues are incorporated in the body of the RNA. Biotinylated RNA and complexes assembled on the RNA are bound to a streptavidin matrix. Affinity chromatography has been used after spliceosomes have been resolved by velocity sedimentation *(18)* or gel filtration *(14)*. Ruby and Abelson *(19)* described an indirect affinity method whereby a nonbiotinylated pre-mRNA forms a double stranded stem with an "anchor" RNA that is biotinylated.

1.1.4. Velocity Sedimentation

Velocity sedimentation in glycerol or sucrose density gradients was a technique first used to identify 60S complexes that were formed upon incubation of pre-mRNA with nuclear extracts *(1–3)*. These complexes were found to contain not only the pre-mRNA but also intermediates (lariat intermediate and 5' exon) and products of pre-mRNA splicing *(2)*.

For a superb explanation of velocity sedimentation we refer the reader to Chapters 10 and 11 of *Biophysical Chemistry* by C. R. Cantor and P. R. Schimmel *(20)*. The principle of sedimentation velocity can be simplified by stating that sedimentation velocity is a function of the size and shape of the sedimenting complex or particle and will always be proportional to the centrifugal force. The proportionality constant is the sedimentation coefficient or *s*, which has units of seconds; however, experimental measurements are given in Svedberg units (S), where $1S = 10^{-13}$ s. The above statement has a mathematical embodiment in the Svedberg equation, which approximates reality

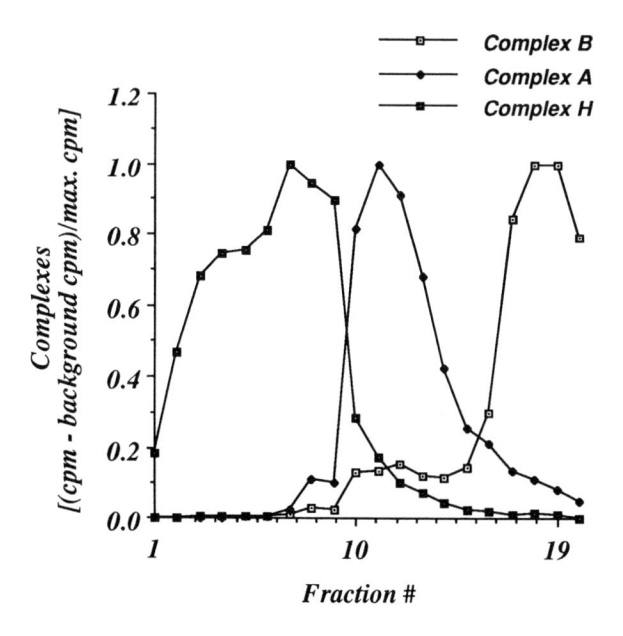

Fig. 2. Analysis of glycerol gradient fractions. Splicing complexes were separated on a glycerol gradient and visualized by native gel electrophoresis and autoradiography. The data were quantified using a Molecular Dynamics Densitometer. The background was subtracted and the values were normalized to the maximum value for each complex.

quite well, $M = sRT/D(1 - V2r)$, where the sedimentation coefficient is a function of both molecular mass (M) and a series of terms that take into account the shape and buoyancy of the particle sedimenting, and the density of the solvent (*20*, pp. 605–610).

Velocity sedimentation using glycerol density gradients has been used to resolve spliceosomes, prespliceosomes, and heterogeneous complexes *(21)* (**Fig. 2**). Using this method it is possible to separate presplicesomes from spliceosomes and it was possible to obtain data that clearly indicated that prespliceosomes were indeed precursors to spliceosomes *(21)*. Glycerol gradient purified prespliceosomes are stable and can be chased to spliceosomes in the presence of excess competitor RNA *(21,22)*. Moreover, the requirements of the prespliceosome to spliceosome transition were studied, and SR proteins were shown to be critical to this reaction *(22)*.

1.2. Purification of Spliceosomes Using Velocity Sedimentation and Affinity Chromatography

We have recently used the combination of velocity sedimentation and affinity chromatography in an attempt to identify mammalian proteins required for the

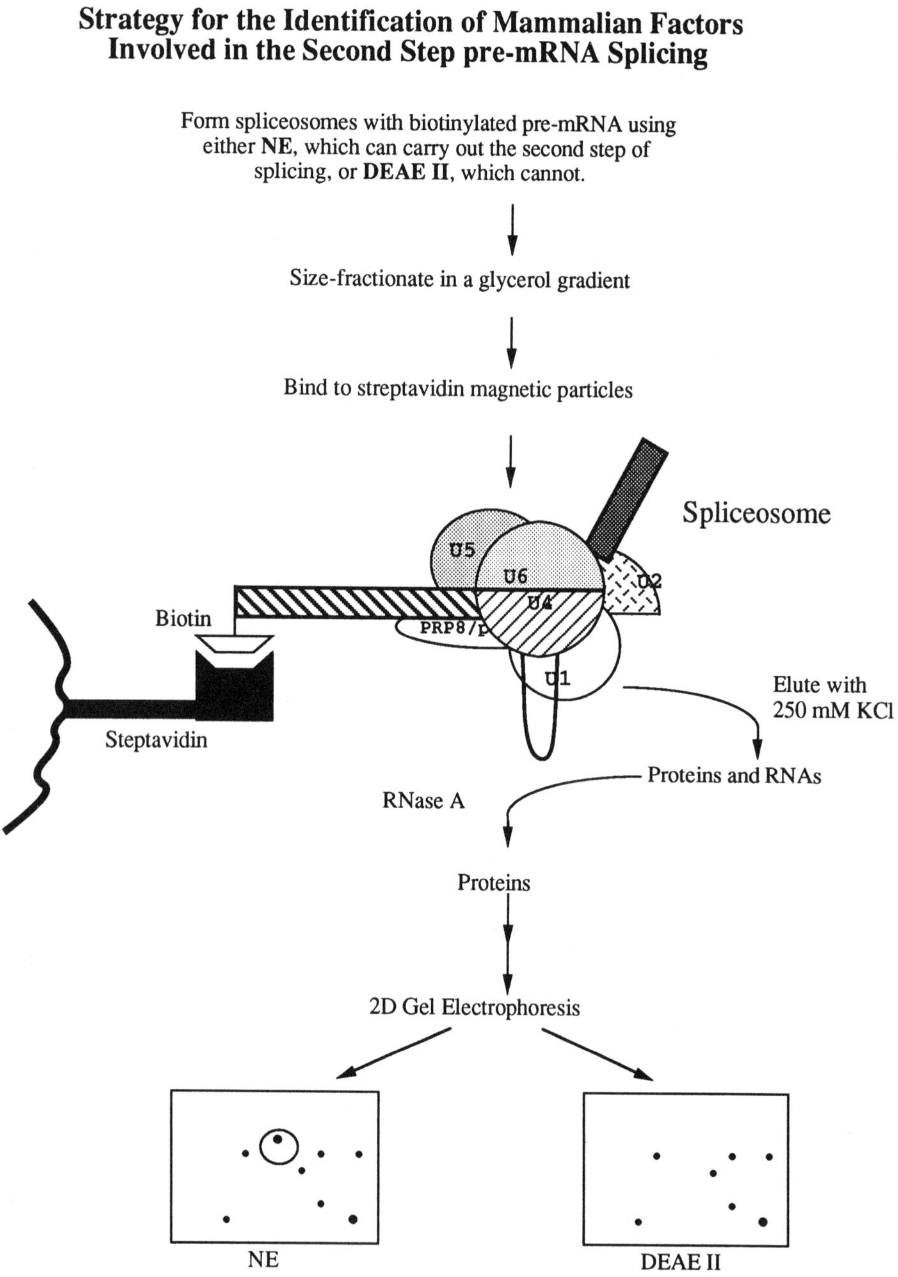

Strategy for the Identification of Mammalian Factors Involved in the Second Step pre-mRNA Splicing

Fig. 3. Schematic for the purification of spliceosomes arrested before the second *trans*-esterification reaction of pre-mRNA splicing.

second step of pre-mRNA splicing (**Fig. 3**). We have fractionated HeLa nuclear extracts (NE) and obtained a fraction, called DEAE II, that is deficient for the second step of splicing *(23)*. To identify proteins required for the second step of splicing, we have purified spliceosomes formed in unfractionated NE which contains all of the components required to carry out the second step of splicing. We also purified spliceosomes formed in the DEAE II fraction, which is deficient for the second step activity, and then compared the protein composition of these two spliceosome preparations using two-dimensional gel electrophoresis *(24)*. Proteins present in spliceosomes formed in the presence of the second step activity, but absent in spliceosomes formed in the absence of the second step activity are good candidates for mammalian second step factors. We have obtained microsequence from two of these proteins, hnRNP H and hnRNP K, and are interested in determining whether these proteins play a role in the second step of splicing.

To purify the spliceosomes, the splicing substrate is transcribed in an in vitro reaction where 50% of the UTPs are biotinylated. The biotinylated RNA is added to a large-scale in vitro splicing reaction with either NE or DEAE II. The spliceosomes are then separated from prespliceosomes and nonspecific H complexes by centrifugation through a glycerol gradient. The glycerol gradient fractions are analyzed using denaturing gels and native gels to identify fractions containing spliceosomes. These fractions are then pooled and the spliceosomes are further purified by binding to streptavidin beads and washing with low salt washes. The RNA is digested with RNase A, and the protein is eluted and analyzed using two-dimensional gel electrophoresis.

2. Materials
2.1. In Vitro Transcription

1. 1 mg/mL *Hin*dIII-digested PIP10 plasmid *(25)*, or other splicing construct.
2. Nucleotides: 10 mM biotin-21-UTP, 10 µCi/mL [α-^{32}P]UTP, 10 mM UTP, 10 mM (ATP, CTP, GTP).
3. 12× Transcription buffer: 480 mM Tris-HCl, pH 8.1, 72 mM MgCl$_2$, 12 mM spermidine, 60 mM dithiothreitol (DTT), 0.12% Triton X-100, 960 g/mL polyethylene glycol (PEG).
4. Pyrophosphatase (0.05 U/mL).
5. RNasin (40 U/µL).
6. T7 RNA polymerase *(26)*.
7. High salt buffer: 10 mM Tris-HCl, pH 7.5, 100 mM LiCl, 10 mM EDTA, 0.5% sodium dodecyl sulfate (SDS), 7 M Urea.
8. 3 M NaOAc, pH 5.2.
9. 100% and 80% ethanol.

2.2. Denaturing Gel for Purification of pre-mRNA and Analysis of Splicing Products

1. Glass plates (27 × 16.5 cm) and standard apparatus for the electrophoresis of polyacrylamide gels.

2. 5× TBE running buffer: 446 mM Tris, 445 mM boric acid, 1 mM EDTA.
3. 15% Acrylamide solution: 1× TBE, 14.5% acrylamide, 0.5% *bis*-acrylamide, 8 M urea.
4. 10% Ammonium persulfate.
5. *N,N,N',N'*-tetramethylenediamine (TEMED).
6. Denaturing gel loading buffer: 98% formamide, 10 mM EDTA, 0.025% xylene cyanol, 0.025% bromophenol blue.
7. Gel elution buffer: 10 mM Tris-HCl, pH 7.5, 0.5 M ammonium acetate, 10 mM magnesium acetate, 0.1 mM EDTA, 0.1% SDS.

2.3. Splicing Reaction

1. PIP 10 pre-mRNA (0.13 mg/mL) (*25*).
2. HeLa NE (*27*) as modified by Miller et al. (*28*) or DEAE II (*23*; for preparation of NE *see also* Chapter 24).
3. 400 mM KCl.
4. 100 mM MgCl$_2$.
5. 10 mM ATP and 50 mM creatine phosphate.
6. 50 mg/mL heparin.

2.4. Glycerol Gradient

1. Gradient solutions: 15% or 30% (v/v) glycerol in 6.4 mM HEPES-KOH, pH 7.9, 50 mM KCl, 2 mM MgCl$_2$.
2. Gradient maker GA.2994 (Gibco-BRL, Grand Island, NY, USA).
3. Beckman polyallomer centrifuge tubes 14 × 95 mm (Beckman, Fullerton, CA, USA).

2.5. Native Gel

1. Glass plates (17 × 16.5 cm).
2. 30% (w/v) acrylamide in dH$_2$O.
3. 1% *Bis*-acrylamide (w/v) in dH$_2$O.
4. 1 M Tris (unbuffered, pH ~ 10).
5. 1 M Glycine (unbuffered, pH ~ 6).
6. 10% Ammonium persulfate.
7. TEMED.
8. Running buffer: 50 mM Tris, 50 mM glycine.

2.6. Affinity Purification

1. Streptavidin magnetic particles 10 mg/mL.
2. Bovine serum albumin (BSA).
3. 5× Binding buffer: 6.4 mM HEPES-KOH, pH 7.9, 50 mM KCl, 12.5 mM EDTA, 0.5% Triton X-100.
4. 1 M DTT.
5. RNasin (40 U/µL).
6. Wash buffer 1: 20 mM Tris, pH 7.9, 250 mM NaCl.

7. Wash buffer 2: 20 mM Tris, pH 7.9, 100 mM NaCl.
8. RNase A (0.2 mg/mL).
9. Elution buffer: 20 mM Tris-HCl, pH 7.9, 20 mM DTT, 2% SDS.
10. Glycogen (20 mg/mL).
11. Acetone.

2.7. Two-Dimensional Electrophoresis of Proteins

1. Ampholyte solution: 22.4 mM Tris-HCl, 17.6 mM Tris base, 7.92 M urea, 0.06% SDS, 1.76% pH 3–10 Ampholytes, 120 mM DTT, 3.2% Triton X-100.
2. Precast pH 3–10 Carrier Ampholyte IEF gel 1 mm × 18 cm (ESA, Chelmsford, MA).
3. IEF running buffer: 10 mM phosphoric acid anode solution; 100 mM NaOH cathode solution.
4. Gel equilibration buffer: 300 mM Tris base, 75 mM Tris-HCl, 50 mM DTT, 3% SDS, 0.01% bromophenol blue.
5. SDS gel: 372 mM Tris, pH 8.8, 9% acrylamide, 0.1% SDS, 0.02% ammonium persulfate, 0.0005% TEMED.
6. SDS running buffer: 25 mM Tris base, 192 mM glycine, 0.1% SDS.
7. Silver stain solutions *(29)*, also commercially available (*see* Chapter 3).

3. Methods
3.1. Transcription

This large-scale 1-mL reaction yields 100 µg of PIP 10 pre-mRNA after gel purification (*see* **Note 1**; *see also* Chapter 1).

1. Add the following components at room temperature (RT) in this order to prevent precipitation of the DNA: 83.3 µL of 12× transcription buffer, 100 µL of 100 mM MgCl$_2$, 200 µL of (10 mM ATP, CTP, GTP), 37.5 µL of 10 mM UTP, 37.5 µL of 10 mM biotin-21-UTP, 30 µL of 10 µCi/µL [α-^{32}P]-UTP, 2 µL of RNasin (40 U/µL), 10 µL of pyrophosphatase (0.05 U/µL), 80 µL of 1 mg/mL PIP10 *Hin*dIII-digested plasmid, 22 µL of T7 RNA polymerase, and dH$_2$O to 1 mL.
2. Incubate at 37°C for 3 h.
3. Add 1/10 vol of 3 M NaOAc, pH 5.2, and 2.5 vol of ethanol (*see* **Note 2**).
4. Precipitate on dry ice for 20 min, centrifuge at 16,000g for 20 min at 4°C, wash with cold 80% ethanol, and air-dry the pellet.

3.2. Denaturing Gel for Purification of pre-mRNA and Analysis of Spliced Products

1. Wash and assemble the glass plates.
2. Add 115 µL of 10% ammonium persulfate and 11.5 µL of TEMED to 21 mL of the 15% acrylamide solution.
3. Pour the mixture between the plates and position the comb.
4. Allow the gel to polymerize >1 h before removing the comb.

5. Resuspend the precipitated RNA in the denaturing gel loading buffer, heat to 100°C for 3 min, and immediately place on ice until ready to load onto the gel.
6. Prerun the gel at 60 V/cm about 15 min until it is warm (see **Note 3**).
7. Rinse the wells and load the sample.
8. Run the gel about 2.5 h until both dyes run off.
9. Remove the top plate and cover the gel with plastic wrap. Place a piece of fluorescent tape on the gel for later positioning of the film.
10. Expose to X-ray film to locate the RNA transcript.
11. Excise the RNA from the gel, crush the excised gel slice, add 500 μL of gel elution buffer, and rock at 4°C overnight or at 37°C for a few hours.
12. Centrifuge the gel solution through a disposable column to remove the gel pieces.
13. Ethanol precipitate the RNA as in **Subheading 3.1.**, steps 3–4.
14. Resuspend the RNA in 500 μL of dH$_2$O and store at –20°C.

3.3. Large-Scale Splicing Reaction (see also Chapter 25)

1. Mix the following components: 56 μL of 100 mM MgCl$_2$, 119 μL of 400 mM KCl, 280 μL of 10 mM ATP and 50 mM creatine phosphate mix, 30 μL of 0.1 μg/μL biotinylated pre-mRNA, 1.12 mL of NE or DEAE II, and dH$_2$O to a final volume of 2.1 mL.
2. Incubate at 30°C for 1 h (see **Note 4**).
3. Add 1/100 vol of 50 mg/mL heparin and incubate for 5 min at 30°C.

3.4. Glycerol Gradient (21)

1. Pour three 13-mL glycerol gradients immediately before use. Place 7 mL of the 30% glycerol gradient solution in the "heavy" chamber and 7 mL of the 15% glycerol gradient solution in the "light" chamber of the gradient maker.
2. Layer 690 μL of the splicing reaction onto each gradient.
3. Centrifuge in a SW40Ti rotor at 40,000 rpm (285,000g) for 8 h at 4°C.
4. Very gently collect 1-mL fractions from top to bottom.

3.5. Native Gel (9)

1. Wash and assemble the glass plates.
2. Mix 5.6 mL of 30% acrylamide, 2 mL of 1% *bis*-acrylamide, 2 mL of 1 M Tris, 2 mL of 1 M glycine, 28 mL of H$_2$O, 0.28 mL of 10% ammonium persulfate, and 28 μL of TEMED.
3. Pour the mixture between the plates and position the comb.
4. Allow the gel to polymerize for 2–4 h before removing the comb.
5. Load 20 μL of each glycerol gradient fraction onto the gel.
6. Run the gel at 160 V (9.4 V/cm).
7. When the gel is done running (see **Note 5**), pick it up using a piece of Whatman 3MM paper, and vacuum dry at 80°C for 45 min.
8. Visualize the splicing complexes by autoradiography.

3.6. Affinity Purification (15,16,18,30,31)

1. Preblock the streptavidin magnetic beads by washing 4× over 2 h with 1× binding buffer plus 0.1% BSA (see **Note 6**).

Presliceosomes and Spliceosomes

2. Pool the glycerol gradient fractions containing spliceosomes (*see* **Note 7**).
3. To the pooled fractions, add 1/5 vol of 5× binding buffer, 2 mM DTT, 3.2 U/mL of RNasin, and 64 μL/mL of streptavidin magnetic beads.
4. Rock overnight at 4°C.
5. Collect the beads and wash with 10 mL of wash buffer 1 for 15 min with rocking at 4°C. Repeat the washes 3×.
6. Wash 1× with wash buffer 2.
7. Resuspend the beads in 0.5 mL of wash buffer 2.
8. Add 2 μL of 0.2 mg/mL RNase A and incubate for 10 min at 30°C.
9. Remove and save the supernatant.
10. Wash the beads 3× with 0.5 mL elution buffer and pool the eluates with the reserved supernatant for a total of 2 mL.
11. Heat the eluates for 5 min at 65°C.
12. Add 2 μL of 20 mg/mL of glycogen and 4 vol of acetone.
13. Incubate at RT for 10 min, and then centrifuge the precipitated proteins for 10 min at 2800g at RT.
14. Remove the supernatant and allow the pellet to dry for 15 min at RT.

3.7. 2-D Gel Electrophoresis (24,32)

1. Assemble the Investigator™ 2-D Electrophoresis System (ESA, Chelmsford, MA, USA) according to the manufacturer's directions.
2. Resuspend the affinity-purified acetone precipitated protein in 15 μL of Ampholyte solution.
3. Centrifuge for 1 min at 16,000g and load the supernatant onto a precast isoelectric focusing (IEF) gel.
4. Run at 100 V for 1 h, 200 V for 1 h, 400 V for 24 h, 650 V for 1 h, 1000 V for 30 min, 1500 V for 10 min, and 2000 V for 5 min, for a total of 11,500 V-h. (*see* **Note 8**).
5. Extrude the gel into gel equilibration buffer and incubate for 5 min at RT.
6. Layer the IEF gel onto the SDS gel and run until the bromophenol blue is 1 cm from the bottom.
7. Silver stain *(29)* the gels to analyze the protein composition (*see* Chapter 3). The amount of protein loaded onto the gel can be scaled up to stain with Coomassie blue *(33)* or copper *(34)* for detection before protein sequencing.

4. Notes

1. Increasing the percentage of biotin-UTP in the RNA above 50% greatly inhibits the splicing reaction.
2. There is loss of biotinylated RNA to the organic phase during phenol:chloroform extractions; therefore to minimize loss, we do not extract before gel purification.
3. The denaturing gel should be warm when running. If it becomes too hot though, the glass plates could break. Therefore monitor the temperature frequently. A metal plate should be clamped onto the front of the glass plate to help distribute the heat and reduce "smiling" of the RNA in the gel.

4. To maximize the accumulation of spliceosomes in the in vitro splicing reaction, a time course should be performed to determine the amount of time needed for formation of spliceosomes.
5. Allow the native gel to polymerize 2–4 h before use. If you have trouble with the gel sticking to the glass plate, treat each plate with Rain-X and/or Photo-flo before assembling the plates. Load TAE loading dye with xylene cyanol and bromophenol blue in one lane of the native gel to determine when you should stop running the gel. Our gels are stopped when both dyes completely run off the gel (about 4 h at 160 V).
6. There are different degrees of nonspecific binding to different streptavidin and avidin beads. We have tried avidin agarose (Vector Labs, Burlingame, CA, USA), streptavidin sepharose (Sigma, St. Louis, MO, USA), neutra avidin (Pierce, Rockford, IL, USA), and streptavidin magnetic particles from Dynal (Lake Success, NY, USA) and Boehringer Mannheim (Indianapolis, IN, USA), and had the lowest amount of nonspecific binding with the magnetic particles (from either company) that had been preblocked with BSA.
7. The glycerol gradient fractions that contain the spliceosomes have to be empirically determined. In our hands they are consistently in fractions 9–13. The yields after the glycerol gradient are 200–300 ng of RNA. Yields after affinity purification are 40–90 ng of RNA.
8. To avoid precipitation of the protein in the IEF gel, the voltage should be "ramped up" such that the current does not exceed 0.1 mA per gel.

Acknowledgments

We thank Sharon Jamison for contributing **Fig. 1A** and **B** (Jamison and Garcia-Blanco, unpublished results), Russ Carstens for his critical reading of this chapter, and Sabina Sager for her help in the preparation of this chapter. We also want to thank the Raymond and Beverly Sackler Foundation for their support.

References

1. Brody, E. and Abelson, J. (1985) The "spliceosome": yeast pre-messenger RNA associates with a 40S complex in a splicing-dependent reaction. *Science* **228**, 963–967.
2. Grabowski, P. J., Seiler, S. R., and Sharp, P. A. (1985) A multicomponent complex is involved in the splicing of messenger RNA precursors. *Cell* **42**, 345–353.
3. Frendewey, D. and Keller, W. (1985) Stepwise assembly of a pre-mRNA splicing complex requires U-snRNPs and specific intron sequences. *Cell* **42**, 355–367.
4. Kramer, A. (1996) The structure and function of proteins involved in mammalian pre-mRNA splicing. *Annu. Rev. Biochem.* **65**, 367–409.
5. Hall, S. L. and Padgett, R. A. (1996) Requirement of U12 snRNA for in vivo splicing of a minor class of eukaryotic nuclear pre-mRNA introns. *Science* **271**, 1716–1718.
6. Tarn, W. Y. and Steitz, J. A. (1996) A novel spliceosome containing U11, U12, and U5 snRNPs excises a minor class (AT-AC) intron in vitro. *Cell* **84**, 801–811.

7. Moore, J. M., Query, C. C., and Sharp, P. A. (1993) Splicing of precursors to messenger RNAs by the spliceosome, in *The RNA World* (Gesteland, R. F. and Atkins, J. F., eds.), Cold Spring Harbor Laboratory Press, Cold Spring Harbor, New York, pp. 303–357.
8. Pikielny, C. W., Rymond, B. C., and Rosbash, M. (1986) Electrophoresis of ribonucleoproteins reveals an ordered assembly pathway of yeast splicing complexes. *Nature* **324**, 341–345.
9. Konarska, M. M. and Sharp, P. A. (1986) Electrophoretic separation of complexes involved in the splicing of precursors to mRNAs. *Cell* **46**, 845–855.
10. Konarska, M. M. and Sharp, P. A. (1987) Interactions between small nuclear ribonucleoprotein particles in formation of spliceosomes. *Cell* **49**, 763–774.
11. Cheng, S.-C. and Abelson, J. (1987) Spliceosome assembly in yeast. *Genes Dev.* **1**, 1014–1027.
12. Reed, R., Griffith, J., and Maniatis, T. (1988) Purification and visualization of native spliceosomes. *Cell* **53**, 949–961.
13. Abmayr, S. M., Reed, R., and Maniatis, T. (1988) Identification of a functional mammalian spliceosome containing unspliced pre-mRNA. *Proc. Natl. Acad. Sci. USA* **85**, 7216–7220.
14. Michaud, S. and Reed, R. (1991) An ATP-independent complex commits pre-mRNA to the mammalian spliceosome assembly pathway. *Genes Dev.* **5**, 2534–2546.
15. Bennett, M., Michaud, S., Kingston, J., and Reed, R. (1992) Protein components specifically associated with prespliceosome and spliceosome complexes. *Genes Dev.* **6**, 1986–2000.
16. Staknis, D. and Reed, R. (1994) Direct interactions between pre-mRNA and six U2 small nuclear ribonucleoproteins during spliceosome assembly. *Mol. Cell. Biol.* **14**, 2994–3005.
17. Gozani, O., Patton, J. G., and Reed, R. (1994) A novel set of spliceosome-associated proteins and the essential splicing factor PSF bind stably to pre-mRNA prior to catalytic step II of the splicing reaction. *EMBO J.* **13**, 3356–3367.
18. Grabowski, P. J. and Sharp, P. A. (1986) Affinity chromatography of splicing complexes: U2, U5, and U4 + U6 small nuclear ribonucleoprotein particles in the spliceosome. *Science* **233**, 1294–1299.
19. Ruby, S. W. and Abelson, J. (1988) An early hierarchic role of U1 small nuclear ribonucleoprotein in spliceosome assembly. *Science* **242**, 1028–1035.
20. Cantor, C. R. and Schimmel, P. R. (1980) *Biophysical Chemistry, Part II: Techniques for the study of biological structure and function*. W. H. Freeman, New York.
21. Jamison, S. F., Crow, A., and Garcia-Blanco, M. A. (1992) The spliceosome assembly pathway in mammalian extracts. *Mol. Cell. Biol.* **12**, 4279–4287.
22. Roscigno, R. F. and Garcia-Blanco, M. A. (1995) SR proteins escort the U4/U6. U5 tri-snRNP to the spliceosome. *RNA* **1**, 692–706.
23. Lindsey, L. A., Crow, A. J., and Garcia-Blanco, M. A. (1995) A mammalian activity required for the second step of pre-messenger RNA splicing. *J. Biol Chem* **270**, 13,415–13,421.
24. O'Farrell, P. H. (1975) High resolution two-dimensional electrophoresis of proteins. *J. Biol. Chem.* **250**, 4007–4021.

25. Kjems, J., Frankel, A. D., and Sharp, P. A. (1991) Specific regulation of mRNA splicing in vitro by a peptide from HIV-1 Rev. *Cell* **67**, 169-178.
26. Grodberg, J. and Dunn, J. J. (1988) OmpT encodes the escherichia coli outer membrane protease that cleaves T7 RNA polymerase during purification. *J. Bacteriol.* **170**, 1245-1253.
27. Dignam, J. D., Lebovitz, R. M., and Roeder, R. G. (1983) Accurate transcription initiation by RNA polymerase II in a soluble extract from isolated mammalian nuclei. *Nucleic Acids Res.* **11**, 1475-1489.
28. Miller, C. R., Jamison, S. F., and Garcia-Blanco, M. A. (1997) HeLa nuclear extract: a modified protocol, in *mRNA Formation and Function* (Richter, J. D., ed.), Academic Press, San Diego, CA, pp. 25-30.
29. Blum, H., Beier, H., and Gross, H. J. (1987) Improved silver staining of plant proteins, RNA and DNA in polyacrylamide gels. *Electrophoresis* **3**, 93-99.
30. Reed, R. (1990) Protein composition of mammalian spliceosomes assembled in vitro. *Proc. Natl. Acad. Sci. USA* **87**, 8031-8035.
31. Calvio, C., Neubauer, G., Mann, M., and Lamond, A. I. (1995) Identification of hnRNP P2 as TLS/FUS using electrospray mass spectrometry. *RNA* **1**, 724-733.
32. Klose, J. and Kobalz, U. (1995) Two-dimensional electrophoresis of proteins: an updated protocol and implications for a functional analysis of the genome. *Electrophoresis* **16**, 1034-1059.
33. Sambrook, J., Fritsch, E. F., and Maniatis, T. (1989) *Molecular Cloning: A Laboratory Manual.* Cold Spring Harbor Laboratory Press, Cold Spring Harbor, New York.
34. Lee, C., Levin, A., and Branton, D. (1987) Copper staining: a five-minute protein stain for sodium dodecyl sulfate-polyacrylamide gels. *Anal. Biochem.* **166**, 308-312.

28

A Yeast Spliceosome Assay

Stephanie W. Ruby

1. Introduction

The spliceosome is a large ribonucleoprotein that splices precursor messenger RNA (pre-mRNA) in the nuclei of eukaryotic cells (reviewed in **refs.** *1–3*). It is composed of five small nuclear ribonucleoproteins (snRNPs) and numerous non-snRNP proteins. Each snRNP consists of a small nuclear RNA (snRNA) and several proteins.

The study of the association of the various components of the spliceosome with the pre-mRNA in vitro has yielded important insights into splicing. It is during the initial association of the spliceosomal components with the pre-mRNA that the intron and exons are first recognized and the splice sites are selected. When a radiolabeled pre-mRNA is added to a splicing extract in vitro, the spliceosome first assembles on the pre-mRNA in an ordered pathway of formation of prespliceosomal and spliceosomal complexes (**Fig. 1**). After the spliceosome is activated, the pre-mRNA is spliced. The steps in spliceosome formation have been defined mainly by the snRNPs that bind to or apparently dissociate from the pre-mRNA-containing complexes. The snRNPs have been the most readily identifiable components of the complexes, and their binding or dissociation can cause substantial changes in the molecular weight of a complex.

1.1. Spliceosome Assays

Four basic methods have been used to study the spliceosome in vitro: centrifugal sedimentation in gradients *(4–6)*, affinity chromatography *(7,8)*, column chromatography *(9)*, and gel electrophoresis *(10–14)*. In the first assays in which the spliceosome was identified *(4–6)*, an in vitro synthesized, radiolabeled pre-mRNA was added to a splicing reaction containing cell extract, and then sedimented in glycerol gradients by centrifugation. The spliceosome sedi-

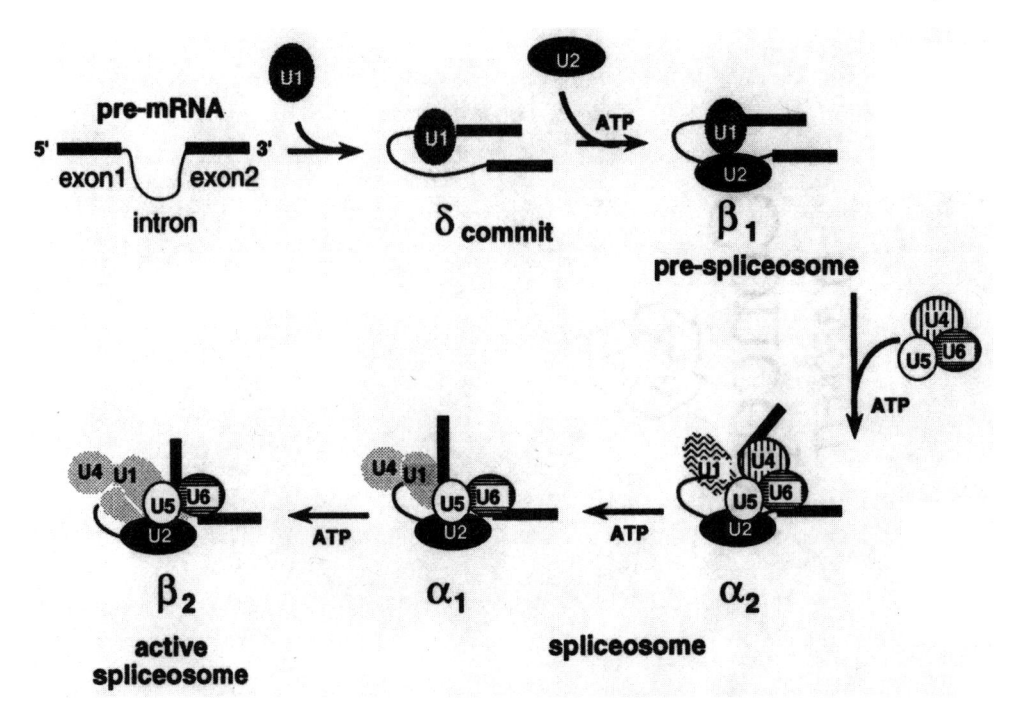

Fig. 1. A current working model of spliceosome assembly is diagrammed here. The U1, U2, U4, U5, and U6 snRNPs are represented by ellipses. Although numerous nonsnRNP proteins are functioning in this pathway, they are not depicted here. The names of the complexes as detected in the electrophoretic gel assay described in this chapter are designated by Greek letters. An initial complex (δ) is formed when U1 snRNP binds to the pre-mRNA; this complex has also been detected in another gel assay system and called the commitment complex *(16)*. U2 snRNP then binds and the prespliceosome (β_1) forms. The U4, U5, and U6 snRNPs form a tri-snRNP particle that binds to the prespliceosome to form the spliceosome. Thereafter, the U4 and U6 snRNPs undergo a conformational change leading to the activation of the spliceosome for splicing. All steps, except the initial one when U1 snRNP binds to pre-mRNA, require ATP. The association of the U1 and U4 snRNPs with the developing spliceosome changes during assembly as evidenced by their loss from some complexes in spliceosome assembly assays and as depicted here by the changes in shading. These two snRNPs may dissociate from the complexes in the splicing reaction or during electrophoresis. The positions and sizes of the snRNPs relative to each other are arbitrarily drawn.

ments as a 40S or 50–60S complex from yeast *(4)* or HeLa *(5,6)* cell extracts respectively, and contains both pre-mRNA and splicing intermediates. Additional analyses revealed other properties of the spliceosome: (1) it is stable enough to be further purified from a glycerol gradient *(7,15)*; and (2) its rate of sedimentation varies with the concentration of salt in the gradient *(4,15)*.

The initial observation that the spliceosome is a relatively stable complex led to other assays to analyze the spliceosome and to identify pre- and postsplicing complexes. Most of these assays consist of adding a radiolabeled pre-mRNA to a splicing reaction with cell extracts, and then fractionating and detecting the pre-mRNA-containing complexes by polyacrylamide gel electrophoresis and autoradiography. Polyacrylamide gel electrophoresis usually gives greater resolution of complexes than glycerol gradient sedimentation.

1.2. Conditions for Native Gel Electrophoretic Assays of the Spliceosome

Various conditions have been used for gel electrophoresis of the spliceosome *(10–14,16,17)*. These electrophoretic gel assays are often called "native" gel assays to indicate that they use conditions that maintain some of the integrity of the splicing-dependent complexes and to distinguish them from the denaturing polyacrylamide gels used to analyze the RNAs from in vitro splicing assays. In addition to preserving the integrity of splicing-dependent complexes, a native gel assay must also resolve and distinguish the splicing-dependent complexes from other complexes including unbound snRNPs, and splicing-independent (nonspecific) complexes that are formed by proteins that can bind to RNAs with or without a functional intron. A plethora of such nonspecifically binding proteins is present in cell extracts.

The native gel assays are based on the same principles as the gel shift assays used in studies of other nucleic acid–protein interactions: the binding of proteins and/or snRNPs retards or shifts the mobility of the pre-mRNA relative to unbound RNA. For native gels, the mobility of a pre-mRNA-containing complex depends on the molecular weight, conformation, shape, and net charge of the complex as well as the effective pore size of the gel and the voltage across the gel. Several conditions can be varied to adapt a native gel electrophoretic assay to the study of a particular RNA–protein interaction. The gel electrophoretic assay described in this chapter was designed to study the early steps of spliceosome formation and especially to detect the U1 snRNP in prespliceosomal complexes *(14)*.

1.2.1. Effective Pore Size

The large molecular weights and sizes of the prespliceosomal and spliceosomal complexes necessitate that the gel used in the assay has large pores. The effective pore size of an acrylamide gel depends on both the concentration of total acrylamide and the number of crosslinks as determined by the ratio of acrylamide to *bis*-acrylamide *(18)*. In the gels described here, both the concentration of acrylamide and the ratio of acrylamide to *bis*-acrylamide were varied to find the lowest concentration and highest ratio that would still enable polymerization and manipulation of the gel. Another means to achieve an effective large pore size is to use a composite gel of agarose and acrylamide as described elsewhere *(11,19)*.

1.2.2. Ionic Conditions

Most native gel and gel shift assays use low ionic strength buffers during electrophoresis (*see* **Note 1**). Such buffers are used for several reasons including to increase the migration of a molecule through the gel. The migration of a molecule during electrophoresis is proportional to the voltage across the gel *(20)*. The gel can be thought of as a simple electrical system in which voltage is defined by Ohm's law, $V = IR$, where voltage, V, is the product of the current, I, and the resistance, R. Because R is inversely proportional to the ionic strength of the buffer, low ionic strength buffers are used to increase the resistance and thus the voltage across the gel. Finally the amount of heat generated in the gel during electrophoresis is proportional to I^2R and is partly reduced by running the gel in low ionic strength buffers. If the heat is not dissipated, there will be temperature differences within the gel that will result in aberrant migrations.

The ionic conditions used in gel electrophoresis can have significant effects on nucleic acid–protein complexes. The conformation, net charge, migration, and stability of a ribonucleoprotein complex can vary with the ionic strength, composition, and pH of the gel buffer. A Tris-phosphate buffer is used for the gel assay described in this chapter. Two other native gel assays of the yeast spliceosome use Tris-acetate *(12)* and Tris-borate *(11)* buffers. A comparison of these two buffers along with a Tris-phosphate buffer showed that the Tris-borate buffer is excellent for fractionation of small DNA fragments in agarose gels, but it causes compaction and relatively slow migration of large DNA fragments (greater than 5 kbp) compared to Tris-acetate and Tris-phosphate buffers *(21)*. The Tris-phosphate buffer offered somewhat better separation of large DNA fragments than the Tris-acetate buffer. Although the basis of these differential effects of these three buffers is not known, a preliminary comparison of mobilities of pre-mRNA-containing complexes in Tris-borate versus Tris-phosphate buffer showed a similar pattern; the large complexes formed on the pre-mRNA in an in vitro splicing reaction migrated faster in Tris-phosphate than in Tris-borate buffer (data not shown).

Additional ions, particularly monovalent or divalent cations, can also be important in a native gel assay. For example, the inclusion of magnesium (Mg) in the gel buffer described here was considered because of four observations: (1) splicing in vitro requires from 1.5 to 3 mM Mg *(22)*; (2) the U1 snRNP had been shown to bind to the pre-mRNA in other assays that had used ionic condi-

A Yeast Spliceosome Assay

tions similar to those in the splicing reaction *(8,23)*; (3) the U1 snRNP is usually absent from complexes fractionated in gel-electrophoretic assays that use EDTA, a chelator of Mg, in the buffer *(12,24)*; (4) bacterial ribosomes migrate as 70S species in gels in 25 mM Tris, 1 mM Mg, but they dissociate into 50S and 30S subunits in gels in 25 mM Tris, 25 mM EDTA *(19)*. A comparison of the complexes fractionated in gels with or without Mg in the buffers shows that Mg affects several prespliceosomal complexes (**Fig. 2**). More radiolabeled pre-mRNA is found in the δ complex (which contains the U1 snRNP bound to the pre-mRNA), and the α and β complexes migrate more slowly and closer to one another with Mg than without it. Additional assays have shown that up to 16-fold more δ complex is detected with Mg than without it *(14)*. Mg may alter the conformations of some components or the binding of a factor to a complex, or both, to increase the stability of the δ complex during electrophoresis.

To adapt the gel assay described here to the study of other RNA–protein interactions, it may be necessary to modify the type and concentration of ions in the gel buffer. In this regard, data from gel shift assays for DNA-binding proteins are also informative. For example, including just 10 mM sodium to increase the ionic strength of the buffer in a gel shift assay allows detection of TFIIIA-induced DNA bending *(25)*.

1.2.3. Competitor RNA or Polyanions

In most spliceosome assays, either competitor RNA *(11,14)* or the polyanion heparin *(10,12)* is added to a reaction to bind the nonspecific factors, thereby allowing the distinction between the splicing-dependent and nonspecific complexes in the gel. A competitor lacking a functional intron and splice sites will usually compete effectively with the pre-mRNA for nonspecific, but not splicing-specific, factors. The result is that mostly splicing-specific complexes bound to the radiolabeled pre-mRNA will be seen in the gel. A competitor can also enhance the resolution of the splicing-specific complexes, so it is routinely added to most assays. A competitor is usually added after the pre-mRNA has been incubated in the splicing reaction and shortly before the sample is loaded onto the gel because it may also inhibit splicing.

The amount of competitor that can be used in the gel assay depends on the specificity and binding constants of both the splicing-specific and and nonspecific factors, but it can be determined empirically by titrating the competitor (**Fig. 3**). Above a certain concentration, however, the competitor will begin to compete with the pre-mRNA for binding to the splicing factors as evidenced by the smearing of the splicing-related complexes and the loss of radiolabeled pre-mRNA from those complexes (**Fig. 3**). In addition to the use of competitors, the specificity of the splicing-dependent complexes needs to be further established by using pre-mRNA substrates with mutations that prevent spliceosome formation and splicing (*see*, e.g., **ref.** *12*).

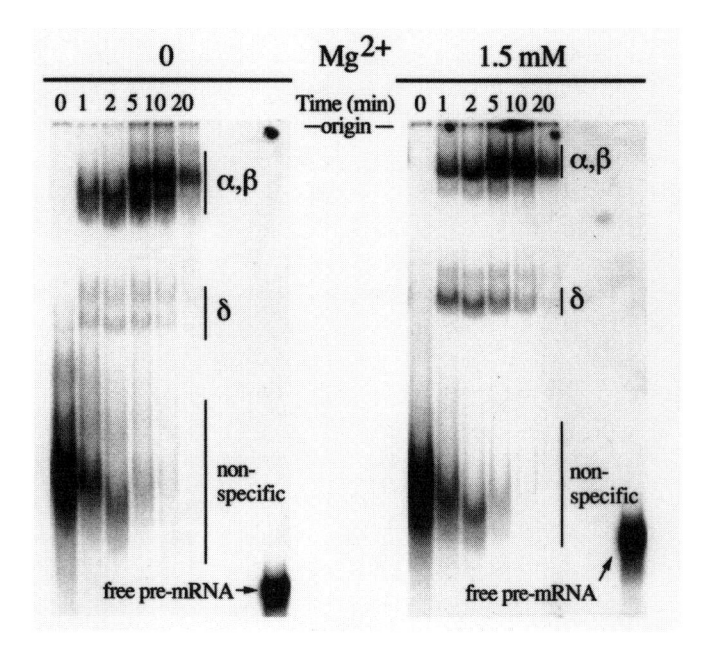

Fig. 2. Presplicing complexes fractionated in gels with or without magnesium. Radiolabeled actin pre-mRNA was incubated in a splicing reaction with added ATP. At the times indicated, samples were withdrawn, and added to ice-cold R buffer and RNA to give a final concentration of 1 µg of competitor mouse intestine RNA per microliter, and incubated for 10 min on ice. After load buffer was added, half the sample was loaded onto a gel with TP8 buffer and half onto a gel with TP8 buffer and 1.5 mM MgOAc. In addition, "free" pre-mRNA in a splicing reaction without extract was similarly treated with competitor RNA and run on the gel. The two gels were run and exposed to film for the same times. The positions of the nonspecific and splicing-dependent complexes (δ, α, and β) as defined in **Fig. 1** are indicated. The individual α and β complexes formed on the radiolabeled pre-mRNA are not clearly resolved in these gels; additional (Northern blot) analyzes and longer electrophoretic times are necessary to distinguish these complexes *(14)*.

The type and preparation of a competitor for routine use in the native gel assay may be critical to the migration and resolution of the splicing-dependent complexes. Competitor RNA may consist of total cellular RNA because the fraction of pre-mRNA in the total cellular RNA is usually very low. Besides being a polyanion, a competitor may have additional properties that are important for the native gel assay. For example, RNA isolated from mouse intestines, mouse liver, and rat liver worked well in the assay shown in **Fig. 4**, but other RNAs and DNA did not. An analysis of these different competitor RNAs by denaturing gel electrophoresis and ethidium bromide staining showed that

A Yeast Spliceosome Assay

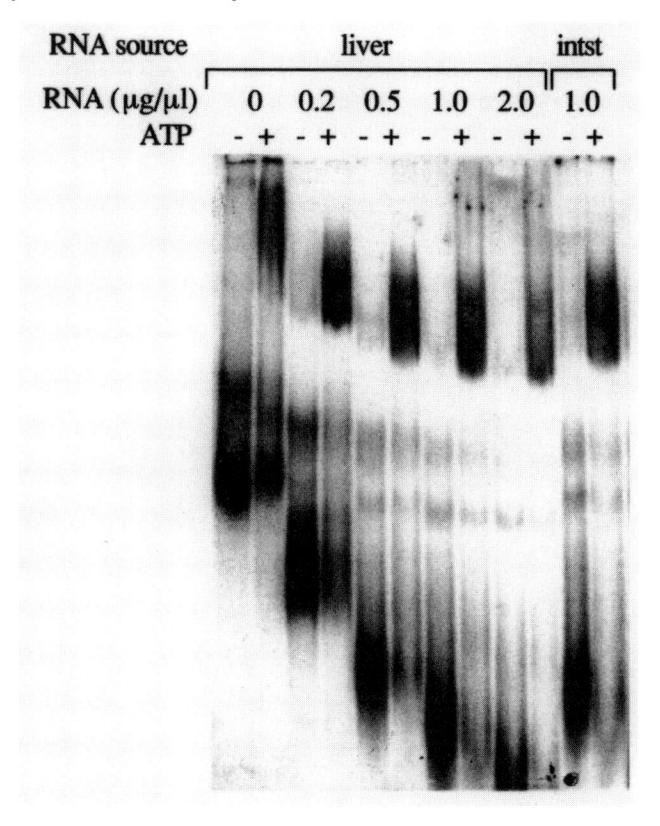

Fig. 3. Titration of competitor RNA. Radiolabeled actin pre-mRNA was incubated in a splicing reaction with (+) or without (–) added ATP. After a 10-min incubation at 23°C, samples were withdrawn from the two reactions and added to ice-cold R buffer containing RNA from mouse liver or mouse intestine (intst) to give the final concentrations of competitor RNA indicated. After 10 min of incubation on ice, load buffer was added and the samples were run on a 3.5% acrylamide gel in TP8 buffer without magnesium.

the good competitors consisted of RNAs mostly larger than 210 nt, whereas the poor competitors (commercial preparations of total yeast RNA and *E. coli* tRNA) consisted of small RNAs, mostly 5S RNA and tRNAs (data not shown). Thus it is worthwhile to try several different RNAs and RNA preparations as well as different polyanions to find one that will work for your specific assay.

2. Materials

2.1. Preparation of Competitor RNA from Mouse Organs

Sterile, distilled water is used for preparing solutions unless otherwise indicated.

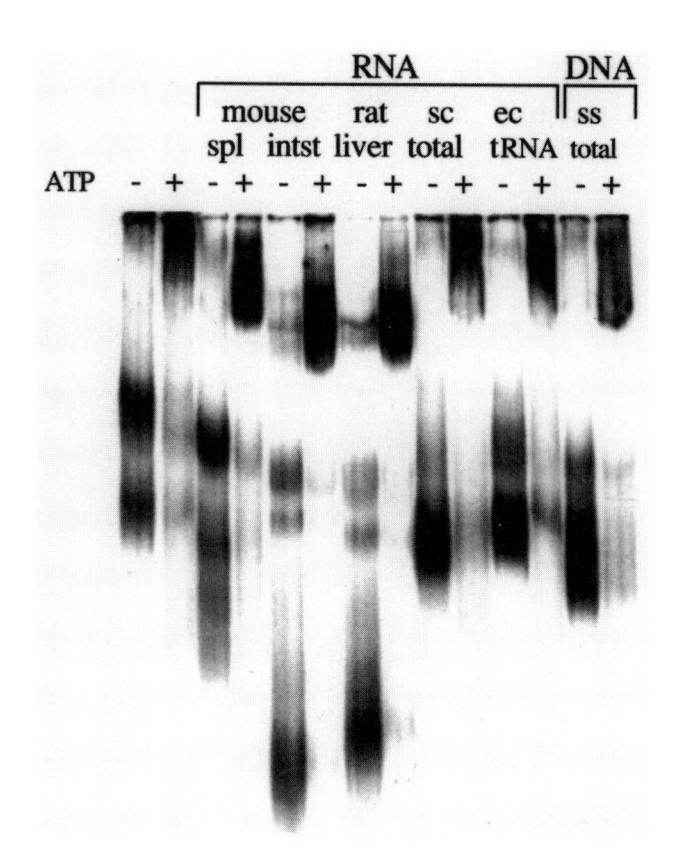

Fig. 4. Effects of various competitor RNAs and DNA on electrophoresis of the complexes. Radiolabeled actin pre-mRNA was incubated in a splicing reaction with (+) or without (–) added ATP. After a 10-min incubation at 23°C, samples were withdrawn from the two reactions and added to ice-cold R buffer containing the competitor RNA or DNA indicated to give a final concentration of 1 µg/µL: total RNA from mouse spleen (spl) or intestine (intst); total RNA from rat liver, total RNA from *Saccharomyces cerevisiae* (sc total); tRNA from *Escherichia coli* (ec tRNA); or single-strand DNA from salmon sperm (ss total). After 10 min of incubation on ice, load buffer was added and the samples were run on a 3.5% acrylamide gel in TP8 buffer without magnesium. The yeast total RNA and salmon sperm DNA were from Sigma Chemical Co., St. Louis, MO, USA, and the *E. coli* tRNA was from Boehringer Mannheim, Indianapolis, MN, USA. Single-strand salmon sperm DNA was prepared as described in **Subheading 2.5., step 7**.

1. Diethylpyrocarbonate (DEPC)-treated water. Under a chemical fume hood (*see* **Note 2**), add 2 mL of DEPC to 2 L of distilled water. Let stand for 12 hour, then autoclave. Autoclaving inactivates DEPC. Store at room temperature.
2. Tris-HCl solutions: 1 *M* Tris-HCl, pH 7.4; 1 *M* Tris-HCl, pH 7.6; 0.1 *M* Tris-HCl, pH 7.4. For a 1 *M* solution, add 30.29 g of Tris base (Tris [hydroxymethyl]

aminomethane, molecular biological grade) to 125 mL of water followed by 92 mL of 1 M HCl. Bring the pH to 7.6 or 7.4 by titrating with 1 M HCl and the volume to 250 mL by adding water. Autoclave then store at room temperature. Dilute 1 M Tris-HCl, pH 7.4, 10-fold with sterile water to make a stock of 0.1 M Tris-HCl, pH 7.4.

3. 0.5 M EDTA, pH 8.0. Add 18.6 g of ethylenediamine tetraacetic acid (EDTA) (disodium, dihydrate form) to 75 mL of water. Dissolve the EDTA and bring the pH to 8.0 by adding 10 N NaOH. Bring the volume to 100 mL by adding water.
4. 10 mM Tris-HCl, pH 7.6, 1 mM EDTA (1× TE). Dilute 1 M Tris-HCl, pH 7.6, and 0.5 M EDTA, pH 8.0, in water.
5. MgCl$_2$ solutions. For 1 M magnesium chloride (MgCl$_2$) add 20.33 g of magnesium chloride hexahydrate to 90 mL of water, and adjust the volume to 100 mL. Autoclave, then store at room temperature. For 0.1 M MgCl$_2$ for splicing reactions, dilute 1 M MgCl$_2$ 10-fold in water and store as 500–1000-µL aliquots at –20°C.
6. 3 M Sodium acetate (NaOAc), pH 5.4. Add 40.82 g of sodium acetate trihydrate to 75 mL of water. Add glacial acetic acid to bring the pH to 5.4, and then bring the volume to 100 mL with water.
7. 10% (w/v) Sodium dodecyl sulfate (SDS). Two different qualities of SDS (high quality [BDH Biochemical from Hoeffer Scientific Instruments, San Francisco, CA, USA] or reagent quality) are used to make 10% solutions. Dissolve 50 g of SDS in 400 mL of distilled water. Bring the volume up to 500 mL and autoclave. Store at room temperature.
8. Homogenization buffer for preparing competitor RNA: 3 M lithium chloride (LiCl), 6 M urea, 20 mM Tris-HCl, pH 7.4, 10 mM MgCl$_2$. Make 500 mL of buffer for extracting RNA from the organs of 16 mice. Use sterile solutions and glassware, and DEPC-treated water for making the buffer, but do not autoclave the buffer. Add 63.6 g of anhydrous LiCl, 180.2 g of urea (ultrapure), 10 mL of 1 M Tris-HCl, pH 7.4, and 5 mL of 1 M MgCl$_2$ to 300 mL of DEPC-treated water. Dissolve the urea with low heat and stirring. When the urea is dissolved, bring the volume to 500 mL with water. Make up fresh within 24 h before use. Store at 4°C.
9. Dissolving buffer for preparing competitor RNA: 10 mM Tris-HCl pH 7.4, 1 mM EDTA, 1% (w/v) SDS. Add 1 mL of 1 M Tris-HCl, pH 7.4, 0.4 mL of EDTA, pH 8.0, and 20 mL of 10% (high-quality) SDS to 79 mL of DEPC-treated water. If kept sterile, the solution can be stored at room temperature for several days before use.
10. 1 M Tris base. Add 121.1 g of Tris base (Tris[hydroxymethyl] aminomethane, molecular biological grade) to 900 mL of warm water. Stir to dissolve then bring the volume up to 1 L with water. Autoclave, then store at room temperature.
11. Phenol equilibrated with Tris/EDTA. Whenever working with phenol, wear gloves, safety glasses, shoes, and a laboratory coat (*see* **Note 2**). Perform the following manipulations in the plastic-coated jar in which the phenol is packaged. Heat 500 g of molecular biological grade phenol at 65°C until it becomes liquid. Add 0.5 g of 8-hydroxyquinoline. Add 250 mL of 1 M Tris base, then 250 mL of water. Mix vigorously by adding a magnetic stir bar to the jar and putting the jar on a magnetic stir plate for about 10 min. Stop stirring to allow the phases to

separate. The lower, phenolic phase will appear bright yellow due to the 8-hydroxyquinoline and will remain slightly cloudy after the phases separate. Remove most of upper aqueous phase by suction. Add two volumes of 1× TE, pH 7.6, stir the solution for 2–5 min, and allow the phases to separate again. Check that the upper aqueous solution is pH 6.8–7.2. Remove most of the aqueous phase to leave about 2 cm of it. Store at 4°C for 1–2 d. For longer storage periods, freeze at –20°C. When the organic phase turns to light orange, discard the solution.

12. $CHCl_3$/*iso*-amyl alcohol (24:1). Working under the fume hood and wearing gloves (*see* **Note 2**), add 20 mL of *iso*-amyl alcohol to 480 mL of chloroform ($CHCl_3$) in a sterile, graduated, glass cylinder. Store in a ground-glass stoppered or screw-capped bottle at room temperature under the fume hood.
13. 7.5 M Ammonium acetate (NH_4OAc). Add 160.84 g of ammonium acetate tetrahydrate to 60 mL of water, stir until dissolved, and adjust volume to 100 mL. Sterilize with a filter of 0.2 μm pore size. Store at 4°C.
14. 70% (v/v) Ethanol. Pour 95% ethanol to 350 mL in a graduated cylinder, then bring the volume to 475 mL with sterile water. Alternatively, pour absolute ethanol to 350 mL in a graduated cylinder then bring the volume to 500 mL with sterile water. Store at room temperature.

2.2. Solutions for Spliceosome Assay

1. Silanizing solution: 15% dichloromethylsilane in chloroform. Wearing gloves and working in a chemical fume hood (*see* **Note 2**), pour 170 mL of chloroform into a 200-mL glass, graduated cylinder. Pour dichloromethylsilane to bring the volume to 200 mL. Store in the fume hood in a glass jar with a ground-glass stopper.
2. 1 M Magnesium acetate (MgOAc). Add 42.9 g of magnesium acetate tetrahydrate to 150 mL of water. Stir to dissolve, then bring up the volume to 200 mL in a graduated cylinder. Sterilize with a filter of 0.2 μm pore size. Store at 4°C.
3. 17× TP8 buffer: 850 mM Tris-phosphate, pH 8.0, at 0°C. Add 108 g of Tris base (Tris[hydroxymethyl] aminomethane, molecular biological grade) and 20.0 mL of 85% phosphoric acid (analytical grade) to 800 mL of sterile water, then adjust the volume to 1 L with water. Check the pH of a small sample diluted to 1× (50 mM Tris-phosphate) and cooled on ice; it should be pH 8.0 at 0°C. Autoclave then store at 4°C.
4. TPM8 buffer: 50 mM Tris-phosphate, pH 8.0, at 0°C, 1.5 mM MgOAc. Make 1 L of TPM8 buffer for one native gel by adding 60 mL of 17× TP8 to 940 mL of sterile water. Slowly add 1.5 mL of 1 M MgOAc while stirring to prevent a precipitate from forming.
5. 20% (w/v) Acrylamide (50:1 acrylamide to *bis*-acrylamide) in water. Dissolve 50 g of acrylamide (Ultra Pure from Boehringer Mannheim, Westbury, NY, USA) and 1 g of *bis*-acrylamide in 200 mL of water and stir until dissolved. Bring the volume up to 250 mL in a sterile graduated cylinder. Filter through a disposable Nalgene filter of 0.2 μm pore size. Store in a sterile, screw-capped bottle at 4°C for up to 1 yr.

A Yeast Spliceosome Assay

6. 10% (w/v) Ammonium persulfate (APS) and N,N,N'N'-tetramethylethylenediamine (TEMED). Put 1 g of APS (molecular biological or electrophoresis grade) into a 15-mL graduated, screw-capped plastic tube and bring volume up to 10 mL by adding water. Store APS for up to 1 wk at 4°C. Store TEMED at 4°C.

7. Yeast whole cell splicing extract. Prepare yeast whole cell splicing extract as described in Chapter 26 of this book. Take care in harvesting the extract after ultracentrifugation at 100,000g. Contamination of the splicing extract in the middle layer with either the upper or bottom layer can lead to loss of resolution of the complexes in the native gel assay as well as retention of significant amounts of the radiolabeled pre-mRNA at the origin of the gel.

8. Radiolabeled pre-mRNA. A radiolabeled, actin pre-mRNA is synthesized in vitro (see Chapters 1 and 26) with the cloned actin DNA template cut with the *Hpa*II restriction endonuclease. The actin pre-mRNA has an exon 1 of 92 nt, an intron of 309 nt, and an exon 2 of 48 nt. It is dissolved at 40,000–80,000 Cerenkov cpm/μL in water and stored at –20°C.

9. Potassium (K) phosphate buffer, pH 7.6. Make a stock of 1 M K phosphate buffer, pH 7.6, by slowly adding 8.85 g of monobasic potassium phosphate anhydrous (KH_2PO_4) alternating with 75.8 g of dibasic potassium phosphate anhydrous (K_2HPO_4) to 400 mL water. When the salts are dissolved, check that the pH is 7.6, then adjust the volume to 500 mL and autoclave. Store at room temperature. For 0.1 M K phosphate buffer, pH 7.6, dilute freshly autoclaved 1 M K phosphate buffer 10-fold and store in 500–1000-μL aliquots at –20°C.

10. 100 mM ATP. Add 5 mL of water to 500 mg of the sodium form of adenosine triphosphate from Amersham Pharmacia Biotech, Piscataway, NJ, USA. Bring the pH carefully to 6.8–7.2 by adding 50 μL of 5 N NaOH followed by about 850 μL of 1 N NaOH in 20 μL aliquots. Monitor the pH by placing a drop of the ribonucleotide solution on pH paper (pH range 6.5–10). Measure the absorbance of a 5000-fold dilution (3 μL into 1.5 mL of water) at λ_{max} (259 nm) in a quartz cuvet in a spectrophotometer. Calculate the concentration of the stock solution; the absorbance coefficient E_{max} ($M^{-1}cm^{-1}$) at pH 7.0 is 1.59×10^4, and absorbance equals $E_{max}M$ in a cuvet with a path length of 1 cm. Prepare 100 mM ATP stocks by dilution in sterile water. Store in 200–500-μL aliquots at –20°C. Thaw the ATP solution quickly in a 37°C water bath and keep on ice while using it at the bench.

11. 30% (w/v) PEG_{8000}. Add 30 g of polyethylene glycol 8000 MW (PEG_{8000}) (Eastman Kodak, Rochester, NY, USA) to 65–70 mL of warm (50°C) water. If necessary, continue stirring on a hot plate to dissolve. After the solution cools, bring volume up to 100 mL with water, then autoclave. Dispense into 1- and 10-mL aliquots and store at –20°C.

12. 1 M potassium HEPES (HEPES-K^+), pH 7.4. Add 2.3 g of N-2-hydroxyethylpiperazine-N'-2-ethanesulfonic acid (Ultrol-grade HEPES acid from Calbiochem, La Jolla, CA, USA) and 6.8 μL of 1 N potassium hydroxide (KOH) to 7 mL of water. Adjust the pH to 7.4 with 1 N KOH and the volume to 10 mL with water. Sterilize with a filter of 0.2 μm pore size and store at –20°C.

13. R buffer: 50 mM HEPES-K^+, pH 7.4, 2 mM MgOAc. Add 0.5 mL of 1 M HEPES-K^+, pH 7.4, and 40 μL of 1 M MgOAc to 9.5 mL of water. Store in 1-mL aliquots at –20 or –70°C.

14. R buffer + competitor RNA: Allow 7.5 μL of R buffer and 15 μg of competitor RNA in 0.5–1 μL of water for 7.5 μL of splicing reaction. Mix the R buffer and competitor RNA fresh before starting the spliceosome assay (*see* **Subheading 3.3., step 2**).
15. Dye solution: 10% (w/v) each bromophenol blue and xylene cyanol in water. Add 1 g of bromophenol blue and 1 g of xylene cyanol to 7 mL of sterile water in a sterile, disposable, polypropylene, screw-capped tube. Bring the volume to 10 mL with sterile water. Store at room temperature. The dyes may not completely dissolve: use as is and vortex vigorously just before removing an aliquot.
16. Load buffer: 217 mM Tris-phosphate, pH 8.0, at 0°C, 40% (v/v) glycerol, and 0.25% (w/v) each bromophenol blue and xylene cyanol. Add 2.5 mL of 17× TP8 buffer, 2.5 mL of autoclaved glycerol (molecular biological), and 250 μL of dye solution to 4.75 mL of water. Store in 1-mL aliquots at –20°C.
17. 15% (v/v) Methanol, 5% (v/v) acetic acid. Pour 150 mL of methanol (analytical grade) into a 1 L graduated cylinder. Add water to 950 mL. Bring the volume to 1 L with glacial acetic acid (*see* **Note 2**). Store at room temperature in a screw-capped bottle.

2.3. Solutions for Northern Blot Analyses of Endogenous snRNPs

1. 10× TBE: 890 mM Tris-borate, 25 mM EDTA, pH 8.3. Dissolve 216 g of Tris base (Tris[hydroxymethyl] aminomethane, molecular biological grade), 110 g of crystalline anhydrous boric acid, and 18.6 g disodium dihydrate EDTA in 1600 mL of water. Check that the pH is 8.3. Bring the volume up to 2 L with water. Store at room temperature in a carboy.
2. 1× TBE, 8 M urea: 89 mM Tris-borate, pH 8.3, 2.5 mM EDTA, 8 M urea. Add 480.4 g of urea (high quality, not ultrapure) to about 500 mL of warm water and 100 mL of 10× TBE. Stir on low heat to dissolve the urea. Pour into a graduated cylinder and bring volume up to 1000 mL with water after the solution has cooled.
3. 0.25× TBE for electroblotting. Make 14 L of 0.25× by diluting the 10× TBE with deionized water. Put most of the 0.25× TBE in the electrotransfer device to cool overnight in the cold room. Store the rest of it in a carboy in the cold room.
4. 20× SSCP: 2.4 M sodium chloride, 0.30 M sodium citrate, 0.40 M sodium phosphate, pH 7.0. Add 140.4 g of sodium chloride (NaCl) and 88.2 g of sodium citrate dihydrate (Na$_3$C$_6$H$_5$O$_7$·2H$_2$O) to 750 mL of water. While stirring the solution, slowly add 18.73 g monobasic sodium phosphate anhydrous (NaH$_2$PO$_4$) alternating with 34.6 g of dibasic sodium phosphate anhydrous (Na$_2$HPO$_4$). When the salts are dissolved, check that the pH is 7.0; if not, the pH can be adjusted slightly by titration with 1 N NaOH or 1 N HCl. Bring the volume up to 1 L with water and store in a carboy at room temperature.
5. Deionized formamide. Some formamide preparations, such as that from Fluka Chemical (Ronkonkoma, NY, USA), may have little formic acid as indicated by a pH above 6.5. Such formamide may be used in the hybridization solution. Store

it at −20°C. However, if the pH is 6.5 or below, then you will need to deionize the formamide as follows. Add 10–50 mL of formamide to a proportional amount (10–50 g) of mixed bed resin AG501-X8 in a sterile beaker and stir for 30 min. at room temperature. Let the beads settle, remove a small sample of formamide and check its pH with a meter to be sure that it is 6.8 or above. Remove the resin by filtering the formamide through a no. 1 Whatman paper in a sterilized buchner funnel into a sterile filter flask. Store in 1–10-mL aliquots at −20°C. Check the pH of aliquots stored longer than 6 mo. If the pH is lower than 6.5, repeat the deionization process.

6. 50× Denhardt's solution: 1% (w/v) Ficoll, 1% (w/v) polyvinylpyrrolidone, 1% (w/v) bovine serum albumin. Add 1 g of Ficoll (average mol wt 400,000), 5 g of polyvinylpyrrolidone (average mol wt 40,000), and 1 g of bovine serum albumin (Pentax Fraction V) to 90 mL of water. Stir to dissolve, then bring the volume to 100 mL with water. Filter through a disposable Nalgene filter with 0.45 μm pore size and store in 20–50-mL aliquots at −20°C.

7. Single-strand, salmon sperm DNA (ssDNA) at 10 mg/mL. Dissolve double-strand salmon sperm DNA at 50 μg/μL in 1× TE. Sonicate it to an average molecular weight of 300 bp as judged by agarose gel electrophoresis. Boil the sonicated (sheared) DNA for 15 min to denature it to single strands, and then dilute it to 10 μg/μL in 1× TE. For **Fig. 4**, the denatured DNA was diluted with water to 20 μg/μL.

8. Northern blot hybridization solution: 50% (v/v) formamide, 0.1% (w/v) SDS, 5× SSCP, 3× Denhardt's, 100 μg ssDNA/mL. Inclusion of phosphate (SSC<u>P</u>) in the hybridization solution makes it unnecessary to separate unincorporated radionucleotides from the radiolabeled probe. For 100 mL of hybridization solution, add the following to a graduated cylinder in the order given: 50 mL of formamide, 25 mL of 20× SSCP, 1 mL of 10% high-quality SDS, and 7 mL of 50× Denhardt's solution. Bring the volume up to 99 mL with water. Add 1 mL of ssDNA. Heat to 60–65°C before adding to the blot. Store at −20°C.

9. 50 mM EDTA, pH 8.0, 0.1% (w/v) SDS. Dilute 0.5 M EDTA and 10% (w/v) high-quality SDS in water.

10. Radiolabeled probe. A double-strand DNA fragment encoding an snRNA gene is used to make the radiolabeled probe for the Northern blot hybridizations described here. The DNA fragment is either isolated by cutting the cloned DNA with restriction endonucleases and purifying the fragment by agarose gel electrophoresis, or amplified by the polymerase chain reaction (*see* **refs.** *14, 26*, and *27*). The DNA fragments for the snRNA genes range in size from about 200 to 1300 bp *(14)*. A DNA fragment (about 500 ng) is radiolabeled by the random hexameric deoxyoligonucleotide priming reaction *(27)* with Klenow polymerase and radionucleotide (2.5 μL [25 μCi] of [α-^{32}P]dATP, 3000 Ci/mmol and 10 mCi/mL) in a 25-μL reaction as described *(27)* except that ATP is used as the radionucleotide because the snRNA genes are 60% or more A–T. The reaction is stopped by the addition of 225 μL of 50 mM EDTA, 0.1% SDS. The probe is denatured by heating the stopped reaction at 100°C for 10 min and then immediately added to 15 mL of hybridization buffer at 65°C.

11. Northern blot wash solutions: 3×: 3× SSCP, 0.1% SDS and 0.1×: 0.1× SSCP, 0.1% SDS. Wash solutions are made by diluting 20× SSCP and 10% (w/v) reagent quality SDS in water.

2.4. Equipment and Supplies

1. Liquid nitrogen (N_2).
2. Centrifuges, homogenizer, and Speed-Vac. Two centrifuges (high-speed and micro), and a Polytronic homogenizer with a sawtooth generator (such as model PT 2000 or 3100 from Brinkmann Instruments, Westbury, NY, USA) are used for the preparation of competitor RNA. A microcentrifuge is needed for the native gel assay.
3. Tubes. 15-mL Corex tubes are filled with chromic acid solution (Chromerge, VWR Scientific Products, West Chester, PA, USA) (*see* **Note 2**), emptied, and then washed with copious amounts of deionized water. The tubes are baked at 252°C for 4 h. Polypropylene Oak Ridge tubes (28 mL, Nalgene Brand no. 3118-0028 available from VWR Scientific Products, West Chester, PA, USA) are filled with DEPC diluted 1000-fold in water just before use, and stood at room temperature for 12 h. The tubes are then rinsed 3× with sterile, distilled water and dried for 30 min at 100°C. Sterile, disposable, polypropylene, graduated, 50-mL, conical centrifuge tubes (Falcon brand no. 2070) are used without pretreatment. Microcentrifuge tubes (1.7 mL) used for the splicing and native gel assays are used as is (not sterilized) as long as they are always handled with gloves.
4. Glass plates. The back gel plate measures 26.5 cm high × 20.2 cm wide and 0.4 cm thick. The "notched" plate has the same dimensions and a notch that is 2.2-cm deep and 16.3-cm wide.
5. Teflon spacers and combs. A sheet of Teflon can be purchased from a local plastics supplier and cut by hand with scissors or a razor blade. The spacers and combs are 0.5 mm thick. The comb is 16 cm wide and 2.7 cm high with 14 teeth separated by 3 mm each. Each tooth is 8-mm wide at the bottom and 11 mm long.
6. Yellow electrical tape (no. 56 from 3M, Minneapolis, MN, USA).
7. Gel box. A plastic gel box that will allow the recirculation of buffer between the two tanks is used (*see* **Note 3**). A simple box with one connector coming out the bottom of the top buffer tank and one coming out the side of the bottom buffer tank is sufficient. Run a piece of tubing between the connectors so that buffer in the top tank will flow by gravity into the bottom tank. The buffer is circulated from the bottom to upper tank with tubing placed in the tanks and connected to a peristaltic pump. Normally the gel box is used only for native gels; however, if it has been used previously with buffers containing SDS or urea, then rinse the box thoroughly with deionized or distilled water before using it for a native gel. If there are traces of RNase from previous uses, the gel box should be treated with DEPC (diluted 1000-fold) in water for 12 h under a fume hood, and then rinsed with copious amounts of sterile, distilled water.
8. Power supply. A power supply that can run up to 200 V and 50 mA, and maintain constant voltage is needed.

A Yeast Spliceosome Assay

9. Pump. A Minipulse-2 peristaltic pump (Gilson Instsruments, Middleton, WI) or the equivalent that can pump at the rate of about 3–5 mL/min per line.
10. Flat-tipped microcapillary pipet tips (no. 53503-189 from VWR Scientific Products, West Chester, PA, USA).
11. Whatman 3MM chromatography paper (46 × 79 cm) available from VWR Scientific Products, West Chester, PA, USA.
12. GeneScreen (NEN Research Products, Boston, MA, USA).
13. Pyrex or stainless steel dishes for soaking gels and washing Northern blots. A set of dishes is used only for native gels and RNA blots. These are initially washed with soap and water and baked at 252°C for 4–5 h. Thereafter they are cleaned by rinsing with deionized water and air-dried.
14. Electrotransfer device. A device described by Church and Gilbert *(28)* is diagrammed in **Fig. 5**. It is attached via standard electrical leads to a bridge line rectifier that converts alternating to direct current. The rectifier is plugged into a 115–120 V grounded electrical outlet. The device is kept in a cold room and is filled with about 14 L of 0.25× TBE that is allowed to cool overnight before the transfer. Two native gels will fit simultaneously on the pad. The buffer can be used twice provided it is cooled overnight between transfers.
15. UV crosslinking device. Three 15-W germicidal UV light bulbs (General Electric Brand G15T8 available from Fotodyne, New Berlin, WI, USA) are separated from each other by 8.5 cm and set 35 cm above the sample platform. These bulbs together give a dose of 5 Joules/s/m^2 at the level of the platform.
16. Microwave oven or hot plate.
17. Hybridization bags and sealer. Scotchpak brand heat-sealable, 8 × 10 in. pouches and Scotchpak brand Pouch Sealer are available from VWR Scientific Products, Westchester, PA, USA.
18. Shaking water bath or hybridization chamber.
19. X-ray film, light-tight film casettes, intensifier screens, and film processing capabilities.
20. Scotchbrite brand pad, 3 × 4 in. (available in any supermarket).
21. Gel dryer (optional).

3. Methods
3.1. Preparation of Competitor RNA for the Spliceosome Assay

1. Wear *powder-free* gloves to avoid RNase contamination from your hands.
2. Remove the small intestines and liver from 8–16 mice. Euthanize one mouse at a time and remove its small intestines and liver. Avoid taking the gall bladder (a small [approx 3 mm diameter], clear, light-yellow, round sac surrounded by liver) and the pancreas (an opaque, cream-colored, 1–2 mm long, ellipsoid) as these organs have high concentrations of nucleases. Drop the liver and intestine into liquid N_2 immediately after they are removed from the mouse. Store overnight at −70°C or for up to 1 yr in liquid N_2.
3. Prepare the organs for homogenization. Thaw several pieces of intestines at once on a clean paper towel at room temperature. With a wet, gloved finger, squeeze

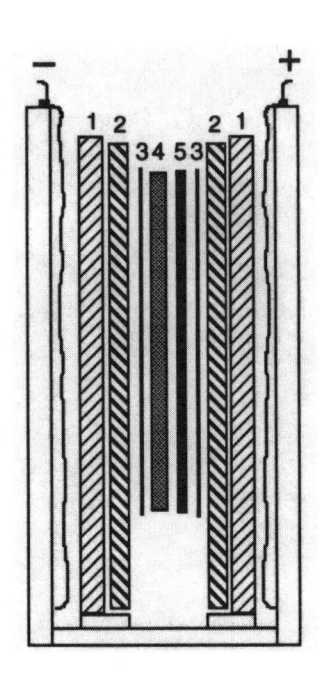

Fig. 5. Electrotransfer device for Northern blots of native gels. An exploded, cross-sectional view of the electrotransfer device is diagrammed. The numbers indicate the various components of the grid: (1) grid plates cut from egg-crate louver panels used in fluorescent light fixtures; (2) pads; (3) Whatmann 3MM chromatography paper; (4) the native gel; and (5) the nylon membrane. The positive (+) and negative (–) electrodes run along the sides of the tank. For a detailed description of this apparatus *see* Church and Gilbert *(28)*.

    out as much of the intestinal contents as possible—usually two or three passes suffice. Cut intestines into 2–3-in. pieces and place in 10 mL ice-cold homogenization buffer in a sterile 50-mL disposable, conical, polypropylene centrifuge tube. Use 10 mL of homogenization buffer per organ-mouse. This ratio of buffer to organ critical; if you use too little buffer, then you will get a very low yield of RNA. Livers can be put right into the buffer.
 4. Homogenize the organs with the Polytron homogenizer. Before using the Polytron homogenizer, wash its probe with DEPC-treated water and ethanol. Homogenize 20–30 mL of sample at a time. Using setting no. 5 (a midrange setting), homogenize a sample of livers for about 1 min and a sample of intestines for 30 s. The time of homogenization is not as critical as the look of the homogenate. It should look thick and have a homogeneous texture with very few fine bits of tissue visible. Let homogenates stand at room temperature while you finish all of them.
 5. Remove large debris by centrifugation. Pour the homogenates into 26-mL Oak Ridge tubes and pellet the large debris by centrifugation at 8500*g* for 15 min at room temperature in a high-speed centrifuge.

A Yeast Spliceosome Assay

6. Precipitate the RNA. Transfer the homogenates to fresh 26-mL Oak Ridge tubes; avoid taking the "slimey goop" in the top 1–2 mL of the supernatant and the pelleted debris at the bottom of the tube. Let the homogenates sit in ice in a coldroom overnight (for at least 6 h and at most 2–3 d). Pellet the precipitated RNA by centrifugation in a high-speed centrifuge at 11,000g for 15 min at 4°C.

7. Dissolve the RNA and extract it with organic solvents. Dissolve each RNA pellet in 2.5 mL of dissolving buffer and transfer to a corex tube. Wearing gloves, a laboratory coat, and safety goggles (*see* **Note 2**), add an equal volume of phenol to the dissolved RNA. Mix the aqueous and phenol solutions by vortexing, and then separate the aqueous and yellow phenolic layers by centrifugation in a high-speed centrifuge at 11,000g for 10 min at 4°C. Remove the upper clear aqueous layer that contains the RNA to a clean Corex tube. Reextract the bottom yellow phenolic phase by mixing it with 1 mL of fresh dissolving buffer and separating the phases by centrifugation. Remove the aqueous layer and add it to the aqueous solution already collected. Extract the combined aqueous solution again with an equal volume of fresh phenol. Remove the aqueous layer into a clean corex tube. Add a one- to twofold volume of chloroform:*iso*-amyl alcohol, mix by vortexing, and separate the aqueous and organic layers by centrifugation. Remove the upper, aqueous phase into a clean Corex tube.

8. Precipitate the RNA with sodium acetate and ethanol. Add 1/15 vol of 3 M NaOAc, pH 5.4, to the aqueous layer, followed by a threefold volume of absolute ethanol. Mix and incubate on dry ice for 10 min or at –20°C overnight. Pellet the precipitated RNA by centrifugation in a high-speed centrifuge at 11,000g at 4°C. Wash the pellet twice with 70% ethanol at room temperature. After the second wash is discarded, wick away any excess ethanol in the tube with a clean paper towel.

9. Dissolve and reprecipitate the RNA. Resuspend the RNA pellet in cold water, allowing 0.25 mL of water per original organ. Keep the RNA on ice. You should have about 3–4 mL of solution for each organ type for 16 mice (*see* **Note 4**). Add cold 7.5 M NH$_4$OAc to the RNA solution to a final concentration of 0.2 M, mix, and aliquot 400 µL per microcentrifuge tube. Add 1.2 mL of absolute ethanol to each tube and store at –70°C.

10. Prepare the competitor RNA for use in the assay. Centrifuge two or three tubes of precipitated RNA at a time for 10 min at 14,000–15,000g in a microcentrifuge at 4°C. Wash the RNA pellets twice with 70% ethanol at room temperature, and dry the pellets in a Speed-Vac. Dissolve the RNA pellets in small volumes of water. Combine the RNA in the tubes for a total volume of 100–150 µL. Determine the concentration and purity of the RNA by measuring the absorbance (OD) at λ_{260} and λ_{280} of a 500- or 1000-fold dilution of the sample in water. Calculate the concentration of RNA using the factor 1 OD unit at λ_{260} = 40 µg of RNA/mL. You want a concentration of 20–30 µg/µL for your stock of competitor RNA. A ratio OD$_{260}$ to OD$_{280}$ of 1.6 or higher is OK for use in the assay. Store the dissolved RNA at –70°C for up to 2 yr.

3.2. Pouring and Setting Up the Native Gel

1. Wash the gel plates, combs, and spacers, and silanize the notched gel plate. While wearing gloves, wash the glass plates and spacers and combs with a mild detergent and rinse with deionized water. Dry the plates and combs and spacers thoroughly with paper towels. Put the notched plate flat and raised (e.g., on an empty pipet tip box) in a chemical fume hood. Pipet about 2 mL of silanizing solution onto the plate (*see* **Note 2**). Spread the solution over the entire surface with a tissue. The solution will evaporate rapidly (within 1–2 min), leaving a glossy film on the plate. Be careful not to let the solution drip over onto the opposite side of the plate or it will be difficult to subsequently seal the tape to the plate. If some of the solution has gotten on the wrong side of the plate, wipe the region with a tissue soaked in 95% ethanol to remove some of the silane. Do not silanize the back plate or the gel may be ruined when the plates are separated after electrophoresis.

2. Assemble the gel plates and spacers together. Lay the back plate flat and raised above the bench. Place spacers along the sides of the plate and with one end of the spacers flush with what will be the bottom of the gel. Lay the notched plate down on top of the spacers with the silanized surface facing inward. Clamp the sides of the plates midway between top and bottom. At this point remove your gloves to tape the plates together (the tape sticks to the gloves). Cut a 20–30-cm piece of yellow electrical tape. Tape the bottom of the plates so that the tape wraps evenly around the bottoms of both plates to form a seal. Wrap part of this tape up about 2 cm along the sides of the plates as well. To ensure a good seal rub your fingernail along the tape. Remove one clip from the side of the plate. Cut a circa 30 cm piece of tape and similarly wrap it along the side and around the corner of the plates. Repeat for the other side.

3. Pour the gel. Lay the assembled plates on an absorbent paper on the bench at a slight angle (with the top of the back plate resting on a 25-mL pipet lying on the bench) to reduce the hydrostatic pressure of the gel when it is poured. For one or two 3.2% gels, add the following to a 150-mL sterile side arm Erlenmeyer flask in the order given: 61.5 mL of water at room temperature, 4.8 mL of 17× TP8, and 12.8 mL of acrylamide stock. Begin gently swirling the flask and slowly add 120 µL of 1 M MgOAc, then 120 µL of TEMED and finally 660 µL of 10% APS while swirling the contents of the flask. Pour the acrylamide mix through the side arm of the flask onto the bottom gel plate and let it slide between the plates until the solution reaches the top of the notched plate. Insert the comb such that the teeth are fully submerged in the gel. Use three clamps along the top of the back plate to press the two plates and gel comb together. Pour any extra gel solution onto the comb to submerge the part of the comb that is just above the notched gel plate. Let the gel polymerize for at least 1 h and up to 4 h at room temperature (*see* **Note 5**).

4. Set up the gel in the gel box with a circulating pump in the cold room at least 1 h before you begin the splicing reaction (*see* **Notes 1**, **3**, and **6**). Pour the gel buffer to submerge the bottom of the gel plate by 2.5 cm and the top of the gel by at least

A Yeast Spliceosome Assay

1 cm. Set and secure two lines of the peristaltic pump into the bottom and top buffer tanks to circulate buffer from the bottom tank to the top tank. Flush the gel wells gently with the buffer in the tank. Cover the top and bottom tanks with plastic wrap to prevent evaporation. Begin prerunning the gel at 160 V one half hour before loading the samples. The current will initially be 12.5 mA and then drop to 11 mA and remain constant.

3.3. Basic Spliceosome Assay

1. The assay for the 1.5 mM Mg^{2+} panel in **Fig. 2** is described here as an example of how to set up a spliceosome assay.
2. Prepare the tubes with R buffer + carrier RNA for the samples. Add 6 μL of competitor RNA stock (20 μg/μL in water) to 60 μL of R buffer. Mix and then add 8.3 μL of the mix to each of six 1.7-mL microcentrifuge tubes. Keep the tubes on ice.
3. Set up the splicing reaction by adding the following solutions to a microcentrifuge tube on ice in the order given: 13.8 μL of H$_2$O, 1 μL of 100 mM dithiothreitol (DTT), 3 μL of 1 M K phosphate buffer, 5.0 μL of 30% PEG$_{8000}$, 1.5 μL of 0.1 M MgCl$_2$, 1 μL of 100 mM ATP, and 20 μL of whole cell extract. Mix by hand and centrifuge for 2 s in a microcentrifuge. For the "0" time sample, remove 6.5 μL of the reaction mix, add it to the first tube with R buffer + competitor, and keep it on ice. (The radiolabeled pre-mRNA will be added later to this "0" time sample.)
4. Start and stop the splicing reaction. Incubate the reaction mix in a 23°C water bath for 2–3 min to bring the reaction mix up to temperature. Start the reaction by adding 5.5 μL of radiolabeled actin pre-mRNA. The concentrations of the components in the reaction are 60 mM phosphate, 3% (w/v) PEG$_{8000}$, 3 mM MgCl$_2$, 2 mM ATP, 20 mM KCl, 8 mM HEPES, 80 μM EDTA, 2.2 mM DTT, 8% (v/v) glycerol, 0.4 nM pre-mRNA, and 40–80 μg of extract. At 1 min after addition of the radiolabeled pre-mRNA, remove 7.5 μL from the reaction and add it to the second tube of R buffer + competitor. Keep the samples in R buffer + competitor on ice while you continue removing aliquots from the reaction for the remaining times. Allow the last sample to incubate in R buffer + competitor on ice for 10 min. Meanwhile, add 1 μL of radiolabeled actin pre-mRNA to the "0" time sample. After the 10 min, add 4 μL of ice-cold load buffer to each sample, mix quickly by hand, and centrifuge for 2 s in a microfuge in the coldroom. Keep the samples on ice until they are loaded onto the gel.
5. Load the samples and run the gel. Just before loading the samples onto the gel, turn off the power supply and the circulating pump. Flush the gel wells vigorously, but carefully so that the gel spacers between the wells are maintained upright. If necessary, readjust the spacers to an upright position with a small piece of used, developed X-ray film. Load 15 μL of each sample using a flat, microcapillary pipet tip as follows: Pull the sample slowly into the pipet tip; avoid getting any air into the tip after pipetting the sample otherwise air will come out as a bubble trapped in the well; position the bottom of the tip next to the bottom of the well; begin pipetting the sample; and as you pipet the sample, pull

the tip to avoid disturbing the sample in the well. Run the sample into the gel for 30 min to 1 h before the turning the circulation pump back on. Run the gel at 150–160 V for 14–22 h (*see* **Notes 6** and **7**) while the pump recirculates the buffer.

6. Remove the gel (*see* **Note 8**). Before you begin to remove the gel, cut two pieces of Whatman 3MM chromatography paper the size of the back plate and set aside. Turn off the power and pump and then disassemble the apparatus at room temperature. Remove the tape and spacers from the gel. Insert a thin spatula between the gel plates and *very slowly* pry apart the gel plates by exerting pressure against the upper notched plate. The gel should be left on the back plate. In a continuous motion, lay one piece of cut 3MM paper on the gel. The paper sticks readily to the gel so that once it contacts the gel, it cannot be removed. Rub a gloved hand over the paper to firmly affix the gel to the paper. Place a second piece of paper on top of the first to give the gel additional support. Turn over the gel and back plate, and pull the gel and papers together down and away from the gel plate.

7. Expose the gel to film. Wrap the gel and papers in a single layer of plastic wrap and expose to film with one or two intensifier screens at –70°C. The gel can be thawed and frozen for one more film exposure without much loss of resolution of the complexes. After the film exposures, the gel can be stored frozen at –20°C and processed later for either drying down the gel or transferring the RNAs to a nylon membrane.

8. For multiple film exposures, or exposure to a phosphor screen, precipitate the RNAs in the gel and dry the gel. Remove plastic wrap and the second piece of Whatman paper, and then soak the gel affixed to the first piece of paper in enough 15% methanol, 5% acetic acid in a dish to submerge the gel by 1 cm. *Very slowly* shake the dish at room temperature for about 20 min. The gel may become detached from the paper, however, it can be guided back onto the paper when the fixer is removed. Slowly remove the fixer by suction with a Pasteur pipet connected to a vacuum line. Gently lift the gel and paper while supporting them underneath with a gloved hand and place them onto two sheets of dry 3MM paper for additional support. Dry the gel in a gel dryer at 60°C for about 45 min to 1 h.

9. Clean the gel box. A precipitate will build up on the platinum wires in the buffer tanks during electrophoresis. To remove the precipitate, rinse the buffer tanks with deionized water. Gently scrub the wires with a Scotchbrite pad in deionized water. Rinse again with copious amounts of deionized water and let air-dry.

3.4. Northern Blot Hybridization Analyses of the Endogenous snRNP Complexes

1. For the initial characterization of the native gel assay for RNA-binding proteins, it is necessary to determine the conditions that will give sufficient resolution of the RNA in a complex from any unbound factors. The pre-mRNA in a splicing-dependent complex, for example, can be distinguished from unbound snRNPs in the extract by Northern blot hybridization analyzes. For the initial tests, the splicing-dependent complexes can be followed by the radiolabeled pre-mRNA, and the positions of the unbound snRNPs can be determined by hybridization with

A Yeast Spliceosome Assay

snRNA-specific probes. For subsequent studies to determine which snRNAs are in the splicing-dependent complexes, a transcript of very low or no radioactivity is used for the native gel assay (*see* **Note 9**), but the gel is treated for the Northern blot analyzes in the same way as described here.

2. Prepare the gel for blotting the RNAs to a membrane. After exposing the gel to film, remove the plastic wrap and the second piece of 3MM paper from the gel. Put the gel attached to the first piece of 3MM paper in a Pyrex dish with enough 1× TBE, 8 M urea to cover the gel by about 2 cm. *Very slowly* shake the gel for 15 min at room temperature. Be careful as the gel will separate from the 3MM paper and it is then easily torn. Remove the paper and then remove the fluid by suction using a Pasteur pipet attached to a vacuum line. Add fresh 1× TBE, 8 M urea and shake for another 15 min. Meanwhile cut four new pieces of 3MM paper to the size of the gel. After the 15 min, slide two of these pieces of 3MM paper underneath the gel to provide additional support to the gel. Remove the fluid by suction, keeping the dish flat so that the gel comes to rest without distortion onto the papers. This soaking in 8 M urea is critical for getting efficient transfer of the snRNAs and pre-mRNA from the complexes to the membrane (*see* **Notes 10–12**).

3. Set up the gel and nylon membrane in the transfer device (*see* **Note 12**). While the gel is soaking in TBE-urea, cut a piece of GeneScreen to the size of the gel. Wet the membrane first in water and then leave it submerged in 0.25× TBE. Lay the grid of the transfer device flat in a tray and open it to separate the pads. Fill the tray with 0.25× TBE to just cover the bottom grid plate and pad.

    Gently pick up the gel with the supporting papers from underneath with a gloved hand. Lay the gel on the buffer-soaked pad. Lay the nylon membrane on the gel and remove any air bubbles by gently rubbing the membrane with a gloved finger. Wet the other two pieces of fresh 3MM paper, lay them on the membrane, and again, remove any air bubbles. Lay the second pad on top of the papers and then the second grid plate on top of the pad. Pour more 0.25× TBE over the grid so that both pads are submerged in buffer. Immediately take the grid to the transfer device in the cold room. While firmly gripping the two grid plates to keep the pads, gel and membrane in place, quickly lift the grid and slide it into the transfer device. The grid should be oriented so that the RNAs will migrate from the gel to the membrane and towards the positive electrode (**Fig. 5**). Run a 2-mL glass pipet between the grid plates and the walls of the transfer device to release trapped air bubbles.

4. Perform the transfer at 115–120 V for 1 h. If you have set the grid into the device as described in the previous step, RNAs will migrate toward the membrane and positive electrode (**Fig. 5**). As soon as power is applied, you will see copious bubbles at the electrodes. By 1 h the buffer in the tank will have warmed to nearly 37°C.

5. Crosslink the RNAs to the membrane. Remove the grid from the transfer device and place it flat in a large tray. Remove the top grid plate and pad. Remove the 3MM paper and gel by rolling them off the membrane. Make a mark on the membrane with a ballpoint pen to indicate the side of the membrane that contacted the gel during the transfer. Quickly rinse the membrane in fresh 0.25× TBE to remove

any gel bits stuck to the membrane and place it on a clean glass plate with the marked side up. Wrap the membrane and plate in a single layer of plastic wrap. Avoid letting the membrane dry, as the crosslinking of the RNAs to the membrane is best done while the membrane is wet (*see* **Note 13**). Lay the plate under the UV lamps at a distance of 35 cm. Turn on the lamps and irradiate for 12 min.

6. Expose the membrane to film to get an image of the radiolabeled pre-mRNA and to check the efficiency of transfer. If you have used a radiolabeled pre-mRNA in the reaction, then you can judge the efficiency of transfer by autoradiography. Let the membrane air-dry for 30–60 min. Put it on a piece of 3MM paper for support with the marked side facing up. Wrap the membrane and paper in a single layer of plastic wrap. Expose it to X-ray film with one or two intensifier screens in a film cassette at –70°C.

7. Pretreat the membrane by heating in low salt and SDS. Put 500–1000 mL of 0.1× blot wash solution in a Pyrex dish and cover with plastic wrap. Bring the solution to boiling in a microwave oven. Carefully uncover the solution to avoid getting burned, submerge the membrane in the solution, put the cover back on, and simmer (use the lowest setting) in the microwave for 10 min. Shake slowly at room temperature for 10 min (*see* **Note 14**).

8. Prehybridize the blot. To make it easy to put the membrane into the hybridization bag, place the membrane between two pieces of 3MM paper cut to the same size as the blot. Slide the "sandwich" into a dry bag, then carefully remove the papers, one at a time, leaving the membrane in the bag. Add about 20 mL of hybridization solution (without the probe) to the bag. Lay the bag flat on the bench and squeeze out any air bubbles, then seal the bag. Incubate for at least 1 h at 42°C while submerged in a slowly shaking water bath.

9. Hybridize the probe to the blot. Prepare the probe for hybridization (*see* **Subheading 2.3., step 10**). Remove the bag with the blot from the water bath, cut off a corner off the bag, and pour out the hybridization solution. Pipette probe in the fresh hybridization solution into the bag, squeeze out any air bubbles, and seal the bag (*see* **Note 2**). It is important to remove air bubbles as they will prevent probe hybridization. Incubate submerged in a shaking water bath at 42°C for 12–24 h.

10. Wash the blot. Cut off a corner of the blot and pour out the radioactive hybridization solution into a radioactive waste can (*see* **Note 2**). Lay the bag flat on several paper towels on top of plastic-backed absorbent paper. Cut through the bag around three sides of the blot, being careful not to cut through the absorbent paper. Peel off one side of the bag and lift out the membrane with blunt-nosed forceps. Put the membrane into a Pyrex dish with about 500 mL of 3× blot wash solution. Shake at room temperature for 10 min. Pour off the wash into a radioactive waste can. Repeat this wash twice. For the fourth wash, add 3× blot wash solution at 55°C and shake in a water bath for 10 min at 55°C. For the final wash, add 500 mL of 0.1× blot wash solution at 55°C and shake in a water bath for 10 min at 55°C.

11. Expose the blot to X-ray film. Remove the membrane from the last wash. Blot it dry with clean 3MM paper and then allow it to air-dry for 30 min to 1 h. Put a

A Yeast Spliceosome Assay

piece of fresh 3MM paper under the membrane for support with the marked side of the membrane facing up, and wrap in a single layer of plastic wrap. Expose the membrane to X-ray film with one or two intensifier screens in a film cassette at –70°C for 12 h to 5 d.

12. Remove the probe from the blot and hybridize with another probe. The blot can be sequentially hybridized with at least six probes. After the blot is exposed to film, remove the first probe by heating the membrane in 0.1× blot wash solution as described in **step 7**. To check the efficiency of probe removal, air-dry the blot and expose it to film before hybridizing it with the next probe.

4. Notes

1. Most gel systems for nucleic acids including the one described here use the same buffer for the running buffer and the gel.
2. Several reagents used in the assay are harmful and suitable safety precautions such as wearing gloves, a laboratory coat, shoes, and safety glasses or goggles should be followed. Acrylamide and *bis*-acrylamide are toxic, dry, aerosolic powders so when they are weighed out, a face mask should also be worn. Once polymerized, acrylamide is less dangerous, but gloves should still be worn when handling the gels. Phenol is caustic and can cause *permanent* damage to the skin and eyes. Phenol, chloroform, excess silanizing solution, and radioactive materials should be disposed of according to environmental health safety regulations.
3. Because of its low ionic strength, the running buffer has to be recirculated or the pH within the gel and in the tanks will change during electrophoresis.
4. You can dilute 50–100 µL of sample into 900–950 µL of water and determine the absorbance at λ_{260} in a spectrophotometer at this point if you want to get an approximate yield (*see* **Subheading 3.1., step 10** for the calculation).
5. The lowest amounts of ammonium persulfate and TEMED to catalyze polymerization at room temperature in a reasonable amount of time were determined experimentally by setting up several tubes of acrylamide solution and adding various amounts of TEMED and 10% ammonium persulfate.
6. It is important that the gel and buffer have reached 4°C before the samples are loaded to avoid potential artifacts caused by temperature; the mobility of a molecule will be faster at higher temperatures. Running the gels at 4°C keeps heat from building up in the gel. Higher voltages than the ones used here can cause gel heating and result in the complexes running as smears with no resolution.
7. The percentage of acrylamide in the gel and the electrophoresis time will vary according to the RNA–protein complexes that one is interested in resolving. For Northern blot analyzes of the snRNAs in the prespliceosome (β_1) and α complexes for example, I usually run a 3.2% gel until the xylene cyanol is about 2.5 cm from the bottom of the gel (about 18 h of electrophoresis). For studying the δ (U1-pre-mRNA) complex with radiolabeled pre-mRNA, I run a 3.5% gel until the xylene cyanol is about 7.5 cm from the bottom of the gel (about 14–16 h of electrophoresis).

8. Low percentage acrylamide gels resemble sticky, viscous slime and are readily stretched or distorted during handling. The gels are most easily handled when they are attached to chromatography paper. Whatmann 3MM chromatography paper is inert to the chemicals used here; however, it does not have high wet strength and can be easily ripped.
9. A pre-mRNA of low specific activity or no radioactivity is made and used at concentrations of 1–4 nM in the splicing reaction *(12,14)*.
10. Most of the snRNAs are not transferred to the membrane without this urea treatment. Presumably the urea is denaturing some components in the complexes, allowing the pre-mRNA and snRNAs to dissociate from at least some proteins and thereby enhancing their transfer. As judged by the radiolabeled pre-mRNA in splicing complexes, the treatment at room temperature is sufficient to transfer 99% of the RNA. Soaking in urea at higher temperatures was not tried because of potentially increasing diffusion of the RNAs; however, higher temperatures may be necessary for some other RNA–protein complexes.
11. It may be possible to alter the gel pretreatment and electrotransfer conditions to transfer the proteins in a complex to a membrane for Western blot analyses. For example, the gel could be soaked in buffer with SDS to allow the proteins to bind SDS, rinsed in buffer without SDS to remove excess SDS, and then electroblotted to a membrane in Towbin buffer *(26)*.
12. Other electrophoretic transfer devices will probably work using the conditions suggested by the supplier. Transfer in phosphate buffer; however, may interfere with the binding of the RNAs to the membrane.
13. The time of UV irradiation for crosslinking nucleic acids to the nylon membrane depends on whether the membrane is wet or dry *(28)*. Too much irradiation will damage the nucleic acids on the blot resulting in decreased hybridization to the probe. You may have to determine the optimal irradiation times and conditions for your particular assay.
14. The pretreatment of the membrane by heating in 0.1× blot wash solution is important for the subsequent hybridization of the probe. The heating in wash solution removes intramolecular basepaired regions in the RNA. It may also release some proteins bound to the RNAs that are also probably transferred to membrane.

Acknowledgments

This work was supported by NSF Grants MCB9104862 and MCB9709915, and a grant from the Dedicated Health Sciences Research Fund of the University of New Mexico Health Sciences Center.

References

1. Guthrie, C. (1991) Messenger RNA splicing in yeast: clues to why the spliceosome is a ribonucleoprotein. *Science* **253**, 157–163.
2. Moore, M. J., Query, C. C., and Sharp, P. A. (1993) Splicing of precursors to mRNA by the spliceosome, in *The RNA World* (Gesteland, R. F. and Atkins, J. F., eds.), Cold Spring Harbor Laboratory Press, Cold Spring Harbor, New York, pp. 303–358.

3. Kramer, A. (1996) The structure and function of proteins involved in mammalian pre-mRNA splicing. *Annu. Rev. Biochem.* **65**, 367–409.
4. Brody, E. and Abelson, J. (1985) The "spliceosome": yeast pre-messenger RNA associates with a 40S complex in a splicing-dependent reaction. *Science* **228**, 963–967.
5. Frendeway, D. and Keller, W. (1985) Stepwise assembly of a pre-mRNA splicing complex requires U-snRNPs and specific intron sequences. *Cell* **42**, 355–367.
6. Grabowski, P. J., Seiler, S. R., and Sharp, P. A. (1985) A multicomponent complex is involved in the splicing of messenger RNA precursors. *Cell* **42**, 345–353.
7. Grabowski, P. J. and Sharp, P. A. (1986) Affinity chromatography of splicing complexes: U2, U5, and U4+U6 small nuclear ribonucleoprotein particles in the spliceosome. *Science* **233**, 1294–1299.
8. Ruby, S. W. and Abelson, J. (1988) An early hierachic role of U1 small nuclear ribonucleoprotein in spliceosome assembly. *Science* **242**, 1028–1035.
9. Reed, R. (1990) Protein composition of mammalian spliceosomes assembled *in vitro*. *Proc. Natl. Acad. Sci. USA* **87**, 8031–8035.
10. Konarska, M. M. and Sharp, P. A. (1986) Electrophoretic separation of complexes involved in the splicing of precursors to mRNAs. *Cell* **46**, 845–855.
11. Pikielny, C. W., Rymond, B. C., and Rosbash, M. (1986) Electrophoresis of ribonucleoproteins reveals an ordered assembly pathway of yeast splicing complexes. *Nature* **324**, 341–345.
12. Cheng, S.-C. and Abelson, J. (1987) Spliceosome assembly in yeast. *Genes Dev.* **1**, 1014–1027.
13. Zillman, M., Rose, S. D., and Berget, S. M. (1987) U1 small nuclear ribonucleoproteins are required early during spliceosome assembly. *Mol. Cell. Biol.* **7**, 2877–2883.
14. Ruby, S. W. (1997) Dynamics of the yeast U1 snRNP during spliceosome assembly. *J. Biol. Chem.* **272**, 17,333–17,341.
15. Clark, M. W., Goelz, S., and Abelson, J. (1988) Electron microscopic identification of the yeast spliceosome. *EMBO J.* **7**, 3829–3836.
16. Seraphin, B. and Rosbash, M. (1989) Identification of functional U1 snRNA-pre-mRNA complexes committed to spliceosome assembly and splicing. *Cell* **59**, 349–358.
17. Company, M., Arenas, J., and Abelson, J. (1991) Requirement of the RNA helicase-like protein PRP22 for release of messenger RNA from spliceosomes. *Nature* **349**, 487–493.
18. Chrambach, A. and Robard, D. (1971) Polyacrylamide gel electrophoresis. *Science* **172**, 440–451.
19. Dahlberg, A., Dingman, C. W., and Peacock, A. C. (1969) Electrophoretic characterization of bacterial polyribosomes in agarose-acrylamide composite gels. *J. Mol. Biol.* **41**, 139–147.
20. Sealey, P. G. and Southern, E. M. (1982) Gel electrophoresis of DNA, in *Gel Electrophoresis of Nucleic Acids, a Practical Approach* (Rickwoods, D. R. and Hames, B. D., eds.), IRL Press, Washington, D.C., pp. 41–151.
21. Zernik, J. and Lichtler, A. (1987) Borate-buffer-related effects on the electrophoretic mobility of linear DNA fragments in agarose. *BioTechniques* **5**, 411–414.

22. Lin, R. J., Newman, A. J., Cheng, S.-C., and Abelson, J. (1985) Yeast mRNA splicing in vitro. *J. Biol. Chem.* **260**, 14780–14792.
23. Michaud, S. and Reed, R. (1991) An ATP-independent complex commits pre-mRNA to the mammalian spliceosome assembly pathway. *Genes Dev.* **5**, 2534–2546.
24. Pikielny, C. W. and Rosbash, M. (1986) Specific small nuclear RNAs are associated with yeast spliceosomes. *Cell* **45**, 869–877.
25. Schroth, G. P., Gottesfeld, J. M., and Bradbury, E. M. (1990) TFIIIA induced DNA bending: effect of low ionic strength electrophoresis buffer conditions. *Nucleic Acids Res.* **19**, 511–516.
26. Sambrook, J., Fritsch, E. F., and Maniatis, T. (1989) *Molecular Cloning*, 2 ed., vol. 3, Cold Spring Harbor Laboratory Press, Cold Spring Harbor, New York.
27. Feinberg, A. P. and Vogelstein, B. (1983) A technique for radiolabeling DNA restriction endonuclease fragments to high specific activity. *Anal. Biochem.* **132**, 6–13.
28. Church, G. M. and Gilbert, W. (1984) Genomic sequencing. *Proc. Natl. Acad. Sci. USA* **81**, 1991–1995.

29

Defining Pre-mRNA *cis* Elements that Regulate Cell-Specific Splicing

Thomas A. Cooper

1. Introduction

A large number of genes express multiple mRNAs that encode diverse protein isoforms via alternative pre-mRNA splicing. For many genes, alternative splicing is regulated according to cell-specific patterns (in this chapter, cell-specific is used as a general term to refer to regulation according to differentiated cell type, developmental stage, gender, or in response to an external stimulus). In many cases, the same pre-mRNA transcript is processed differently in different cells, indicating that elements within the pre-mRNA direct cell-specific splicing via interactions with cell-specific factors or a cell-specific combination of ubiquitous factors. Results from a large number of vertebrate experimental systems have identified elements that affect splice site choice; however, only a few of these have been shown to direct cell-specific splicing (*1,4,5,7,14,15,19*). While ubiquitously recognized elements are likely to participate in cell-specific splicing events, this chapter focuses on experimental strategies to identify *cis* elements that promote cell-specific splicing in vertebrates and to distinguish cell-specific and general splicing elements.

From an experimental point of view, cell-specific splicing can be mediated from three types of *cis* elements: (1) auxiliary elements that are distinct from the consensus splicing elements (e.g., *1,14*); (2) auxiliary elements that are integral components of the splice sites (e.g., *5,19*); (3) the splice sites themselves in which regulation is mediated without auxiliary elements (e.g., *15,16*). This chapter focuses on auxiliary splicing elements (1 and 2).

Identification of elements that regulate splicing can provide insights into the mechanism of regulation. Once localized, the *cis* element becomes a reagent for in vitro investigations to isolate regulatory factors. Point mutants used to

From: *Methods in Molecular Biology, Vol. 118: RNA-Protein Interaction Protocols*
Edited by: S. Haynes © Humana Press Inc., Totowa, NJ

localize nucleotides required for regulated splicing in vivo can also be used to demonstrate sequence specificity in RNA binding studies and in vitro splicing assays. These mutants will provide the strongest evidence that what is observed in vitro is relevant to the regulation observed in vivo. Characterization of *cis* elements should be approached with this future goal in mind.

Elements that are required for a cell-specific splicing event are not necessarily binding sites for factors that are cell-specific. Results from *Drosophila* and vertebrate experimental systems indicate that cell-specific splicing requires multiple and often repeated elements located upstream and downstream from the regulated splice sites *(1,2,4,6,10,13,14)*. Regulated splicing most likely involves assembly of a multicomponent complex containing constitutive factors in combination with cell-specific factors. Mutations that disrupt complex assembly will knock out cell-specific regulation, whether the mutation is in a binding site for a constitutive or a cell-specific factor. Regulation of some genes may not involve cell-specific factors but rather is mediated by combinatorial effects of multiple repeats of different elements that have different binding affinities for constitutive splicing factors combined with what may be subtle differences in the nuclear concentrations of these factors. Whether or not a cell-specific factor is involved, identification of the *cis* elements and *trans* factors that regulate splicing is a significant step toward understanding the mechanism of regulated splicing.

Cell-specific splicing elements have been difficult to identify, particularly in vertebrate systems. The major goal of this chapter is to outline experimental strategies to identify cell-specific splicing elements by noting general principles from the splicing elements identified to date and by describing ways to circumvent the pitfalls of earlier investigations. Potential explanations for the difficulties encountered in these investigations are listed below along with a summary of the solutions presented in this chapter. Many aspects of this strategy are applicable to investigation of other RNA processing signals.

1. Like cell-specific transcription, a cell-specific splicing event is likely to require multiple regulatory elements each of which can be present in multiple copies. Some elements are located close to the regulated splice sites while others may be several thousand nucleotides away. The functional redundancy of these elements can make them difficult to identify experimentally. Mutations in one element are compensated by an immediately adjacent element or deletions in one set of regulatory elements may simply bring distal elements sufficiently close to permit regulated splicing. Solution 1: Define a minimal genomic fragment that is necessary and sufficient for regulated splicing in a transient transfection assay. Focus on the proximal elements; the distal elements can be defined later. Solution 2: Define minimal elements within this fragment with the aim of performing a gain of function experiment in which the element (or most likely concatemers of the element) direct regulated splicing of a heterologous exon.

2. Variability is a hallmark of vertebrate RNA processing signals, making repeated and conserved elements difficult to identify on the basis of nucleotide sequence. Solution: Generate a database of intron and exon sequences from as many examples of similarly regulated genes as possible to identify potential consensus sequences.
3. The nonspecific effects of changing intron size can be striking. In addition, there is no such thing as a "splicing-neutral" stuffer sequence. Stuffer fragments and even substitutions of only a few nucleotides can have effects independent of the sequence being replaced. This clearly complicates interpretation of mutations and can lead to conflicting results from different minigene constructs. Solution: Get a consensus result from several stuffer fragments when performing deletions (use both orientations of two stuffer fragments). In some cases it is necessary to test more than one set of point mutations in a putative regulatory sequence.
4. It is often difficult to determine whether a mutation disrupts cell-specific splicing or merely changes general splicing efficiency. Solution: Construct minigenes such that the splicing pathways are balanced to produce both spliced products in the cell culture that normally uses the default splicing pattern. Mutations in a cell-specific splicing element should alter the cell-specific splicing pattern toward the default in cell cultures that regulate splicing but should have little effect on the ratio of the RNAs expressed in the default cell type. Mutations that affect general recognition of the regulated splice sites will affect splicing in both the default and regulated cell cultures.
5. Regulation in some systems may be mediated by the nuclear concentrations and binding affinities of ubiquitous components of the splicing machinery rather than cell-specific splicing elements. This mechanism of regulated splicing will be the most difficult to dissect using the standard reductionist approach. Solution: The pattern of regulated splicing may provide hints as to the mechanism of regulation. The scenario most likely to involve cell-specific regulatory factors is a gene that is expressed in all cell types but includes a cassette type alternative exon in a cell-specific manner. A mechanism that balances splice site and polyadenylation site recognition has been shown to use a balance of constitutive processing factors for cell-specific regulation, suggesting that this may be a mechanism common to this gene architecture *(16)*.

2. Steps for Analysis of Elements that Regulate Alternative Splicing
2.1. Sequence Analysis

Results from several laboratories provide some reasonable assumptions as to the nature of auxiliary regulatory elements:

1. With some exceptions (e.g., *19*), cell-specific regulatory elements are located within introns. While the presence of splicing enhancers within alternative and constitutive exons is well established, the vertebrate enhancers identified thus far are not cell-specific *(17)*.
2. Multiple elements with different sequence motifs can contribute to regulation.

3. Multiple repeats of each element are likely.
4. Repeated elements are usually located upstream and downstream of the regulated exon.
5. Elements are relatively close to the exon (within 200–300 nucleotides) but additional distal elements are likely.
6. Regulation may involve elements with both positive and negative effects.
7. Elements are often conserved between species (if the regulated splicing is conserved) and may be common to different genes that undergo similar regulated splicing.

The first step toward identifying elements that regulate a cell-specific splicing event is to perform sequence analysis based on these assumptions. This can streamline identification of regulatory elements by providing potential targets for mutagenesis.

Collect a database of genomic sequence flanking alternatively spliced exons that are regulated similarly. Within each gene look for relatively short (8–12 nucleotides) repeated sequences upstream and downstream of the alternative exon. Also compare similarly regulated genes for short regions of homology. If the introns are large, the initial searches can be limited to the exon and 300 nt upstream and 300 nt downstream of the exon to reduce the complexity of the analysis. Because the repeats are often imperfect, they are easiest to find using a matrix comparison (available in DNA Strider and DNASTAR, for example). The sequences that come out of this analysis can be ranked according to their potential significance to regulated splicing. For example, sequences that are repeated within each gene and are common to similarly regulated genes are most likely to be functionally important and should be the first targets for mutagenesis in transfection studies.

2.2. Transfection System

An ideal transfection system to define cell-specific regulatory elements is a cell line in which a transition in the splicing pattern can be induced by differentiation (e.g., induction of neuron-specific splicing *[8,15]*) or by external cues *(3)*. A common alternative is to compare splicing patterns in two different cell cultures. One is a primary culture or differentiated cell line in which splicing of the endogenous and transfected pre-mRNAs follows the cell-specific splicing pattern. A second cell line that does not express the cell-specific splicing pattern is used for comparison. Often this is a heterologous cell type that does not express the gene of interest. Because these cells do not express the gene, it is assumed that they do not express the appropriate regulatory factors and the pre-mRNA is spliced according to a default splicing pattern. The default splicing pathway is determined by the efficiency with which the constitutive splicing machinery recognizes the constitutive splicing signals. This depends on several features such as splice site strength, exon size, presence of general splicing enhancers or repressors in exons and introns, relative strength of competing splice sites, and secondary structure. Regulation could involve a negative mechanism (blockage of strong splice sites) or a positive mechanism (activation of suboptimal splice sites). Defining the default splicing pathway therefore defines whether alternative splicing is positively or negatively regulated.

Defining splicing patterns in different cells as default vs regulated has been a useful working model in *Drosophila* where regulation of several genes has been shown to involve true auxiliary splicing factors that modulate a default splicing pattern via interactions with auxiliary splicing elements. The discussion in this chapter is based on a default vs regulated model. However, it is important to note that default vs regulated is a working model subject to revision as it is usually not clear which splicing pattern is the default and which is regulated until auxiliary regulatory elements are identified. In addition, this model will not be appropriate for all alternatively spliced genes. For some genes, there may not be an unregulated default splicing pattern. Splice site selection may be determined by a balance of constitutive splicing factors and/ or a combination of positive and negative acting elements such that splicing is repressed in some cells and activated in others.

Ideally the differentiated phenotype of the cells should be easy to maintain and the cells should take up DNA reasonably well. Transient transfection is preferable to stable transfection because of the rapid turnaround time of the experiments. However, there are examples in which regulation was reproduced only in stably transfected cells (e.g., *9*).

Once cell lines that express differentially regulated splicing patterns are identified, the next step is to determine whether the preferred splicing patterns are dependent on gene-specific elements. If the regulated splicing pattern is gene-specific, a heterologous pre-mRNA of the same architecture should undergo the default splicing pattern in both the default and regulated cell types. The elements responsible for cell-specific regulation can be mapped as described below using these cell cultures. If a heterologous minigene of the same architecture expresses the same cell-specific splicing patterns, it is possible that differential regulation based on differences in constitutive factors is the mechanism of regulation. This has been nicely demonstrated for IgM *(11,16)*. It is also possible that the cell cultures do not reproduce the mechanism of regulated splicing. It is easy to imagine that splicing patterns in stable cell lines derived from a differentiated tissue have little to do with the regulatory molecules that exist in the tissue. The cell cultures used for the regulated splicing pattern should be as close to the in vivo phenotype as possible (e.g., express cell-specific markers, morphology, and cell-specific alternative splicing of other pre-mRNAs). In our laboratory, for example, the best muscle-specific regulation by far is obtained using primary skeletal muscle cultures.

Determine whether the regulated splicing pattern is saturable by transfecting increasing amounts of minigene DNA. Confirm that increasing amounts of pre-mRNA and mRNA are expressed. As a control, cotransfect the minigene with increasing amounts of a heterologous minigene that lacks putative regulatory elements. This should have no effect on the ratio of splice products in the regulated cell line. If regulation is reduced as more RNA is expressed, this result suggests that the regulated splicing pathway requires titratable *trans*-acting factors. In addition, transfections to map regulatory elements should use significantly less DNA than this threshold level to avoid saturating regulatory factors. If the regulated splicing pattern is not saturable by this simple experiment, this result does not indicate that saturable *trans*-acting factors are not involved because there are multiple alternative explanations. This result does mean that regulation will not be affected by high levels of RNA expressed from minigenes.

2.3. Quantitation of Results

mRNA splicing patterns can be analyzed by S1 nuclease, RNase protection, primer extension, or reverse transcriptase-polymerase chain reaction (RT-PCR). RT-PCR is often the method of choice since its high sensitivity allows detection of products even from cells that transfect poorly. However, RT-PCR is not necessarily a quantitative assay as small differences in the efficiency of reverse transcription of different mRNAs or in the amplification of the different cDNAs can accumulate into huge disparities after multiple cycles. In addition, RT-PCR analysis of alternatively spliced mRNAs can produce artifactual bands *(18,20)*. To avoid potential problems, RT-PCR can be optimized using samples containing known ratios of in vitro synthesized RNAs. The RNA products are synthesized in vitro from RT-PCR products cloned downstream from a viral RNA polymerase promoter. Test samples containing different ratios of the in vitro synthesized RNAs are used to optimize conditions such that the ratio of the RT-PCR products matches the ratios of the input RNAs.

Mutations that affect one of many elements can result in relatively small changes in splicing patterns. Therefore, multiple independent transfections with statistical analysis are likely to be necessary. Small changes are not unexpected but it is important that they be shown to be consistent from transfection to transfection.

Results are usually quantitated as the percent mRNA spliced in the regulated splicing pattern (calculated as amount of mRNA spliced in the regulated pattern divided by the total amount of mRNA spliced in all patterns × 100). Because percentages have an upper limit of 100, this is a nonlinear scale and comparing the effects of different mutations is not necessarily straightforward. For example, if a cassette exon has a default level of inclusion of 50%, it is not possible to have more than a twofold activation in the level of exon inclusion. Constructs that have an intrinsically lower level of default inclusion due to features unrelated to regulatory elements (small exon, weak 5' splice site) will have higher potential -fold inclusion. In addition, many mutations affect the general recognition of splice sites in both the default and regulated cell cultures, which can have the appearance of altered regulation. For example, if a mutation causes a drop in the levels of inclusion of a neuron-specific alternative exon in both the default and regulated cell cultures by the same number of percentage points (e.g., from 25% to 5% default inclusion and from 75% to 55% inclusion in neurons), the -fold inclusion will increase (from 3 to 11). It is unclear whether the -fold difference or the arithmetic difference between the default and regulated cell types is a more accurate measure of changes in regulated splicing. Because so many mutations affect basal splicing efficiency, we use the arithmetic difference in our laboratory.

2.4. Minigene Constructs

When screening cell lines for appropriate regulation, the minigene serves as an indicator of the regulatory potential of the cells and it is important to have all possible elements for maximum regulation. Therefore, the minigene used to screen cell lines for regulation should contain a genomic fragment with the alternative exon plus several upstream and downstream exons and introns in case regulatory elements are located quite distal to the exon. For example, regulation of a neuron-specific alternative exon in the nonmuscle myosin II heavy chain mRNA requires an element located two introns downstream from the exon 1.5 kb away *(8)*.

Once cell cultures that demonstrate regulated splicing have been chosen and regulation by gene-specific elements has been demonstrated (*see* **Subheading 2.2.**), a different minigene is used as a starting point to define regulatory elements. This minigene should contain the minimal genomic fragment that is necessary and sufficient for regulated splicing. Based on results from several experimental systems, a genomic fragment containing the alternative exon(s) plus 200–300 nucleotides of the upstream and downstream introns is likely to contain most of the information required for regulated splicing. This genomic fragment is inserted into an intron of a heterologous minigene and tested for appropriately regulated splicing. Further deletions from the 5' and 3' ends of the genomic fragment can be tested until the minimal genomic fragment that is necessary and sufficient for regulated splicing is identified. This approach will reduce the incidence of conflicting results from different constructs due to functionally redundant distal elements. In addition, if the genomic fragment contains the minimal number of required regulatory elements, any mutation that disrupts a regulatory element should have an effect. If the results from the smaller genomic fragment indicate that regulatory elements are missing, gain of function add-back experiments can be used to localize the distal elements.

2.5. Loss of Function Analysis

Theoretically, mutation of a cell-specific element will change the regulated splicing pattern toward the default and have little effect on the ratio of splicing pathways in the default cell cultures. Mutations in constitutive splicing elements are expected to affect both regulated and default splicing patterns. To distinguish cell-specific from constitutive effects, the splicing patterns should be balanced such that all splicing pathways are expressed at some level in the default cell culture. This is the only way that the effects of the mutation on both the regulated and basal splicing patterns can be evaluated. For example, if a minigene containing a neuron-specific exon expresses 30% inclusion in a neuronal cell line and is completely skipped in nonneuronal cells, it is impossible to know whether a mutation that results in complete skipping in neurons disrupted a general splicing element or a cell-specific splicing element. An exon that is constitutively skipped or constitutively included can be modified to restore balanced splicing by improving or reducing general splicing efficiency (*see* **Subheading 2.6.**). For example, increasing the size of the neuron-specific exon in the example above by 12 nucleotides might result in 15% inclusion in nonneuronal cells and 55% inclusion in neurons. If the mutation in a putative regulatory element has no effect on the level of inclusion in nonneuronal cells but the level of inclusion in neurons drops to 15%, this mutation clearly had a neuron-specific effect.

Because deletions can have effects unrelated to sequence being tested, deletions need to be "stuffed" with heterologous sequence. Experience has shown that there are no "splicing-neutral" sequences. To be certain that the effects of a stuffed deletion are due to the absence of the gene segment and not to the introduced heterologous sequence, get a consensus result from testing both orientations of at least two stuffer fragments. Otherwise, the effects of altered spacing (if the deletion is not stuffed) or the introduced sequence (if only one stuffer sequence is tested) may be incorrectly attributed to the absence of the sequence being tested. The best stuffer fragments restore the intron size and come from introns of constitutively spliced genes.

Even point mutant substitutions can have effects independent of the sequence being tested. In our laboratory, we have found that different substitutions of the same five nucleotides gave different results. We concluded that some mutations inadvertently introduced active sequences. This was not an isolated event as a similar result was obtained in a different four-nucleotide region. These effects were detected only because the original mutation had unexpected effects, so additional substitutions of the same nucleotides were tested. It is impractical to test all nucleotides of interest with multiple mutations. However, it is reasonable to test key elements with more than one mutation.

As noted in the introduction, mutation analysis of putative regulatory elements should be designed with future applications to in vitro assays in mind. RNA binding assays are likely to be the primary assay to identify regulatory factors once a cell-specific splicing element has been identified. A particularly useful strategy is to generate a series of mutations that increase, decrease, and completely inactivate the regulatory event in transfection assays. The RNA binding factors that are relevant to regulation should have lower and higher affinity for mutations that decrease and increase regulatory activity in vivo, respectively. Mutants with increased activity can be generated by concatemerizing a minimal sequence or from consensus motifs derived from comparisons of similarly regulated genes. This approach was used to establish the relevance of binding of a subset of SR proteins to an exonic splicing enhancer *(12)*.

The possibility that exonic elements contribute to cell-specific regulation should be tested early on by substituting a heterologous exon. Exon splicing enhancers are often found in alternative exons, so a heterologous exon of the same size may result in a lower default level of exon recognition. It may necessary to compensate for the absence of the splicing enhancer by increasing the size of the exon or improving a splice site sequence (*see* **Subheading 2.6.**). This approach was used to demonstrate that the cTNT exon 5 exon enhancer was not required for cell-specific splicing *(17)*.

Determine whether the natural splice sites are required for regulated splicing. If both splice sites can be replaced by heterologous splice sites while maintaining regulated splicing, the elements responsible for regulation are distinct from the splice sites. If the sequence of a splice site is required for regulation, the target for regulation is either the splice sites themselves (implying that constitutive factors alone regulate splicing) or an auxiliary element buried within the splice site (implying a requirement for a splicing factor not required for other splice sites). Splice site substitutions can be accomplished by using either splice sites from unregulated exons or by introducing mutations that change the sequence of the splice sites without changing splice site strength. When using the latter approach, it is most convenient to generate multiple mutant constructs using a doped oligonucleotide in which mutated positions contain more than one nucleotide change. A series of splice site mutants is more likely to provide at least one efficiently spliced construct to determine whether or not regulation is intact. If necessary, the default level of exon inclusion can be "titrated" using a series of exon sizes (*see* **Subheading 2.6.**).

2.6. Tricks to Modulate the Default Splicing Pathway

As noted previously, mutations that disrupt an element required for cell-specific splicing should affect only the regulated splicing pattern in the regulating cell cultures and have little effect on the ratio of splicing pathways in the default cell cultures. However, some mutations will alter splicing efficiency in both default and regulated cell cultures. Those mutations that result in constitutive inclusion or constitutive skipping in both regulated and default cell cultures are not informative with regard to cell-specific regulation. In many instances, splicing still has the potential to be regulated but the basal recognition of the splice sites is either too weak or too strong to be regulated. This is a problem particularly when auxiliary regulatory elements are components of a splice site. Most alternative exons are flanked by suboptimal splice sites. Mutations that improve a splice site may lead to constitutive splicing of the alternative exon in all cells tested. This result indicates that the splice site was functionally suboptimal, but it does not address whether the splice site is a target for cell-specific regulation. When a mutation results in constitutive splicing, minigene constructs that change the basal recognition of the exon should be tested so that the effect of the mutation on the regulated and default splicing patterns can be evaluated. There are several ways to restore balanced splicing patterns by "tweaking" basal recognition. The best way to modulate the default is by changing exon size (assuming that the exon has been demonstrated not to contain a cell-specific regulatory element). One approach to this is to develop an "accordion exon" that can be used to generate a series of exons of different sizes (**Fig. 1**). This approach is a versatile and relatively fast way to determine whether a mutation that changes overall splicing efficiency also affects cell-specific regulation. An alternative strategy is to change the strength of a splice site sequence. This can be done by changing the exonic or intronic components of the splice site sequence. For example, recognition of an exon can be substantially improved by changing the last three nucleotides of the exon to the consensus "CAG." As noted previously, it is usually worthwhile to obtain multiple mutations using a doped oligonucleotide to obtain a range of splice site strengths.

2.7. Gain of Function Experiments

A gain of function experiment provides the strongest evidence that a specific element mediates a cell-specific splicing event: Transfer the element to a heterologous gene and activate the cell-specific splicing pattern. This is often a difficult experiment because of the multicomponent nature of the elements that regulate alternative splicing. However, because these elements are often repeated and different elements are functionally redundant, strategies that are often successful are to replace one element with another or to use concatemers of single elements to push regulation. In our investigations of the elements that regulate cardiac troponin T exon 5, we found that three copies of one element or six copies of a second element can replace the three elements found in the

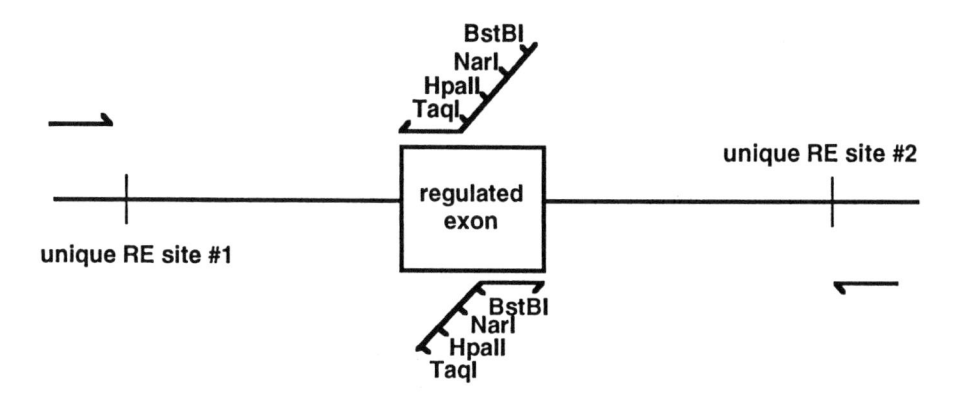

Fig. 1. Accordion exon approach to generating a series of exon sizes. Oligos that prime within the first and last 12–15 nucleotides of the exon are designed to contain "tails" of compatible restriction sites. The two PCR products containing the 3' and 5' ends of the exon are generated. Each PCR fragment is cut at the restriction site that is unique to the plasmid (unique RE site nos. 1 or 2) and the multiple compatible restriction sites. Exons of different sizes are reassembled in three-way ligation reactions containing the minigene cut at both unique restriction enzyme sites, and combinations of the 5' and 3' end exon PCR products that were cut at different sites in the exon oligos. Although using the compatible sticky ends aids in the efficiency of ligation, it is also possible to blunt end ligate filled-in exon sites to generate additional exon sizes.

downstream intron. In addition, we have been able to reproduce robust cell-specific regulation by placing three copies of a 43 nucleotide element upstream and downstream of a heterologous exon *(2a)*.

Conclusions

The *cis*-acting elements that mediate regulated splicing are the key to identifying the regulatory factors, whether the mechanism involves a cell-specific factor or cell-specific differences in the nuclear concentrations and affinities of constitutive factors. The success of in vitro experiments to define regulatory factors depends on the quality of the *cis* element as a reagent. Ultimately, understanding the mechanism of cell-specific splicing requires identification of the molecular link between the cell-specific element and the constitutive splicing machinery.

Acknowledgments

This work was supported by the National Institutes of Health and the Muscular Dystrophy Association.

References

1. Black, D. L. (1992) Activation of c-*src* neuron-specific splicing by an unusual RNA element in vivo and in vitro. *Cell* **69**, 795–807.
2. Chan, R. C. and Black, D. L. (1995) Conserved intron elements repress splicing of a neuron-specific c-*src* exon in vitro. *Mol. Cell. Biol.* **15**, 6377–6385.
2a. Cooper, T. A. (1998) Muscle-specific splicing of a heterologous exon mediated by a single muscle-specific splicing enhancer from the cardiac troponin T gene. *Mol. Cell. Biol.* **18**, 4519–1525.
3. Endo, H., Matsuda, C., and Kagawa, Y. (1994) Exclusion of an alternatively spliced exon in human ATP synthase gamma-subunit pre-mRNA requires de novo protein synthesis. *J. Biol. Chem.* **269**, 12,488–12,493.
4. Gooding, C., Roberts, G. C., Moreau, G., Nadal-Ginard, B., and Smith, C. W. J. (1994) Smooth muscle-specific switching of alpha-tropomyosin mutually exclusive exon selection by specific inhibition of the strong default exon. *EMBO J.* **13**, 3861–3872.
5. Guo, W. and Helfman, D. M. (1993) *Cis*-elements involved in alternative splicing in the rat beta-tropomyosin gene — the 3'-splice site of the skeletal muscle exon-7 is the major site of blockage in nonmuscle cells. *Nucleic Acids Res.* **21**, 4762–4768.
6. Horabin, J. I. and Schedl, P. (1993) Sex-lethal autoregulation requires multiple *cis*-acting elements upstream and downstream of the male exon and appears to depend largely on controlling the use of the male exon 5' splice site. *Mol. Cell. Biol.* **13**, 7734–7746.
7. Huh, G. S. and Hynes, R. O. (1994) Regulation of alternative pre-mRNA splicing by a novel repeated hexanucleotide element. *Genes Dev.* **8**, 1561–1574.
8. Kawamoto, S. (1996) Neuron-specific alternative splicing of nonmuscle myosin II heavy chain-B pre-mRNA requires a *cis*-acting intron sequence. *J. Biol. Chem.* **271**, 17,613–17,616.
9. Libri, D., Marie, J., Brody, E., and Fiszman, M. Y. (1989) A subfragment of the beta tropomyosin gene is alternatively spliced when transfected into differentiating muscle cells. *Nucleic Acids Res.* **17**, 6449–6462.
10. Lynch, K. W. and Maniatis, T. (1995) Synergistic interactions between two distinct elements of a regulated splicing enhancer. *Genes Dev.* **9**, 284–293.
11. Peterson, M. L. (1994) Regulated immunoglobulin (Ig) RNA processing does not require specific *cis*-acting sequences: non-Ig RNA can be alternatively processed in B cells and plasma cells. *Mol. Cell. Biol.* **14**, 7891–7898.
12. Ramchatesingh, J., Zahler, A. M., Neugebauer, K. M., Roth, M. B., and Cooper, T. A. (1995) A subset of SR proteins activates splicing of the cardiac troponin T alternative exon by direct interactions with an exonic enhancer. *Mol. Cell. Biol.* **15**, 4898–4907.
13. Rio, D. (1993) Splicing of pre-mRNA: mechanism, regulation and role in development. *Curr. Opin. Genet. Dev.* **3**, 574–584.
14. Ryan, K. R. and Cooper, T. A. (1996) Muscle-specific splicing enhancers regulate inclusion of the cardiac troponin T alternative exon in embryonic skeletal muscle. *Mol. Cell. Biol.* **16**, 4014–4023.

15. Tacke, R. and Goridis, C. (1991) Alternative splicing in the neural cell adhesion molecule pre-mRNA: regulation of exon 18 skipping depends on the 5'-splice site. *Genes Dev.* **5,** 1416–1429.
16. Takagaki, Y., Seipelt, R. L., Peterson, M. L., and Manley, J. L. (1996) The polyadenylation factor CstF-64 regulates alternative processing of IgM heavy chain pre-mRNA during B cell differentiation. Cell **87,** 941–952.
17. Xu, R., Teng, J., and Cooper, T. A. (1993) The cardiac troponin T alternative exon contains a novel purine-rich positive splicing element. *Mol. Cell. Biol.* **13,** 3660–3674.
18. Zacharias, D. A., Garamszegi, N., and Strehler, E. E. (1994) Characterization of persistent artifacts resulting from RT-PCR of alternatively spliced mRNAs. *Biotechniques* **17,** 652–655.
19. Zhang, L., Ashiya, M., Sherman, T. G., and Grabowski, P. J. (1996) Essential nucleotides direct neuron-specific splicing of gamma(2) pre-mRNA. *RNA* **2,** 682–698.
20. Zorn, A. M. and Krieg, P. A. (1991) PCR analysis of alternative splicing: identification of artifacts generated by heteroduplex formation. *BioTechniques* **11,** 181–183.

30

In Vivo SELEX in Vertebrate Cells

Thomas A. Cooper

1. Introduction

Iterative selection strategies have been widely used to enrich specific RNA molecules from randomized pools based on binding affinities or an RNA-mediated activity *(6,8)*. The vast majority of these procedures have been performed in cell-free systems. Of particular use would be iterative selection within cell cultures to enrich RNA sequences that mediate a specific function in vivo. This approach is readily applied to selection of exonic sequences that enhance exon inclusion; these sequences are known as exonic splicing enhancers *(2)*. This chapter describes a procedure that uses cycles of transient transfection of minigene plasmids containing an alternative exon with randomized sequence and selective reverse transcriptase-polymerase chain reaction (RT-PCR) to enrich exon sequences that enhance inclusion of the alternative exon. The selection scheme is outlined in **Fig. 1A**. A DNA cassette containing a randomized region (10–15 nucleotides) is directionally ligated into a weakly recognized exon of a minigene. The ligation reaction is transiently transfected directly into cultured cells and total RNA is extracted after 40–48 h. Selectable exons that are spliced into the mRNA are selectively amplified by RT-PCR using oligos that prime only on spliced mRNAs that include the randomized exon. The randomized cassette is excised from the PCR product by the restriction enzymes used for cloning. This digestion product contains a selected population of sequences that enhance exon inclusion. It is then ligated into the minigene exon to begin additional cycles of selection.

The exons obtained from the procedure are evaluated by two criteria: sequence and enhancer activity. To determine whether individual sequence motifs are enriched by repeated rounds of selection, an aliquot of the ligation reaction from each round is transformed into bacteria and the sequences of 20–

From: *Methods in Molecular Biology, Vol. 118: RNA-Protein Interaction Protocols*
Edited by: S. Haynes © Humana Press Inc., Totowa, NJ

A

```
5'GGACGTAGGGTCGACGTTNNNNNNNNNNNNNGAATGGATCCGTCGTGACTGGGAAAAC 3'
```

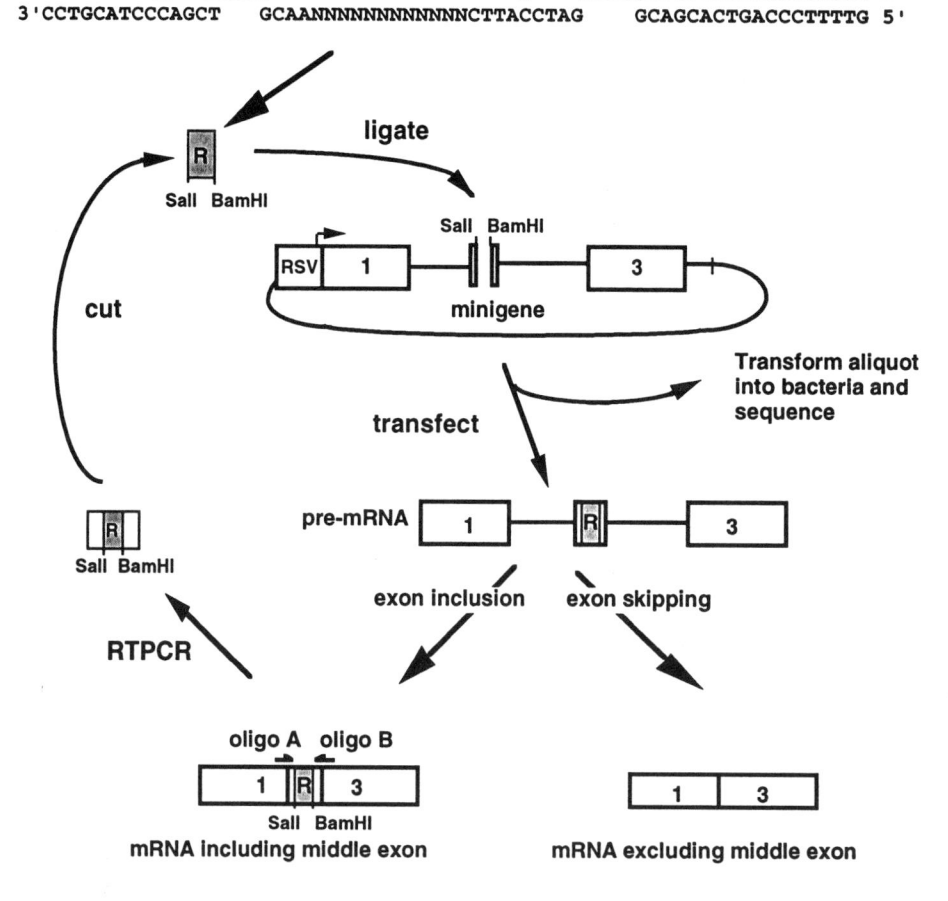

synthesize second strand

```
         SalI                            BamHI
5'GGACGTAGGGTCGACGTTNNNNNNNNNNNNNGAATGGATCCGTCGTGACTGGGAAAAC 3'
3'CCTGCATCCCAGCTGCAANNNNNNNNNNNNNCTTACCTAGGCAGCACTGACCCTTTTG 5'
```

cut

```
5'GGACGTAGGG   TCGACGTTNNNNNNNNNNNNNGAATG    GATCCGTCGTGACTGGGAAAAC 3'
3'CCTGCATCCCAGCT    GCAANNNNNNNNNNNNNCTTACCTAG    GCAGCACTGACCCTTTTG 5'
```

ligate

cut

Transform aliquot into bacteria and sequence

transfect

exon inclusion   **exon skipping**

RTPCR

oligo A   oligo B

mRNA including middle exon   mRNA excluding middle exon

B

```
5' TAATACGACTCACTATA 3'                           BamHI
   5' AATACGACTCACTATAGGTCGACGTTNNNNNNNNNNNNNGAATGGATCCGTACGT 3'
                       SalI                  3' CTTACCTAGGCATGCAG 5'
```

30 individual clones are obtained from miniprep DNA. Enhancer activity of individual clones is tested by transient transfection and determination of the level of exon inclusion. To determine whether the procedure enriched for splicing enhancers, the level of inclusion of selected exons is compared to that of nonselected exons. To determine whether the selected sequence has intrinsic enhancer activity, independent of the minigene used for selection, its ability to enhance splicing of a different alternative exon should be tested.

This procedure has been performed in a fibroblast cell line to identify an A/C-rich motif that enhances splicing (2). Now that the feasibility of the approach is established, it would be of interest to perform selection in different cell types to identify cell specific exonic enhancers. This procedure could also be performed in cells that overexpress a protein known to mediate enhancer activity (such as SR proteins) to identify enhancer sequences that are preferred by individual proteins in vivo.

2. Materials

1. 10× Ligation buffer (supplied by manufacturer: New England Biolabs, Beverly, MA, USA): 500 mM Tris-HCl, pH 7.8, 100 mM MgCl$_2$, 100 mM dithiothreitol (DTT), 10 mM ATP, 500 μg/mL bovine serum albumin (BSA). Store 200-μL aliquots at –20°C.
2. TEE: 10 mM Tris-HCl, pH 7.5, 0.1 mM EDTA.
3. 0.5 M EDTA.
4. 5.0 M NaCl.
5. Acrylamide gel elution buffer (10 mL): 0.5 M NH$_4$OAc, 10 mM EDTA. Store at room temperature.
6. BES-buffered saline (BBS) (300 mL): 50 mM N,N-bis(2-hydroxyethyl)-2-aminoethanesulfonic acid (BES; Calbiochem, San Diego, CA, USA), 280 mM NaCl, 1.5 mM Na$_2$HPO$_4$. Bring to 250 mL with ddH$_2$O, adjust the pH to 6.96 (exactly) with 1 M NaOH (about 6 mL), and add ddH$_2$O to 300 mL. Filter sterilize (prerinse filter twice with ~20 mL of ddH$_2$O) and store at –20°C in 15-mL aliquots for 1 yr. This buffer can be thawed and reused for a second time only (*see* **Note 5**).
7. 2.5 M CaCl$_2$ (50 mL). Add solid CaCl$_2$ to a 100-mL graduated cylinder with 35 mL of ddH$_2$O. Once it is completely dissolved, fill to 50 mL with ddH$_2$O. Filter sterilize (prerinse filter twice with ~20 mL of ddH$_2$O) and aliquot (1.5 mL) and store at –20°C for 1 yr.
8. RNA extraction solution A (52.5 mL): 25 mL of phenol (H$_2$O-saturated); 25 mL of solution B; 2.5 mL of 2 M NaOAc, pH 4.0, and 180 μL of β-mercaptoethanol. Can be stored at 4°C for >4 mo.

Fig. 1. Strategy for in vivo selection. (**A**) Selection scheme. (**B**) Amplification of single stranded oligonucleotide containing randomized region for first round of selection. Double-stranded copies are generated by PCR using oligonucleotides that anneal to constant sequences that flank the randomized region. The *Sal*I and *Bam*HI restriction sites are indicated by underlining.

9. RNA extraction solution B (50 mL): 4 M guanidinium thiocyanate, 25 mM sodium citrate, and ddH$_2$O to 45 mL. Adjust the pH to 7.0 with 1 N HCl. Add ddH$_2$O to 50 mL. Can be stored at 4°C for >4 mo.
10. 5× In vitro transcription buffer (supplied by manufacturer: Promega, Madison, WI, USA): 200 mM Tris-HCl, pH 7.5 at 37°C, 30 mM MgCl$_2$, 10 mM spermidine, 50 mM NaCl.
11. DNase: Worthington (Freehold, NJ, USA) DPFF DNase at 1.0 mg/mL (2 U/µL) in 10 mM Tris-HCl, pH 7.5, in 10-µL aliquots stored at –80°C.
12. 10× Vent polymerase buffer (supplied by manufacturer: New England Biolabs): 10 mM KCl, 10 mM (NH$_4$)$_2$SO$_4$, 20 mM Tris-HCl, pH 8.8 at 25°C, 2 mM MgSO$_4$, 0.1% Triton X-100.
13. 10× *Taq* DNA polymerase buffer (supplied by manufacturer: Promega, Madison, WI, USA): 50 mM KCl, 10 mM Tris-HCl, pH 9.0 at 25°C, 0.1% Triton X-100.
14. 25 mM MgCl$_2$ (supplied by manufacturer: Promega, Madison, WI, USA).
15. 25 mM dGATC: This contains 25 mM dGTP, 25 mM dATP, 25 mM dTTP, 25 mM dCTP and is made by mixing equal volumes of 100 mM stocks from Pharmacia (Piscataway NJ, USA).
16. Vent DNA polymerase (New England Biolabs).
17. *Taq* DNA polymerase.
18. AMV reverse transcriptase (Life Sciences, St. Petersburg, FL, USA).
19. RNasin, 40 U/µL (Promega).
20. 100 mM DTT.
21. Acetylated BSA (Promega).
22. Random hexamers (Pharmacia).
23. 25:25:1 Phenol:chloroform:isoamyl alcohol.
24. 25:1 Chloroform:isoamyl alcohol.
25. Ethanol
26. 40% Acrylamide:*bis*-acrylamide (20:1)
27. 10× TBE: 89 mM Tris base, 89 mM boric acid, 2 mM EDTA.
28. GeneClean (Bio101, Vista, CA, USA).
29. T4 DNA ligase.
30. Carrier DNA (any plasmid that lacks a eukaryotic transcription promoter).
31. 10× phosphate-buffered saline (PBS): 136.9 mM NaCl, 2.7 mM KCl, 10.0 mM Na$_2$HPO$_4$, 1.4 mM KH$_2$PO$_4$. Bring pH to 7.4 with HCl. Dilute 1:10 with ddH$_2$O and filter sterilize to make 1× working stock.
32. Isopropanol

3. Methods

3.1. Selectable Exon Design

The double stranded cassette containing the randomized region is generated from a single stranded oligonucleotide. This oligo must contain the following features (illustrated in **Fig. 1B**):

1. Priming sites for oligonucleotides located upstream and downstream of the randomized region to allow PCR amplification of a double-stranded cassette.

Table 1
Inverse Relationship of the Number of Variable Nucleotides and Sequence Copy Number

Variable nucleotides	Potential combinations	Copies of each sequence in 1 µg plasmid
0	1	1.9×10^{11}
5	1024	1.9×10^{8}
10	1.0×10^{6}	1.9×10^{5}
11	4.2×10^{6}	4.6×10^{4}
12	1.7×10^{7}	1.2×10^{4}
13	6.7×10^{7}	2.0×10^{3}
14	2.7×10^{8}	7.4×10^{2}
15	1.0×10^{9}	1.9×10^{2}
16	4.3×10^{9}	46
18	6.9×10^{10}	2
20	1.1×10^{12}	1 (only 17% of sequences represented)

2. Two different restriction endonuclease sites on either side of the randomized region for directional cloning (underlined in **Fig. 1B**, *Sal*I and *Bam*HI sites). These should be unique to the vector (*see* **Subheading 3.2.1.**), have incompatible overhangs, and should cut and ligate efficiently. It is most convenient if both enzymes are optimally active in the same buffer.
3. As the number of randomized nucleotides increases, the number of molecules containing a particular sequence decreases in a constant amount of DNA (*see* **Table 1**). When designing this approach, we used a relatively low number of random positions due to the concern that a large variable region would result in insufficient copies of any one sequence to allow detection by RT-PCR. Now that feasibility of the approach has been demonstrated, it would be worthwhile testing large randomized sequences.
4. Position the restriction sites within the oligonucleotide such that the fragment containing the randomized region will be a different size than the other two fragments generated from digestion with both restriction enzymes. The restriction endonuclease digestion product containing the randomized region is gel isolated and quantitated.
5. Make certain that the constant regions of the selected exon do not contain sequences that can affect splicing such as potential cryptic splice sites, known splicing enhancers, or in-frame translation stop codons.

3.2. Minigene Vector Design

1. The exon that is to receive the randomized region should have the appropriate restriction sites for directional cloning of the cassette into an exon. These restriction sites must be unique to plasmid.

2. It is useful to clone a large stuffer fragment into these sites so that plasmids that are cut with both enzymes can be distinguished from single cut and uncut plasmid. Otherwise, low levels of contaminating plasmid containing a potentially "spliceable" sequence will generate a major contaminant of the selected sequences.
3. The stringency of selection can be adjusted by adjusting the basal level of exon inclusion in the absence of enhancer sequences (*see* **Note 1**). A selectable exon that is completely skipped will theoretically select for stronger exonic enhancers than one that has a low basal level of inclusion. Establishing the desired balance of exon inclusion and exclusion may require modifying features such as exon size and strength of the 5' and 3' splice sites.
4. Because premature stop codons can result in skipping of the resident exon *(3)*, it is best not to have a natural open reading frame in the minigene mRNA.
5. Ubiquitously active transcription enhancers (such as Rous sarcoma virus [RSV] or cytomegalovirus [CMV]) are most useful as they allow use of the minigene in almost any cell type.

3.3. Preparation of the Selectable Cassette

1. Take 100 ng of the single stranded oligonucleotide containing the randomized region and convert to double stranded DNA in a standard PCR reaction using oligos that flank the randomized region (**Fig. 1B**). Prepare a 100-µL reaction containing 500 ng of each flanking oligo, 1× Vent DNA polymerase buffer, 0.2 mM dGATC, and 1 U of Vent DNA polymerase. The PCR conditions will depend on the T_m of the oligonucleotides and should be adjusted based on the quality of the final product as assayed on an acrylamide gel.
2. Following the PCR reaction, add EDTA to 2 mM and NaCl to 0.2 M, and extract the PCR product once with an equal volume of 25:25:1 phenol:chloroform:isoamyl alcohol and once with an equal volume of 25:1 chloroform:isoamyl alcohol. Precipitate with 2.5 vol of ethanol and wash the pellet once in 70% ethanol. Vacuum dry the pellet and dissolve in 20 µL of TEE.
3. Digest the PCR product with the appropriate restriction enzymes. Plan on using approx 0.5 pmol of the selectable cassette DNA (7.4 ng of a 24-bp fragment) for every 1 µg of vector (of 6500 bp) in the ligation, and 1 µg of vector per round of selection. The restriction digest is loaded directly onto a 7% nondenaturing polyacrylamide gel (20:1 acryl:*bis*) in 1× TBE. Visualize the digestion products by staining in ethidium bromide.
4. Use a razor blade to cut a piece of acrylamide containing the band. Transfer to a 1.5-mL Eppendorf tube and cut into pieces about 1 mm square. Add 400 µL of acrylamide elution buffer and incubate at 37°C with shaking overnight. Centrifuge at 12,000g for 5 min to pellet gel fragments and transfer the supernatant (without gel fragments) to a clean tube and add 16 µL of 5 M NaCl and 1 mL of 100% ethanol, mix well, and centrifuge (12,000g) for 15 min. Residual soluble acrylamide from the gel precipitates in ethanol and acts as a carrier so there is no need to add glycogen. Redissolve the pellet in 10–20 µL of TEE and run an ali-

quot on an acrylamide gel alongside known amounts of size marker to estimate the DNA concentration.

3.4. Vector Preparation

1. Cut 20 μg of plasmid DNA with the appropriate enzymes. Load the restriction digest directly onto a 0.9% agarose gel (*see* **Note 2**).
2. Isolate the DNA using GeneClean or a comparable product (*see* **Note 3**).
3. Check the recovery and approximate concentration of the isolated DNA by running an aliquot on a minigel beside known amounts of marker DNA.

3.5. Ligation of Selectable Exon Cassette into the Minigene Vector

1. Ligate 1 μg of gel isolated vector and a twofold molar excess of insert at 15°C overnight in a 100-μL ligation reaction using 1× ligation buffer and 400 U of T4 DNA ligase (*see* **Note 4**).
2. Transfect the ligation reaction directly into two 60 mm plates of cells containing 10^6 cells per plate (*see* **Subheading 3.6.2.**).

3.6. Calcium Phosphate Transfection Protocol (modified from ref. 1)

1. It is important to express as many pre-mRNA molecules as possible so that a large pool of sequences is available for selection. The number of expressed RNAs is determined primarily by the ligation efficiency (the fraction of isolated vector that recircularizes with insert) and the transfection efficiency (the fraction of ligated molecules that make it to the nucleus and are transcribed). Before performing the selection procedure, ligation efficiency and transfection efficiency should be optimized (*see* **Notes 4** and **5**)
2. The day prior to transfection (the day of the ligation), plate the cells at 10^6 cells per 60 mm plate.
3. Three hours prior to transfection, change the medium and feed the cells with fresh medium. Also at this time, remove tubes of BBS and 2.5 M $CaCl_2$ from the freezer and leave at room temperature. The purpose is to thaw the solutions and to bring them to room temperature (see **Note 5**)
4. For each DNA to be transfected, set up a 15-mL sterile tube in the sterile cell culture hood and make the transfection cocktail. To transfect two 60-mm plates make 1 mL of cocktail: add 0.095 mL of the ligation reaction (the remaining 5 μL is transformed into bacteria; *see* **Subheading 3.8.**), 0.05 mL of 2.5 M $CaCl_2$, and 5–10 μg of carrier DNA and sterile filtered ddH_2O to 0.5 mL. To this, add 0.5 mL of BBS and gently swirl to mix. Incubate at room temperature for 20 min.
5. Add 0.5 mL of the transfection cocktail dropwise to each 60 mm plate (do not remove media). Agitate the plate gently and return to the incubator. Do not swirl the plates as this causes the precipitate to collect in the center of the plate. Incubate for 15–18 h, then remove the media (*see* **Note 6**). Wash the plate once with sterile 1× PBS to remove the precipitate. Add growth medium and return the plates to the incubator.
6. Harvest the cells 40–48 h. after the start of transfection (when the precipitate was added to plates).

3.7. RNA Extraction and DNase Treatment (9)

1. Wash the plates once with 2 mL of cold 1× PBS. Let the plates drain at a 45°C angle for 1 min and aspirate the remaining liquid at the bottom edge of the plate.
2. Add 650 µL of solution A to each of the two plates, then scrape the cells off with a policeman and pool both plates into one 1.5-mL Eppendorf tube. Add 210 µL of chloroform:isoamyl alcohol (25:1). Vortex for at least 20 s, making sure that the phases mix with great agitation.
3. Place the tubes on ice for 20 min, then centrifuge (12,000g) for 20 min at 4°C.
4. Transfer the upper aqueous phase to a new Eppendorf tube containing 870 µL of isopropanol. Vortex to mix well and store at –20°C for at least 1 h.
5. Centrifuge tubes (12,000g) for 20 min at 4°C. Discard the supernatant and wash the pellet by vortexing with 1 mL of 75% ethanol (stored at –20°C). Centrifuge for 10 min. Pour off the supernatant immediately after the centrifuge stops; otherwise the pellet may dislodge from the tube and pour out. Repeat this washing procedure a second time, then vacuum dry the pellet.
6. Redissolve the pellet in 50 µL of the DNase cocktail by gently vortexing for about 30 s, being careful to introduce few bubbles. Incubate at 37°C for 30 min. For each sample, the DNase cocktail contains: 10 µL of 5× in vitro transcription buffer, 2.5 µL of 100 mM DTT, 0.2 µL of RNasin, 0.4 µL of DNase, and 36.8 µL of ddH$_2$O.
7. Following the incubation, add 2 µL of 0.5 M EDTA, 4 µL of 5 M NaCl, and 44 µL of ddH$_2$O. Extract once with phenol:chloroform:isoamyl alcohol and once with chloroform/isoamyl alcohol. Ethanol precipitate the aqueous layer using 260 µL of ethanol. Redissolve the pellet in 384 µL of ddH$_2$O then add 16 µL of 5 M NaCl and 1 mL of 100% ethanol and vortex to mix. Store this precipitated RNA at –20°C as a suspension in 70% ethanol; do not pellet the RNA.

3.8. RT-PCR (modified from ref. 5)

The RT-PCR procedure is designed to perform two functions: (1) detect low amounts of RNA with very little background, and (2) selectively amplify only those mRNAs that contain the alternative exon (*see* **Note 7**). These goals are accomplished by nested PCR (**Fig. 2**). First, cDNA is synthesized using random hexamers (*see* **Note 8**), then cDNAs from mRNAs that include or exclude the alternative exon are amplified using oligonucleotides that anneal to the upstream and downstream exons (external 5' and external 3' oligos, respectively). Finally, a nested PCR reaction is performed using the internal oligos. Both internal oligos prime the polymerase reaction from within the alternative exon but these oligos will anneal only to correctly spliced mRNAs (*see* **Fig. 2**), selectively amplifying exons that are spliced into the mRNA. PCR products from mRNAs that skip the randomized exon or from the pre-mRNA are not amplified. In addition, the specificity of nested primers means that a large number of cycles can be used to amplify small amounts of RNA. Because the restriction sites are maintained in the RT-PCR product, the PCR product can

In Vivo SELEX in Vertebrate Cells

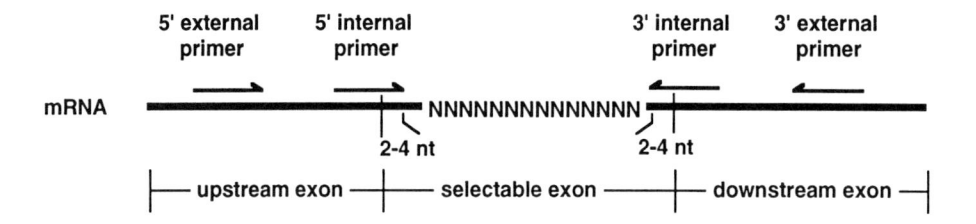

Fig. 2. Nested RT-PCR. The mRNA sequence is represented as a thick line. Exon boundaries are indicated by thin vertical lines. The randomized region is represented by *N*s. Following reverse transcription, PCR no. 1 is performed using the external primers. An aliquot of PCR no. 1 is then amplified using the internal primers. Note that the internal primers are designed such that the last two to four nucleotides anneal within the selectable exon. These oligonucleotides will prime DNA synthesis only on correctly spliced mRNAs and not on plasmid DNA, unspliced pre-mRNA, or mRNAs that skip the randomized exon.

be digested and cycled through multiple rounds of ligation, transfection, and amplification.

3.8.1. Reverse Transcription (RT) Reaction

1. Vortex each RNA sample to resuspend the RNA precipitate and remove 40 µL. This works out to be about 1/5 of a 60-mm plate or 10–15 µg of total RNA. Centrifuge the RNA and dry the pellet under a vacuum.
2. Make up the RT cocktail and use a 20-µL aliquot to dissolve each RNA pellet. For each reaction combine 2.0 µL of 10× PCR buffer, 2.4 µL of 25 m*M* MgCl$_2$, 2.0 µL of 1 µg/µL acetylated BSA, 2.0 µL of 10 m*M* dGATC, 1.0 µL of 100 pmol/µL hexamers, 1.0 µL of 100 m*M* DTT, 5 U of RNasin, 2 U of AMV reverse transcriptase, and H$_2$O to 20 µL.
3. Once the RNA is dissolved, let it sit at room temperature for 10 min, then transfer the tubes to a 42°C waterbath for 1 h. Following the RT reaction, heat the tubes at 95°C for 5 min in a heating block, then immediately plunge into an ice/water slurry.

3.8.2. PCR No. 1

1. Make up a cocktail for 80 µL PCR reactions (per reaction): 8.0 µL of 10× PCR buffer, 6.4 µL of 25 m*M* MgCl$_2$, 1.6 µL of BSA (1 µg/µL), 0.5 µL of *Taq* polymerase, 200 ng of external 5' oligo, 200 ng of external 3' oligo, and H$_2$O to 80 µL. Note that additional dGATC is not necessary because the final concentration is 0.2 m*M* from the RT reaction.
2. Mix the cocktail well. Add 80 µL of the cocktail to each of the 20-µL RT reactions using a fresh pipet tip for each tube; transfer the reactions to 500-µL tubes. Add two drops of light mineral oil and run a PCR program using appropriate temperatures for 20 cycles (*see* **Note 9**).

3.8.3. PCR No. 2 (Nested PCR Reaction)

1. Make up cocktail for 80 µL PCR reactions (per reaction): 8.0 µL of 10× PCR buffer, 6.4 µL of 25 mM MgCl$_2$, 0.64 µL of 25 mM dGATC, 1.6 µL of 1 µg/µL BSA stock, 200 ng of internal 5' oligo, 200 ng of internal 3' oligo, 2.5 U of *Taq* polymerase, and H$_2$O to 79 µL.
2. Mix the cocktail well. Add 79 µL of the cocktail to 500 µL tubes. Add 1 µL of PCR no. 1 and two drops of light mineral oil and run a PCR program using appropriate temperatures and number of cycles (*see* **Note 9**).
3. Add EDTA to 2 mM and NaCl to 0.2 M, and extract the PCR product once with an equal volume of 25:25:1 phenol:chloroform:isoamyl alcohol and once with an equal volume of 25:1 chloroform:isoamyl alcohol. Precipitate with 2.5 vol of ethanol and wash the pellet once in 70% ethanol.
4. Digest the PCR product with the appropriate restriction enzymes, and isolate the fragment from a 7% nondenaturing polyacrylamide gel as in **steps 3** and **4** of **Subheading 3.3**.
5. Repeat the steps in **Subheadings 3.3.–3.8.** for the desired number of rounds of selection.

3.9. Monitor for Enrichment of Exon Sequences that Function as Splicing Enhancers

There are several approaches to monitor the success of the selection. The easiest is to determine whether identifiable sequence motifs are enriched after several rounds of selection. We found clear enrichment after two and three rounds *(2)*. To obtain individual clones from each round of selection, 5 µL of the ligation reaction containing randomized exon and minigene vector is transformed into competent bacteria. Twenty to thirty individual colonies are inoculated into 1-mL cultures for preparation of plasmid DNA according to standard procedures. Miniprep plasmid DNA is sequenced using an oligonucleotide that primes within a convenient distance from the exon (30–50 nucleotides).

To determine whether the procedure is enriching for bona fide splicing enhancers, individual clones are tested by transient transfection and the level of exon inclusion is determined by a quantitative RT-PCR assay, primer extension, or RNase protection.

Alternatively, the levels of exon inclusion can be determined directly on the pools of RNA from each round using a quantitative RT-PCR assay. However, because cells express a high background level of mRNAs that lack the exon even in the absence of insert (*see* **Note 10**), this assay does not give reliable results.

4. Notes

1. It is necessary to determine the background level of exon inclusion to know whether the selected sequences enhance inclusion. This is not as straightforward

as one might expect owing to the strong effects that exon sequences have on the level of exon inclusion. We initially picked 10 clones containing random sequence in the exon intending to transfect them individually and average the level of exon inclusion for a background level. Surprisingly, there were large variations in the level of exon inclusion between the clones. Therefore, a better estimate of the "background" level of exon inclusion of nonselected exon sequences comes from using a large number of pooled nonselected exons. This is accomplished by transforming the initial ligation into high efficiency competent bacteria according to standard procedures (*4*) to the point of adding SOC and incubating for 1 h. Instead of plating out the transformed cells on agar plates, use this 1 mL to inoculate 100 mL of media including the appropriate antibiotic to prepare plasmid DNA for transfection.

2. Use a large well such as a single-well comb or make a large well from a standard comb by placing tape over several teeth. It is important to use a large well so as not to overload the gel. Overloading can lead to smearing of the bands and contamination of the desired double-cut DNA with uncut plasmid DNA.

3. When using the Bio101 gene clean kit, it is particularly important to remove the residual wash solution before suspending the resin in TEE for elution of the DNA. This is done by a quick centrifuging and removal using a pipet tip. The wash solution contains ethanol which will interfere with the ligation.

4. Ligation efficiencies can be optimized using a bacterial transformation assay that contains 50 ng of vector and one- to fivefold molar excess of insert in a 10-µL reaction. Ligation efficiency is defined as the number of colonies obtained per nanogram of vector DNA. A particularly important reagent in the ligation reaction is ATP. Ligase buffer supplied by companies often includes ATP and it is best to store aliquots at –20°C. Each aliquot is used for 1 mo. We chose 1 µg of vector DNA for the ligation as a convenient amount. This amount can be increased to increase the number of sequences expressed and available for selection. The amount of DNA used is limited by the ability to isolate large amounts of cut DNA and the sensitivity of the cells to transfected DNA. We have not determined the effect of increasing amounts of DNA on ligation efficiency.

5. The transfection buffers should be at room temperature at the time of transfection. If the transfection cocktail is too warm or cool, the character of the precipitate is altered. The pH of the BBS is also a critical factor in determining transfection efficiency. We find pH 6.96 to work best; however, day-to-day variability in pH meters may necessitate testing different pHs. To optimize the pH, divide one large batch of BBS into three aliquots and pH to 6.93, 6.96, and 6.99. Each aliquot is tested for transfection efficiency using plasmids expressing a marker such as green fluorescent protein or β-galactosidase. The BBS buffer loses potency if stored at 4°C or after repeated freeze–thaw cycles. Both the 2.5 *M* $CaCl_2$ and BBS are aliquotted, stored at –20°C, and should be discarded after the second thawing. The transfection buffer can be stored >1 yr at –20°C. Choose cell cultures that take up and express the minigene efficiently. Several alternative methods of transfection are available and can give strikingly different efficien-

cies in different cell lines. It is often useful to talk to the suppliers to determine optimal conditions for individual cell lines.
6. Individual cell lines and primary cultures will differ as to their sensitivity to the calcium phosphate precipitate. Some primary cultures may not tolerate an overnight incubation. It might be necessary to determine empirically the time of exposure to the precipitate that provides a balance between a high level of expression and cell viability.
7. The extremely high sensitivity of PCR makes it necessary to establish the rules listed below (*see also* **ref. 7**).
   a. Use dedicated equipment for PCR including pipetman with aerosol resistant tips, tube racks, microfuges, and vortexers. Do not use these for plasmid DNA.
   b. Include a "no RNA" and a "no DNA" control for each experiment to detect contamination.
   c. Water is a common source of plasmid contamination. Purchase bottled water if necessary.
   d. Aliquot reagents, mark one tube as "in use," and use only that tube until it is finished.
   e. If oligos are suspected to be contaminated with plasmid DNA, they can be cleaned up by gel isolation on 10% nondenaturing acrylamide gel. Visualize the oligos by UV shadowing, cut out the piece of acrylamide containing the band, place it in a 1.5-mL Eppendorf tube, and grind it using a blue tip. Add 1 mL of water, and incubate with shaking overnight at 37°C. Filter the solution through a 0.2-μm Millex filter (Millipore, Bedford, MA, USA) using a 3-mL syringe.
8. An alternative is to use an mRNA-specific primer to synthesize cDNA. We have found this to be a more sensitive approach than using random hexamers for some mRNAs. To increase the specificity, it is best to use a primer that primes cDNA synthesis from downstream of the external PCR primer pair; *see* **Fig. 2**.
9. RT-PCR conditions must be optimized empirically for sensitivity with low background by varying the $MgCl_2$ concentration, the number of cycles, and the annealing and reaction temperatures. Design oligos that avoid internal or interprimer annealing that involves the last three nucleotides of the oligos. For our nested PCR *(2)*, it was necessary to use 79°C for annealing and polymerization, because at 76°C, the reaction generated significant background bands and low amounts of the specific band after 60 cycles. At 81°C, no PCR product was formed, probably due to an absence of annealing. At 79°C, the reaction produced large amounts of only the specific band.
10. Transfected linear plasmid molecules will become blunt-ended and circularized by endogenous enzymes in cultured cells. Therefore, transfection of only vector DNA without ligase results in expression a high level of mRNAs that lack the middle exon. This makes it difficult to reliably quantitate the level of exon inclusion from cells that were transfected directly using ligation reactions.

Acknowledgments

This work was supported by the American Cancer Society, the Moran Foundation, and the National Institutes of Health.

References

1. Chen, C. and Okayama, H. (1987) High-efficiency transformation of mammalian cells by plasmid DNA. *Mol. Cell. Biol.* **7,** 2745–2752.
2. Coulter, L. R., Landree, M. A., and Cooper, T. A. (1997) Identification of a new class of exonic splicing enhancers by in vivo selection. *Mol. Cell. Biol.* **17,** 2143–2150.
3. Dietz, H. C., Valle, D., Francomano, C. A., Kendzior, R. J., Pyeritz, R. E., and Cutting, G. R. (1993) The skipping of constitutive exons in vivo induced by nonsense mutations. *Science* **259,** 680–683.
4. Hanahan, D. (1983) Studies on transformation of *Escherichia coli* with plasmids. *J. Mol. Biol.* **155,** 557–508.
5. Kawasaki, E. S. (1990) Amplification of RNA, in *PCR Protocols. A Guide to Methods and Applications* (Innis, M. A., Gelfand, D. H., Sninsky, J. J., and White, T. J., eds.), Academic Press, San Diego, pp. 21–27.
6. Klug, S. J. and Famulok, M. (1994) All you wanted to know about SELEX. *Mol. Biol. Rep.* **20,** 97–107.
7. Kwok, S. and Higuchi, R. (1989) Avoiding false positive with PCR. *Nature* **339,** 237–238.
8. Tuerk, C. (1997) Using the SELEX combinatorial chemistry process to find high affinity nucleic acid ligands to target molecules. *Methods Mol. Biol.* **67,** 219–230.
9. Xie, W. and Rothblum, L. I. (1991) Rapid, small-scale RNA isolation from tissue culture cells. *BioTechniques* **11,** 325–327.

31

Purification of SR Protein Splicing Factors

Alan M. Zahler

1. Introduction

In recent years, the SR protein family of precursor messenger RNA splicing factors has emerged as a key player in the assembly of the spliceosomal machinery onto pre-mRNA. The SR proteins are essential splicing factors and different family members can direct usage of alternative splice sites in vitro and in vivo (reviewed in **refs.** *1* and *2*). SR proteins are required for the earliest spliceosomal interactions in recruitment of U1snRNP *(3,4)*, they can interact with the splicing factors that recruit U2 snRNP *(5)*, and they appear to be essential for recruitment of the tri-snRNP to the pre-mRNA *(6)*. When present in excess in a U1snRNP-depleted extract, they can function to bypass a U1 snRNP requirement for splicing *(7,8)*. They have been identified as *trans*-acting factors that interact with *cis* regulatory sequences in alternatively spliced exons *(3,9–13)*. Because they are involved in activating, enhancing, and repressing splicing for constitutive and alternatively spliced messages, they have become the focus of a great number of recent studies.

SR proteins are found at conserved molecular weights — 20, 30 (three family members), 40, 55, and 75 kDa in all metazoans — and they have also been identified in plants (reviewed in **refs.** *1* and *2*). Structurally, the SR proteins are very similar *(14)*. They each contain a similar N-terminal RNA recognition motif (RRM) domain common to many RNA binding proteins *(15)*. SR proteins all contain a distinctive C-terminal domain consisting mainly of alternating serine and arginine residues. Many serines in this domain are phosphorylated and this is recognized by a phosphoepitope-specific monoclonal antibody, mAb104 *(14,16,17)*. The hybridoma line that produces this antibody is available from the American Type Culture Collection (Manassas, VA, USA). A subset of the SR proteins (SRp30a [SF2/ASF], SRp40, SRp55,

From: *Methods in Molecular Biology, Vol. 118: RNA-Protein Interaction Protocols*
Edited by: S. Haynes © Humana Press Inc., Totowa, NJ

and SRp75) contain a domain between the RRM and the SR domain that appears to be a direct repeat of a subset of the amino acids in the RRM. By itself, this domain does not score strongly as an RRM and is referred to as an RRM homology (RRMH) or pseudo RRM (ΨRRM) *(14,17,18)*. The size differences between the family members is based on the presence or absence of the RRMH region and the length of the SR domain, which varies from 50 to 350 amino acids among the different family members *(19)*.

The ability to study SR protein function has been enhanced by the fact that it is relatively easy to purify the family members from any tissue or cell type by taking advantage of the fact that these proteins are insoluble in magnesium *(14,17)*. This observation was made by Mark Roth as the result of mAb104 immunostaining of lampbrush chromosomes from *Xenopus* oocytes prepared in the presence or absence of magnesium *(16)*. He observed that the SR proteins were detectable only on the lampbrush chromosomes prepared in the presence of magnesium. He reasoned that the ability of these proteins to interact with the RNA or other proteins depended on magnesium. This observation led to an experiment to test solubility of proteins containing this phosphoepitope in the presence of magnesium. This is one of the best examples of the use of cytological observations to advance molecular biology. In addition to being easy to purify, it is easy to assay for SR protein activity by determining whether purified proteins can complement a standard SR protein-deficient, splicing-deficient HeLa cell S100 extract *(20*; *see* Chapters 24 and 25) to restore splicing activity *(14,21–23)*. This was the biochemical characterization that led to the functional identification of the first SR protein, splicing factor 2 (SF2) *(21,22)*. This same protein was also identified as alternative splicing factor (ASF) by its ability to affect 5' splice site choice *(24)*.

In this chapter, techniques to easily isolate SR proteins to near homogeneity from any tissue or cell line using only two salt precipitations steps are discussed. Although this will easily and quantitatively yield all the SR proteins from a source, it has proven much more difficult to isolate individual SR proteins because they are all found in a narrow size range and contain many of the same structural characteristics. I will also discuss a somewhat inefficient but reliable method for the purification of individual SR protein family members from one another using recovery and renaturation of the proteins from preparative sodium dodecyl sulfate (SDS)-polyacrylamide gels. This technique is based on a modification of one developed by Hager and Burgess *(25)*. **Figure 1** shows a Coomassie-stained SDS-polyacrylamide gel electrophoresis (SDS-PAGE) gel containing representative examples of SR proteins purified from HeLa and *Drosophila* Kc cells as well as individual SR proteins gel purified from calf thymus.

Purification of SR Protein Splicing Factors

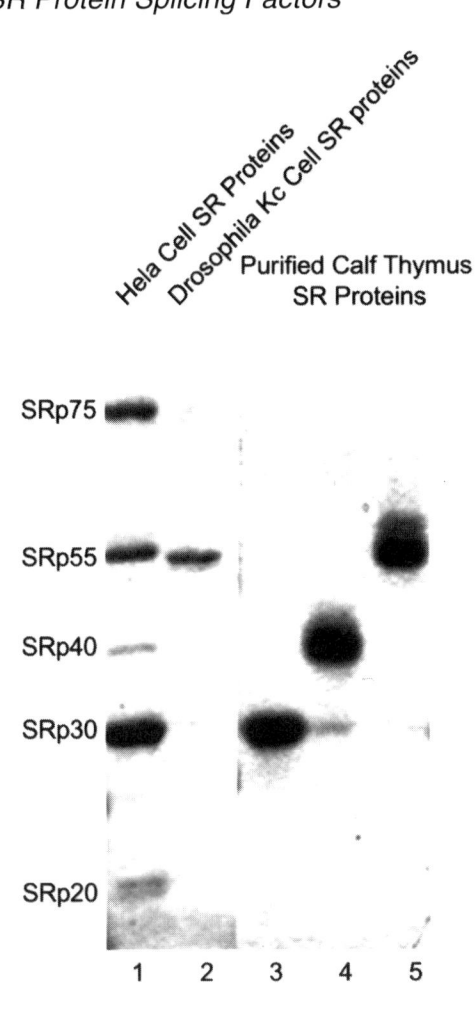

Fig. 1. Coomassie stained SDS-PAGE gel showing different representative SR protein purifications. **Lane 1**: Total SR proteins purified from HeLa cells. **Lane 2**: Total SR proteins purified from *Drosophila* Kc cells. **Lanes 3–5**: Individual calf thymus SR proteins purified from a preparative SDS-PAGE gel. SRp30b, **Lane 3**; SRp40, **Lane 4**; and SRp55, **Lane 5**.

1.1. Choice of Starting Material for SR Protein Purification

SR proteins can be easily prepared from any tissue culture cell line or animal tissue or organ. While tissue culture cells yield cleaner protein and are easier to work with, organs have the advantage of allowing a larger quantity of starting material more cheaply. Starting material choice can also depend on the goal of the purification. HeLa cells yield large amounts of all the family members. *Drosophila* Kc and SL2 cells contain SRp55 almost exclusively. Bovine brain contains mostly SRp75. Calf thymus seems to have the most nuclei per gram of tissue and has large amounts of SRp30b (SC35) along with SRp40 and SRp55 but does not appear to contain significant amounts of SRp30a (SF2/ASF) (A. M. Zahler and M. B. Roth, unpublished observations). Thus, thymus has the advantage of allowing one to isolate more SR proteins per gram of tissue and for allowing isolation of SRp30b enriched away from SRp30a. Methods for purification of SR proteins from both tissue culture cells and organs are described.

1.2. Purification of Individual SR Proteins

Several techniques have been described for the purification of individual SR proteins, including recombinant SR proteins. Functional recombinant proteins made in *E. coli* have been described, but they have the disadvantage of lacking the posttranslational phosphorylation found in native proteins and of difficulty with solubility in the absence of chaotropic reagents *(24,26)*. In addition, *E. coli* codon bias prevents expression of the longer SR domains (serine and arginine are both sixfold degenerate) of some of the family members (A. M. Zahler and M. B. Roth, unpublished observations). Recombinant baculovirus expression systems for several SR protein family members have been described *(5,27)*. These proteins are phosphorylated and soluble. High levels of recombinant protein expression are required so that when SR proteins are purified from infected cells, the vast majority of SR proteins isolated are the recombinant ones and endogenous SF9 cell SR proteins are only a minor contaminant. The techniques described here for purification of SR proteins from tissue culture cells can be adapted for the purification of recombinant SR proteins from baculovirus-infected SF9 cells. It is very difficult to use column chromatography to purify individual SR proteins because the family members occur in a narrow size range and are very similar in structure. A method that is quite useful for purification of individual SR proteins albeit with poor yields is to separate magnesium-purified SR proteins on preparative SDS-PAGE gels and extract the proteins from the gel and renature them *(14,28)*. This technique is described in this chapter.

2. Materials
2.1. Purification of Total SR Proteins from Cells and Tissues

1. Tissue culture cells are grown to mid log phase. Cells are harvested by low-speed centrifugation, washed 2× in 4°C phosphate-buffered saline (PBS) and pellets are stored at −80°C until needed. Preps described below typically begin with 6–8 × 10^9 HeLa cells or 4–5 × 10^{10} *Drosophila* Kc cells. Cell pellets are ground to a fine powder under liquid N_2 in a mortar and pestle packed in dry ice prior to addition to the extraction buffer to ensure uniform thawing.

Purification of SR Protein Splicing Factors

2. Powdered organs. Animal organs and tissues can be used in the methods described below. It is important to work with a supplier who can assure you that the animal parts have been frozen in dry ice or liquid N_2 within minutes of the animals' death. These should be shipped on dry ice and stored at –80°C. A large mortar and pestle is used to turn these frozen tissues into a powder. The mortar is packed into an ice bucket filled with crushed dry ice and liquid nitrogen is used to cool down the inside surface. Animal tissues are pounded and ground to a fine powder with frequent addition of liquid nitrogen and transferred to a beaker on dry ice. Pulverized organs can be stored at –80°C until the preparation day.
3. SR isolation buffer: 10 mM HEPES-KOH, pH 7.6, 67 mM KCl, 13 mM NaCl, 10 mM EDTA, 5 mM dithiothreitol (DTT), 5 mM KF, 5 mM β-glycerophosphate, 0.2 mM phenylmethylsulfonylfluoride (PMSF). Make 500 mL.
4. SR dialysis buffer: 10 mM HEPES-KOH, pH 7.6, 67 mM KCl, 13 mM NaCl, 1 mM EDTA, 2 mM DTT, 5 mM KF, 5 mM β-glycerophosphate, 0.2 mM PMSF. Make 4 L.
5. Buffer D: 20 mM HEPES-KOH, pH 7.6, 5% v/v glycerol, 0.1 M KCl, 0.2 mM EDTA, 0.5 mM PMSF, 0.5 mM DTT. Make 50 mL.
6. Sonicator with a large probe tip: Branson sonicator 450 with a 0.5-in. diameter disruptor tip or equivalent. Tissue culture cells and ground tissues are disrupted in a 600-mL beaker containing isolation buffer.
7. Phase-contrast microscope: used for monitoring the progress of cell disruption by the sonicator.
8. Centrifuges, rotors, and tubes: Beckman J2 or equivalent centrifuge with a JA-17 or equivalent rotor and lipped 50-mL polypropylene centrifuge tubes.
9. Ultracentrifuge, rotors, and tubes: Beckman L7 or equivalent ultracentrifuge with an SW-28 or equivalent large volume swinging bucket rotor and appropriate tubes.
10. Ultrapure ammonium sulfate ground to a fine powder in a mortar and pestle: grinding is done just prior to addition to extract.
11. 90% Ammonium sulfate SR dialysis buffer: prepared by adding 66.2 g of ammonium sulfate to 100-mL SR dialysis buffer.
12. 6000–8000 mol wt cutoff (MWCO) dialysis membrane tubing. Tubing is prepared by boiling in 10 mM Na-bicarbonate for 10 min followed by boiling in 10 mM EDTA for 10 min. The membrane is washed extensively in deionized dH_2O and stored in 50% ethanol at 4°C until use. The membrane is then rinsed in dH_2O and used for dialyzing ammonium sulfate away from extracts.
13. Eppendorf 5415C or equivalent benchtop microcentrifuge.
14. Silanized 1.7-mL microcentrifuge tubes.
15. 1 M MgCl$_2$ solution.
16. Cuphorn sonicator probe: 2-in. diameter model manufactured by Branson Ultrasonics (Danbury, CT, USA) cat. no. 101-147-047 or equivalent. Mg^{2+} precipitated SR proteins are resuspended in buffer D by holding the microcentrifuge tubes in a cuphorn sonicator probe bath.

2.2. Purification of Individual SR Proteins from Denaturing Polyacrylamide Gels

1. A vertical polyacrylamide gel electrophoresis apparatus for running gels of 16 cm in height, including 0.75-mm thick spacers and a preparative comb.

2. SDS-PAGE solutions for running a classic Laemmli 10% acrylamide (29.2:0.8 acrylamide:*bis*-acrylamide ratio) discontinuous protein gel. Tank buffer: 0.025 *M* Tris base, 0.192 *M* glycine, 0.1% SDS. Resolving gel acrylamide solution: 10% acrylamide (29.2:0.8 monomer:*bis*-acrylamide ratio), 0.375 *M* Tris-HCl, pH 8.8, 0.1% SDS. Stacking gel acrylamide solution: 4% acrylamide (29.2:0.8 monomer:*bis*-acrylamide ratio), 0.125 *M* Tris-HCl, pH 6.8, 0.1% SDS.
3. 0.25 *M* KCl chilled to 4°C.
4. 2× Protein sample buffer: 0.125 *M* Tris-HCl, pH 6.8, 4% SDS, 20% glycerol, 10% 2-mercaptoethanol.
5. Protein elution buffer (make fresh before use): 0.1% SDS, 50 m*M* Tris-HCl, pH 8.0, 5 m*M* DTT, 0.1 m*M* EDTA, 0.15 *M* NaCl, 0.1 mg/mL acetylated bovine serum albumin (BSA).
6. Acetone chilled to –20°C.
7. Microcentrifuge chilled to and operated at –20°C.
8. 6 *M* Guanidine HCl made in buffer D.
9. Microdialysis chamber with 6000–8000 MWCO dialysis membrane and peristaltic pump set up to pump buffer D through chamber. Spectrum Spectra/por (Houston, TX, USA) Microdialyzer cat. no. 132321 or equivalent apparatus with chambers capable of dialyzing volumes of 100–200 µL.
10. Coomassie staining solutions: Staining solution is 0.125% Brilliant Blue R250 in 50% methanol, 10% acetic acid. Stain the gel for 20 min and then destain the gel, first in 50% methanol, 10% acetic acid for 15 min followed by destaining to completion with several changes of 5% methanol, 7% acetic acid.

3. Methods

3.1. Purification of Total SR Proteins - Preparation of 65–90% Ammonium Sulfate Fraction

1. Mix the appropriate amount of cells, either approx 5×10^9 mammalian tissue culture cells, 4×10^{10} insect tissue culture cells, or 100 g of powdered tissue such as calf thymus (from –80°C freezer) with 330 mL of 4°C SR isolation buffer in a 600-mL beaker on ice. Place a stir bar in the beaker. Use a spoonula initially to mix the frozen powder and buffer (*see* **Note 1**).
2. Stir at 4°C for approx 5 min. Place the beaker in an ice bucket and remove the stir bar.
3. Sonicate in a 4°C room using a large probe and sonicator power setting near maximum. The cells are still in the 600-mL beaker on ice. Stir every 30–60 s with a spatula to ensure uniform sonication. Check for lysis with a microscope every 5 min. Tissues could take as long as 15 min to lyse and tissue culture cells could take as little as 5 min. Prepare a slide of cells before sonication as a reference (*see* **Note 2**).
4. Centrifuge lysed cells at 10,000 rpms (10,000*g*) in a JA-17 rotor at 4°C for 20 min in chilled 50-mL polypropylene centrifuge tubes.
5. After centrifugation, remove tubes from the rotor and place on ice. A lipid layer is often visible on top of the extract in the tube. Remove lipids with a spoonula (*see* **Note 3**). Pool all supernatants in a chilled 600-mL beaker on ice. Measure

the exact volume of the extract with a chilled 500-mL graduated cylinder and return the extract to the 600-mL beaker on ice.
6. Add a magnetic stir bar to the beaker and stir slowly at 4°C. Add fine powdered ammonium sulfate to 65% saturation over a period of 15–20 min, a little bit at a time. Just before use, make sure that the salt is ground completely to a fine powder, with a mortar and pestle (*see* **Note 4**). To obtain 65% saturation with ammonium sulfate, add 0.43 g of ammonium sulfate per milliliter of extract.
7. After all the salt is added, stir at 4°C for 1.5 h.
8. Transfer the extract to chilled 50 mL polypropylene centrifuge tubes. Centrifuge at 10,000 rpm in a JA-17 rotor (10,000*g*) for 20 min at 4°C.
9. After centrifugation, remove tubes from the rotor and place on ice. Remove any visible lipid layer on top of the extract with a spoonula.
10. Pour the supernatant from each tube into a clean chilled 50-mL polypropylene centrifuge tube. Centrifuge again at 10,000 rpm in a JA-17 rotor (10,000*g*) for 20 min at 4°C (*see* **Note 5**).
11. After centrifugation, remove tubes from the rotor and place on ice. A lipid layer may still appear on top of the extract in the tube. If so, remove this with a spoonula.
12. Pool all supernatants in a chilled 600-mL beaker on ice. Measure the exact volume of the extract in a chilled 500-mL graduated cylinder and return the extract to the 600-mL beaker on ice.
13. Add a magnetic stir bar to the beaker and stir slowly at 4°C. Add fine powdered ammonium sulfate to 90% saturation over a period of 15–20 min as described in **step 6**. To obtain 90% saturation with ammonium sulfate, add 0.19 g ammonium sulfate per milliliter of 65% ammonium sulfate-containing extract.
14. Stir the 90% ammonium sulfate extract overnight at 4°C; cover the beaker with plastic wrap.
15. Next morning, aliquot the sample into chilled ultracentrifuge tubes for the SW28 rotor on ice.
16. Centrifuge in a Beckman L7 ultracentrifuge at 25,000 rpm (85,000*g*) at 8°C for 1 h in the SW28 (swinging bucket) rotor (*see* **Note 6**).
17. Gently pour the supernatants out of the tubes.
18. Gently pipet in 3 mL of 4°C 90% ammonium sulfate SR dialysis buffer. Allow the solution to run down the side of the tube to the pellet and then pour it off gently. Be careful not to disturb the pellet.
19. Pipet 0.5 mL of SR dialysis buffer (no ammonium sulfate) into each tube on ice. I usually put the ice bucket containing all the tubes on a benchtop rotator and allow it to rotate for 10–15 min to help in resuspending the pellets.
20. Pool all resuspended pellets into one tube on ice and then serially rinse the other tubes with 1 mL of SR dialysis buffer and pool this with the rest of the sample. The total volume at this step should be roughly 6 mL.
21. Rinse 6–7 in. of 6000–8000 MWCO dialysis tubing with dH$_2$O and clamp the bottom with a dialysis clamp.
22. Put the resuspended pellet into the tubing using a transfer pipet. Clamp the top.
23. Dialyze against 1300 mL of SR dialysis buffer for 1 h at 4°C.

24. Change the dialysis buffer and dialyze against another 1300 mL for 2 h at 4°C.
25. Change the dialysis buffer again and dialyze against another 1300 mL overnight at 4°C.
26. The next morning, remove the dialyzed 65–90% ammonium sulfate cut from the dialysis tubing and place it into silanized microcentrifuge tubes in 1-mL aliquots.
27. Place the tubes on dry ice for 1 min or until frozen.
28. The dialyzed 65–90% ammonium sulfate cut can be stored frozen at –80°C indefinitely.

3.2. Magnesium Precipitation of SR Proteins

1. Thaw the dialyzed 65–90% ammonium sulfate fractions on ice.
2. Perform a clearing centrifugation to remove precipitated proteins (*see* **Note 7**), which can be done right in the microcentrifuge tubes at 14,000g (13,000 rpm) for 30 min in a 4°C microcentrifuge. Alternatively for larger volume SR protein preparations, pool the thawed dialyzed 65–90% ammonium sulfate fractions into a baked 15-mL Corex centrifuge tube and centrifuge at 10,000 rpm (10,000g) at 4°C for 30 min in a JA17 or equivalent rotor using rubber tube adaptors.
3. Using a transfer pipet, transfer the cleared supernatants to new chilled tubes on ice (Corex tubes or silanized microcentrifuge tubes). Be careful to avoid any pellets you may see. It is better to sacrifice some supernatant than to contaminate it with precipitated proteins.
4. Add 1 M MgCl$_2$ solution to the cleared extract to a final concentration of 15 mM MgCl$_2$. Mix and allow the tubes to sit on ice for 1 h.
5. Centrifuge the Mg^{2+} treated extracts as described in **step 2**.
6. Remove the supernatants. Gently pipet a small amount of 20 mM MgCl$_2$ SR dialysis buffer into each tube as a wash. You only need to add a small amount, enough to cover the pellets. Remove this wash buffer completely. Centrifuge briefly to ensure that all this buffer is at the bottom of the tube where it can be removed.
7. Resuspend the pellets in buffer D. For pellets in Corex tubes, resuspend the pellet by gently pipeting up and down with 200–400 µL of buffer D. Keep pipeting to a minimum to avoid sample loss to the pipet tip. Transfer the resuspended pellet to a silanized microcentrifuge tube. Rinse the corex tube with an additional 100 µL of buffer D and pool with the previous sample. For microcentrifuge tubes, add 20–100 µL of buffer D to pellets derived from 1 mL of 65–90% dialyzed ammonium sulfate cut. Do not pipet up and down to resuspend. In both cases, resuspend the protein pellets in the microcentrifuge tubes by holding the tubes in a cuphorn sonicator bath at 4°C for 30 s with the cuphorn probe at the highest setting. If necessary, repeat several times taking care that the water in the cuphorn sonicator is kept cool, either by adding ice or replacing the water with 4°C water (*see* **Note 8**). (*See* **step 2** of **Subheading 3.3.** if individual SR proteins will be purified from these resuspended SR proteins.)
8. The purity of resuspended SR proteins can be checked by running the SR proteins on an SDS-PAGE gel and comparing the Coomassie staining of proteins

with the mAb104 immunoblot of the same proteins. In our experience, SR proteins are not detected by silver staining techniques (*see* **Note 9**).

3.3. Purification of Individual SR Proteins with SDS-PAGE Followed by Renaturation

1. Prepare a preparative 10% acrylamide SDS-PAGE gel using 0.75-mm thick spacers, a 5–8-cm wide preparatory well, and an approx 4-cm tall stacking gel. Allow the acrylamide in the gel several hours to polymerize prior to use.
2. Add an equal volume of 2× protein sample buffer to the magnesium precipitates of SR proteins already resuspended in buffer D. For these preparative gels we find that the SR proteins derived from 100 g of calf thymus starting material work best. Care should have been taken in the resuspension step to keep the volumes to a minimum. Do not resuspend the magnesium precipitate pellets directly into protein sample buffer or add protein sample buffer to resuspended proteins in the tube in which the magnesium precipitation step was performed, as this will transport any contaminants that have stuck to the sides of the centrifuge tube.
3. Heat the sample to 90°C for 5 min.
4. Load the sample into the preparative well of the gel. Position a prestained protein molecular weight marker next to the preparative lane to orient you. Save 1% of your sample and run it in a single well next to a prestained marker.
5. Run gel at 100 V for 7 h or until the dye front reaches the bottom of the gel.
6. Disassemble the gel. Coomassie stain the single lane of 1% of the sample and the marker next to it to keep as a record of the purification.
7. Soak the gel containing the preparative lanes and the prestained marker in a dish containing 500 mL of prechilled 0.25 M KCl for 10 min at 4°C.
8. Place the gel on a glass plate at room temperature and put the plate over a black background. The KCl will interact with the SDS on the proteins to form a visible light white precipitate where proteins are present (*see* **Note 10**).
9. Cut out the individual protein bands with a razor blade. Place each piece of acrylamide into a disposable 15-mL conical tube.
10. Add 10 mL of 4°C 1 mM DTT to each tube. Soak bands for 3 min at room temperature and remove and discard solution from the tubes.
11. Place a small mortar and pestle in an ice bucket filled with crushed dry ice. Add liquid N_2 to the mortar. Using tweezers, place a piece of acrylamide containing an SR protein into the liquid nitrogen. Gently grind the band to a fine powder in the mortar and pestle.
12. Using a piece of weighing paper chilled on dry ice, remove powdered acrylamide from the mortar and place the powder into a 15-mL conical tube or a silanized 1.7-mL microcentrifuge tube.
13. Add 0.5–2.0 mL of freshly made protein elution buffer. This volume is determined by the volume of acrylamide powder. A good rule is to use one volume of elution buffer per volume of acrylamide powder.
14. Elute the proteins from the acrylamide by rotating the tubes gently at room temperature overnight.

15. Separate the eluate from the acrylamide by centrifuging the tubes in which the proteins were eluting for 5 min at high speed at room temperature either in a clinical centrifuge for the conical tube elution or a microcentrifuge for the microcentrifuge tube elution.
16. Using a transfer pipet, remove the supernatants to new silanized microcentrifuge tubes. For larger volumes, divide the supernatant into multiple silanized microcentrifuge tubes and be careful not to exceed 800 µL per tube.
17. Centrifuge the supernatants again for 5 min to pellet out any acrylamide that came across.
18. Remove the supernatants to new silanized microcentrifuge tubes. Be very careful to avoid any acrylamide in the pellets. Determine the volume of supernatant in each tube using a micropipetor as a measuring device.
19. Place these tubes in a salt ice bath (5 M NaCl solution mixed with ice such that ice is wet but tubes can still stand up in it) at –20°C for 10 min to chill them.
20. In the coldroom, add 0.9 vol of acetone chilled to –20°C to each tube. Quickly mix the tube by inversion several times and immediately return the tube to the salt ice bath (*see* **Note 11**).
21. Incubate the acetone precipitations in the salt ice bath for 20 min in the –20°C freezer.
22. Place the tubes into a microcentrifuge prechilled to –20°C and centrifuge them at 13,000g for 20 min at –20°C.
23. Quickly remove the tubes from the microcentrifuge and place in a salt ice bath. Go to the coldroom immediately.
24. In the coldroom, remove as much of the supernatant as possible, as quickly as possible using a fine tip transfer pipet or Pasteur pipet. Immediately add 100 µL of 6 M guanidine HCl in buffer D (*see* **Note 12**).
25. Incubate the tubes at 4°C for 5 min.
26. Transfer the tubes to room temperature and incubate for 15 min with the caps open to allow residual acetone to evaporate.
27. Cap the tubes and sonicate each tube in a cuphorn sonicator bath for 30 s to aid in resuspending the pellet.
28. Incubate the tubes for an additional 15 min at room temperature with caps open.
29. Centrifuge the tubes for 10 min at 13,000g in a room temperature microcentrifuge to remove any proteins that were not solubilized in guanidine HCl.
30. Place each supernatant into a well of a microdialysis device that is designed for 100-µL samples. Dialyze against buffer D containing 5 mM β-glycerophosphate (a phosphatase inhibitor) at 4°C with a 6000–8000 MWCO dialysis membrane. Pump buffer D through the device at the rate of 1 mL/minute with a peristaltic pump. Dialyze samples for a minimum of 3 h and a maximum of 7 h.
31. Remove the dialysate from each well to a silanized microcentrifuge tube. Rinse each well with 50 µL of buffer D and pool with the dialysate for each protein.
32. Centrifuge the dialysates at 13,000g in a microcentrifuge for 30 min at 4°C to remove any proteins that became insoluble when the guanidine HCl was dialyzed away.
33. Put the supernatant into a new silanized microcentrifuge tube. Add 1 M MgCl$_2$ to a final concentration of 20 mM MgCl$_2$.

Purification of SR Protein Splicing Factors

34. Store the tubes on ice at 4°C overnight.
35. Centrifuge the tubes in a microcentrifuge at 4°C for 30 min at 13,000g.
36. Remove the supernatants completely and add 20–60 µL of buffer D to each pellet (*see* **Note 13**).
37. Resuspend the pellets with the use of a cuphorn sonicator bath set at maximum power for 30 s. Do this in the coldroom and take care to ensure that the sonicator bath water is kept ice cold.
38. The purity of SR proteins can be checked by running 2–4 µL on an SDS-PAGE gel and visualizing by Coomassie staining. We often run a dilution series of a known amount of protein (most times we use 0.1–1.0 µg of BSA) on the same gel and use the intensity of staining of this protein by Coomassie to determine the concentration of our SR proteins.

4. Notes

1. The volumes and amounts of cells and SR isolation buffer described can be scaled up or down according to the amount of starting material present.
2. Place 5 µL of cells onto a slide and cover with a coverslip. Prepare comparison slides of cells before and after sonication. More than 95% of the cells should be lysed.
3. Lipids and other membranous components of the disrupted cells are fairly easy to peel off from the top of the supernatant using a spoonula. It is important to remove as much of this as possible because it seems to interfere with later steps of the preparation if too much is carried over.
4. Having the ammonium sulfate ground to finer crystals than obtained from the supplier seems to be important for increasing the rate at which they dissolve into the extract. Slow addition is important. Place a few grams into the extract at a time about once per minute for 15 min.
5. One of the keys to getting clean SR proteins is to be certain that the proteins that are insoluble in 65% ammonium sulfate are completely cleared from the extract, and the second centrifugation step helps accomplish this.
6. We have found that doing this ultracentrifugation step at 8°C rather than 4°C helps to ensure that the ammonium sulfate stays in solution.
7. It is important to make sure that any proteins that have precipitated be removed prior to addition of magnesium so that they do not contaminate the SR proteins in the magnesium precipitate.
8. A cuphorn sonicator is a great tool for this type of protein work. It helps to prevent loss of protein pellets to the pipet tip during resuspension. We use a 2-in. diameter cuphorn probe made by Branson and our sonication is done for 30 s. The procedure works best if you can see cavitation inside the tube. It is important that you do not allow the cuphorn probe's water to heat up. We do this procedure in the cold room and add ice or change the water in the bath every 30 s to keep it cool.
9. We find that tissue culture cells have SR proteins at apparent homogeneity at this step. For preparations starting from animal tissues, some Coomassie staining contaminating bands are visible, but they are present in lower quantities than the SR

proteins and can be removed in the gel purification step described here.
10. Visibility of the protein bands may on occasion be poor. Sometimes we destain the gel by soaking in 4°C deionized H_2O for 2 min if the bands are hard to see above background. The black background also aids the visualization. Sometimes we put the glass plate onto feet made from microcentrifuge tube lids on top of a black benchtop. This helps prevent liquid from getting trapped under the glass plate which can affect visualization of the bands.
11. The role of the acetone is to precipitate the proteins and leave the SDS soluble. The 0.9 vol of acetone that is added has been determined empirically as the least amount of acetone required to precipitate the highly charged SR proteins. When working with acetone precipitation of proteins that one wants to renature, we have found that it is very important to maintain the acetone at cold temperatures. Any warming of the protein in the acetone leads to irreversible denaturation of the protein. For this reason acetone is kept at –20°C, acetone precipitations are kept on salt ice and at –20°C, and the microcentrifuge is kept and operated at –20°C, which is a highly unusual precaution but is important for this purification.
12. The 6 M guanidine HCl in buffer D is made fresh every time by weighing out an appropriate amount of guanidine HCl and adding buffer D until a final concentration of 6 M is achieved. It should be noted that this method of preparing this solution gives a buffer that is 6 M guanidine HCl and only about 0.6× buffer D.
13. At this point faint pellets should be barely visible to the naked eye.

References

1. Fu, X. D. (1995) The superfamily of arginine/serine-rich splicing factors. *RNA* **1**, 663–680.
2. Manley, J. L. and Tacke, R. (1996) SR proteins and splicing control. *Genes Dev.* **10**, 1569–1579.
3. Staknis, D. and Reed, R. (1994) SR proteins promote the first specific recognition of the pre-mRNA and are present together with the U1 small nuclear ribonucleoprotein particle in a general splicing enhancer complex. *Mol. Cell. Biol.* **14**, 7670–7682.
4. Zahler, A. M. and Roth, M. B. (1995) Distinct functions of SR proteins in recruitment of U1 small nuclear ribonucleoprotein to alternative 5' splice sites. *Proc. Natl. Acad. Sci. USA* **92**, 2642–2646.
5. Wu, J. Y. and Maniatis, T. (1993) Specific interactions between proteins implicated in splice site selection and regulated alternative splicing. *Cell* **75**, 1061–1070.
6. Roscigno, R. F. and Garcia-Blanco, M. A. (1995) SR proteins escort the U4/U6. U5 tri-snRNP to the spliceosome. *RNA* **1**, 692–706.
7. Crispino, J. D., Blencowe, B. J., and Sharp, P. A. (1994) Complementation by SR proteins of pre-mRNA splicing reactions depleted of U1 snRNP. *Science* **265**, 1866–1869.
8. Tarn, W. Y. and Steitz, J. A. (1994) SR proteins can compensate for the loss of U1 snRNP functions in vitro. *Genes Dev.* **8**, 2704–2717.
9. Sun, Q., Mayeda, A., Hampson, R. K., Krainer, A. R., and Rottman, F. M. (1993) General splicing factor SF2/ASF promotes alternative splicing by binding to an exonic splicing enhancer. *Genes Dev.* **7**, 2598–2608.

10. Lavigueur, A., La Branche, H., Kornblihtt, A. R., and Chabot, B. (1993) A splicing enhancer in the human fibronectin alternate ED1 exon interacts with SR proteins and stimulates U2 snRNP binding. *Genes Dev.* **7,** 2405–1417.
11. Ramchatesingh, J., Zahler, A. M., Neugebauer, K. M., Roth, M. B. and Cooper, T. A. (1995) A subset of SR proteins activates splicing of the cardiac troponin T alternative exon by direct interactions with an exonic enhancer. *Mol. Cell. Biol.* **15,** 4898–4907.
12. Kanopka, A., Muhlemann, O., and Akusjarvi, G. (1996) Inhibition by SR proteins of splicing of a regulated adenovirus pre-mRNA. *Nature* **381,** 535–538.
13. Gontarek, R. R. and Derse, D. (1996) Interactions among SR proteins, an exonic splicing enhancer, and a lentivirus rev protein regulate alternative splicing. *Mol. Cell. Biol.* **16,** 2325–2331.
14. Zahler, A. M., Lane, W. S., Stolk, J. A. and Roth, M. B. (1992) SR proteins: a conserved family of pre-mRNA splicing factors. *Genes Dev.* **6,** 837–847.
15. Birney, E., Kumar, S., and Krainer, A. R. (1993) Analysis of the RNA-recognition motif and RS and RGG domains: conservation in metazoan pre-mRNA splicing factors. *Nucleic Acids Res.* **21,** 5803–5816.
16. Roth, M. B., Murphy, C., and Gall, J. G. (1990) A monoclonal antibody that recognizes a phosphorylated epitope stains lampbrush chromosome loops and small granules in the amphibian germinal vesicle. *J. Cell Biol.* **111,** 2217–2223.
17. Roth, M. B., Zahler, A. M., and Stolk, J. A. (1991) A conserved family of nuclear phosphoproteins localized to sites of polymerase II transcription. *J. Cell Biol.* **115,** 587–596.
18. Caceres, J. F. and Krainer, A. R. (1993) Functional analysis of pre-mRNA splicing factor SF2/ASF structural domains. *EMBO J.* **12,** 4715–4726.
19. Zahler, A. M., Neugebauer, K. M., Stolk, J. A., and Roth, M. B. (1993) Human SR proteins and isolation of a cDNA encoding SRp75. *Mol. Cell. Biol.* **13,** 4023–4028.
20. Dignam, J. D., Lebovitz, R. M., and Roeder, R. G. (1983) Accurate transcription initiation by RNA polymerase II in a soluble extract from isolated mammalian nuclei. *Nucleic Acids Res.* **11,** 1475–1489.
21. Krainer, A. R. and Maniatis, T. (1985) Multiple factors including the small nuclear ribonucleoproteins U1 and U2 are necessary for pre-mRNA splicing in vitro. *Cell* **42,** 725–736.
22. Krainer, A. R., Conway, G. C., and Kozak, D. (1990) Purification and characterization of pre-mRNA splicing factor SF2 from HeLa cells. *Genes Dev.* **4,** 1158–1171.
23. Mayeda, A., Zahler, A. M., Krainer, A. R., and Roth, M. B. (1992) Two members of a conserved family of nuclear phosphoproteins are involved in general and alternative pre-mRNA splicing. *Proc. Natl. Acad. Sci. USA* **89,** 1301–1304.
24. Ge, H., Zuo, P., and Manley, J. L. (1991) Primary structure of the human splicing factor ASF reveals similarities with *Drosophila* regulators. *Cell* **66,** 373–382.
25. Hager, D. A. and Burgess, R. R. (1980) Elution of proteins from sodium dodecyl sulfate-polyacrylamide gels, removal of sodium dodecyl sulfate, and renaturation of enzymatic activity: results with sigma subunit of *Escherichia coli* RNA polymerase, wheat germ DNA topoisomerase, and other enzymes. *Anal. Biochem.* **109,** 76–86.

26. Krainer, A. R., Mayeda, A., Kozak, D., and Binns, G. (1991) Functional expression of cloned human splicing factor SF2: homology to RNA binding proteins, U1 70K, and *Drosophila* splicing regulators. *Cell* **66,** 383–394.
27. Tian, M. and Maniatis, T. (1993) A splicing enhancer complex controls alternative splicing of doublesex pre-mRNA. *Cell* **74,** 105–114.
28. Zahler, A. M., Neugebauer, K. M., Lane, W. S., and Roth, M. B. (1993) Distinct functions of SR proteins in alternative pre-mRNA splicing. *Science* **260,** 219–222.

32

Processing mRNA 3' Ends In Vitro

Michael J. Imperiale

1. Introduction

The 3' ends of most nonhistone mRNAs in mammalian cells are generated by the endonucleolytic cleavage of an mRNA precursor (pre-mRNA) followed by the addition of a polyadenylate (poly[A]) tail (see **ref. *1*** for a recent review on mRNA 3' end processing). The ability to process pre-mRNAs in vitro has contributed greatly to our understanding of the biochemical details of the reaction and, more recently, the regulation of this processing event. The in vitro conditions are such that the same *cis*-acting signals on the pre-mRNA that are required for processing in vivo are also required in vitro. The assay can be performed in one of two ways. In the first, a plasmid encoding a gene containing the poly(A) site of interest is incubated in a cell-free extract under conditions that allow transcription of the gene by RNA polymerase II and processing of the nascent transcript *(2)*. In the second, the pre-mRNA is synthesized in a separate reaction using standard bacteriophage polymerase-driven systems, and then incubated in the cell-free extract *(3)*. These are referred to as transcription-processing and processing reactions, respectively. The reaction conditions can be adjusted so as to examine only cleavage of the precursor, to examine cleavage and addition of the poly(A) tail, or to examine the addition of a poly(A) tail to a precleaved RNA molecule. The products of the reaction are separated from the precursor by standard denaturing gel electrophoresis, allowing for detection using autoradiographic techniques.

3' end processing requires three *cis*-acting signals on the pre-mRNA and a number of *trans*-acting processing factors. The *cis*-acting signals are the cleavage (or polyadenylation) site itself, the highly conserved AAUAAA sequence, which is usually found approximately 10–35 nt upstream of the cleavage site, and a less well-conserved G/U or U-rich sequence just downstream of the cleav-

From: *Methods in Molecular Biology, Vol. 118: RNA-Protein Interaction Protocols*
Edited by: S. Haynes © Humana Press Inc., Totowa, NJ

age site. The processing factors that are provided in the nuclear extract include cleavage and polyadenylation specificity factor (CPSF); cleavage stimulatory factor (CstF); poly(A) polymerase, which is required for both cleavage and the addition of the poly(A) tail; and other factors including CF1, CF2, and poly(A) binding protein II. Processing occurs after the ordered assembly of these factors into a large, multicomponent complex. First, CPSF recognizes the pre-mRNA through an interaction with the AAUAAA signal. This unstable complex is converted to a stable complex by the binding of CstF, which recognizes the downstream element. It is thought that this complex is now committed toward processing, and the rest of the factors then associate with the complex and the reaction proceeds. In this chapter, the methods for studying these reactions in vitro are described.

2. Materials
2.1. Preparation of Nuclear Extracts

1. Exponentially growing HeLa cells.
2. Phosphate-buffered saline (PBS): 0.2 g/L KCl, 0.2 g/L KH_2PO_4, 8.0 g/L NaCl, 2.16 g/L $Na_2HPO_4 \cdot 7H_2O$.

The following three buffers are stored at 4°C, and the dithiothreitol (DTT) is added from a 1 M stock just prior to use.

3. Buffer A: 10 mM HEPES, pH 7.9 at 4°C, 10 mM KCl, 1.5 mM $MgCl_2$, 0.5 mM DTT.
4. Buffer C: 20 mM HEPES, pH 7.9 at 4°C, 420 mM NaCl, 0.2 mM EDTA, 25% glycerol (USP grade), 1.5 mM $MgCl_2$, 0.5 mM DTT.
5. Buffer D: 20 mM HEPES, pH 7.9 at 4°C, 100 mM KCl, 0.2 mM EDTA, 20% glycerol, 0.5 mM DTT.
6. Spectropore MWCO 3500 (1 mL/cm) dialysis tubing.
7. 0.04% Trypan blue.
8. Dounce homogenizer with tight-fitting pestle.
9. Corex centrifuge tubes.
10. 15-mL Falcon or Corning centrifuge tubes.

2.2. Preparation of pre-mRNA

1. Reaction buffer: 40 mM Tris-HCl, pH 8.0, 25 mM NaCl, 8 mM $MgCl_2$, 2 mM spermidine.
2. 280 mM Cap analog (m7G[5']ppp[5']G, Pharmacia, Piscataway, NJ, USA)
3. 1 M DTT.
4. 10 mM GTP.
5. A solution containing 100 mM each UTP, ATP, and CTP.
6. [^{3}H]UTP, 30–60 Ci/mmol.
7. RNA polymerase, appropriate for the promoter being used.
8. RNase-free DNase (e.g., RQ1 DNase from Promega, Madison, WI, USA).

2.3. In Vitro Processing

1. 50 mM MgCl$_2$.
2. 10% Polyvinyl alcohol.
3. 0.5 M Phosphocreatine.
4. A solution containing 10 mM each GTP, ATP, UTP, and CTP.
5. 25 mM ATP.
6. Stop buffer: 50 mM Tris-HCl, pH 7.4, 50 mM EDTA, 0.1% SDS.
7. Proteinase K.
8. Phenol saturated with 10 mM Tris-HCl, pH 7.4, 10 mM EDTA.
9. Chloroform/isoamyl alcohol (24:1).
10. 7.5 M NH$_4$OAc.
11. 10 mg/mL of yeast tRNA.
12. Ethanol.

2.4. Oligo-dT Cellulose Chromatography

1. 5 M NaCl.
2. Oligo-dT cellulose.
3. TE: 10 mM Tris-HCl, pH 7.4, 1 mM EDTA.
4. Binding buffer: TE + 0.5 M NaCl.
5. Wash buffer: TE + 0.1 M NaCl.
6. 3 M NaOAc.
7. Econo-column (Bio-Rad, Hercules, CA, USA), or other small disposable chromatography support.

2.5. S1 Analysis

1. S1 hybridization buffer: 80% deionized, ultrapure formamide, 10 mM PIPES, pH 6.4, 0.4 M NaCl.
2. 10 mg/mL Sonicated salmon sperm DNA.
3. 5× S1 digestion buffer: 0.15 M NaOAc, pH 4.5, 2 M NaCl, 5 mM ZnSO$_4$.
5. S1 nuclease.
6. 40% Acrylamide solution (19 parts acrylamide:1 part *bis*-acrylamide).
7. Gel loading buffer: 80% deionized formamide, 0.1% bromophenol blue, 0.1% xylene cyanol, 10 mM NaOH, 1× TBE.
8. 10× TBE (per liter): 108 g of Tris base, 55 g of boric acid, 9.3 g of Na$_2$EDTA.
9. X-ray film and cassettes for exposing the film. To quantitate results, a scanning densitometer or other quantitation device (e.g., PhosphorImager) will be required.
10. [^{32}P]dNTP at 3000 Ci/mmol. The dNTP to be used depends on the sequence of the probe. *See* **Subheading 3.6., step 1**.
11. Klenow fragment of DNA polymerase.

3. Methods

3.1. Preparation of the Nuclear Extract

The standard procedure for nuclear extract preparation is based on that of Dignam et al. *(4)* for in vitro transcription reactions, as modified by Moore and

Sharp *(3)*. HeLa cells are commonly used, but it should be possible to make an extract from any cell line. The keys to preparing high potency extracts are to ensure that the cells are in the exponential growth phase at the time of harvest, and to work as quickly as possible in the cold room. Our laboratory typically grows the HeLa cells in spinner cultures to facilitate growth of large numbers of cells, allowing us to prepare large quantities of extract, but the protocol can be used for attached cells as well. In this case, the cells must be detached from the dish either mechanically or using trypsin prior to the first step (*see* **Notes 2** and **3**).

1. Centrifuge the cells at 200*g* at 4°C for 5 min. Wash the cells with cold, sterile PBS 3× by resuspending in the PBS and centrifuging as described previously.
2. Resuspend the cell pellet in five packed cell volumes (PCVs) of buffer A (to which you have remembered to add the DTT). Perform this step and all subsequent steps in the cold room. Resuspend the cells thoroughly and incubate on ice for 10 min. Transfer the cells to a cold, sterile Corex tube and centrifuge at 600*g* at 4°C for 10 min. The pelleted cells should be quite swollen (i.e., the volume of the pellet should be greater than the original PCV). Remove the supernatant with a pipet and add twice the original PCV of buffer A. Resuspend thoroughly.
3. Transfer the mixture to a cold, sterile Dounce homogenizer and perform 15 strokes (once down and once up = one stroke) with a tight-fitting pestle. Check for cell lysis by mixing a small aliquot with an equal volume of trypan blue. Place on a slide and view under the microscope. Nuclei will take up the stain and appear blue while intact cells will remain colorless. Greater than 90% of the cells should be lysed (unlysed cells will be roughly twice as large as the nuclei). If fewer than 90% of the cells are lysed, perform a few more Dounce strokes. Transfer the lysed cells to a cold, sterile Corex tube and centrifuge at 1500*g* at 4°C for 10 min to pellet the nuclei. During this centrifugation rinse the Dounce homogenizer with cold dH$_2$O and then with buffer C. (Remember to add the DTT to the buffer C.)
4. **This next step is very important.** After the nuclei are pelleted, remove the supernatant and add 3 mL of buffer C for every 1×10^9 cells with which you started. (An alternative is to add buffer C in proportion to the size of the original PCV. Typically, use 3.8 mL of buffer C for a starting PCV of 5 mL. The total volume after the pellet is resuspended is roughly 5.5 mL). The amount of buffer C is critical: if too little buffer C is used the nuclear proteins are not extracted, and if too much buffer C is added the extract will be too dilute.
5. After the pellet is thoroughly resuspended, transfer the nuclei to the washed homogenizer and perform 15 Dounce strokes. Transfer the nuclei to a cold, sterile 15 mL Falcon centrifuge tube and gently mix at 4°C for 1 h. This can be performed on a rocking platform or a Nutator.
6. During the extraction step rinse the dialysis tubing in dH$_2$O and then let it soak in dH$_2$O at 4°C. Also add DTT to 1 L of buffer D in a sterile 2-L Erlenmeyer flask containing a sterile stir bar, and keep it at 4°C.
7. After the extraction step is complete, transfer the solution containing the nuclei and the extracted proteins to a cold, sterile Corex tube and centrifuge at 10,000*g*

at 4°C for 30 min. Transfer the supernatant to the dialysis tubing and clamp it. Put this into the buffer D and dialyze in the cold room for 2–2.5 h. Some protocols dialyze for as little as 1 h. Once this is complete aliquot the extract into microfuge tubes that have been placed on crushed dry ice. Typically, aliquot 50–100 µL into each tube. Immediately place in a –70°C freezer. Measure the protein concentration of the extract using any standard assay, such as the Bio-Rad (Hercules, CA, USA) assay.

3.2. Synthesis of Pre-mRNA Using Bacteriophage RNA Polymerase

To prepare pre-mRNA for the processing reaction, a plasmid containing a promoter that is recognized by a bacteriophage RNA polymerase such as T7, T3, or SP6 and encoding the poly(A) site is used. The plasmid is linearized with a restriction enzyme that cuts downstream of all the 3' processing signals, extracted with phenol and chloroform, precipitated with ethanol, and used in the synthesis reaction. The RNA can be synthesized in the presence of [^{32}P]UTP, generating a molecule that can be detected directly, or in the absence of ^{32}P, generating a molecule that can be detected using a ^{32}P-labeled S1 probe. The conditions for synthesizing ^{32}P-labeled RNA are given elsewhere in this volume (*see* Chapter 1). To prepare a precleaved precursor, use a restriction enzyme that linearizes the plasmid template at the poly(A) site. Described below is a procedure for preparing "cold" pre-mRNA (trace labeled with ^{3}H), whose processing can be measured using S1 nuclease analysis.

1. To prepare cold pre-mRNA, incubate 2 µg of linearized template DNA with 150 U of T7 RNA polymerase in a 50-µL reaction containing 40 mM Tris, pH 8.0, 25 mM NaCl, 8 mM MgCl$_2$, 2 mM spermidine, 2.8 mM cap analog, 5 mM DTT, 5 µCi lyophilized [^{3}H]UTP, 100 mM GTP, and 1 mM each ATP, UTP, and CTP at 18°C for 3 h. Stop the reaction by adding 2 U of RNase-free DNase and incubating at 37°C for 15 min.
2. Determine the RNA yield by measuring the amount of tritiated UTP incorporated into trichloroacetic acid-precipitable counts. Because the processing of these pre-mRNAs will be assayed by S1 hybridization to a 3' end-labeled DNA probe, there is no need to purify full-length molecules.

3.3. Processing Reactions

1. Pre-mRNAs are processed by incubation with the nuclear extract as follows. Assemble a reaction containing 2 mM MgCl$_2$, 3% polyvinyl alcohol, 20 mM phosphocreatine, 1 mM ATP, 100 µg of nuclear extract, and 2 nM (100 fmol) pre-mRNA in a 50-µL volume. The amount of nuclear extract will vary from preparation to preparation and should be titrated, but 100 µg is usually excess. Similarly, the amount of pre-mRNA needs to be determined empirically; the optimum is usually in the 1–10 nM range (see **Note 5**.)
2. Incubate the reaction for 2 h at 30°C.

3. Stop the reaction by adding an equal volume of stop buffer plus 30 μg of proteinase K and incubating at 37°C for 15 min. Extract the RNA once with phenol and chloroform:isoamyl alcohol and once with chloroform:isoamyl alcohol, and precipitate by the addition of ammonium acetate to 2.5 M, 25 μg of yeast tRNA, and 2 vol of ethanol.
4. When ^{32}P-labeled RNA is used, assay the products directly by gel electrophoresis. When cold RNA is used, purify the products by oligo-dT chromatography prior to S1 analysis (*see* **Subheadings 3.5. and 3.6.**).

3.4. Transcription-Processing Reactions

1. Incubate 1 μg of supercoiled substrate DNA at 30°C for 2 h in a 50-μL reaction containing 100 μg of nuclear extract, 3.0% polyvinyl alcohol, 20 mM phosphocreatine, 2 mM MgCl$_2$, and 600 mM each GTP, ATP, UTP, and CTP.
2. Stop the reaction by adding an equal volume of stop buffer plus 30 μg of proteinase K and incubating at 37°C for 15 min. Extract the RNA once with phenol and chloroform:isoamyl alcohol and once with chloroform:isoamyl alcohol, and precipitate it by adding ammonium acetate to 2.5 M, 25 μg of yeast tRNA, and 2 vol of ethanol. Isolate the poly(A)$^+$ RNA on oligo(dT)-cellulose columns (see **Note 6**.)

3.5. Oligo-dT Cellulose Chromatography

1. Pour a small Econo-column with 0.1-mL bed volume of oligo-dT cellulose (*see* **Note 1**). Equilibrate with 5 column volumes of binding buffer.
2. Dissolve the RNA in TE, heat to 60°C for 5 min to denature, put on ice, then add NaCl to 0.5 M to prevent renaturation. Apply the RNA to the column, collecting the flow-through. Reapply the flow through, including the liquid in the exit port of the column.
3. Wash with 5 column volumes of binding buffer, collecting the effluent. This, together with what flowed through during loading, is the poly(A)$^-$ fraction, which can be discarded or saved and assayed as needed.
4. Wash with 3 column volumes of TE + 0.1 M NaCl. Discard the wash.
5. Elute the poly(A)$^+$ RNA with 3 column volumes of TE.
6. Precipitate the RNA by adding NaOAc to 0.3 M and 2 vol of ethanol.

3.6. S1 Analysis of Processed RNA

1. Prepare a DNA probe for S1 analysis by digesting a plasmid (usually the same plasmid used for the transcription-processing reaction or for the production of the pre-mRNA) at a site upstream of the poly(A) site with a restriction enzyme that leaves a 5' overhang. Fill in this end with the appropriate ^{32}P-labeled NTP (depending on the sequence at the restriction site) using the Klenow fragment of DNA polymerase. For most restriction digests, add 50–100 μCi α-^{32}P-radiolabeled nucleotide (3000 Ci/mmol) and 1 U of Klenow enzyme directly to the restriction digest once the digest is complete, and incubate on ice for 30 min to incorporate the label. If this doesn't work, the DNA must be extracted and precipitated prior to the fill-in reaction. After labeling, extract the DNA with phenol

and chloroform, ethanol precipitate it, and digest it with a second enzyme that cuts downstream of the poly(A) site. Gel purify the labeled fragment using standard methods.
2. To set up the hybridization, coprecipitate the probe and the RNA and dissolve them in 50 µL of S1 hybridization buffer. The amount of probe to use must be determined empirically to ensure that the probe is in excess to the products of the reaction. Heat the mixture to 90–95°C for 3 min and then **rapidly** immerse the tube in a water bath at the appropriate temperature, which is determined empirically (*see* **Note 4**). This must be performed rapidly as the temperature should never drop below the hybridization temperature, and the tube should be surrounded with water to a level even with the bottom of the lip to prevent condensation on the underside of the lid. Hybridize the samples overnight.
3. Prepare the following mix on ice for each reaction: 100 µL of 5× S1 digestion buffer, 350 µL of dH$_2$O, 10 µg of freshly denatured (by boiling for 5 min) sonicated salmon sperm DNA, and 250 U of S1 nuclease. This amount of S1 nuclease is generally enough for most applications but should be tested empirically. It is useful to prepare enough of this reaction mix for the number of samples one has, plus one. To begin the digestion, carefully open the tube containing the hybridization mix while it is still mostly submerged in the water bath. Take 450 µL of the ice-cold digestion mix and squirt it into the tube. Immediately remove the tube from the water bath, cap it, invert it a few time to mix the contents, and place it on ice. Continue until all the tubes have been processed, then quickly centrifuge the tubes briefly (2 s) in a microfuge and transfer them to a water bath at the appropriate digestion temperature. This is usually between 20–25°C, but a lower temperature may be needed if the region of hybridization is A + T rich, to avoid digestion of the probe at regions in which the duplex is "breathing." Digestion is usually complete in 1–2 h, but this should be determined empirically for each probe.
4. Stop the reaction by extracting with 250 µL of phenol and 250 µL of chloroform:isoamyl alcohol. Remove the aqueous layer to a new tube, add 25 µg of yeast tRNA, and 1 mL of 95% ethanol to precipitate. Centrifuge and then wash the pellet with 70% ethanol. Dry the pellet and resuspend in 5 µL of gel loading buffer.

3.7. Electrophoresis of Reaction Products

The products of the reactions using labeled RNA, or the products of the S1 analysis, are separated using polyacrylamide-urea sequencing gels. When using labeled RNA, one will see the input RNA and a smear that runs slower than the input, which represents the cleaved and polyadenylated products. It runs as a smear because of the variable length of the poly(A) tail. If one inhibits the addition of the poly(A) tail, one will see the 5' cleavage product, which will run faster than the input RNA. In the S1 analysis, one will see undigested probe and the digestion product due to protection by the cleaved RNA (*see* **refs. 2** and **3** for examples).

4. Notes

1. To recycle oligo-dT cellulose, wash with H_2O, then with 0.1 M NaOH + 5 mM EDTA, then H_2O again. Wash well with binding buffer to eliminate all the NaOH (check the pH with pH paper). The oligo-dT cellulose can usually be used three times.
2. We generally do not treat our RNA reagents with any RNase inhibitors such as diethylpyrocarbonate (DEPC), and find that autoclaving is enough if one takes care during the procedures. On the other hand, we do have pipetors that are used for RNA work only.
3. As stated previously, the key to preparing active extracts is to have healthy cells and to work as quickly as possible during the preparation of the extract.
4. There are two key steps in performing S1 nuclease analysis, to ensure that the hybridization mix never falls below the hybridization temperature. First, the tube must be transferred from the denaturing step into the hybridization bath very rapidly. Second, the digestion mix must be added to the reaction while the tube is still submerged in the hybridization bath. The optimal hybridization temperature is one that gives the cleanest signal and the lowest background. For most probes, we have found that temperatures between 55 and 60°C are optimal, but for some probes, lower or higher temperatures are required.
5. To assay for cleavage only of the pre-mRNA, add 0.5 mM 3' dATP to the reaction. This will be incorporated at the site of cleavage by the poly(A) polymerase in the extract and cause chain termination.
6. We have not attempted to assay the products of the transcription/processing reactions directly by adding labeled UTP to the reaction, but there is no theoretical reason why this shouldn't work assuming that one is obtaining promoter-specific initiation of transcription.

References

1. Wahle, E. and Kuhn, U. (1997) The mechanism of 3' cleavage and polyadenylation of eukaryotic pre-mRNA. *Prog. Nucleic Acid. Res. Mol. Biol.* **57,** 41–71.
2. Wilson-Gunn, S. I., Kilpatrick, J. E., and Imperiale, M. J. (1992) Regulated adenovirus mRNA 3' end formation in a coupled in vitro transcription/processing system. *J. Virol.* **66,** 5418–5424.
3. Moore, C. L. and Sharp, P. A. (1985) Accurate cleavage and polyadenylation of exogenous RNA substrate. *Cell* **41,** 845–855.
4. Dignam, J. D., Lebovitz, R. M., and Roeder, R. G. (1983) Accurate transcription initiation by RNA polymerase II in a soluble extract from isolated mammalian nuclei. *Nucleic Acids Res.* **11,** 1475–1489.

33

Analysis of Poly(A) Tail Lengths by PCR: *The PAT Assay*

Fernando J. Sallés and Sidney Strickland

1. Introduction

The procedure described here, the PCR poly(A) test (PAT), allows the analysis of the poly(A) tail on a specific mRNA within a pool of total RNA in a rapid (1 day) and sensitive (subnanogram total RNA) fashion. The assay also provides quantitative estimates of the RNAs poly(A) tail length *(1)*. RNA manipulations up to and including cDNA synthesis are carried out in a single tube, minimizing handling and possible RNase contamination. Most importantly, once the PAT cDNAs are synthesized they are stable and can be used repeatedly to study any mRNA within the population by a PCR reaction. The ease and sensitivity of this technique make it amenable to the analysis of poly(A) tail lengths in virtually any system. Current and past applications include poly(A) analysis in mouse oocytes *(1)*, *Drosophila* oocytes and embryos *(2–4)*, *C. elegans* (B. Goodwin, personal communication; M. Gallegos and J. Kimble, personal communication), yeast (J. Engebrecht, personal communication), and cultured cells *(5)*. A previous simpler version of this assay *(6)* has been used to study cytoplasmic polyadenylation in mouse *(7)* and zebrafish (M. O'Connell, personal communication) embryos, as well as nuclear polyadenylation of nascent transcripts (mouse *[8]*; *Xenopus [9]*) and polyadenylation in mouse submaxillary glands *(10)*.

1.1. Theory

PAT cDNAs are synthesized as follows: RNA is isolated from cells or tissue of interest and heat denatured at 65°C in the presence of phosphorylated oligo(dT)$_{12-18}$ [p(dT)$_{12-18}$] in an Eppendorf tube (**Fig. 1**). The tube is then transferred to a 42°C water bath and a high concentration T4 DNA ligase cocktail is

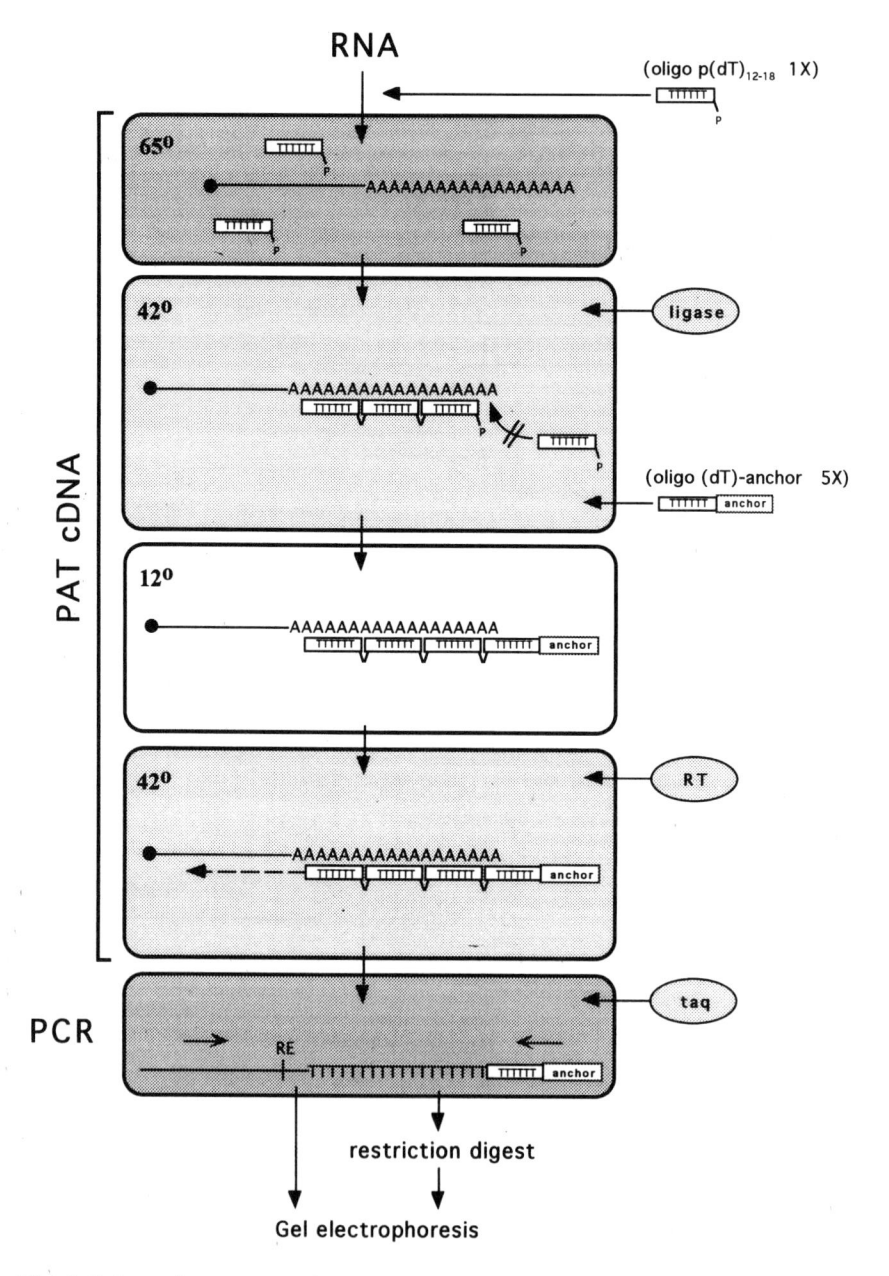

Fig. 1. Schematic representation of the PAT assay. RE, restriction endonuclease cleavage site; RT, reverse transcriptase; taq, *Taq* DNA polymerase.

added to ligate the p(dT)$_{12-18}$ subunits as they anneal on the poly(A) tails. When the p(dT)$_{12-18}$ subunits reach the 5' and 3' ends of the poly(A) tail a small unannealed stretch of poly(A) (≈<10 residues) should remain due to unfavorable base pairing at 42°C. At the end of the incubation period, a fivefold molar excess (over p[dT]$_{12-18}$) of an oligo(dT)-primer linked to a GC-rich anchor (oligo[dT]-anchor) is added to the reaction and the temperature lowered to 12°C. At the lower temperature and with the increased oligo(dT) concentration (both p[dT]$_{12-18}$ and oligo[dT]-anchor), annealing and subsequent ligation at the 3' poly(A) overhang is more favorable and the excess oligo(dT)-anchor ensures that a greater percentage of molecules will end with the anchor sequence (the anchor sequence serves to maintain the length of the poly[A] stretch during subsequent PCR amplification). After 2 h, the reaction is returned to the high temperature and reverse transcriptase is added to synthesize first strand cDNAs using the newly created poly(dT)-anchor as the primer. These PAT cDNAs are very stable.

To study the polyadenylation state of any mRNA within the pool, an aliquot is amplified by polymerase chain reaction (PCR) using a message-specific primer and the oligo(dT)-anchor primer. PCR products are analyzed by standard electrophoretic techniques. As a control to ensure that amplification was specific to the mRNA of interest and that any observed heterogeneity is due to 3' polyadenylation, the amplified products can be digested with a restriction endonuclease allowing visualization of a constant 5' portion and the heterogeneous 3' ends (**Fig. 2A**). This process can also aid in allowing a more accurate quantitation of the size of the poly(A) tail by allowing the analysis of a smaller fragment of DNA. If the mRNA contains alternative polyadenylation sites, they will appear as a double (or triple, etc.) profile (**Fig. 2B**).

2. Materials
2.1. PAT cDNAs

(All of the following enzymes and solutions are stable at –20°C).

1. Diethylpyrocarbonate (DEPC)-treated dH$_2$O.
2. T4 DNA ligase high concentration (approx 10 Weiss U/μL) (*see* **Note 1**).
3. Superscript II RNase H⁻ reverse transcriptase (RT) (Life Technologies, Gaithersburg, MD, USA) (200 U/μL) (*see* **Note 2**).
4. 5× Superscript II RNase H⁻ RT buffer (supplied with the enzyme).
5. 0.1 *M* Dithiothreitol (DTT) in DEPC-dH$_2$O (supplied with Superscript II RNase H⁻ RT).
6. 10 m*M* ATP in DEPC-dH$_2$O (required by the ligase).
7. 40 m*M* dNTP solution in DEPC-dH$_2$O: 10 m*M* each dATP, dTTP, dCTP, and dGTP. This solution will also be used for the PCR reaction.
8. Phosphorylated oligo(dT)$_{12-18}$ (p[dT]$_{12-18}$) at 10 ng/μL in DEPC-dH$_2$O. It is crucial for the oligo(dT)$_{12-18}$ to be phosphorylated or the ligation step will not work.

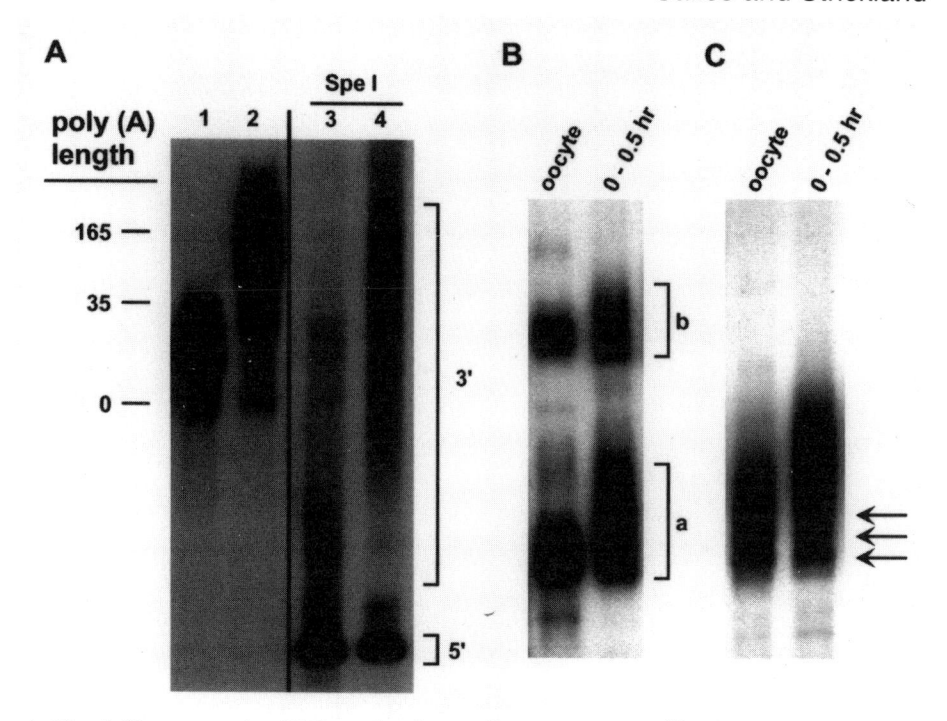

Fig. 2. Representative PAT results. Autoradiograms of trace-^{32}P-labeled PAT products electrophoresed on 5% nondenaturing (**A,B**) or 6% denaturing (**C**) polyacrylamide gels. (**A**) PAT analysis of tPA mRNA in mouse oocytes isolated 1 or 6 h after germinal vesicle breakdown (GVBD) (*lanes 1* and *2*, respectively). tPA mRNA has a short poly(A) tail in 1 h GVBD oocytes (<35 nt, *lane 1*) that undergoes cytoplasmic polyadenylation during the completion of meiosis I (approx 165 nt, *lane 2*). Estimates of poly(A) tail length agree with those derived from Northern analysis *(14)*. PAT cDNA synthesis was performed on 20 oocytes (0.8 ng total RNA). One oocyte equivalent was added to the PCR reactions (27 cycles: 90°C, 30 s; 62°C, 1 min; 72°C, 1 min). To compensate for changes in RNA concentration due to differences in RNA stability between time points, one fourth of the PCR products was loaded in *lane 1* and one half of the PCR products in *lane 2*; *see* **Note 9**. *Lanes 3* and *4* show the same samples as *lanes 1* and *2* after restriction digestion with *Spe*I. This digest reveals the constant 5' (55 bp) and heterogeneous 3' (130 + As) portion of the amplified products. (tPA specific primer [5'-ACT CTA TAG ATG GTT GGG AG]; expected minimum size {poly[A] length = 0} is 185 bp). (**B**) PATs for hunchback mRNA in *Drosophila* oocytes and embryos. The gel clearly shows two 3' termini (labeled a and b) both of which show a similar change in polyadenylation and correspond to different nuclear polyadenylation signal sequences within the 3' UTR of the mRNA. RNA was isolated from total ovary (oocyte) or 0–0.5 h embryos (0–0.5 h). (Hunchback specific primer [5'-AGC CAC CTT TCA ATC TGT CTC CTA T]; expected minimum sizes are [a] 244 bp and [b] 403 bp). (**C**) PAT analysis of bicoid mRNA in *Drosophila* oocytes and embryos. This shows a clear example of the "laddering" effect (arrows) sometimes seen under poor annealing/Ligation conditions during PAT cDNA synthesis. (Bicoid specific primer [5'-CAT TTG CGC ATT CTT TGA CC]; expected minimum size is 286 bp).

Poly(A) Tail Analysis

9. An oligo(dT)-anchor primer (5'-GCG AGC TCC GCG GCC GCG T$_{12}$) at 200 ng/μL for a 30-mer, in DEPC-dH$_2$O. Any oligo with a 3' (dT) stretch (n > 10) and a 5' GC-rich region that can serve as an anchor for PCR can substitute for this primer.

2.2. PCR

1. 40 mM dNTP solution in dH$_2$O (same as **item 7** above).
2. Oligo(dT)-anchor primer (same as **item 9** above).
3. Message specific primer in dH$_2$O (*see* **Note 3**).
4. *Taq* DNA polymerase (variety of suppliers).
5. 10× *Taq* polymerase buffer (generally supplied with the enzyme).

3. Methods
3.1. PAT cDNAs

1. Isolate total RNA from cells or tissues and resuspend at the appropriate concentration(s) in DEPC-dH$_2$O. Place 5 μL in a sterile RNase-free microfuge tube (*see* **Notes 4** and **5**).
2. Add 2 μL (20 ng) of p(dT)$_{12-18}$ to each RNA sample. The final volume of the RNA and p(dT)$_{12-18}$ should be 7 μL.
3. Heat denature at 65°C for 5–10 min.
4. Immediately transfer the tube to a 42°C water bath without an ice-quenching step. The immediate transfer to 42°C theoretically aids in preventing p(dT)$_{12-18}$ annealing to the extreme ends of the poly(A) tail.
5. Add 13 μL of the following prewarmed (42°C) mixture, mix by pipetting up and down, and incubate at 42°C for 30 min. Prepare a master mix with enough reagents for all of the PAT reactions. For each reaction add: 4 μL of 5× Superscript II H⁻ RT buffer, 2 μL of 0.1 M DTT, 1 μL of 40 mM dNTP mixture, 1 μL of 10 mM ATP, 4 μL of DEPC-dH$_2$O, and 1 μL of T4 DNA ligase at 10 Weiss U/μL (*see* **Note 1**). This step allows annealing and ligation of the p(dT)$_{12-18}$ along the poly(A) tail.
6. At the end of the incubation, while still at 42°C, add 1 μL (200 ng) of oligo(dT)-anchor. Vortex, quickly microfuge, and incubate at 12°C for 2 h (*see* **Note 6**).
7. Transfer the samples to 42°C for 2 min (*see* **Note 7**).
8. Add 1 μL of Superscript II RNase H⁻ RT (*see* **Note 2**). Vortex. Incubate at 42°C for 1 h. During this step the RNA is reverse transcribed using poly(dT)-anchor as a primer.
9. Heat inactivate RT and Ligase by incubating at 65°C for 20 min. If necessary, the PAT cDNAs can be diluted.

3.2. PCR

1. Set up a standard 25–50 μL PCR reaction containing 1× buffer (final component concentrations: 50 mM KCl, 10 mM Tris-HCl, pH 9.0 at 25°C, 0.1% Triton X-100, 1.5 mM MgCl$_2$), 0.2 mM each dNTP, 0.5 μM message specific primer (*see* **Note 3**), 0.5 μM oligo(dT)-anchor, 0.5–1 U *Taq* DNA polymerase, and 0.5–1 μL of each PAT cDNA as template.

2. Suggested initial cycling conditions for amplification of up to a 500 bp fragment are as follows: initial denaturation of 93°C for 5 min; then 30–35 cycles of 93°C for 30 s, 57–65°C (annealing temperature of message specific oligo) for 1 min, and 72°C for 1 min; followed by a final extension of 72°C for 7 min (*see* **Note 8**).
3. If desired, take an aliquot of the amplified products and digest them to completion with a restriction endonuclease to aid in analysis of the 3' ends (*see* **Note 3**).
4. Analyze the amplified products by gel electrophoresis (*see* **Note 9**).

4. Notes

1. Ligase concentration: The large amount of T4 DNA ligase is used to offset the enzyme's half-life during the initial 42°C incubation. A minimum required amount has not been established; however, enzyme from a variety of suppliers at different concentrations (6–40 Weiss U/μL using 1 μL/reaction) has been used successfully.
2. Choice of reverse transcriptase: Superscript II reverse transcriptase yields excellent results. However, in certain situations where even greater sensitivity is required or for RNAs with large amounts of secondary structure AMV reverse transcriptase (variety of suppliers) may work better and can be substituted along with its appropriate buffer *(11)*.
3. PCR message specific primer choice: A major variable in PAT optimization is the choice of message specific primer. Any PCR primer should theoretically work; however, certain primers never give clean results and a primer to a different region should be tested. In general, you should avoid very AT-rich sequences whenever possible and approximate the GC-content of the oligo(dT)-anchor (53% for the one listed here). For maximal electrophoretic separation the primer should be 100–400 nt 5' of the 3' end of the mRNA. In addition, it is convenient to have a unique restriction site (within the amplified region) approx 50–100 bp from the 5' primer to test amplification specificity (*see* **Fig. 2A**). For the analysis of hnRNA, the primer should lie in an intronic region and amplification carried out with ^{32}P-end-labeled oligo(dT)-anchor followed by a restriction digest to allow visualization of only the 3' end of the molecules *(8,9)*.
4. RNA isolation and treatment: Many different RNA isolation techniques have been used and any technique leaving the RNA in dH$_2$O should be compatible with the PAT. We routinely use the acid phenol method *(12)*. We rarely add carrier RNA even for small RNA preparations (<10 ng); however, the use of carrier tRNA should not interfere with the assay. For most applications the isolated RNA does not have to be treated with DNase. However, we have found that for certain mRNAs some background amplification was removed by the prior treatment of the RNA with DNase. DNase treatment can be carried out as follows: resuspend RNA in dH$_2$O and 1/10 volume of 10× DNase buffer (500 m*M* Tris-HCl, pH 7.5, 10 m*M* MgCl$_2$, 1 mg/mL bovine serum albumin [BSA]; this solution should be RNase free) or an RNA polymerase buffer (commercially supplied RNA polymerase buffers work well) and add 1 μL of RNase-free DNase. In addition, one can add 30–40 U (1 μL) of an RNase inhibitor. Incubate at 37°C for 30 min.

Phenol extract and precipitate the RNA in 0.3 M NaOAC and 2.5 vol of ethanol *(13)*. Resuspend in dH$_2$O.
5. RNA concentration: One of the major variables in optimizing the PAT results is the proper saturation of the poly(A) tail with p(dT)$_{12-18}$. If too much input RNA is used complete saturation and/or annealing of the p(dT)$_{12-18}$ subunits will not occur and could lead to a "laddering" of the products (**Fig. 2C**), which usually does not interfere with the interpretation of the results, or very poor amplification. Interestingly, a broad range of RNA concentrations (20 ng–1 µg) has been used successfully with the concentration of oligo p(dT) used here (20 ng). However, it may be helpful (although not required) to perform several PAT reactions for each sample at different RNA concentrations, e.g., 0.001, 0.01, 0.1, and 1 µg total (Betsy Goodwin, personal communication). Then test each PAT cDNA reaction for optimal PCR results with your specific mRNA. Altering RNA concentrations should eliminate the need to alter any other reagent concentration.
6. PAT annealing temperature: We routinely use 12°C as the annealing temperature for this step; however, we have successfully used temperatures as high as 25°C.
7. First strand synthesis: This short incubation before adding the RT should decrease, but not eliminate, annealing of the oligo(dT)-anchor to the 5' end of the poly(A) tail, which will preferentially prime first-strand synthesis over the ligated poly(dT)-anchor. This 5' end priming serves as a good control because it allows visualization of the 5' border of the poly(A) tail (*see* **Fig. 1A**; *lanes 1* and *2*, poly(A) length = 0).
8. PCR cycling conditions: In general one should use no more cycles than are necessary to properly visualize the results. Too many cycles can increase background bands. Unfortunately this has to be determined empirically but 30–35 cycles is a good starting point for ethidium bromide stained agarose gels with fewer cycles necessary for radiolabeled PCR analysis (*see* **Note 9**).
9. PAT analysis: The initial electrophoretic analysis can be conducted using agarose gels stained with ethidium bromide. This allows the visualization of even small changes (<50 nt) in poly(A) length *(1)* and is also the easiest method. However, for greater resolution and quantitation of size, the use of nondenaturing or denaturing polyacrylamide gels (generally 4–6% acrylamide) is recommended. The best way to visualize PCR products using these gel systems is by spiking the PCR reaction with a small amount (0.5–1 µL) of [α-^{32}P]dATP. After amplification, the samples should be precipitated using NH$_4$OAc and ethanol (to remove unincorporated radionucleotide) and resuspended in dH$_2$O *(13)*. For some RNAs, changes in polyadenylation alter their stability and hence concentration; therefore, when analyzing PCR products from two samples it is sometimes necessary to adjust the amounts of each sample used for electrophoresis (*see* **Fig. 2A**, *ref. 1*).

Acknowledgments

The authors acknowledge the excellent technical assistance of M. Lieberfarb and C. Wreden. The authors are indebted to and graciously thank the following individuals for sharing their unpublished experiences and modifications of the original assay: B. Conne, T. Chu, M. Gallegos, E. Gavis, B. Goodwin, M. Hentze,

J. Kimble, M. Lieberfarb, Z.-J. Liu, Y. Minami, S. Moody, M. Muekenthaler, M. O'Connell, J. Richter, J. Schisa, R. Schultz, A. Stutz, S. Sullivan, J.-D. Vassalli, A. Verrotti, P. Webster, M. Wickens, C. Wreden, and L. Wu. This work was supported by NIH Grants GM51584 and HD25922 to S.S.

References

1. Sallés, F. J. and Strickland, S. (1995) Rapid and sensitive analysis of mRNA polyadenylation states by PCR. *PCR Methods Appl.* **4**, 317–321.
2. Gavis, E. R., Curtis, D., and Lehmann, R. (1996) Identification of *cis*-acting sequences that control nanos RNA localization. *Dev. Biol.* **176**, 36–50.
3. Lieberfarb, M. E., Chu, T., Wreden, C., Theurkauf, W., Gergen, J. P., and Strickland, S. (1996) Mutations that perturb poly(A)-dependent maternal mRNA activation block the initiation of development. *Development* **122**, 579–588.
4. Sallés, F. J., Lieberfarb, M. E., Wreden, C., Gergen, J. P., and Strickland, S. (1994) Coordinate initiation of *Drosophila* development by regulated polyadenylation of maternal messenger RNAs. *Science* **266**, 1996–1999.
5. Muckenthaler, M., Gunkel, N., Stripecke, R., and Hentze, M. W. (1997) Regulated poly(A) tail shortening in somatic cells mediated by Cap-proximal translational repression proteins and ribosome association. *RNA* **3(9)**, 983–995.
6. Sallés, F. J., Darrow, A. L., O'Connell, M. L., and Strickland, S. (1992) Isolation of novel murine maternal mRNAs regulated by cytoplasmic polyadenylation. *Genes Dev.* **6**, 1202–1212.
7. Temeles, G. L., and Schultz, R. M. (1997) Transient polyadenylation of a maternal mRNA following fertilization of mouse eggs. *J. Reprod. Fertil.* **109(2)**, 223–228.
8. Huarte, J., Stutz, A., O'Connell, M. L., Gubler, P., Belin, D., Darrow, A. L., Strickland, S., and Vassalli, J.-D. (1992) Transient translational silencing by reversible mRNA deadenylation. *Cell* **69**, 1021–1030.
9. Rao, M. N., Chernokalskaya, E., and Schoenberg, D. R. (1996) Regulated nuclear polyadenylation of *Xenopus* albumin pre-mRNA. *Nucleic Acids Res.* **24(20)**, 4078–4083.
10. Sheflin, L. G., Brooks, E. M., Keegan, B. P., and Spaulding, S. W. (1996) Increased epidermal growth factor expression produced by testosterone in the submaxillary gland of female mice is accompanied by changes in poly-A tail length and periodicity. *Endocrinology* **137(5)**, 2085–2092.
11. Brooks, E. M., Sheflin, L. G., and Spaulding, S. W. (1995) Secondary structure in the 3' UTR of EGF and the choice of reverse transcriptases affect the detection of message diversity by RT-PCR. *Biotechniques* **19**, 806–815.
12. Chomczynski, P. and Sacchi, N. (1987) Single step method of RNA isolation by acid guanidinium thiocyanate phenol–chloroform extraction. *Anal. Biochem.* **162**, 156–159.
13. Sambrook, J., Fritsch, E. F., and Maniatis, T. (1989) *Molecular Cloning: A Laboratory Manual*, 2nd ed. Cold Spring Harbor Laboratory Press, Cold Spring Harbor, New York.
14. Huarte, J., Belin, D., Vassalli, A., Strickland, S., and Vassalli, J.-D. (1987) Meiotic maturation of mouse oocytes triggers the translation and polyadenylation of dormant tissue-type plasminogen activator mRNA. *Genes Dev.* **1(10)**, 1201–1211.

34

In Vitro Translation Extracts from Tissue Culture Cells

Kazuko Shiroki and Akio Nomoto

1. Introduction

The in vitro translation system provides an important means of identifying mRNA species of a gene of interest, characterizing the protein products, and investigating translational control. Rabbit reticulocyte lysate (RRL) or wheat germ lysate have been used to successfully translate protein in response to mRNAs from a variety of species including mammals, plants and phage *(1,2)*. The translation of viral mRNA in cell-free extracts is one of the most powerful tools available for elucidation of virus-specific translational control. The RNA genomes of some viruses including poliovirus are not capped, having instead a long noncoding region at their 5' end. The translation of such viral mRNAs is initiated by entry of the ribosome within the noncoding region in a cap-independent manner; the site is called the internal ribosome entry site, IRES *(3)*. Poliovirus is not translated in a cell-free translation system prepared from wheat germ, and is translated inefficiently, and usually incorrectly, in RRL. However, poliovirus mRNA is translated efficiently in HeLa cell S10 extract, and in RRL upon adding HeLa cell S100 extract (*see* **Note 1**) *(4–7)*. The cellular factors necessary for poliovirus translation are present in sufficient amounts in HeLa cell S10 extract but not in RRL. Thus, the in vitro system is useful for identification of the factors necessary for poliovirus cap independent translation, such as the La protein *(8)*, poly (rC) binding protein 1 (PCBP 1), and PCBP2 *(9,10)*.

Here, we describe the techniques currently in use in our laboratory for the study of cap-independent translation in mammalian cell S10 extracts using poliovirus and hepatitis C virus mRNAs (**Fig. 1**). We recently isolated poliovi-

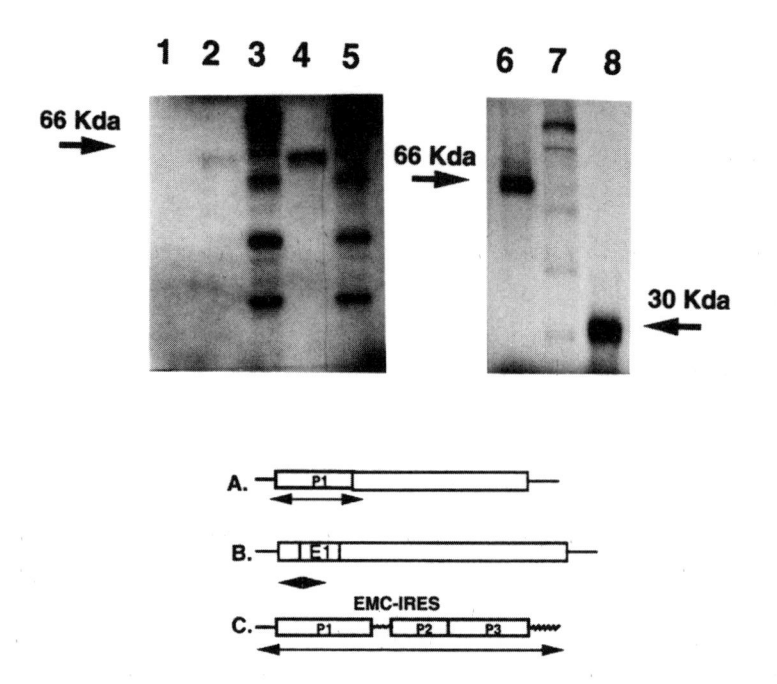

Fig. 1. Cell-free translation products of poliovirus, encephalomyocarditis virus (EMCV), and hepatitis C virus (HCV) mRNA in TgSVA and HeLa cell S10. Template RNAs (shown by *arrowhead* at both ends) were transcribed by T7 RNA polymerase from poliovirus (**A**), HCV (**B**), and polio and EMCV dicistronic (**C**) IRES containing cDNAs. The in vitro translation mixtures were incubated with poliovirus (*lanes 2, 4, and 6*), EMCV (*lanes 3, 5, and 7*) and HCV (*lane 8*) RNAs at 32°C for 1 h in TgSVA cell S10 (*lanes 1, 2, and 3*), TgSVA cell S10 with HeLa S100 (*lanes 4 and 5*) and HeLa cells S10 (*lanes 6, 7, and 8*). The products (poliovirus-66 kDa, HCV-30 kDa and EMCV-several bands) are shown.

rus host range mutants carrying mutations in the IRES region that grow efficiently in human HeLa cells but not in mouse TgSVA cells *(11,12)*. We also showed that the host range phenotype between HeLa and mouse TgSVA cells was consistent with the translation efficiencies in cell-free systems using S10 extracts from HeLa and mouse TgSVA cells *(11,12)* (**Fig. 2**).

2. Materials

2.1. Cells

1. HeLa S3 cells are maintained in suspension cultures at 37°C in RPMI 1640 medium supplemented with 5% newborn calf serum (NCS) and 0.15% NaHCO3 using spinner flasks. HeLa cells are diluted every day and should be maintained between 2 and 5×10^5 cells/mL *(11–14)*.

A. HeLa cell S10
(Human)

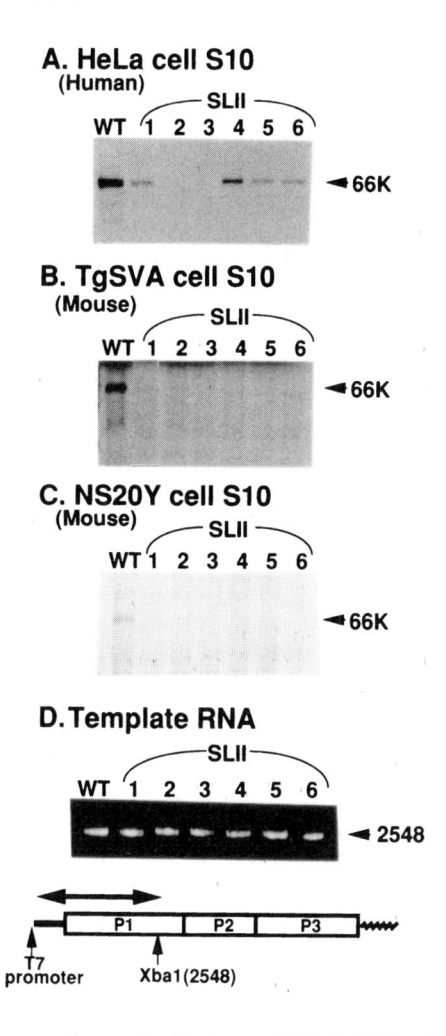

B. TgSVA cell S10
(Mouse)

C. NS20Y cell S10
(Mouse)

D. Template RNA

Fig. 2. Cell-free translation products of poliovirus mRNA in HeLa, TgSVA, and NS20Y cell S10 extracts. Template RNAs (**D**) were transcribed by T7 RNA polymerase from poliovirus WT cDNA and mutants of IRES region (pSLII-1, pSLII-2, pSLII-3, pSLII-4, pSLII-5, and pSLII-6) cDNAs that had been cleaved by XbaI. Length of template RNAs (2548 nt) is shown on the right of **D** and indicated by lines with *arrowheads* at both ends in **D**. The in vitro transcripts were incubated at 32°C for 1 h in S10 extracts from HeLa cells (**A**), TgSVA cells (**B**), and N20Y cells (**C**). Positions of the product (66 kDa) are indicated by arrowheads on the right side of the figures.

2. TgSVA cells were derived from the transgenic mouse kidney which expressed the poliovirus receptor. Maintain as a monolayer using Dulbecco's modified Eagle medium (DMEM) supplemented with 5% fetal calf serum (FCS) and 0.15% $NaHCO_3$ *(15)* and passaged every 2 d.

3. Lx cells are Ltk⁻ cells that express the human poliovirus receptor, and are maintained in a monolayer using DMEM supplemented with 5% FCS, 10^{-4} M hypoxanthine, 4×10^{-7} M aminopterin, 8×10^{-4} M thymidine, and 0.15% $NaHCO_3$ *(16)*.
4. NS20Y (mouse neuroblastoma) cells are maintained in high glucose (4.5 g/mL) DMEM supplemented with 10% FCS and 0.15% $NaHCO_3$ *(11)*.
5. HepG2 cells (from ATCC, Manassas, VA, USA) originated from human hepatoma, and are maintained in DMEM supplemented with 10% FCS and 0.15% $NaHCO_3$.
6. All media are sterilized by filtration or autoclaving. The cells in suspension culture are maintained in a spinner flask (BELLCO BIOTECHNOLOGY, Vineland, NJ, USA) on a magnetic stirrer at 37°C and all monolayer cells are maintained in a CO_2 incubator.

2.2. Reagents and Materials for Preparation of S10 Extracts

RNase free reagents, H_2O, and sterilized disposable plasticware should be used (*see* **Note 2**).

1. 1 M HEPES-KOH, pH 7.5: Autoclave and store at room temperature.
2. 1 M Tris-HCl, pH 7.5: Autoclave and store at room temperature.
3. Isotonic buffer: 35 mM HEPES-KOH, pH 7.5, 146 mM NaCl, 11 mM glucose. Sterilize by autoclaving and store at 4°C.
4. Hypotonic buffer: 10 mM HEPES-KOH, pH 7.5, 15 mM KCl, 1.5 mM $Mg(CH_3COO)_2$, 6 mM 2-mercaptoethanol. Store at –20°C.
5. 10× Salt buffer: 200 mM HEPES-KOH, pH 7.5, 1.2 M KCl, 50 mM $Mg(CH_3COO)_2$, 6 mM 2-mercaptoethanol. Store at –20°C.
6. Dialysis buffer: 10 mM HEPES-KOH, pH 7.5, 120 mM KCl, 1.5 mM $Mg(CH_3COO)_2$, 1 mM dithiothreitol (DTT).
7. 0.2 M Creatine phosphate (CP): CP in RNase-free H_2O. Freeze in 0.2 to 0.3-mL aliquots at –20°C.
8. 10 mg/mL creatine phosphokinase (CPK): CPK in 20 mM HEPES, pH 7.5, 50% glycerol solution. Freeze in 0.2-mL aliquots at –20°C.
9. 15K U/mL nuclease: Dissolve 15,000 U (=0.667 mg) of Nuclease S7 (Boehringer, Mannheim GmbH, Germany) in 1 mL of RNase-free H_2O. Store at –20°C.
10. 15 mg/mL tRNA: Dissolve 15 mg tRNA (ultrapure yeast tRNA) in 1 mL of RNase-free H_2O, phenol extract several times, then ethanol precipitate. Dissolve the precipitate in RNase-free H_2O and make up to 15 mg/mL. Store at –20°C (*see* **Note 2**).
11. 1 mM Hemin: Dissolve 6.5 mg Hemin (mol wt: 651.95) in 0.5 mL of 1 N KOH, add 0.5 mL of 1 M Tris-HCl, pH 7.5; adjust the pH to 7.8 with 1 N HCl, and add 8.5 mL of ethylene glycol. Centrifuge at 18,000g for 5 min and store the supernatant at –20°C (*see* **Note 3**).
12. 100 mM Spermidine: Dissolve in RNase-free H_2O. Store at –20°C.
13. 0.2 M EGTA: Dissolve EGTA in RNase-free H_2O and adjust the pH to 7.5 with 5 N KOH. Store at 4°C.

In Vitro Translation Extracts

14. 20 mM ATP: Dissolve 122 mg of ATP (RNase free) in 10 mL of RNase-free H$_2$O, then adjust the pH to 7.0 with 1 M NaOH. Store in 0.2 to 0.5-mL aliquots at –20°C (*see* **Note 2**).
15. 20 mM GTP: Dissolve 96 mg of GTP (RNase free) in 10 mL of RNase-free H$_2$O, then adjust pH as described for preparation of 20 mM ATP (*see* **Note 2**).
16. High-speed centrifuge such as Beckman Avanti TM-J-25 (Beckman, Palo Alto, CA, USA).
17. Tight-fitting Dounce homogenizer (A pestle).

2.3 Reagents and Materials for In Vitro Transcription

See Chapter 1.

1. Plasmid DNA: plasmid DNA containing the T7 or T3 promoter, the IRES from poliovirus type 1 Mahoney strain or hepatitis C virus (HCV), genotype 1b and coding region *(14,15,17)*.
2. DNase I: RNase-free DNase I (2 U/µL).
3. LiCl Precipitation solution: 7.5 M LiCl, 50 mM EDTA.
4. TE: 10 mM Tris-HCl, pH 7.6, 1 mM EDTA.

2.4. Reagents and Materials for In Vitro Translation.

1. S10 translation mixture: 40 µL of 0.5 M HEPES-KOH, pH 7.5, 26 µL of 5 M K(CH$_3$COO), 0.8 µL of 1 M Mg(CH$_3$COO)$_2$, 50 µL of 20 mM ATP, 2.7 µL of 20 mM GTP, 45 µL of 0.2 M CP, 2 µL of 1 M DTT, 2 µL of 0.1 M spermidine, and 11 µL of 1 mM amino acid mixture (methionine-free, Promega, Madison, WI, USA). Make up to 180 µL with RNase-free H$_2$O. Freeze in 30 to 50-µL aliquots at –80°C.
2. [^{35}S]Methionine: [^{35}S]Methionine (1200 Ci/mmol) at 10 mCi/mL, translational grade such as Amersham [^{35}S]methionine (Arlington Heights, IL, USA; cat. no. SJ1515 or SJ1015). Store at –80°C.
3. Creatine phosphokinase (CPK): 0.5 µg/µL of CPK; make fresh as required from 10 mg/mL CPK with RNase-free H$_2$O.
4. Sodium dodecyl sulfate (SDS) gel apparatus.
5. 10% SDS-polyacrylamide Gel (PAG): 9.8% acrylamide, 0.2% *N,N'*-methylene-*bis*-acrylamide, 1% SDS, 0.375 M Tris-HCl, pH 8.8.
6. Electrophoresis buffer: 3 g of Tris base, 14.4 g of glycine, 1 g of SDS, made up to 1 L with H$_2$O.
7. 2× Loading buffer: 100 mM Tris-HCl, pH 6.8, 200 mM DTT, 4% SDS, 0.2% bromophenol blue, 20% glycerol.
8. Reagents for fixation of PAG: 50% methanol, 10% CH$_3$COOH in H$_2$O.

3. Methods
3.1. Preparation of In Vitro Translation Extracts
3.1.1. Uninfected HeLa Cells

1. All procedures should be carried out at 0–4°C unless otherwise specified. Collect HeLa cells from 2 or more liters of suspension culture (a 5-L suspension culture

contains about 2×10^9 cells) by centrifugation at 300g for 7 min in a refrigerated centrifuge.
2. Wash 3× in isotonic buffer. After this point, the procedure is identical for all types of cells or infected cells.
3. After the last wash, resuspend the cells in a small volume of isotonic buffer and transfer them to a graduated conical centrifuge tube.
4. Pack the cells by centrifugation at 300g for 7 min. Note the cell volume (2×10^9 cell = 7–8 mL), resuspend the cells in 1.5 cell volume (10 mL) of hypotonic buffer, and let stand for 10 min at 0°C.
5. Break the swollen cells with 50 strokes of a cooled tight-fitting Dounce homogenizer (*see* **Note 4**). Check the homogenate microscopically for cell lysis.
6. Add 5/18 vol of 10× salt buffer, centrifuge the homogenate at 18,000g for 15 min, and discard the pellet (the pellet contains nuclei, mitochondria, and membrane-like cell components). This supernatant is the crude S10 extract.
7. To run off the endogenous mRNA from ribosomes in the crude S10 extract, add ATP to a final concentration of 1 m*M*, GTP to 0.2 m*M*, CP to 8 m*M*, and CPK to 0.2 mg/mL, and incubate at 37°C for 30 min.
8. Dialyze against 100 vol of dialysis buffer at 4°C for 1 h with one change of buffer. Alternatively, instead of dialysis, pass the extract through a G-25 column.
9. If the S10 extract becomes turbid, centrifuge at 6600g for 5 min, and use the opalescent supernatant for further steps.
10. To destroy the endogenous mRNA, add 0.2 mL of 1 m*M* hemin, 50 µL of 10 mg/mL CPK, 0.1 mL of 0.1 *M* CaCl$_2$, and 0.1 mL of 15K U/mL nuclease S7 and incubate for 5–15 min at 15–20°C.
11. Add 0.1 mL of 0.2 *M* EGTA and 40 µL of 15 mg/mL tRNA to stop nuclease activity. If necessary, centrifuge again at 6600g for 5 min and freeze in 0.2 to 0.3-mL aliquots at –80°C (*see* **Note 5**).

3.1.2. Poliovirus-Infected HeLa Cells

1. For preparation of poliovirus infected HeLa cell S10 extract, collect HeLa cells (5 L, 2×10^9 cells) and suspend in 1/20 vol (250 mL) of serum-free RPMI 1640.
2. Infect with poliovirus at a multiplicity of infection (MOI) of 20 (4×10^{10} plaque-forming units [pfu]).
3. Adsorb at room temperature for 20 min, and then at 37°C for 40 min in suspension.
4. Add 3 vol (750 mL) of RPMI 1640 containing 7.5% NCS. Incubate the infected cells at 37°C for 4 h in a spinner bottle. Preparation of S10 extracts from infected cells is by the same method as for uninfected HeLa cells.

3.1.3. Infected and Uninfected Monolayer Cells

1. For preparation of S10 extract from monolayer cells (TgSVA, Lx, NS20Y, and HepG2), the cells should be in good condition. Collect growing cells (2×10^9 cells) from 60–100 plates (150 m*M* in diameter) by treatment with 0.02 *M* EDTA and 0.1% trypsin, suspend in DMEM-containing serum (growth medium), and culture in a spinner flask for 5–8 h (*see* **Notes 6** and **7**).

2. For infection with poliovirus, add poliovirus at an MOI of 20 after growth in suspension culture for 5 h. Adsorb by the same method as for HeLa cells.
3. After infection, maintain the cells in suspension culture for a further 4 h. Preparation of S10 extracts from these cells is by the same method as for HeLa cells (*see* **Notes 8 and 9**).

3.2. In Vitro RNA Synthesis

1. Linearize plasmid DNAs with the appropriate restriction enzyme and purify by phenol:chloroform extraction. It is not necessary to separate the DNA fragments. Prepare transcripts by standard in vitro transcription methods (*see* Chapter 1) or with commercially available kits (MEGAscript™ in vitro Transcription Kits for Large Scale Synthesis of RNA: T7 and T3 Kits, Ambion, Austin, TX, USA).
2. For digestion of the template DNA after the RNA has been transcribed, add 1 U/sample of RNase free DNase I and incubate at 37°C for 15 min.
3. Add a half volume of 7.5 M LiCl solution and chill at −20°C for 30 min.
4. Centrifuge at 10,000g for 10 min, wash the precipitated RNA with 80% ethanol, dry the pellet and dissolve it in 0.5 mL RNase free TE (final concentration 1 μg/μL). Usually, 50–100 μg RNA is transcribed from 1 μg of template DNA. The RNA is stored 10–20-μL aliquots at −80°C.
5. Determine the size of RNA products on an agarose gel and the yield using spectrophotometry.

3.3. Cell-Free Translation

1. S10 extracts are generally frozen at −80°C in 0.3-mL aliquots and thawed only once. The optimum concentrations of K^+ and Mg^{2+} ions are different for each extract and each mRNA (*see* **Note 10**).
2. Assemble a mix containing all ingredients common to all tubes in the experiment. Final reactions contain 20 (or 10) μL of mix, additional RNAs, and RNase-free H_2O to make the final volume up to 25 (12.5) μL.
3. As an example, the mixture for each reaction of poliovirus mRNA in HeLa S10 contains: 13 μL of S10, 4.5 μL of translation mixture, 1 μL of 0.5 μg/μL CPK, 0.1 μL of RNasin, 0.5 μCi of ^{35}S-methionine, template RNA (0.5 μg), made up to 25 μL with RNase-free H_2O. The final concentration of each reagent is: 1.0 mM ATP, 57 μM GTP, 9 mM creatine phosphate, 0.02 μg/μL creatine phosphokinase, 2 mM DTT, 0.2 mM spermidine, 20 mM HEPES, 120 mM (150 mM for HCV mRNA) K(CH$_3$COO), 0.8 mM (1.5 mM for HCV mRNA), Mg(CH$_3$COO)$_2$, 60 μM each of 19 amino acids, 0.5 μCi [^{35}S]methionine, and 0.5 μg of RNA.
4. After incubation at 30–32°C for variable periods of time (generally 60 min) add sample buffer to the translation mixtures, boil for 2 min, and analyze by electrophoresis on 10% SDS-PAGE.
5. If necessary, treat the gels with autoradiography enhancer (ENLIGHTNING™ or EN^{3}HANCE; Dupont, Wilmington, DE, USA) before drying and exposing to X-ray film.

4. Notes

1. S10 or S100 extract is supernatant after centrifugation of cytoplasm at 18,000g for 15 min, or 100,000g for 120 min, respectively.

2. RNase-free and highest grade reagents should be used. We have had problems with GTP and tRNA being contaminated with RNase. Glassware is made RNase-free by washing 3× with 2 N NaOH or soaking in 0.05 N NaOH overnight and then thoroughly rinsing with freshly deionized H_2O.
3. Hemin is used to block phosphorylation of eIF 2α. The preparation should be carefully adjusted for pH and should be RNase free.
4. To prepare S10 extract from TgSVA cells, the swollen cells are broken with more than 80 strokes. Cell homogenate preparation is different for each cell type. The methods for S10 extract preparation from suspension cultures of HeLa, L, CHO, NGP (neuroblastoma) cells and ascitic Krebs-2 cells have been described previously *(7,11–14,17–19)*.
5. The activity of the S10 extract is reduced by nuclease treatment. Check the activity of the parental extract and the nuclease treated extract. The concentration of EGTA and Ca^{2+}, the incubation time (5–15 min), and the temperature (15–20°C) are important. Failure is most likely caused by making up the $CaCl_2$ and EGTA solutions incorrectly, as residual calcium ion allows the nuclease to remain active.
6. For the preparation of efficient S10 extracts, the best growing cells should be used. To obtain cells in optimum condition, actively growing monolayer cells should be trypsinized, transferred into suspension culture, and incubated for 5–8 h in a spinner flask. At present, we have obtained good preparations of S10 extract from the monolayer cell lines TgSVa, Lx, NS20Y, and HepG2, but could not prepare efficient S10 extract from a human neuroblastoma cell line, IMR32, and a human glioma cell line using this protocol. The major reason for this could be that these cell lines can not grow well in suspension for more than 10 h.
7. The success rate for preparation of active S10 extract is about 70%. It is not clear at present why some preparations fail.
8. The preparation of S10 extract from HeLa cells infected with poliovirus or coxsackie virus has an usually high success rate. In the S10 extract from infected cells, cap-independent translation of poliovirus mRNA occurs efficiently, but cap-dependent translation does not. Recently, it was reported that the eIF4G C-terminal domain, produced by cleavage of eIF4G with viral 2A protease, fulfilled the function of eIF4G in cap-independent translation *(20)*.
9. Translationally active extract is difficult to prepare from uninfected monolayer cells, and the success rate is about 30–50%.
10. In cell free translation, the optimal concentration of K^+ and Mg^{2+} ions varies considerably, depending on each extract and each mRNA species. It is essential to determine the optimal concentration of both ions for each extract and each mRNA. For example, the optimal concentration in HeLa S10 is, in our hands, around 100 mM K^+ and 0.8 mM Mg^{2+} for poliovirus mRNA, and 176 mM K^+ and 1.6 mM Mg^{2+} for HCV mRNA. The optimal concentration for poliovirus mRNA is 0.5 µg in HeLa S10 under these ionic conditions.

Acknowledgments

We are grateful to N. Kamoshita, T. Fujimaki, and N. Iizuka for critical reading and helpful suggestions. This work was supported in part by a grant-

References

1. Walter, P. and Blobel, G. (1983) Cell-free translation of messenger RNA in a wheat germ system. *Methods Enzymol.* **96**, 38–50.
2. Jackson, R. J. and Hunt, T. (1983) Preparation and use of nuclease-treated rabbit reticulocyte lysates for the translation of eukaryotic messenger RNA. *Methods Enzymol.* **96**, 50–74.
3. Pelletier, J. and Sonenberg, N. (1988) Internal initiation of translation of eukaryotic mRNA directed by a sequence derived from poliovirus RNA. *Nature* **334**, 320–325.
4. Jackson, R. J., Howell M. T., and Kaminski, A. (1990) The novel mechanism of initiation of picornavirus RNA translation. *Trends Biochem. Sci.* **15**, 477–488.
5. Brown, B. A. and Ehrenfeld, E. (1979) Translation of poliovirus RNA in vitro: changes in cleavage pattern and initiation sites by ribosomal salt wash. *Virology* **97**: 396–405.
6. Dorner, A. J., Semler, R. J., Jackson, R., Duprey, E., and Wimmer, E. (1984) In vitro translation of poliovirus RNA: utilization of internal initiation sites in reticulocyte lysate. *J. Virol.* **50**, 507–514.
7. Villa-Komaroff, L., McDowell, M., Baltimore, D., and Lodish, H. F. (1974) Translation of reovirus mRNA, poliovirus RNA, and bacteriophage QB RNA in cell free extracts of mammalian cells. *Methods Enzymol.* **30**, 709–723.
8. Meerovitch, K., Pelletier, J., and Sonenberg, N. (1989) A cellular protein that binds to the 5'-noncoding region of poliovirus RNA: implication for internal translation initiation. *Genes Dev.* **3**, 217–227.
9. Blyn, L. B., Towner, J. S., Semler, B. L., and Ehrenfeld, E. (1997) Requirement of poly(rC) binding protein 2 for translation of poliovirus RNA. *J. Virol.* **71**, 6243–6246.
10. Gamarnik, A. V. and Andino, P. (1997) Two functional complexes formed by KH domain containing proteins with the 5' noncoding region of poliovirus RNA. *RNA* **3**, 882–892.
11. Shiroki, K., Ishii, T., Aoki, T., Ota, Y., Yang, W.-Y., Komatsu, T., Ami, Y., Arita, M., Abe, S., Hashizume, S., and Nomoto, A. (1997) Host range phenotype induced by mutations in the internal ribosomal entry site of poliovirus RNA. *J. Virol.* **71**, 1–81.
12. Ishii, T., Shiroki, K., Hong, D.-K., Aoki, T., Ohta, Y., Abe, S., Hashizume, S., and Nomoto, A. (1998) A new internal ribosomal entry site 5' boundary is required for poliovirus translation initiation in a mouse system. *J. Virol.* **72**, 2398–2405.
13. Iizuka, N., Yonekawa, H., and Nomoto, A. (1991) Nucleotide sequences important for translation initiation of enterovirus RNA. *J. Virol.* **63**, 5354–5363.
14. Tsukiyama-Kohara, K., Iizuka, N., Kohara, K., and Nomoto, A. (1992) Internal ribosome entry site within hepatitis C virus RNA. *J. Virol.* **66**, 1476–83
15. Shiroki, K., Kato, H., Koike, S., Odaka, T., and Nomoto, A. (1993) Temperature-sensitive mouse cell factors for strand-specific initiation of poliovirus RNA synthesis. *J. Virol.* **67**, 3989–3996.

16. Koike, S., Horie, H., Ise, H., Okitsu, A., Yoshida, M., Iizuka, N., Takeuchi, T., Takegami, T., and Nomoto, A. (1990) The poliovirus receptor protein is produced both as membrane-bound and secreted forms. *EMBO J.* **9,** 3217–3224.
17. Kamoshita, N., Tsukiyama-Kohara, K., Kohara, M., and Nomoto, A. (1997) Genetic analysis of internal ribosomal entry site on hepatitis C virus RNA: implication for involvement of the highly-ordered structure and cell type specific trans-acting factors. *Virology* **223,** 9–18.
18. Haller, A. A., Stewart, S. R., and Semler, B. L. (1996) Attenuation stem-loop lesions in the 5' noncoding region of poliovirus RNA: neuronal cell-specific translation defects. *J. Virol.* **70,** 1467–1474.
19. Svitkin, Y. V., Yu, V. and Agol, V. I. (1978) Complete translation of encephalomyocarditis virus RNA and faithful cleavage of virus-specific proteins in a cell-free system from Krebs-2 cells. *FEBS Lett.* **87,** 7–11.
20. Haghighat, A., Svitkin, Y., Novoa, I., Kuechler, E., Skern, T., and Sonenberg, N. (1996) The eIF4G-eIF4E complex is the target for direct cleavage by the rhinovirus 2A proteinase. *J. Virol.* **70,** 8444–8450.

35

Messenger RNA Turnover in Cell-Free Extracts from Higher Eukaryotes

Jeff Ross

1. Introduction
1.1. mRNA Stability in Cells and In Vitro

Two groups of observations support the notion that mRNA turnover influences gene expression in virtually all cells *(1–3)*. (1) The steady-state levels of many mRNAs are determined more by their half-lives than by their gene transcription rates. In other words, mRNA levels often fluctuate without any measurable change in transcription. (2) The half-lives of some mRNAs change when the environment of the cell, its replication cycle status, or its stage of differentiation changes. The sequences and structures that determine the half-lives of many mRNAs have been mapped in great detail. In contrast, little is known about the RNases that specifically degrade mRNAs (mRNases) and the regulatory factors that influence mRNA stability. How many mRNases are expressed in each cell? Do different cells express different mRNases with different specificities? How do these enzymes function? Are they endonucleases or exonucleases, and where do they first begin to attack the mRNA molecule? What are the major pathways of mRNA decay, and what sorts of degradation intermediates are generated as mRNAs are destroyed? How does translation affect mRNA turnover? What *trans*-acting factors regulate mRNA turnover? How do they function — by binding to the mRNA molecules they affect, by up- or down-regulating mRNase activity, or by indirect mechanisms?

These questions have been particularly difficult to address in higher eukaryotes, in which genetic approaches to investigate mRNA stability have not been forthcoming. mRNA decay mutants in bacterial and yeast systems have revealed several of the enzymes and cofactors that affect mRNA stability (e.g., *refs. 4* and *5*). In contrast, most of what is known about the "players" involved in mRNA stability in higher eukaryotes has come from biochemical and gene transfection studies.

In vitro mRNA decay systems offer another approach for identifying, characterizing, and purifying mRNases and regulatory factors in higher eukaryotes. Two properties of the mRNA decay systems described in this chapter are particularly important (**Table 1**). (1) They can be exploited to identify putative mRNA decay intermediates. Many of these intermediates are very short-lived and thus quite difficult to detect in intact eukaryotic cells. The intracellular half-life of a relatively stable mRNA might be 20 h, but the degradation intermediates generated when each mRNA molecule is actually undergoing degradation are rapidly destroyed. If the intermediates were long-lived, they would accumulate in cells and be easy to detect, which, with few exceptions, is not the case. Because degradation rates are slower in vitro than in cells *(6)*, it is feasible to identify decay intermediates in vitro. Once identified, the investigator can construct specific probes to search for the same intermediates in cells and thereby confirm that what was observed in vitro occurs also in cells. In other words, it is easier to detect a decay intermediate if you know what to look for. (2) In vitro mRNA decay systems can be exploited to detect and purify mRNases, cofactors, and *trans*-acting regulators.

This chapter describes several in vitro methods currently used to investigate mRNA turnover in higher eukaryotes. Improvements on these and other in vitro systems will surely continue to be made, but their capacity to identify putative mRNases and mRNA decay regulatory factors is already well established *(7–9)*. On the one hand, it remains to be proved that all the factors thus far identified using cell-free systems actually affect mRNA stability in intact cells. On the other hand, several properties of these systems indicate that they mimic authentic intracellular mRNA decay in many respects and are therefore valid models of mRNA decay.

The most important characteristic of the systems described here is the observed correlation between relative mRNA half-lives in vitro and in cells. Some mRNAs that are known to be stable in cells are also relatively stable in vitro. Conversely, some unstable mRNAs are relatively unstable in cells and in vitro. It is important to stress the notion of relative in vitro mRNA stability, because mRNA decay rates are slower in vitro than in intact cells *(3)*. (In this respect, cell-free mRNA decay is similar to in vitro transcription and translation.) As a result, mRNA half-lives in cells are usually shorter than those measured in vitro. Nevertheless, if the decay rates of two mRNAs differ by 30-fold in cells, they should differ to approximately the same extent in vitro. In other words, their relative decay rates in vitro should reflect their relative decay rates in intact cells, regardless of their absolute intracellular half-lives. The half-lives of only a handful of mRNAs have thus far been determined in vitro. How-

Table 1
Advantages of In Vitro mRNA Decay Systems

1. mRNases and *trans*-acting regulators can be assayed and purified.
2. Relative mRNA turnover rates can be measured without toxic transcription inhibitors.
3. mRNA decay intermediates that are very short-lived in cells can be more easily detected in vitro.

Table 2
Use of Endogenous and Exogenous Substrates in Cell-Free mRNA Decay

Endogenous substrate (free or polysome-associated mRNP)
Advantages
1. These are "authentic" substrates made in intact cells.
2. Published experiments validate using mRNPs as substrates.

Disadvantages
1. Low abundance mRNP is difficult to detect.
2. mRNP can be degraded during isolation.

Exogenous substrate (deproteinized cell mRNA or in vitro transcribed RNA)
Advantages
1. Radiolabeled substrates are easy to synthesize, detect, and quantitate.
2. Modified forms of in vitro substrates are readily synthesized.

Disadvantages
1. These substrates are not as "authentic" as endogenous mRNP.
2. Deproteinized substrates might not respond to regulatory factors.

ever, the correlation between in vitro and intracellular relative mRNA half-life is excellent for these mRNAs. In summary, in vitro mRNA decay systems appear to reflect mRNA stability in cells. Therefore, it is reasonable to assume that they can be used to identify and characterize at least some of the enzymes, cofactors, and RNA-binding proteins that determine mRNA half-life.

1.2. mRNA Substrates for In Vitro Decay

One major issue in any cell-free mRNA decay assay is whether the substrate should be endogenous (polysome-associated) or exogenous mRNA. The answer depends, in part, on the questions being asked, the number of available cells, the abundance of the mRNA of interest, and the ease with which undegraded endogenous substrate can be prepared (**Table 2**). The methods described in this chapter use either endogenous or exogenous substrates.

The two endogenous substrates are messenger ribonucleoprotein (mRNP) associated with polysomes and free mRNP. Because polysomes contain at least some mRNases *(6,10,11)*, polysomes can degrade the mRNAs with which they are associated. There are several advantages to using endogenous mRNP substrates. (1) Polysomal mRNP was synthesized by the cell and was being translated. Therefore, it can be considered more or less authentic. (2) Some endogenous mRNPs are known to respond in vitro to mRNA stability regulatory factors *(12–15)*.

There are two major disadvantages of endogenous substrates. (1) Low abundance mRNA can be difficult to detect and quantify. (2) Polysomes containing undegraded mRNA can be difficult to prepare. mRNA degradation during polysome isolation is a major problem when using tissues from animals such as mice, rats, or frogs. In these cases, special techniques are required to prepare undegraded polysomes.

Exogenous substrates have two major advantages. (1) They are easy to prepare and to analyze. Radiolabeled substrate RNA is relatively easy to make by in vitro transcription. It is added to a cell extract, usually polysomes, postpolysomal supernatant, or ribosomal salt wash. Following incubation, the RNA is extracted and analyzed directly by gel electrophoresis and autoradiography. In contrast, endogenous substrates must be assayed by more cumbersome techniques such as Northern blotting, RNase protection, or polymerase chain reaction (PCR). (2) Exogenous substrates can be readily modified by mutating the transcription substrate (cDNA). Therefore, exogenous substrates can be prepared with or without caps or poly(A) tracts, with deletions, substitutions, additions, etc.

The major disadvantage of exogenous substrates is that they are not as "authentic" as endogenous ones. They are not made in cells, and there is no reason a priori to think they will associate in vitro with relevant RNA-binding proteins to generate functional mRNP complexes. For this reason, it is particularly important to compare the decay rates of different exogenous substrates. Do they reflect the relative decay rates of their mRNA counterparts in intact cells? If so, the system would seem to be a valid measure of mRNA stability, as demonstrated by several investigators *(12,16–18)*. If not, it is difficult to justify using the exogenous substrates.

1.3. Some Caveats About In Vitro mRNA Decay

It is well known that mRNA stability and translation are linked *(3,19)*. For example, inhibiting translation in higher eukaryotes invariably causes many unstable mRNAs to be stabilized at least fourfold. However, most extract preparations from nucleated mammalian cells translate endogenous and exogenous mRNAs poorly, if at all. Therefore, any links between mRNA stability and translation might be lost with these extracts. Prolonged translation has been observed in some extracts prepared from higher eukaryotic cells *(20)*, and it might therefore be useful to characterize the capacity of these extracts to translate and degrade mRNAs.

Extraction procedures disrupt cell architecture. Therefore, any potential links between cell structure, mRNA localization, and mRNA stability can be destroyed. For example, histone mRNA is normally translated on unbound polysomes, and its half-life is regulated as a function of DNA replication *(21)*. However, it is not regulated in the same way if it is targeted to membrane-bound polysomes *(22)*. Cell extraction/disruption procedures also dilute potential *trans*-acting regulatory factors. It is unclear whether these factors freely diffuse throughout the cytosol or are associated with ribosomes, endoplasmic reticulum, the mRNA itself, etc. Therefore, extract preparation might destroy interactions between regulatory factors and cell structures. It might also dilute the factors so much that they are inactivated or ineffectual. In summary, the same constraints that apply to any in vitro assay system apply also to mRNA decay systems, and it is important to recognize these limitations.

1.4. Precautions About Extract Preparation and Incubation

The methods described in this chapter use polysomes, a polysome extract, or a reticulocyte translation system. When using polysomes, the primary requirements for successful results are an appropriate cell or tissue source for preparing undegraded polysomes plus careful, consistent technique to maintain enzyme activity. Some tissues and cell lines contain more nonspecific RNases than others, making it difficult to isolate undegraded polysomes from them. For example, in spite of numerous attempts, we have never isolated undegraded polysomes from rodent spleen. Therefore, we recommend isolating polysomes from tissue-culture cells, to avoid the high level of nonspecific RNase activity present in spleen and other animal tissues. K562 human erythroleukemia cells are an excellent source, being easy to grow, low in endogenous RNases, and easy to lyse without detergents. All procedures should be performed at 2–4°C, from the time the cells are harvested until the extracts are placed in the –70°C freezer. For most applications, extracts can be thawed at least twice.

If necessary, various RNase inhibitors are available and can be added to all buffers. They include placental or liver RNase A inhibitor, which is used in the methods described here, vanadyl ribonucleoside complex (4 mM), and heparin (0.2 mg/mL). Sometimes, a combination or cocktail of RNase inhibitors will work best *(23)*. It is also essential to avoid contaminating the buffers and extract samples with RNases. Use sterile plasticware and wear disposable plastic gloves at all times, to avoid contact between the cell extract and body surfaces. Fingertips, e.g., contain active RNases. Glassware should be prerinsed with DEPC and baked at least 1 d before use (*see* **Protocol 1**). We recommend diethylpyrocarbonate (DEPC) rather than autoclaving, because autoclaving might not inactivate all RNase contaminants in the glassware. Moreover, DEPC is an excellent sterilizing agent. Solutions should be filtered and stored at 4°C or –20°C. To avoid subsequent contamination of solutions with RNase-containing microorganisms, sterile solutions should be opened only in a sterile hood. Dispense what is needed, then seal the container and place it back in the refrigerator or freezer. Tape a note on the container indicating that it is to be opened only under sterile conditions. Glass-distilled water can be sterilized by filtration or by adding DEPC to a final concentration of 0.02% (v/v) and then autoclaving.

1.5. Data Interpretation and Troubleshooting

Since mRNA decay rates are slower in vitro than in cells, cell-free mRNA decay systems should be used to quantify relative, as opposed to absolute, decay rates. To assess whether the system is functioning properly, the decay rates of mRNAs known to have different intracellular half-lives should be determined. mRNAs that are unstable in cells should also be unstable in vitro, relative to stable mRNAs. Some caution must be exercized in plotting and interpreting cell-free mRNA decay data. In cells, mRNA is thought to be degraded stochastically and logarithmically according to the following formula:

$$C / C_o = e^{-k_d t}$$

C and C_o are mRNA levels at time t and time 0, respectively, and k_d is the decay constant (reviewed in **ref. 24**). However, there is a problem when attempting to apply this formula to in vitro data, because decay rates decrease over time in vitro. The basis for this decrease has not been investigated but presumably involves depletion of one or more essential components or inactivation of mRNases. For this reason, we suggest that you can plot in vitro mRNA decay data as either a linear or logarithmic function, provided you are consistent and aware of the situation.

Below is a list of some trouble spots and some suggestions on how to deal with them.

- Excessive RNA degradation occurs during polysome and S130 preparation. Make sure all glassware and reagents are RNase-free. Work in the cold, and keep all solutions, test tubes, centrifuges, and glassware cold. Pick a different source for preparing extracts. For example, use a tissue-culture cell line instead of primary animal cells, or pick a different cell line. Some cell lines contain less "nonspecific" nuclease activity than others. By screening several cell lines, you might find one that makes the mRNA(s) of interest and contains minimal nonspecific RNase activity.
- mRNAs are degraded in vitro rapidly and at similar rates, regardless of their intracellular half-lives. Nonspecific RNase activity might be overwhelming the system. Try adding more RNase inhibitor, using combinations of inhibitors *(23)*, or modifying the reaction conditions. For example, nonspecific RNase activity

might be depressed by changing the monovalent or divalent cation concentration, the pH, or the reaction temperature. Add carrier single-stranded RNA to act as a competitor, e.g., poly(C) at 1 µg/25 µL of reaction. Additional competitor RNA should be avoided, because it can inhibit mRNA decay.

- No mRNA degradation occurs. One explanation for this observation is that mRNases were inactivated during extract preparation. Monitor the composition and pH of all extraction buffers and work in the cold. Another possibility is that all RNase activity, including that of mRNases, is blocked by the RNase inhibitor. We consider this explanation unlikely but recommend, nevertheless, that you try adding less inhibitor to the reactions.

- Regulated decay is not observed. For example, polysomes are isolated from a cell line that is either untreated or treated with a hormone that stabilizes a particular mRNA. Yet, this mRNA is degraded in vitro at the same rate with polysomes from treated and untreated cells. There could be many explanations for this result. One is that the hormone-responsive factors responsible for regulating mRNA decay are located in the S130. Try mixing polysomes with S130s from treated and untreated cells. Try altering the reaction conditions by changing the pH, salt, and incubation temperature. Another explanation is that regulation is coupled to translation, in which case it might be informative to incubate the mRNA of interest in an extract that permits some translation (*21*; **Protocol 7**).

1.6. Outline of Procedures Provided

1.6.1. Protocol 1: DEPC Treatment of Glassware and Solutions

This protocol is straightforward. Use caution when handling DEPC and dispose of undiluted DEPC in the fume hood.

1.6.2. Protocol 2: Preparation of Polysomes and Postpolysomal Supernatant (S130) from Exponentially Growing Tissue Culture Cells

There is no single, universally accepted method for preparing polysomes or for analyzing mRNA decay in vitro. The method described here involves lysing cells by homogenization in a buffer without detergents (*6*). The nuclei are removed by low-speed centrifugation. The cytosol is then fractionated by ultracentrifugation at 130,000g into two fractions, polysomes and postpolysomal supernatant or S130. The S130 seems to have little mRNase activity but does contain factors capable of regulating mRNA stability (*13,25*). These factors can be assayed in reactions containing mixtures of S130 and polysomes (*see* **Protocol 4**).

One advantage of this polysome preparation method is its flexibility. Polysomes contain both mRNA and mRNases. Therefore, the turnover of polysome-associated mRNAs can be analyzed simply by incubating polysomes under appropriate conditions (**Protocol 4**). Regulatory factors in the S130 are assayed by incubating polysomes and S130. Exogenous, radiolabeled mRNA can also be assayed in reactions containing polysomes or polysomes plus S130 (**Protocol 5**). Ribosome-bound mRNases and perhaps certain regulatory factors can be separated from polysomes by exposing the polysomes to high salt, pelleting them in an ultracentrifuge, and harvesting the supernatant or ribosomal salt wash (RSW; **Protocol 3**). mRNases and regulatory factors in the RSW can then be assayed by incubating RSW with in vitro synthesized mRNA substrates.

As noted previously, one disadvantage of using polysomes from higher eukaryotes is that they translate endogenous and exogenous mRNAs poorly, if at all. Moreover, many variables can affect the quality and performance of the extract. Aside from variables introduced by the investigator, the general health of the cells, their growth rate, and their stage of differentiation can vary among experiments and with cell passage time. Whenever possible, it is therefore important to eliminate these and other variables. When harvesting cells, make certain that the cell density is the same in each experiment. Lyse the cells and prepare the extracts in the same way, keeping the extracts ice-cold at all times, the centrifugation times constant, and the glassware and plasticware clean and cold.

Preparing polysomes from animal tissues might require special care to reduce nonspecific RNase activity. The following types and amounts of RNase inhibitors might be useful:

- Placental or liver RNase inhibitor. Appropriate inhibitor concentrations must be determined empirically, but a reasonable starting point is 100 U/mL.
- Ribonucleoside–vanadyl complexes. Starting point: 2.5 mM.
- Buffers can also contain RNase inhibitor combinations, which might be more effective than a single inhibitor. One buffer used to isolate intact polysomes from primary amphibian liver cell cultures included 4 mM vanadyl–ribonucleoside complex, 1 mM EGTA, 0.2 mg/mL heparin, 100 U/mL placental RNase inhibitor, pH 7.85 *(23)*.

1.6.3. Protocol 3: Preparation of Ribosomal Salt Wash (RSW) from Polysomes

Polysomal proteins include core ribosomal and mRNA components as well as proteins that are bound to ribosomes and mRNAs but are not necessarily integral components of the ribosome or mRNP complex. Noncore proteins include some translation initiation factors, RNases, and cytosolic proteins that bind to ribosomes during extract preparation. Relatively loosely bound proteins can be eluted from and separated from polysomes in two steps. First, the polysomes are exposed to high salt (0.5–1.0 M). Then, they are pelleted, and the supernatant (RSW) is harvested. RSW has been used as the starting material to purify putative mRNases *(8)*.

1.6.4. Protocol 4: Turnover of Endogenous Polysomal mRNA

Polysomes are incubated in a simple buffer at an appropriate temperature, usually 37°C. At various times, reactions are harvested, and the RNA is extracted and analyzed. Polysomes, S130, and RSW can be mixed in various combinations to assess whether the stability of a polysome-associated mRNA is influenced by soluble regulatory factors in the S130 or the RSW. If excessive nonspecific RNA degradation is observed, combinations of RNase inhibitors should be considered (*see* **Subheading 1.6.2.**). It is important to stress that this protocol should be used as a general guide on how to proceed and might need to be modified, depending on experimental parameters. For example, the S130 from cells infected with Herpes simplex virus 1 contains an activity that destabilizes polysome-associated mRNAs *(9)*. This destabilizer is relatively inactive under the standard reaction conditions described in **Protocol 4** but is fully active in reactions containing approximately one third lower potassium and magnesium. It is also important to stress the role of the placental/liver RNase inhibitor in these reactions. We have not observed differential mRNA decay in vitro in the absence of this inhibitor. Without it, all mRNAs, regardless of their stability in cells, are rapidly degraded. One tentative conclusion from these observations is that at least some mRNases are resistant to this class of inhibitor, while "nonspecific" RNases are efficiently inhibited. It is not known whether some mRNases are also inhibitor-sensitive and, therefore, are not detected in this system under these circumstances.

1.6.5. Protocol 5: Turnover of Exogenous mRNA Substrates Using Polysomes as a Source of mRNase Activity

Reaction mixtures should contain an mRNA substrate plus a source of mRNase activity, either crude postnuclear supernatant (S8), polysomes, or RSW. Two common substrate sources are either deproteinized cellular mRNA or in vitro synthesized RNA. Investigators should also consider using purified mRNP as a substrate. Cellular RNA should be prepared from a cytoplasmic extract and can be further purified by affinity chromatography to select poly(A)$^+$ mRNA. Synthetic RNA substrates can be made to contain a cap, phosphate, or hydroxyl group at the 5'-terminus. They can be poly- or oligo-adenylated at the 3'-terminus. In fact, all sorts of "tails" can be added to the 3'-terminus. At least in preliminary experiments, we recommend preparing mRNA substrates that are as similar as possible to authentic mRNA and contain both a cap and a poly(A), to ensure that the stability of each deproteinized mRNA substrate mimics that of its mRNP counterpart in cells. Reaction mixtures containing S8, polysomes, or RSW can be modified to include S130, purified RNA-binding proteins, or any potential mRNA-stability regulating factors.

1.6.6. Protocol 6: Turnover of Exogenous mRNA Substrates Using Ribosomal Salt Wash (RSW) as a Source of mRNase Activity

This protocol involves a minor modification of **Protocol 5**, primarily to account for the high salt in the RSW. Some mRNases are apparently inactive at salt concentrations above 0.3 M and lose activity above 0.2 M *(26)*. Therefore, the salt concentration in the reactions should be below 0.2 M and preferably at 0.1 M or lower. It is unclear whether differential mRNA stability with exogenous mRNA substrates is better studied using polysomes or RSW as a source of mRNase activity. RSW has two potential advantages. (1) Polysomes contain an activity that rapidly degrades uncapped substrates, but RSW apparently lacks this enzyme *(27)*. (2) RSW contains soluble RNases and thus can serve as the starting material for mRNase purification *(8)*. Regardless of the source of mRNase, it is important to ensure that the relative mRNA decay rates of the exogenous substrates reflect those in intact cells. Our experience has been that exogenous substrates are more likely than endogenous mRNP to fail this test.

1.6.7. Protocol 7. Turnover of Exogenous mRNA in an mRNA-Dependent Rabbit Reticulocyte Translation System

This protocol describes a simple system with three important advantages. (1) The extract can be purchased commercially and contains minimal nonspecific RNase activity. (2) mRNA can be translated for some time. (3) A variety of substrates can be tested. Following incubation, the RNA is extracted and analyzed by gel electrophoresis, Northern blotting, or any suitable technique. Translation can be monitored by including appropriate labeled amino acid(s) in the reactions.

2. Materials
2.1. Protocol 1
1. DEPC (store at –20°C) (*see* **Note 1**).
2. High-temperature oven.

2.2. Protocol 2
1. Exponentially growing tissue culture cells (*see* **Note 2**).
2. Solution A (*see* **Note 3**): 1 mM potassium acetate, 2 mM magnesium acetate, 2 mM dithiothreitol (DTT), 10 mM Tris-acetate, pH 7.6 at room temperature.
3. Solution B: solution A containing 30% (w/v) RNase-free sucrose and sterilized as per **Note 3**.
4. Tissue culture medium without serum or antibiotics (*see* **Note 4**).
5. Clean Potter–Elvehjem homogenizer with a Teflon pestle and a clearance of approx 0.1 mm (*see* **Note 5**).
6. 1- or 2-mL Dounce-type glass homogenizer with a tight-fitting pestle (clearance 0.06 mm).
7. Low-speed centrifuge, ultracentrifuge, and appropriate rotors and tubes (*see* **Note 6**).
8. Single-edged razor blades.

2.3. Protocol 3

1. Polysomes isolated as per **Protocol 2**.
2. Solutions A and B plus ultracentrifugation equipment, as per **Protocol 2**.
3. DEPC.
4. Magnetic stirring instrument and stirring bar.
5. 4.0 M Potassium acetate.

2.4. Protocol 4

1. Solution C stock, prepared as follows and stored in small (0.2- to 0.4-mL) aliquots at −70°C: 200 µL of sterile, deionized water, 200 µL of 1.0 M creatine phosphate, 40 µL of 1 M DTT, 100 µL of 0.2 M ATP (Tris salt), 20 µL of 0.2 M GTP (Tris salt), 1 mL of 2.0 M potassium acetate, 800 µL of 2.5 mM spermine, 200 µl of 1.0 M Tris-acetate, pH 7.5.
2. Solution A and polysomes at a concentration of 10^6 cells' worth per milliliter prepared as described in **Protocol 2**.
3. Solution D, prepared on ice just prior to performing the reaction (*see* **Note 7**). The volumes listed are for each 25 µL of reaction required in **step 1** of this protocol: 1.1 µL of sterile, deionized water, 3.2 µL of solution C, 0.25 µL of 4 mg/mL creatine kinase, 0.25 µL of 40 U/µL RNase inhibitor, 16 µL of solution A, 0.2 µL of 0.1 M magnesium acetate.

2.5. Protocol 5

1. Solution A and polysomes (10^6 cells' worth of polysomes per microliter, as described in **Protocol 2**).
2. Solution C, as described in **Protocol 4**.
3. Either a ^{32}P-labeled RNA substrate (~10^7 cpm/µg; 5000–20,000 cpm/µL in water) or unlabeled RNA (0.1–2.0 µg/µL in water; *see* Chapter 1).
4. Solution E, prepared on ice just prior to performing the reaction. The volumes listed are for each 25 µL of reaction used in **step 1** of the protocol: 3.2 µL of solution C, 0.25 µL of 4 mg/mL creatine kinase, 0.25 µL of 40 U/µL RNase inhibitor, 16 µL of solution A, 0.2 µL of 0.1 M magnesium acetate, and 1.1 µL of RNA substrate.

2.6. Protocol 6

1. Solution A, as per **Protocol 2**.
2. RNA substrate, as described in **Protocol 5**.
3. RSW (*see* **Protocol 3**).
4. Solution F stock, prepared as follows and stored at −70°C in multiple small (0.2–0.4 mL) aliquots: 1.2 mL of deionized, sterile water, 200 µL of 1.0 M creatine phosphate, 40 µL of 1.0 M DTT, 100 µL of 0.2 M ATP (Tris salt), 20 µL of 0.2 M GTP (Tris salt), 800 µL of 2.5 mM spermine, 200 µL of 1.0 M Tris-acetate, pH 7.5.
5. Solution G, prepared on ice just prior to performing the reaction. The volumes listed are for each 25-µL reaction required in **step 1**: 3.2 µL of solution F, 0.25 µL of 4 mg/mL creatine kinase, 0.25 µL of RNase inhibitor (*see* **Note 8**), 15 µL of solution A, 0.2 µL of 0.1 M magnesium acetate, and 1.1 µL of RNA substrate.

2.7. Protocol 7

1. Commercial source of rabbit reticulocyte lysate pretreated with micrococcal nuclease.
2. mRNA dissolved in sterile, RNase-free water at 0.1–5.0 µg/µL, either from cells or from in vitro transcriptions (*see* **Note 9**).
3. Placental or recombinant RNase inhibitor, 40 U/µL.
4. 1 mM Amino acid mixture (*see* **Note 10**).
5. Radioactive amino acids(s), if desired (*see* **Note 10**).

3. Methods
3.1. DEPC Treatment of Glassware

1. Add a small volume of DEPC to each item of glassware.
2. Rotate the item, taking care to expose all the inner surfaces to the DEPC.
3. Pour off the DEPC and rinse the item thoroughly with deionized, distilled water until all traces of the distinctive DEPC odor have disappeared. A minimum of 10 rinses should be sufficient.
5. Cover the item with aluminum foil and bake it in a 220°C oven for at least 1 h, and preferably overnight.

3.2. Preparation of Polysomes and Postpolysomal Supernatant (S130) from Exponentially Growing Tissue Culture Cells (see Notes 2 and 3)

1. Count the cells (*see* **Note 11**).
2. Centrifuge the cells for 4 min at 4°C, 34g (average), corresponding to 500 rpm in a low-speed centrifuge with a swinging bucket radius of 8 cm.
3. Aspirate and discard the supernatant.
4. Gently resuspend the cells in ice-cold tissue-culture medium without serum or antibiotics.
5. Centrifuge as in **step 2**.
6. Resuspend the cells in solution A (*see* **Note 12**).
7. Transfer the cells to the prechilled Potter–Elvehjem homogenizer.
8. Place the homogenizer vessel in an ice-water bath and break the cells by moving the pestle rapidly up and down and twisting it to the left and right at the same time (*see* **Note 13**).
9. Centrifuge the homogenate for 10 min at 8300g (average), which corresponds to 9000 rpm in a swinging bucket rotor, radius 9.1 cm.
10. Remove and save the supernatant (S8), taking care to avoid harvesting any of the pelleted material (nuclei and membranes). Leave behind a small amount of supernatant so that none of the pelleted material contaminates the S8.
11. Gently layer the S8 over a cushion of cold solution B in a prechilled ultracentrifuge tube. The volume of solution B should be approx 25% of the tube capacity. If the tube is insufficiently filled with S8, add solution A.

12. Centrifuge at 130,000g (average), 2°C, 2.5 h, which corresponds to 36,000 rpm in a Beckman SW60 rotor. If you intend to save the supernatant (S130), allow the rotor to decelerate without the brake.
13. Remove the S130 above the sucrose cushion, taking care not to harvest the whitish material located on top of the cushion. Leave a small amount of S130 above the cushion, if necessary. Do not be concerned if some of the sticky white lipid material at the top of the tube is harvested along with the S130. Set aside the S130 in an ice bucket.
14. Aspirate the remaining liquid, including the sucrose, and discard it. A bluish-white pellet of polysomes should be visible in the center of the tube bottom.
15. Working in the cold room, lay the centrifuge tube on its side on a piece of plastic wrap and cut it across with a razor blade approximately two thirds from the top. Discard the top piece.
16. Gently add approx 1 mL of solution A down the sides of the remaining (bottom) part of the centrifuge tube. Gently move the tube back and forth to wash any contaminating material, including sucrose, from the upper surface of the polysome pellet. Aspirate or pour off the liquid.
17. Repeat this washing step once. It is not necessary to perform these washing steps quickly, as the polysomes will not dissolve in the liquid, but it is necessary to add the liquid slowly and gently to avoid dislodging the polysome pellet from the bottom of the tube.
18. Resuspend the polysomes in a small volume of solution A. The volume will depend on the quantity of polysomes recovered and the subsequent experiments. As a rule, use a volume such that the polysomes from 10^6 cells will be resuspended in 1 µL (*see* **Note 14**).
19. Measure the optical density of the polysomes and S130 at 260 nm and 280 nm and store both in multiple aliquots of 100–200 µL each at –70°C (*see* **Note 15**).
20. To assess polysome integrity, RNA from a small amount of the preparation can be extracted and analyzed by agarose gel electrophoresis. The rRNA can be visualized by ethidium bromide staining. The rRNA is intact if the 28S and 18S rRNA bands are discrete and if the band intensity is approx 2:1 (28S:18S).

3.3. Preparation of Ribosomal Salt Wash (RSW) from Polysomes

1. Working in the cold room, if possible, resuspend the polysomes in Solution A. Freshly prepared polysomes are not required. You can use frozen polysomes as starting material. The polysomes need not be fully dispersed but should be more concentrated than the 10^6 cells' worth per microliter suggested in **Protocol 2**. A reasonable starting point is $2–5 \times 10^9$ cells' worth of polysomes per milliliter, but higher concentrations can be used, if desired.
2. Transfer the polysomes to a small test tube or beaker containing a stirring bar. Prior to use, the stirring bar should be soaked for 3–5 min in DEPC and rinsed thoroughly with water.
3. Add sufficient 4.0 *M* potassium acetate dropwise and with gentle stirring to bring the final concentration to 0.5 *M* potassium acetate.

4. Stir gently for an additional 15 min in the cold. Avoid bubbles.
5. Place Solution B in a polyallomer ultracentrifuge tube so that it is ~25% of the tube capacity, carefully layer the salt-washed polysomes on top of the solution B, fill the tube to within several millimeters of the top, and centrifuge for 2.5 h at 130,000g (average), 2°C.
6. Using a pipet, harvest the RSW located above the sucrose cushion. Retain the salt-washed polysome pellet (*see* **step 9**).
7. Mix the RSW gently but thoroughly and determine the protein concentration (*see* **Note 16**).
8. Store the RSW in small aliquots (100–200 μL) at –70°C.
9. If desired, the salt-washed polysomes in the pellet at the bottom of the tube (**step 6**) can also be harvested, as described in **Protocol 2**.

3.4. Turnover of Endogenous Polysomal mRNA

1. Add the following to the required number of microcentrifuge tubes on ice: 21 μL of solution D and 4 μL of polysomes (*see* **Note 17**).
2. Mix gently without vortexing and incubate the reactions for the desired time at an appropriate temperature, usually 37°C (*see* **Note 18**).
3. Remove the reactions and either store them at –70°C or extract the RNA immediately. When performing a time course, the most practical method is to place each time point in dry ice immediately after you remove it from the water bath. Any RNA extraction method can be used, as long as care is taken to avoid nonspecific RNase contamination.
4. Analyze and quantitate the endogenous mRNAs of interest (*see* **Note 19**).

3.5. Turnover of Exogenous mRNA Substrates Using Polysomes as a Source of mRNase Activity

1. Add the following to the required number of microcentrifuge tubes on ice: 21 μL of solution E and 4 μL of polysomes.
2. Incubate the reactions at the desired temperature for the desired time (*see* **Note 18**).
3. Either store the reaction mixtures at –70°C or extract the RNA immediately. When performing a time course, the most practical method is to place each time point in dry ice immediately after you remove it from the water bath. Any RNA extraction method can be used, as long as care is taken to avoid nonspecific RNase contamination.
4. Analyze and quantitate the endogenous mRNAs of interest (*see* **Note 19**).

3.6. Turnover of Exogenous mRNA Substrates Using Ribosomal Salt Wash (RSW) as a Source of mRNase Activity

1. Add the following to the required number of microcentrifuge tubes on ice: 20 μL of solution G and 5 μL of RSW.
2. Incubate at the desired temperature for the desired time (*see* **Note 18**).
3. Either store the reaction mixtures at –70°C or extract the RNA immediately. When performing a time course, the most practical method is to place each time

point in dry ice immediately after you remove it from the water bath. Any RNA extraction method can be used, as long as care is taken to avoid nonspecific RNase contamination.
4. Analyze and quantitate the endogenous mRNAs of interest (*see* **Note 19**).

3.7. Turnover of Exogenous mRNA in an mRNA-Dependent Rabbit Reticulocyte Translation System

1. Add the reagents in the order listed below to microcentrifuge tubes on ice, with the final reaction volume being 50 µL: 35 µL of nuclease-treated lysate, 1 µL of RNase inhibitor, 1 µL of amino acid mixture, radioactive amino acid(s) and/or water, as necessary, mRNA substrate in water, as necessary.
2. Mix gently and incubate at 30°C for the desired time.
3. Either store the samples at –70°C or extract the RNA immediately.

4. Notes

1. DEPC is volatile and potentially carcinogenic. Dispose of undiluted DEPC in the fume hood and avoid contact with skin and eyes.
2. This method was designed specifically for K562 human erythroleukemia cells. Other cells can be used, but you will need to monitor their suitability with respect to the amount of substrate (mRNA) available and the ease with which cell lysates containing undegraded polysomes and mRNA can be prepared. As a general rule for this and other protocols, all glassware and buffers should be nuclease-free. Unless otherwise noted, all steps should be performed at 2–4°C and the buffers, solutions, centrifuges, and centrifuge tubes should be ice-cold. Therefore, ice and ice buckets and, preferably, a coldroom should be available.
3. Solution A should be sterilized using a 0.22-µm filter and stored at 4°C. Open the solution A stock only in a tissue-culture hood under sterile conditions. Pour out an aliquot and return the sterile stock bottle to the 4°C refrigerator.
4. Ham's F 12 and RPMI-1640 media (Flow, Gibco-BRL, HyClone) have been used in our laboratory, but any isotonic growth medium will suffice.
5. The homogenizer should hold a volume two- to threefold greater than the amount of solution A used to lyse the cells. If the capacity of the homgenizer is too small, lysate can splash out during homogenization.
6. Polyallomer ultracentrifuge tubes are preferable. They need not be baked or treated with DEPC, but they should be thoroughly rinsed with RNase-free water and then drained dry immediately prior to use.
7. It is unknown whether the order of addition of these components is important. However, it is essential that solution C be added *before* nonrecombinant RNase inhibitor, because tissue-derived (nonrecombinant) RNase inhibitor contains RNase that is activated in the absence of a reducing agent.
8. You can add more RNase inhibitor if the nonspecific RNase activity in the extract is not fully blocked. We have not determined the upper limit of RNase inhibitor that can be added to these reactions.
9. If using cellular RNA, it might be necessary to prepare poly(A)$^+$ mRNA, because excess ribosomal RNA can interfere with translation and mRNA decay.

10. Various amino acid mixtures are available from the companies that sell the lysate. Some of these mixtures contain all 20 of the common amino acids. Others lack one amino acid, so that it can be replaced with a radioactive amino acid, which is useful for monitoring both translation and mRNA decay.
11. The cells should be in exponential phase. Using slowly dividing or nongrowing cells generally results in poor polysome yields. There is no hard and fast rule for how many cells to use. If a large stock of polysomes is to be made, at least 10^9 cells are recommended.
12. The volume of solution A will depend primarily on the cell number and the cell type. Some cells do not lyse easily when homogenized in a buffer without detergent, especially if they are too concentrated. The maximum concentration of cells for high-efficiency breakage (95% or greater cell lysis) must be determined in a pilot experiment by counting the percentage of broken cells using a hemocytometer and phase contrast microscope. In most cases, 10^8 cells/mL is a reasonable starting point for homogenization, but the optimum number might be 10-fold lower. Save a prehomogenization sample as a control. Homogenize with 20 strokes, and compare the pre and posthomogenization samples in the microscope by counting the number of nuclei and unbroken cells to determine the percentage of lysed cells. It is acceptable to have cytoplasmic "tags" surrounding the nuclei of lysed cells.
13. This step should be hard work! Push, pull, and turn the pestle vigorously. Depending on your condition, your hand and arm should feel tired by the time you are finished. The number of up and down strokes required for efficient cell lysis will vary, depending upon the cell type and the cell concentration (*see* **Note 12**). As a starting point, 20 strokes should be sufficient. Avoid generating bubbles by keeping the head of the homogenizer beneath the upper surface of the liquid.
14. Polysomes can be difficult to resuspend. They tend to be "sticky" and to form clumps that are difficult to disperse. First, try resuspending them by pipetting the liquid up and down with a micropipettor with a 200 μL or 1-mL disposable tip. If visible clumps remain, transfer the material to the tight-fitting, 1 or 2 mL Dounce-type glass homogenizer with a clearance of 0.06 mm. Several up and down strokes with this homogenizer should break up the clumps.
15. As a guide, when polysomes isolated from a logarithmically growing human erythroleukemia cell line (K562) are resuspended at 10^6 cells' worth of polysomes per microliter, the absorbance reading at 260 nm is 220–260 through a 1 cm light path.
16. Kits available from Bio-Rad (Hercules, CA, USA) and Pierce (Rockford, IL, USA) contain reagents for the Bradford and BCA protein quantitation assays.
17. The components of the reaction mixture prepared in step 1 can be modified in many ways to suit different experimental conditions and to address various questions. For example, factors that regulate mRNA stability might be present in the S130 and can be assayed by adding S130 to the reactions. Simply substitute S130 for an equivalent volume of solution A when preparing solution D. For example, to analyze the effect of 5 μL of S130 on the decay of polysomal mRNP, prepare solution D with 11 μL per reaction of solution A, instead of 16 μL. Each final 25 μL reaction mix should then contain 16 μL of modified solution D, 5 μL of S130, and

4 μL of polysomes. You can also analyze mRNP decay in S8 simply by substituting S8 for the polysomes and solution A. For any such substitution, make sure that the magnesium ion concentration is maintained at 2 mM.
18. mRNA decay reactions can be performed at a variety of temperatures, depending on the particulars of the experiment. The standard incubation temperature is 37°C, but situations may arise in which lower temperatures are preferred. For example, in vitro mRNA decay reactions are useful for identifying mRNA decay products, the structures of which provide a snapshot of the decay pathway: its direction (5' to 3' or 3' to 5') and the mRNases involved (exo- or endoribonucleases). However, some mRNA decay products are so unstable at 37°C that they are difficult to detect. If so, try incubating at temperatures between 20°C and 32°C. Reaction rates are slower at lower temperatures, permitting easier detection of decay products.
19. Many methods are available for analyzing and quantifying endogenous mRNA levels. They include Northern blotting, nuclease S1 mapping, RNase protection, primer extension, and PCR. The method of choice will depend on the quantity per cell of the particular mRNA. Blotting and nuclease mapping/protection are preferred for moderately abundant mRNAs. For very scarce mRNAs, more sensitive techniques such as PCR might be required. As a time zero standard, prepare a complete reaction mix and keep it on ice or place it in dry ice.

References

1. Belasco, J. G. (1993) mRNA degradation in prokaryotic cells: an overview, in *Control of Messenger RNA Stability* (Belasco, J. and Brawerman, G., eds.), Academic Press, New York, pp. 3–12.
2. Caponigro, G. and Parker, R. (1996) Mechanisms and control of mRNA turnover in *Saccharomyces cerevisiae. Microbiol. Rev.* **60**, 233–249.
3. Ross, J. (1995) mRNA stability in mammalian cells. *Microbiol. Rev.* **59**, 423–450.
4. Hajnsdorf, E., Carpousis, A. J., and Régnier, P. (1994) Nucleolytic inactivation and degradation of RNase III processed *pnp* message encoding polynucleotide phosphorylase of *Escherichia coli. J. Mol. Biol.* **239**, 439–454.
5. Beelman, C. A., Stevens, A., Caponigro, G., LaGrandeur, T. E., Hatfield, L., Fortner, D. M., and Parker, R. (1996) An essential component of the decapping enzyme required for normal rates of mRNA turnover. *Nature* **382**, 642–646.
6. Ross, J. and Kobs, G. (1986) H4 histone mRNA decay in cell-free extracts initiates at or near the 3' terminus and proceeds 3' to 5'. *J. Mol. Biol.* **188**, 579–593.
7. Bernstein, P. L., Herrick, D. J., Prokipcak, R. D., and Ross, J. (1992) Control of c-*myc* mRNA half-life *in vitro* by a protein capable of binding to a coding region stability determinant. *Genes Devel.* **6**, 642–654.
8. Caruccio, N. and Ross, J. (1994) Purification and characterization of a ribosome-associated 3' to 5' exoribonuclease from human cells. *J. Biol. Chem.* **269**, 31,814–31,821.
9. Zelus, B. D., Stewart, R. S., and Ross, J. (1996) The virion host shutoff protein of Herpes simplex virus type 1: messenger ribonucleolytic activity *in vitro. J. Virol.* **70**, 2411–2419.
10. Amara, F. M., Chen, F. Y., and Wright, J. A. (1994) Phorbol ester modulation of a novel cytoplasmic protein binding activity at the 3'-untranslated region of mam-

malian ribonucleotide reductase R2 mRNA and role in message stability. *J. Biol. Chem.* **269**, 6709–6715.
11. Dompenciel, R. E., Garnepudi, V. R., and Schoenberg, D. R. (1995) Purification and characterization of an estrogen-regulated *Xenopus* liver polysomal nuclease involved in the selective destabilization of albumin mRNA. *J. Biol. Chem.* **270**, 6108–6118.
12. Krikorian, C. R. and Read, G. S. (1990) In vitro mRNA degradation system to study the virion host shutoff function of herpes simplex virus. *J. Virol.* **65**, 112–122.
13. Brewer, G. (1991) An A + U-rich element RNA-binding factor regulates c-*myc* mRNA stability in vitro. *Mol. Cell. Biol.* **11**, 2460–2466.
14. Amara, F. M., Chen, F. Y., and Wright, J. A. (1993) A novel transforming growth factor-β1 responsive cytoplasmic *trans*-acting factor binds selectively to the 3'-untranslated region of mammalian ribonucleotide reductase R2 mRNA: role in message stability. *Nucleic Acids Res.* **21**, 4803–4809.
15. Brown, B. D., Zipkin, I. D., and Harland, R. M. (1993) Sequence-specific endonucleolytic cleavage and protection of mRNA in *Xenopus* and *Drosophila*. *Genes Devel.* **7**, 1620–1631.
16. Wreschner, D. H. and Rechavi, G. (1988) Differential mRNA stability to reticulocyte ribonucleases correlates with 3' non-coding $(U)_nA$ sequences. *Eur. J. Biochem.* **172**, 333–340.
17. Hepler, J. E., Van Wyk, J. J., and Lund, P. K. (1990) Different half-lives of insulin-like growth factor I mRNAs that differ in length of 3'-untranslated sequence. *Endocrinology* **127**, 1550–1552.
18. Gorospe, M. and Baglioni, C. (1994) Degradation of unstable interleukin-1α mRNA in a rabbit reticulocyte cell-free system. Localization of an instability determinant to a cluster of AUUUA motifs. *J. Biol. Chem.* **269**, 11845–11851.
19. Maquat, L. E. (1995) When cells stop making sense: effects of nonsense codons on RNA metabolism in vertebrate cells. *RNA* **1**, 453–465.
20. Molla, M., Paul, A. V., and Wimmer E. (1991) Cell-free, de novo synthesis of poliovirus. *Science* **254**, 1647–1651.
21. Osley, M. A. (1991) The regulation of histone synthesis in the cell cycle. *Annu. Rev. Biochem.* **60**, 827–861.
22. Zambetti, G., Stein, J., and Stein, G. (1990) Role of messenger RNA subcellular localization in the posttranscriptional regulation of human histone gene expression. *J. Cell. Physiol.* **144**, 175–182.
23. Blume, J. E. and Shapiro, D. J. (1989) Ribosome loading, but not protein synthesis, is required for estrogen stabilization of *Xenopus laevis* vitellogenin mRNA. *Nucleic Acids Res.* **17**, 9003–9014.
24. Hargrove, J. L., Hulsey, M. G., and Beale, E. G. (1991) The kinetics of mammalian gene expression. *BioEssays* **13**, 667–674.
25. Brewer, G. and Ross, J. (1989) Regulation of c-*myc* mRNA stability in vitro by a labile destabilizer with an essential nucleic acid component. *Mol. Cell. Biol.* **9**, 1996–2006.
26. Ross, J., Kobs, G. Brewer, G., and Peltz, S. W. (1987) Properties of the exonuclease activity that degrades H4 histone mRNA. *J. Biol. Chem.* **262**, 9374–9381.
27. Peltz. S. W., Brewer, G., Kobs, G., and Ross, J. (1987) Substrate specificity of the exonuclease activity that degrades H4 histone mRNA. *J. Biol. Chem.* **262**, 9382–9388.

Index

Acrylamide, precautions, 29
Affinity chromatography, *see* Chromatography
Aptamer, 217
Ascorbic acid, *see* Photochemical crosslinking
5-Azido(phenacylthio)-CTP (5-APAS-CTP), 22–23
5-Azido(phenacylthio)-UTP (5-APAS-UTP), 22–23
Bacteriophage,
    filamentous, 190
    growth, 199–201
    λ and N protein, 177–179
    M13K07, use as helper phage, 195, 214
β-galactosidase assay,
    colony color, 179–180, 183
    solution, 167, 180, 183
Biacore, *see* Surface plasma resonance
cDNA,
    first strand synthesis, 236–237
    tailing reaction, 237–238
cDNA library, *see* Library
Cell extract, for immunoprecipitation, 269, 272–273
Cell line,
    for splicing assay, 39 4–396
    for translation extract, 456
Chromatography,
    anti-m3G, 293–295, 297
    antisense affinity, 275
    DEAE-sepharose, 97
    H386 antibody, 297
    mono Q, 295
    oligo-dT cellulose, 438, 440
    size exclusion, 68
    streptavidin agarose, 278–279, 282
        regeneration, 279
    streptavidin paramagnetic beads, 206–208

Combinatorial library, *see* Library
Crosslinking, *see* Photochemical crosslinking
Diethyl pyrocarbonate (DEPC), use as RNase inhibitor,
    treatment of glassware, 470
    treatment of solutions, 28
DNA,
    gel purification, 225, 410–411
    mutagenesis, 168–170
DNase treatment of RNA, 446–447
3' End processing of mRNA, 433–434
    in vitro assay, 433, 437–438
Endoprotease digestion of ribonucleoprotein complexes, 69, 70
Expression, *see* Library
Footprinting of RNA and RNA-protein complexes, 73–77
    base specific probes, 84–86
    chemical probes, 82–84
    phosphorothioate-containing transcripts, 84
    RNase, 80–82
Gel purification, *see* DNA; Protein; RNA
Gel retardation, 116
    dissociation constants, 116–117, 120–121, 124
    equilibrium binding constants, 115–118, 120
HeLa cells, suspension culture, 311, 313
Heparin, use in transcription reactions, 29
Heterogeneous nuclear ribonucleoprotein (hnRNP) complex purification, 299–300
Immunoaffinity chromatography, *see* Chromatography
Immunoblotting, *see* Western blotting

477

Immunoprecipitation of ribonucleoprotein
        complexes, 226–227, 239, 242,
        265–267, 269–270, 273
    hnRNP complexes, 304, 305
In vitro transcription, *see* Transcription, in
        vitro
In vitro translation, *see* Translation, in vitro
Lac repressor, 163
lacZ reporter, 162–163, 164
Library,
    cDNA, preparation, 271–272
    cDNA expression, 248–249, 258–259
        plating 251, 254, 255
        protein extract preparation, 261–262
        solution-based screening assay, 257–260
    combinatorial,
        preparation of peptide expressing,
            180, 184, 208–210
        preparation of transcription template,
            224–225
        sampling considerations, 180–181,
            208, 214, 226, 230, 409
        screening, 184–185
        selection of RNA binding proteins,
            169–170, 180–181, 189–190
        selection of RNA ligands, 217–221
        selection of splicing enhancers, 410–411
Ligase, bacteriophage T4 DNA, 12, 13
Ligase, bacteriophage T4 RNA, 12
Ligation, 411, 415
Matrix-assisted laser desorption/ionization
        mass spectrometry (MALDI-MS), 63–66, 69–71
5-Mercapto-CTP (5-SH-CTP), 23
5-Mercapto-UTP (5-SH-UTP), 23
Methylene blue, *see* Photochemical
        crosslinking
Modification interference analysis, 74–77, 86
mRNA decay, 459–460
    data analysis, 464
    endogenous mRNA, 466–467, 472, 474–475
    exogenous mRNA, 467–468, 472–473
    in vitro assays, 459–465

Nitrocellulose filter binding, 105–107
    assay conditions, 105–107, 110–111
    dissociation constants, 105–107, 110–111
    dot blot apparatus, 107–108
    isotherms, 106, 109–110, 113
    stoichiometry of RNA-protein complexes, 112–113
Nonequilibrium pH gel electrophoresis
        (NEPHGE), 306–307
Northern blot, 98–99, 101, 384–387, 388
Northwestern screening, 245–248
    plaque lifts, 252
    protein induction, 252
    RNA probe binding and washing, 252–253
Nuclear extract preparation, 292–293, 303–304, 311–312, 435–437
    preclearing on streptavidin agarose,
            277–278
    preparation of high salt, 279–280, 284–285
2'-*O*-methyl RNA oligonucleotides, 276,
        282, 284
Oligo-dT cellulose, *see* Chromatography
Oligonucleotides, annealing to RNA, 98
PACE, *see* Polyacrylamide
        coelectrophoresis
PAGE, *see* Polyacrylamide gel electrophoresis
PCR, *see* Polymerase chain reaction
PCR poly(A) test (PAT), 441–443
    electrophoresis of products, 447
    optimizing, 446–447
Phage display, 189–194
    binding phage to beads, 207–208
    binding reaction, 210-211
    pDISPLAY-B, 193, 195
    growth, 200–201
    pDISPLAYblue-B, 193, 194–195
Phagemid library production, 208–210
Phenylmethylsulfonylfluoride (PMSF), safe
        handling, 261
*pho*A, 194, 199, 215
Phosphodiesterase, 5'-3', 70

Index

Photochemical crosslinking, 21–23, 26–28
  dsRNA to protein, 35–38
  methylene blue, 36–38
    ascorbic acid as quenching agent, 42
    base specificity, 44
    crosslinking of spliceosome components, 37–38
    concentration for crosslinking, 41
    protein-protein crosslinking, 41–42
    RNase digestion of crosslinked material, 44–45
  nucleotide analogs, 21–23
  psoralen, 49–51
    adduct structure, 50
    conjugation to protein, 53
    crosslinking reaction, 53–54
    synthesis of crosslinking agent, 52
  2-iminothiolane, 67, 70, 71
  UV irradiation, 68
PMSF, *see* Phenylmethylsulfonylfluoride
Poly(A) tail, 441
Polyacrylamide coelectrophoresis (PACE), 129–130
  dissociation constant (K_d) measurement, 136–137
  equipment, 129–131
  gel running time, 139–140
  preparation of gel, 132–135, 138–139
Polyacrylamide gel electrophoresis (PAGE),
  denaturing (urea), 6–7, 25, 27, 53–54, 98–99, 205, 318–319, 339–341, 359–360
    mobility of RNA, 7, 27
  nondenaturing (native), 100, 101, 122–123, 125–126, 225
  SDS- (SDS-PAGE), 25, 202, 305–306, 427
    silver staining, 26, 28
    staining proteins with KCl, 427, 430
  spliceosomes, 382–383
    acrylamide concentration, 387–388
    conditions, 367–371
  two-dimensional, 306–307

Polyadenylation, 433–434
Polymerase chain reaction (PCR), 227, 445–446
  cDNA library amplification, 238
  chase reaction, 271, 273
  colony screening, 272
  random, 270–271
Polysome preparation, 465–466, 470–471, 473–474
Primer extension,
  mapping cleavage fragments, 99
  mapping crosslink sites, 56
  RNA sequencing, 56, 61
Protein, gel purification of, 427–429, 430
Psoralen, *see* Photochemical crosslinking
Rabbit reticulocyte lysate, 449, 468, 473
Reverse phase–high performance liquid chromatography (RP-HPLC), 69, 70
Reverse transcription, 227, 239–240, 270, 445
Reverse transcription-polymerase chain reaction (RT-PCR), 412–414
  optimization, 416
  preventing contamination, 416
Ribonuclease (RNase),
  A, 45
  B. cereus, 55, 60
  cleavage specificities, 81
  cobra venom V1, 45
  endogenous, 463, 464
  H,
    cleavage specificity, 93
    oligonucleotide-targeted cleavage, 93–94
    protection analysis, 93–95, 98, 101
  inhibitors, 463–464, 466
  T1, 45, 55, 60, 69
Ribosomal salt wash, preparation, 471–472
RNA, *see also* Transcription in vitro
  alkaline hydrolysis, 55, 70, 80–81
  aniline hydrolysis, 86
  electroblotting, 99, 385–386, 388
  extraction from animal tissues, 379–381
  extraction from cells, 412

RNA *(cont.)*,
  extraction from ribonucleoprotein complexes, 97
  gel purification, 6–7, 205–206, 341–342, 360
  labeling,
    end labeling 5–6
    internal, 5, 9
    nonisotopic, 9, 26–27
    site-specific, 14–16
  ligation, 11–12, 16
  modification
    chemical probes, 79, 82–87
    diethyl pyrocarbonate, 84–85
    dimethylsulfate, 85–86
    hydrazine, 86–87
    hydroxyl radicals, 82–83, 89
    N-ethyl-*N*-nitrosourea, 83
    nucleotide analogs, 22–23
    phosphorylation, 15
    site specific labeling, 14–16
    UV shadowing, 7, 184
RNA ligands,
  selection from combinatorial libraries, 217–221
  selection from libraries of natural sequences, 234–235
RNA-protein complexes, electroblotting to nitrocellulose, 28
RNA recognition motif (RRM), 234, 419
rPCR, *see* PCR
RP-HPLC, *see* Reverse phase–high performance liquid chromatography
RT-PCR, *see* Reverse transcription-polymerase chain reaction
S1 analysis, 438–439, 440
Selex, *see* Library, combinatorial
Shine-Dalgarno sequence, 161–162
Silver staining, *see* Polyacrylamide gel electrophoresis
Small nuclear ribonucleoprotein particles, *see* snRNP
snRNP,
  depletion by antisense affinity chromatography, 280–281, 284–286

snRNP *(cont.)*,
  protein composition, 290
  purification, 289–291
    anti-m3G immunoaffinity chromatography, 293–295
    antisense affinity chromatography, 278–279, 282
    glycerol gradient centrifugation, 293
    mono Q chromatography, 295
    17S U2 snRNP, 296
    tri-snRNP complex, 295–296
Spliceosome,
  polyacrylamide gel electrophoresis,
    native, 360
    two dimensional, 361
  purification,
    competitor, 369–371
    glycerol gradient, 360
    mammalian nuclear extract, 351–357
    streptavidin affinity purification, 360–361
    splicing reaction, 360–361, 383
    yeast, 365–371
Splicing,
  alternative, 391–393
  cis regulatory elements, 391–394, 405–407
  extract preparation from HeLa cells, 309–310
    nuclear extract, 311–312
    S100, 313
  extract preparation from yeast, 323–326, 345
    freeze-fracture method, 337–338
    spheroplast homogenate, 335–337, 345–346
  mammalian in vitro, 315, 317–318
    minigene constructs, 397–401, 408–410
    optimization, 319
    substrates, 316, 319–320
  quantization, 396–397
  yeast in vitro, 325–326, 342–344, 348
    purification of spliceosomes, 383
SPR, *see* Surface plasmon resonance

Index

SR proteins, 419–420
  ammonium sulfate precipitation, 424–426, 429
  magnesium precipitation, 426–427
  purification of individual proteins, 422
  sources for purification, 421–422
Surface plasmon resonance (SPR), 143–145
  detection of RNA-protein interactions, 145–147
  enzyme kinetics, 154
  sensor chips,
    immobilization, 147–151, 156–157
    regeneration, 152
    stoichiometry, 152–154
T7 RNA polymerase, purification, 223–224
Transcription antitermination assay, 177–181
  competent cell preparation, 185
Transcription in vitro, 1–2, 4–5, 26, 238–239, 338–339, 347–348, 359, 437
  bacteriophage RNA polymerase, 1, 8
  combinatorial library, 225–226, 227–228
  double-stranded RNA production, 7–8

Transcription in vitro *(cont.)*,
  E. coli RNA polymerase, 26
  heparin, 29
  labeling RNA,
    internal, 5, 9
    nonisotopic, 9, 26–27
    phosphorothioate nucleotide, 84
Transfection of tissue culture cells, 411, 415–416
Translation in vitro, 449–450, 455
  cap-independent, 449
  extract preparation from tissue culture cells, 453–455, 456
Translational repression, 161–163
  assay, 166–168
  dissociation constants, 171–173
  placZ-Rep, 162–165
Two dimensional polyacrylamide gel electrophoresis, *see* Polyacrylamide gel electrophoresis
UV crosslinking, *see* Photochemical crosslinking
UV shadowing, *see* RNA
Western blotting, 202–203